The Interaction of Food Industry and Environment

The Interaction of Food Industry and Environment

Edited by

Charis Galanakis

Department of Research & Innovation, Galanakis Laboratories, Chania, Greece

Food Waste Recovery Group, ISEKI Food Association, Vienna, Austria

ELSEVIER

ACADEMIC PRESS

An imprint of Elsevier

Academic Press is an imprint of Elsevier
125 London Wall, London EC2Y 5AS, United Kingdom
525 B Street, Suite 1650, San Diego, CA 92101, United States
50 Hampshire Street, 5th Floor, Cambridge, MA 02139, United States
The Boulevard, Langford Lane, Kidlington, Oxford OX5 1GB, United Kingdom

Library of Congress Cataloging-in-Publication Data
A catalog record for this book is available from the Library of Congress

British Library Cataloguing-in-Publication Data
A catalogue record for this book is available from the British Library

ISBN: 978-0-12-816449-5

For information on all Academic Press publications visit our website at https://www.elsevier.com/books-and-journals

Publisher: Charlotte Cockle
Acquisition Editor: Megan R. Ball
Editorial Project Manager: Laura Okidi
Production Project Manager: Sojan P. Pazhayattil
Cover Designer: Greg Harris

Typeset by TNQ Technologies

Contents

Contributors xi
Preface xiii

1 Cleaner production strategies for the food industry 1
Saioa Ramos, Susana Etxebarria, Maite Cidad, Mônica Gutierrez, David San Martin, Bruno Iñarra, Idoia Olabarrieta, Ángela Melado-Herreros and Jaime Zufia
1 Introduction 2
2 Cleaner production: principles and methodology for food industry 3
2.1 Planning and organizing cleaner production 5
2.2 Preassessment 6
2.3 Assessment 7
2.4 Evaluation and feasibility study 8
2.5 Implementation of viable cleaner production opportunities 9
3 Environmental improvement in the food industry 9
3.1 Water efficiency 10
3.2 Energy efficiency 14
3.3 Raw material efficiency 18
4 Performance and environmental indicators 22
4.1 Key Performance Indicators 22
4.2 Environmental impact indicator 23
5 Future perspectives: food industry 4.0 26
5.1 The use of information in the food industry 4.0: smart data 27
5.2 Food industry 4.0: a change in the cleaner production paradigm 28
References 30

2 Green supply chain 35
Stella Despoudi
1 Introduction 35
2 Definition of green food and green food supply chain 39
3 Green farming 40
4 Green food product design 42
4.1 Food product stewardship 42
4.2 Eco-design of food products or design for the environment (DfE) 43
4.3 Eco-labeling of food products' and product certification 44
4.4 Life Cycle Assessment and life cycle costing 47

5 Green food sourcing and procurement **47**
6 Green warehousing **49**
7 Green food logistics **51**
8 Green food supply chains and circular food supply chains **52**
8.1 Internal environmental management **52**
8.2 Waste management and food waste management **53**
8.3 Reverse logistics and circular supply chains **54**
9 Conclusion **55**
References **57**

3 Life cycle assessment in the food industry **63**
Adriana Del Borghi, L. Moreschi and M. Gallo
1 Introduction **63**
2 Life cycle assessment (LCA) **65**
2.1 The application of LCA to food sector **68**
2.2 LCA databases **81**
3 Labeling schemes **90**
4 EU organic logo and other standards on organic farming **98**
5 Product environmental footprint (PEF) **99**
6 Environmental product declaration (EPD) **100**
7 Examples of product category rules (PCR) related to food **107**
8 Conclusions **111**
References **112**

4 Microbes and the environment: fermented products **119**
Maria Aspri and Dimitrios Tsaltas
1 Introduction **119**
2 Interaction of cheese with the dairy environment **120**
3 House microbiota of dairy industry **121**
4 Examples of cheeses with characteristic in house microbiota **122**
4.1 Pit cheese **122**
4.2 Cheeses ripening in wooden vats **122**
4.3 Cheese ripening in goat's skin bag **130**
4.4 Cheese ripening in caves **130**
5 Interaction of fermented sausages with meat industry **131**
6 House microbiota of meat industry **131**
7 Examples of traditional sausages with characteristic house microbiota **134**
7.1 Serbia **134**
7.2 Portugal **134**
7.3 Spain **134**
7.4 Greece **138**
7.5 Italy **138**
7.6 Methods to study the microbial ecology of fermented foods **139**
8 Conclusion **141**
Funding **141**
References **141**

5 **Biorefineries for the valorization of food processing waste** **155**
Antonio D. Moreno, Mercedes Ballesteros and María José Negro
1 Introduction **156**
2 The food industry: wastes and valuable by-products **157**
2.1 Animal-derived waste **157**
2.2 Plant-derived waste **158**
2.3 Market potentials from food-derived by-products **159**
3 Biorefineries based on food processing waste **160**
3.1 Biorefinery based on wine grape waste **160**
3.2 Biorefinery based on brewer's spent grain **165**
3.3 Biorefinery based on olive waste **169**
3.4 Biorefinery based on potato peels **174**
3.5 Other biowaste-based biorefineries **176**
4 Legislation on waste management **178**
5 Concluding remarks **179**
Acknowledgments **180**
References **180**

6 **Packaging and food sustainability** **191**
Fabio Licciardello and Luciano Piergiovanni
1 Introduction **191**
2 Environmental aspects of packaging materials for food use **192**
3 Impact of food wastes on the overall sustainability **194**
4 The contribution of packaging to the sustainability of food chains **196**
5 Sustainability-oriented strategies through the improvement of packaging efficiency **200**
5.1 Packaging lightweighting **200**
5.2 The choice of packaging materials with a focus on recyclability **201**
5.3 The shift to biobased plastic materials **202**
6 Sustainability-oriented strategies through the improvement of packaging effectiveness **208**
6.1 Redesign product-packaging configurations **210**
6.2 Maximize packaging protective role **210**
6.3 Nanotechnologies for packaging improvement **211**
6.4 Correct shelf life assignment **214**
7 Conclusion **215**
References **216**

7 **The concept of (virtual) water in the food industry** **223**
Camelia Adriana Bucatariu
1 Introduction **223**
1.1 Water (blue, green, gray): all virtual for the global food and agricultural system **224**
1.2 Water on the global policy agenda **225**

1.3 Data needs: blue, green, and gray (virtual) water 227
1.4 Water: a global vital flow impacted by climate change 231
1.5 Governance: from virtual water to a global food and agricultural system water requirement 233
2 Conclusions 236
References 239
Further reading 241

8 Wastewater treatment and water reuse in the food industry 245
A.G. Tekerlekopoulou, Ch.N. Economou, T.I. Tatoulis, C.S. Akratos and D.V. Vayenas
1 Introduction 246
1.1 Food industry wastewaters 246
1.2 Characteristics of food industry wastewaters 247
2 Treatment methods 247
2.1 Biological methods 250
2.2 Physicochemical methods 255
2.3 Electrochemical methods 257
2.4 Constructed wetlands 258
2.5 Hybrid systems 259
3 Case studies of food industry wastewater treatment 260
3.1 Biological trickling filters 261
3.2 Constructed wetlands 265
3.3 Electrocoagulation treatment 270
4 Reuse of treated food industry wastewaters 272
5 Conclusions 273
Acknowledgments 274
References 274

9 Control of odors in the food industry 281
Nastaein Qamaruz-Zaman, Nurashikin Yaacof and Fatin Farhana Kamarzaman
1 Introduction 282
2 Example sources of odor in the food industry 283
2.1 Brewery industry 284
2.2 Baking industry 285
2.3 Palm oil industry 286
2.4 Poultry industry 288
3 Control of odor through odor management plan 289
3.1 Odor management plan documentation 289
3.2 Identification of odor sources from processing activities 290
3.3 Validation of odor exposure at the surroundings (ambient) 294
4 Odor control technologies 295
4.1 Adsorption 295
4.2 Absorption 296

4.3 Thermal oxidation 297
4.4 Catalytic oxidation 297
4.5 Nonthermal plasma injection 298
4.6 Biotechnologies 299
5 Modifications of biofilter technology for improved odor removal 301
5.1 Activated carbon assisted biofilters 301
5.2 Nonthermal plasma assisted biofilters 302
5.3 Multistaged high performance modules 302
5.4 Engineered biofilter media 303
6 The selection of odor control systems 304
6.1 Selection based on investment cost 304
6.2 Selection based on operational requirements 305
6.3 Maintenance requirement 305
6.4 Selection based on preceding environmental impacts 306
7 Conclusions 307
References 307

10 Innovation management and sustainability in the food industry: concepts and models 315
Barbara Bigliardi and Charis Galanakis
1 Introduction 315
2 How to classify innovation in the food industry 317
3 Models of innovation in the food industry 321
3.1 Food incremental innovation: the low-fat margarine and me-too food products 323
3.2 Food market breakthrough: the soy-based food 324
3.3 Food technological breakthrough: 3D food printing, the digital gastronomy and foodomics 324
3.4 Food radical innovation: functional foods and foodomics 326
4 Sustainability driven foods innovation: the case of food waste recovery 327
4.1 Other sustainability-driven foods innovation cases 329
5 Concluding remarks 331
References 333
Further reading 340

11 Food waste valorization opportunities for different food industries 341
Charis Galanakis
1 Introduction 341
2 The Universal Recovery Strategy 343
3 Innovation challenges and commercialization aspects of food waste recovery 344
4 Cereal processing by-products 347
5 Coffee processing by-products 358
6 Meat processing co-products and by-products 366

7 Olive mill processing by-products **370**
8 Grape processing by-products **377**
8.1 Fruit processing by-products **387**
9 Conclusion **402**
References **403**
Further reading **421**

Index **423**

Contributors

C.S. Akratos Department of Civil Engineering, Democritus University of Thrace, University Campus, Xanthi, Greece

Maria Aspri Department of Agricultural Sciences, Biotechnology and Food Science, Cyprus University of Technology, Lemesos, Cyprus

Mercedes Ballesteros CIEMAT, Department of Energy, Biofuels Unit, Madrid, Spain

Barbara Bigliardi Department of Engineering and Architecture, University of Parma, Parma, Italy

Camelia Adriana Bucatariu Independent Researcher, Rome, Italy

Maite Cidad Department of Efficient and Sustainable Processes, AZTI, Parque Tecnológico de Bizkaia, Astondo Bidea, Derio, Spain

Adriana Del Borghi Department of Civil, Chemical and Environmental Engineering (DICCA), University of Genoa, Genova, Italy

Ch.N. Economou Department of Chemical Engineering, University of Patras, Patras, Greece

Susana Etxebarria Department of Efficient and Sustainable Processes, AZTI, Parque Tecnológico de Bizkaia, Astondo Bidea, Derio, Spain

Charis Galanakis Department of Research & Innovation, Galanakis Laboratories, Chania, Greece; Food Waste Recovery Group, ISEKI Food Association, Vienna, Austria

M. Gallo Department of Civil, Chemical and Environmental Engineering (DICCA), University of Genoa, Genova, Italy

Mónica Gutierrez Department of Efficient and Sustainable Processes, AZTI, Parque Tecnológico de Bizkaia, Astondo Bidea, Derio, Spain

Bruno Iñarra Department of Efficient and Sustainable Processes, AZTI, Parque Tecnológico de Bizkaia, Astondo Bidea, Derio, Spain

Fatin Farhana Kamarzaman School of Civil Engineering, Universiti Sains Malaysia, Engineering Campus, Nibong Tebal, Penang, Malaysia

Fabio Licciardello DSV, University of Modena and Reggio Emilia, Reggio Emilia, Italy; GSICA, Italian Scientific Group of Food Packaging, Italy

Ángela Melado-Herreros Department of Efficient and Sustainable Processes, AZTI, Parque Tecnológico de Bizkaia, Astondo Bidea, Derio, Spain

Antonio D. Moreno CIEMAT, Department of Energy, Biofuels Unit, Madrid, Spain

L. Moreschi Department of Civil, Chemical and Environmental Engineering (DICCA), University of Genoa, Genova, Italy

María José Negro CIEMAT, Department of Energy, Biofuels Unit, Madrid, Spain

Idoia Olabarrieta Department of Efficient and Sustainable Processes, AZTI, Parque Tecnológico de Bizkaia, Astondo Bidea, Derio, Spain

Luciano Piergiovanni DeFENS, University of Milan, Milan, Italy; GSICA, Italian Scientific Group of Food Packaging, Italy

Nastaein Qamaruz-Zaman School of Civil Engineering, Universiti Sains Malaysia, Engineering Campus, Nibong Tebal, Penang, Malaysia

Saioa Ramos Department of Efficient and Sustainable Processes, AZTI, Parque Tecnológico de Bizkaia, Astondo Bidea, Derio, Spain

David San Martin Department of Efficient and Sustainable Processes, AZTI, Parque Tecnológico de Bizkaia, Astondo Bidea, Derio, Spain

Stella Despoudi Aston Business School, Aston University, Birmingham, United Kingdom

T.I. Tatoulis Department of Environmental and Natural Resources Management, University of Patras, Agrinio, Greece

A.G. Tekerlekopoulou Department of Environmental and Natural Resources Management, University of Patras, Agrinio, Greece

Dimitrios Tsaltas Department of Agricultural Sciences, Biotechnology and Food Science, Cyprus University of Technology, Lemesos, Cyprus

D.V. Vayenas Department of Chemical Engineering, University of Patras, Patras, Greece; Institute of Chemical Engineering Sciences (ICE-HT), Patras, Greece

Nurashikin Yaacof School of Civil Engineering, Universiti Sains Malaysia, Engineering Campus, Nibong Tebal, Penang, Malaysia

Jaime Zufia Department of Efficient and Sustainable Processes, AZTI, Parque Tecnológico de Bizkaia, Astondo Bidea, Derio, Spain

Preface

The food industry contributes significantly to the economic growth of many nations. Therefore, as resources become more restricted, demand grows, and environmental problems increase, sustainability and environmental impacts begin to affect the food industry around the world. In particular, this sector is associated with various environmental problems including high water consumption (foodstuff ingredients, cleaning, sanitation, heating, refrigeration, etc.) and the production of large volumes of wastewaters. To minimize environmental impact and achieve sustainability in the food sector, efforts include optimization of food manufacturing, the maximum utilization of energy, water and resources (from farm to folk), the valorization of processing by-products and the recycle of wastewater, but also the development of innovations within restricting environmental legislative frameworks. Subsequently, there is a need for a new guide covering the latest developments in the interface between food industry and environment.

Food Waste Recovery Group (www.foodwasterecovery.group of ISEKI Food Association) has published books dealing food waste recovery technologies, different food processing by-products' valorization (e.g., from olive, grape, cereals, coffee, meat, etc.), sustainable food systems and water processing, innovations in the food industry and traditional foods, nutraceuticals and nonthermal processing, shelf-life and food quality, personalized nutrition as well as targeting applications of functional compounds like polyphenols, proteins, carotenoids, and dietary fiber. Following these efforts, the current book aims to cover the interaction between the food industry and environment in all possible levels. The ultimate goal is to support professionals and enterprises that aspire to improve efficiency of the food industry and diminishing its environmental impact by denoting the available option for the responsible operational excellence in food production and processing.

The book consists of 11 Chapters. Due to increases in world population, minimizing the environmental impact associated with food production is key to ensuring the sustainability of the planet. One of the most implemented strategies to improve the environmental performance of industries and manufacturing companies is the Cleaner Production methodology. The main objective of this methodology is to identify and avoid operational inefficiency and optimize production to the maximum. Chapter 1 describes the phases necessary to implement this methodology in the food sector, as well as the main strategies to reduce the consumption of water, energy and raw materials.

Food companies and food supply chains face more significant challenges as they are highly affected by climatic change, food regulations, and resource limitations. Food companies need to incorporate environmental thinking into their core strategies and extend it across the whole supply chain. To this line, Chapter 2 provides an overview of how a green food supply chain can be achieved. Different elements of the green food supply chain are discussed which are as follows: green food product, green farming, green food product design, green food sourcing and procurement, green warehousing, green food logistics, and green food supply chains.

Chapter 3 presents and discusses the implementation of Life Cycle Assessment (LCA) in different food industries. Specific methodological issues related to the food sector are discussed with particular attention to functional unit, system boundary, allocation rules, crop cultivation, manure management and packaging. In addition, different product categories (e.g., products of agriculture, live animals and animal products, fish and other fishing products and beverages) are considered. A further step in applying LCA to food industry would be the spread of environmental certification and labeling schemes utilization for Business to Business and Business to Consumer communication. For this reason, applications of environmental labels to food products are presented, too.

Food processing environments can play a major role in the development of microbial consortia of the final food product. For instance, once microbes are introduced into food-industry facilities they may occupy specific niches, persisting and affecting the physicochemical characteristics of the food product. Chapter 4 describes the microbial diversity of naturally fermented dairy and meat products, and the impact of processing environment and equipment on its growth.

Chapter 5 deals with the use of the waste resulting from the food manufacturing industry, taking grape-derived waste, brewer's spent grain, olive-derived waste, potato-derived waste, and dairy by-products as biorefinery examples. These processes are presented showing the feasibility and constraints of applying industrial symbiosis toward the implementation of a circular bioeconomy. Furthermore, value-added products with especial interest for the nutraceutical and pharmaceutical industry are highlighted, including some antioxidants and phenolic compounds with anticancer activity.

Chapter 6 presents a deep investigation into the role of packaging within the sustainability of the food chains. In this context, food becomes a key element and the life cycle impacts related to both the packaging and the food, including the food that is not consumed, must be considered. Food waste should be taken into great account since in many cases its environmental impact is higher than that arising from packaging. The chapter presents various sustainability-oriented strategies based on the improvement of packaging efficiency and effectiveness, i.e., lightweighting, shift to more recyclable materials and to biopolymers, redesign of packaging configuration, adoption of active packaging solutions, and potential of nanotechnologies.

Chapter 7 argues that all water (blue, green, and gray) utilized by the food and agricultural system is virtual. Once extracted, all water becomes invisible to the food supply chain stakeholders, including the end consumers and relevant waste management actors. Water security (availability, accessibility, stability, utilization) is directly linked with global population and the dynamics of diets. Therefore, in order to reflect this global viewpoint, the concept of a global food and agricultural system water requirement is presented, linking consumption and production in multiactor demand and supply cycle.

Chapter 8 provides an overview of the types and characteristics of wastewaters that originate from different agri-food industrial activities such as olive oil and table olive production, dairies and wineries, summarizes the technologies referred to in the literature for the treatment of such wastewaters, and presents different case studies on the treatment of dairy, table olive, and olive mill wastewaters with emphasis on biological methods, electrochemical processes and constructed wetlands. Finally, the quality of posttreatment wastewaters is demonstrated and their potential for reuse is discussed.

In the last decades, large-scale agricultural operations and food industries have increased. These operations generate numerous types of odors. To develop environmentally sound, sustainable agricultural and food industrial operations, it is necessary to focus research and its applications toward approaches that are innovative, effective, and reliable enough to meet the challenging demands of the industrial emissions. Chapter 9 revises the available information regarding odor from agricultural operations and food industries by giving an overview about odor problems and suggesting solutions to control odor pollution.

Innovation in the food industry comes in different forms: incremental, product or process, social, and so on. To date, no classification of innovation in this context exists. Chapter 10 reviews the main classification proposed in the extant literature and identifies an innovation model suitable for the food industry, following the urgent need for sustainability and reduced environmental impact of our times. In addition, it provides examples for each kind of innovation giving emphasis on sustainability driven innovations as well as market, consumer, and technology aspects.

Chapter 11 revises current aspects of food waste recovery in the food industry, giving emphasis on different cases of processing by-products. At first, food waste–related and newly introduced definitions are provided, whereas the *Universal Recovery Strategy* is analyzed to explore how it can help overcome innovation obstacles and improve the sustainability of food systems. Thereafter, it discusses current valorization trends of by-products obtained in different food industries, particularly those generated during processing of cereals, coffee, meat, olives, grapes, and other fruit.

Conclusively, the book addresses food and environmental science and technology researchers, academics and consultants working in food processing, as well as those who are interested in the environmental management of food processes. It could be used by university libraries and institutes all around the world as a textbook and/or ancillary reading in undergraduate- and postgraduate-level multidisciplinary courses dealing with food and environmental science and technology.

At this point it is important to thank all authors for contributing in this book. Their acceptance of my invitation, adaption to editorial guidelines and respect of timeline are highly appreciated. I would also like to acknowledge the acquisition editor Megan Ball, the book manager Katerina Zaliva, and Elsevier's production team for their help during editing and publication process. Finally, a message for you, the reader. This collaborative reference is a scientific effort integrating different approaches and concepts in one integral test of hundreds of thousands of words. Therefore, it may contain some errors or gaps. Instructive comments, questions, or even criticism are always welcome, so please do not hesitate to contact me in order to discuss any relevant issues.

Charis Galanakis
Food Waste Recovery Group
ISEKI Food Association
Vienna, Austria
foodwasterecoverygroup@gmail.com
Research and Innovation Department
Galanakis Laboratories
Chania, Greece
cgalanakis@chemlab.gr

Cleaner production strategies for the food industry

1

Saioa Ramos, Susana Etxebarria, Maite Cidad, Mónica Gutierrez, David San Martin, Bruno Iñarra, Idoia Olabarrieta, Ángela Melado-Herreros, Jaime Zufia
Department of Efficient and Sustainable Processes, AZTI, Parque Tecnológico de Bizkaia, Astondo Bidea, Derio, Spain

Chapter outline

1. Introduction 2
2. Cleaner production: principles and methodology for food industry 3
2.1 Planning and organizing cleaner production 5
2.2 Preassessment 6
2.3 Assessment 7
2.4 Evaluation and feasibility study 8
2.5 Implementation of viable cleaner production opportunities 9
3. Environmental improvement in the food industry 9
3.1 Water efficiency 10
3.1.1 Water consumption 10
3.1.2 Wastewater generation 12
3.2 Energy efficiency 14
3.2.1 Reducing energy consumption at source 16
3.2.2 Energy or heat recovery 16
3.2.3 Alternative energy sources 17
3.3 Raw material efficiency 18
3.3.1 Food waste reduction 20
3.3.2 Reusing food by-products 20
3.3.3 Recovery and recycling 20
4. Performance and environmental indicators 22
4.1 Key Performance Indicators 22
4.2 Environmental impact indicator 23
5. Future perspectives: food industry 4.0 26
5.1 The use of information in the food industry 4.0: smart data 27
5.2 Food industry 4.0: a change in the cleaner production paradigm 28
References 30

The Interaction of Food Industry and Environment. https://doi.org/10.1016/B978-0-12-816449-5.00001-1

1. Introduction

The world population is expected to increase by more than 1 billion people in the next 15 years (UN-DESA, 2017), which generates a growing demand for food globally. Furthermore, scientific and technological developments are revolutionizing the production systems, communication, and consumption worldwide. This situation has triggered a series of global challenges which compromises the future of our planet. Climate change, loss of biodiversity, accumulation of waste, resource depletion (water, fossil fuels, etc.), and the modification of natural ecosystems, are some of the consequences of main anthropogenic activities (Foster et al., 2006).

The agrifood industry is main activity of the European manufacturing industry, representing 14.6% of its total turnover and a value exceeding €1,048,000 (Food Drink Europe, 2013). It has about 286,000 companies, most of them (99.1% of the total) Small and Medium Enterprises (SMEs) with less than 250 workers.

The global food production, processing, transport and consumption systems are responsible for 25% of the total emissions of greenhouse gases, 60% of the loss of terrestrial biodiversity, 33% of the degraded soils, overexploitation of 20% of aquifer resources, as well as overexploitation of 90% of commercial marine species (Tilman and Clark, 2014; WWF, 2017; FAO and PAR, 2011; Scherr, 1999; Dalin et al., 2017; Kituyi and Thomson, 2018) In addition, 70% of fresh water is consumed for food production (WWAP, 2012) and the 32% of food waste every year (FAO, 2011b). Thus, improving food production systems is one of the main actions that urgently needed to be carried out to reduce the global impacts on the environment.

This is visualized through the 2030 sustainable development goals established by the UN in 2015. Many of the objectives are directly related to food, specifically objective two, "to end hunger, achieve food security and improve nutrition and promote sustainability agriculture" and Goal 14, "conserve and sustainably use oceans, seas and marine resources for sustainable development." Other objectives are also indirectly related to food, such as Goal 12 "responsible production and consumption." Likewise, specific objective 12.3 intends, by the year 2030, to "halve per capita global food waste at the retail and consumer levels and reduce the food losses along production and supply chains." This situation leads to the need to develop new production and consumption models based in ecological foundations and sustainable development.

Before proceeding to look for approaches to reduce the environmental impact of food production, it is necessary to understand main cause and origins of environmental impacts caused by food production systems. Depending on the raw material acquisition method, distribution, processing, packaging materials, use and final disposal, the environmental impact would be also different because the overall environmental impacts are accumulated **along the entire food chain** (Andersson et al., 1994).

In all the production processes of a given food or beverage product, there is a consumption of resources and materials coming directly from the nature or from the technosphere (water, energy, fertilizers, chemicals, packaging material, or fossil fuels). Likewise, losses of raw materials, wastes, wastewater and atmosphere emissions are also accumulated along the production process. They are the so-called "inputs" and "outputs" of the system, respectively.

These inputs and outputs interact with the environment and have negative effects in human health, ecosystems and resource depletion. In 2010 the EC approved the ILCD method to measure the environmental impact along the value chain of products and services (EC-JRC, 2011). This method stablishes the methodologies to transform those inputs and outputs to a series of environmental impacts, such as global warming or eutrophication potential. Therefore, to reduce the pressure on the environment, it is essential to consider the inputs and outputs of the whole food chain, focusing on the identification of main origins and causes of the impact in each stages of the chain, from raw material production to final disposal of the product.

To improve the environmental performance of foods and beverages, it is important to take into account the potential environmental impact from the beginning of the product design (Ramos, 2015). In this phase main causes of the environmental are determined, from raw material, energy and water requirement for the processing to processing requirement. The so-called ecodesign introduces the environmental variable as one more factor when developing new foods, reducing the environmental impacts linked to all stages, from the raw material production, to the final use and disposal. Currently the design of new food products is subject to other factors that can influence the environmental behavior of a food. Taste and other organoleptic characteristics, cultural preferences, quality, brand image, food safety, costs and price, market situation should be considered together with the sustainability when a product is ecodesigned (Ramos, 2015).

Finally, when a food processing plant is already in operation, and food products are on the market, one the most effective strategies to reduce environmental impacts is to follow a **Cleaner Production (CP) methodology**. The concept was developed during the preparation of the Rio Summit as a program of United Nations Environmental Program (UNEP) and United Nations Industrial Development Organization (UNIDO). The main objective of this methodology is to reduce the company-specific inputs and outputs to finally reduce the related environmental impacts (UNEP, 1993). Thus, in food industries main strategies are focused to minimize wastes and emissions and increase the productivity. Implementing CP programs along the entire value chain of a given food and drink products, could results in great environmental reduction.

2. Cleaner production: principles and methodology for food industry

CP was defined by UNEP in 1990 as: "The continuous application of an integrated environmental strategy to processes, products and services to increase efficiency and reduce risks to humans and the environment." This definition has been widely used as the working definition in all contexts related to the promotion of CP. It applies to:

- production processes: preserving raw materials and energy, eliminating toxic raw materials and reducing the quantity and toxicity of all emissions and wastes,

- products: reducing negative impacts along the life cycle of a product from raw materials extraction to its ultimate disposal,
- services: incorporating environmental concerns into designing and delivering services.

Nonetheless, the confluence of global economic and environmental crisis that has occurred in recent years has consolidated the understanding of the dependence between our economic and environmental systems and has provided new motivation to promote transition toward more sustainable industrial systems. This fact has required the extension of the definition of CP to include **resource efficiency** which is a key element to that transition.

CP is an approach to environmental management that aims to improve the environmental performance of products, processes and services by focusing on the causes of environmental problems rather than the symptoms. It is not a "react and treat" approach, CP is based a preventive philosophy to face inefficiencies with the ai to reduce "end-of-pipe" solutions, or in some cases, even eliminate completely (UNEP, 1994). By preventing inefficient use of resources and avoiding unnecessary generation of waste, an organization can benefit from reduced operating or waste disposal costs.

CP is most commonly applied to specific company or production processes; however, it can also be applied throughout the life cycle of a product, from the initial design phase through to the consumption and disposal phase. Techniques for implementing CP include improved housekeeping practices, process optimization, raw material substitution, new technology and new product design. It is important to emphasize that CP is about attitudinal as well as technological change (UNEP, 1994). In many cases, the most significant CP benefits can be gained through lateral thinking, without adopting technological solutions. A change in attitude on the part of company directors, managers and employees is crucial to gaining the most from CP.

As stated above, industrial food processing plants are characterized by significant water, energy and raw material consumption, while disposing also large amounts of organic bioproducts, wastes and wastewater to the environment. Therefore, CP in food industries focuses on minimization of resource consumption, reduction of the waste generation, a better use of food by-products looking for increasing the process efficiency.

However, food production has specific characteristics which makes the application of the CP more challenging than in other manufacturing sectors. For instance, most of the food and drink products are perishable products, implying the need to include preservation systems and technologies that allow a safe commercialization of the foodstuffs. Indeed, the conservation processes used to be the stage with the highest water and energy consumption all the food processing industries.

Other key element in the food production chain is the maintenance of the food safety in all the steps of the food production chain. Food contamination happens when food is corrupted with another substance, like hair, plant stalks or pieces of plastic and metal (physical contamination) or chemical substances like pesticides, migration agents from packaging materials and toxins or biological agents like bacteria, virus or parasites. To avoid any kind of contamination the maintenance of strict hygiene requirements, among other practices, is key aspect in all the food

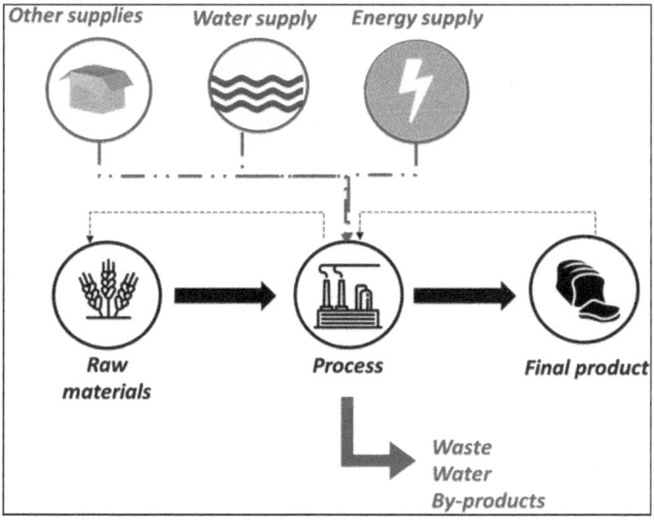

Figure 1.1 Cleaner production scheme.

commercialization chain. The maintenance of the hygiene requirements is always fundamental stage when water usage is accounted and most of the times this use could not be minimized.

Finally, as the whole sector relays on the production of natural living resources this sector also generates large amounts of organic wastes. The generation of a large percentage of these wastes could be avoided with good manufacturing and handling practices, while some of them, like skin, scales or bones, could not be avoided and are specific to the type of raw material being processing.

Overall, although challenging, the implementation of CP strategies can be used as a competitive advantage in the Food industry (Parfitt et al, 2010). The methodology described in this chapter is based on the guide developed by UNEP (1993, 1994), and consists on the following basic steps (Fig. 1.1):

1. Planning and organizing CP
2. Preassessment (overview of the production and environmental aspects of the company)
3. Assessment (collect data and evaluate the environmental performance and production efficiency of the company)
4. Evaluation and feasibility study
5. Implementation of viable CP opportunities and developing a plan for the continuation of CP efforts

2.1 Planning and organizing cleaner production

The following steps are imperative for a successful CP Assessment:

(i) to obtain the management commitment,
(ii) to organize and involve the project team,
(iii) to identify barriers and solutions, and
(iv) to set the milestones and goals.

Once the management of the company recognizes the need for implementing a CP assessment, the project team must be set. The project team should include not only the management but also the workers involved in everyday operations and maintenance, because they have a better understanding of the process and they are able to come up with suggestions for improvement. The success of the implementation largely depends on the collaboration of the staff, so it is a key factor to involve main staff responsible of operational and auxiliary activities. In food industries, suitable project team should be comprised by the quality manager, production manager, maintenance manager, and some employees responsible for the facility cleaning or management of the wastewater treatment plant. In addition, the team should be supported by consultants from outside the company where necessary.

Once the project team is established, and all the staff members are aware of the CP benefits, the team should detail objectives of the CP assessment. At best scenarios, all processes and operations should be assessed, however, time and resource constraints may make necessary to select the most important aspects or process areas. It is common for CP assessments in food industry to focus on those processes that

(i) generate a large quantity of waste and emissions,
(ii) generate a large consumption of natural resources (water, energy, raw materials), ii) entail a high financial loss,
(iii) has clear potential for improvement or
(iv) are considered as a problem by everyone involved.

2.2 Preassessment

Once the basis of the CP is established, a preassessment should be performed. The objective of the preassessment is to obtain an overview of the production and environmental aspects of the food company.

Producing a **flow chart of the production activity** identifying main inputs (raw materials, ingredients, water, energy, packaging or preservation elements) outputs (organic residues, by-products, wastewaters or packaging residues) and environmental problem areas is key step in the preassessment. All the information gathered in the flow chart will be the basis for the material and energy balances which should be detailed later in the assessment. To create a complete process flow chart, the team should also pay attention to the auxiliary processes that go hand in hand with the main process. Some of them are:

- Equipment and facility cleaning
- Material storage and handling
- Cooling, steam and compressed air production
- Equipment maintenance and repair
- Auxiliary materials (catalysts, lubricants etc.)
- By-products released to the environment

On-site visit is the most effective technique for getting first-hand information about a production operation in a short time. Indeed, most of the information needed in the preassessment may be obtained during this visit. The production site inspection should follow the process from the reception of the raw materials to the delivery of the final food products, focusing on those areas where products, wastes and emissions are produced.

2.3 Assessment

Afterward, detailed assessment should be performed to collect data about the food process production, auxiliary processes and evaluate the production efficiency of the company.

Amounts of resources consumed and wastes and emissions generated in each step of the activity should be quantified, to identify main processes to improve. For instance, to calculate the water consumption, water consuming activities should be identified, and the amount of water consumed should be calculated or estimated on direct measurements if necessary. Data collection is the base to do the material balance which allows the identification and quantification of uncontrolled losses, wastes or emissions, and suggest main sources and causes of those inefficiencies. A simple calculation of the material balance could be done with the following formula (Eq. 1.1):

$$\text{Raw Material}_{\text{in}} = \text{Product}_{\text{output}} + \text{Wastes}_{\text{output}} + \text{Emissions}_{\text{output}} + \text{Accumulated} \tag{1.1}$$

The material balance and the analysis of main causes and origin allow the project team to know the production efficiency and environmental performance of the process under study. All this information will be further presented as **environmental or ecoefficiency analysis** of the company, where performance indicators for the process are included.

These performance indicators are achieved by dividing the quantity of a material input or waste stream by the production over the same period. Performance indicators may be used to identify over consumption of resources or excessive waste generation by comparing them with those of other companies or figures quoted in the literature. Those indicators are also valuable to analyze and track the ecoefficiency performance along the years (see Section 4.1 of current chapter for more information regarding Key Performance Indicators).

After the analysis of the information collected during preassessment and assessment steps, the project team should identify as much as possible improvement opportunities. Indeed, some of them should have already been identified during the diagnosis and the on-site visits, but other ideas may come from literature review, workers' knowledge, discussions with suppliers, examples established in other companies or additional research and development projects.

Most common alternative to create improvement ideas for CP opportunities is to run a brainstorming session with all the CP project team. Brainstorming sessions have proved to be the most effective technique when managers, technicians, process operators and other employees as well as some outside consultants are required to work together without hierarchical constraints. During the brainstorming sessions participants should carefully analyze main causes and origin of the identified inefficiencies and will provide ideas for improve them. Some of the main alternatives to improve ecoefficiency could be aggregated through following criteria:

- Material changes or substitution
- Technology change or modifications
- Good operating practices
- Product changes or modifications
- Reuse of by-product or waste recycling

Finally, the CP team should undertake a preliminary screening of the CP improvements identified to decide on implementation priorities. In such a screening exercise, the improvements will be categorized into two groups: those that can be implemented directly and those that require further feasibility study because of the technical difficulty or the cost of investment.

2.4 Evaluation and feasibility study

The objective of the evaluation and feasibility study phase is to evaluate the proposed CP opportunities and to select those suitable for implementation. Those opportunities that required further investigation should be evaluated according to their technical, economic and environmental criteria.

For performing a complete **technical evaluation** the following changes should be assessed for each improvement opportunity: new material requirement, water and energy balances, variations in the quality or sensory characteristics of the products, alteration of main characteristics of the packaging (temperature resistance or portion size), modification of production methods, modification of cleaning and maintenance instructions, changes in the requirement of human resources such as more labor or new skills required, modification on the spaces or zones of the production facility and finally, time and easiness of the implementation should be also addressed.

The objective of the **economic evaluation** is to evaluate the cost effectiveness of each improvement strategy. Economic viability is often the key parameter that determines whether an opportunity will be implemented. Required capital investment, amortization of the equipment, expected financial savings or costs due to changes in water and energy consumption, reduction or increase of new material costs or changes in supplier requirements are main criteria to take into account for this evaluation.

Least but not last, the **environmental performance evaluation** of an improvement action should take into account the consumption of raw materials, energy and water and outputs to the environment of the activity under study. For a good environmental evaluation, the following expected changes should be addressed: energy and water requirements, amount of raw material, packaging material and auxiliary materials required, amount and toxicity of wastes or emissions, changes in degradability of wastes and other emissions or amount of renewable raw materials used.

The three evaluation criteria allow the selection and prioritization of previously identified improvement opportunities. Depending on the financial situation or the environmental awareness and commitment of the company, each criterion will be weighted generating a comparative ranking to prioritize opportunities for implementation (Fig. 1.2). The options with the highest scores will probably be best suited for implementation. However, the results of this analysis should not be blindly accepted, they should form a starting point for a discussion. The more promising strategies must be selected in close collaboration with the management of the company and those selected ones should be deeply analyzed in the next step "the cleaner production implementation plan."

2.5 Implementation of viable cleaner production opportunities

The objective of the last phase of the assessment is to ensure that the selected strategies or actions are implemented, and that the resulting reductions in resource consumption and waste generation are maintained over time. To ensure fruitful implementation of the selected options an action plan should be developed detailing the activities that needs to be carried out, human and financial resource requirements, person and position responsible for undertaking those activities and the time frame for completion with intermediate milestones.

To ensure positive changes and achievement of previously defined targets, **periodic monitoring** is a compulsory and very important practice. Therefore, to evaluate the effectiveness over the time some performance indicators are used. Reduction in wastes and emissions per unit of production, reduction in resource consumption (including energy) per unit of production or improved profitability are Key Performance Indicators in CP of food industries.

By providing technical information on CP opportunities individuals and organizations within the food industry will be able to take advantage of the benefits that CP has to offer.

3. Environmental improvement in the food industry

As stated before, one of the most important steps when carrying out a CP assessment is the data acquisition for the assessment of the environmental status of a given company. Main inputs and outputs that are used in the industrial stages associated with food products should be considered and summarized to stablish a complete overview of the activity. The consumption of natural resources, such as water, energy or fossil resources, as well as emissions to water, atmosphere or soils from waste and discharges, contribute to current environmental impacts. Thus, reducing the amounts of resource requirement is, always, one of the main strategies to reduce the impact. In this section, water, energy and resource efficiency are deeply analyzed as key aspects to take into account when performing a CP assessment in food industries.

Water consumption is a key aspect for the food industry, since it is used as a raw material for the product, but especially for the cleaning processes of industrial facilities and equipment (Foster et al., 2006). The consumption of large amounts of water generates impacts in the basins and watersheds where the catchments are produced, as well as in the groundwater (Boulay et al., 2011). Today, many of the rivers and aquifers of industrial zones are in a water stress status, that is, the contribution of natural water does not reach the necessary levels to maintain the ecosystems and at the same time have enough water for the anthropogenic activities demand (consumption in households, industry, agriculture, etc.).

In addition, water not incorporated into the product drags a large amount of discharges with high organic load that also have an impact on the environment. Organic loads of nitrogen and phosphorus compounds poured into the natural channels have great potential for eutrophication, and greenhouse gases that contribute to global warming are emitted from the nitrogen decomposition cycle (Smith et al., 1999). Likewise, the discharges of chemical substances with the cleaning agents, lubricants, paints or other auxiliaries generate toxicity in the ecosystems and in human health, with potential carcinogenic effects in the population.

Regarding energy, most energy sources are based on fossil resources such as oil or coal, which generate global warming potential and depletion of abiotic resources. According to Foster et al. (2006) 25% of climate change is due to activities related to food production, and much of that impact is directly related to energy consumption, whether in the form of electricity or related to transport or generation of heat or steam.

Another of the most important aspects that will be discussed in this section includes the use of raw materials and the generation of waste. It was not until 1990 when the term circular economy was published for the first time, which bases are established in the principles of reduce, reuse and recycle. This model must be the basis to reduce the amount of food waste and make the most of the current waste streams, since in many cases they are sending to landfill molecules or substances with a great potential to be used in the food industry or in other sectors.

3.1 Water efficiency

Food and beverage industry are one of the major contributors to the growth of all economies (CED, 2017). However, the sector has been associated with various environmental issues including high levels of water consumption and wastewater production (Valta et al., 2013).

Efficient water use is important to the success of the food industry. Ecoefficiency is based on the idea of "doing more with less," and efficient water use is an important factor. When referring to water efficiency, two main aspects should be considered, on the one hand the water consumption and on the other hand, wastewater generation and organic and toxic loads.

3.1.1 Water consumption

Total water abstraction in Europe is about 350 km^3/year, in other words, approximately 10% of Europe's total freshwater resource is abstracted annually (EEA, 2018). The three main users of water are agriculture, industry and the domestic sector, e.g., households. On average, 44% of total water abstraction in Europe is used for agriculture, 40% for industry and energy production, and 15% for public water supply.

The food and drink sectors are particularly reliant on a supply of freshwater as raw material and also for operational needs. Water uses are classified depending on the final use, distinguishing following categories: general purpose water, process water, cooling water and boiler feed water. In food processing facilities, water is used from the conditioning of raw materials such as soaking, cleaning, blanching and chilling, continues with cooling, sanitizing, steam generation for sterilization, power and heating, and, finally, direct "in-process" use (Table 1.1). Moreover, to ensure sanitary conditions and standards, high amounts of water have always been used in most facilities for cleaning operations.

Depending on the final use or purpose different water quality is required. When **water is used as a food ingredient**, its quality (e.g., impurities) can affect the properties of the food, including texture, shelf stability, appearance, aroma and flavor. In this case, the quality of water can have significant impact on the safety, quality and taste of products. Besides, **as a processing aid**, water may be used for conveying, heating, cooling, rinsing, dissolving, dispersing, blanketing, diluting, separating, steam generation and other activities. In each case, purity of the water will affect its performance. Finally, the **water used for cleaning** the equipment, facilities, and ancillary equipment should be potable, clear, colorless, and free of contaminants that affect taste or odor.

In the recent years, being aware of the need to guarantee the sustainability of resources, important efforts have been made to reduce water consumption in the food and drink industries, reporting significant savings. In fact, controlling all water uses, improving production plan, reusing and recycling water in the facility or improving layout designs are some premises to reduce water consumption in food industries.

According to Schug (2016) water consumption can be reduced by up to 30% by simple cultural and operational changes with low capital investment. Examples include awareness and monitoring programs and taps with presence control. Similar savings could be achieved by reusing water flows in the same facility or designing

Table 1.1 Main processes and operations consuming water resources in food and drink industry.

Process	Operation
Product preparation	Washing, cleaning, peeling, cutting
Processing	Mixing, steaming, pumping/transferring, water-in-product
Preservation	Freezing, heating, boiling
Packing	Canning, bottling
Equipment cleaning	Cleaning-in-place (CIP), rising, washing, sterilization
Site facilities	Site maintenance, vehicle washing, staff facilities

improvement programs, however capital investment could be higher, and a deep analysis should be done due to potential consequences on the quality and safety on the final products (Kirby et al., 2003).

3.1.2 Wastewater generation

Most of the water not used as ingredient or steam generation, ultimately is disposal at the waste water stream. Although the food sector is an extremely diverse, certain sources of wastewaters are common. Washing and steeping of raw material, cleaning facilities or equipment, cleaning packaging or containers, cooling or cooking water, water used for transporting raw material, products or wastes and finally storm-water run-off are most common sources of waste water in most food and drink subsectors.

The wastewater profile and organic load is largely dependent on production and cleaning patterns. Moreover, some food sectors have season dependent production which affects the effluent flowrate and characteristics along the year. Overall, effluents could be extremely variable in composition, but typically, untreated wastewater has the following common characteristics:

- High COD and BOD levels, with amounts 10 to 100 times higher than in domestic wastewater. This characteristic has high potential to deplete the oxygen in receiving streams.
- Significant presence of materials in suspension and, in a lesser degree, colloidal matter.
- Usually physical parameters are easily perceptible: turbid waters, colored, fouled and/or very hot effluents.
- High loads of organic chemicals such as oils and fats which are very difficult to remove.
- Inorganic matter is also common due to active substances of the cleaning agents. Sulfates, chlorides, phosphates or ammonium ions could potentially vary the pH of the water effluents.

Due to the earlier-mentioned characteristics and the amounts of water used in some subsectors, food and drink processing facilities used to have their own wastewater treatment or pretreatment plants. When the wastewater is treated, it is commonly discharged to sewers and finally a municipal wastewater treatment plant (WWTP). In that cases, the degree of treatment required is determined by the specific limitations defined in the municipal ordinances for discharging to sewers (Waldron, 2009). However, depending on the treatment applied or the location of the facility, those wastewaters could be discharged directly to water bodies. In that cases, high regulatory and legislative standards should be compliment.

In both cases, the quantity and characteristics of wastewater final streams has economic and environmental consequences with respect to treatment and disposal. Overall, there are two approaches to reduce wastewater amounts and pollution.

On the one hand, CP methodology could be implemented to reduce water consumption and pollution. The following approaches are the most common practices to achieve water efficiency in the food and drink industry:

- Promotion of best environmental practices: Good practices should be applied in each stage of the production chain. Dry floor cleaning, reducing losses (as leakage) or training of employees are main actions to consider.

- Maintenance of equipment: Regular maintenance of equipment and facilities is required to minimize water spills and leakage of contaminants.
- On-site reuse of water: Depending on the characteristics of the wastewater, it could be reused within the same process or in other one, such as the use of cooling water for floor cleaning. However, depending on the regulation nonexhaustive sanitization is usually required to close water loops.
- Control and substitution of materials: The use of less hazardous auxiliaries or cleaning agents, and materials with longer lifetime could potentially reduce the amounts of contaminates on the wastewaters.
- Optimization or substitution of processing technology: Sometimes old equipment or inappropriate design of the production chain could lead to large amount of wasted water. These amounts could be easily reduced by controlling water consumption along the whole process, acquiring more efficient technologies or equipment or redesigning processes and facility layout.
- Optimization of the product characteristics and design: Improving product characteristic such as the lifetime, raw material reception conditions or packaging properties could reduce the use of water the minimize the organic loads of the effluents.

On the other hand, when the wastewater amounts are already reduced to minimum, end-of-pipe techniques should be applied. In this case, on-site wastewater treatment plants are usually the most effective methods.

To ensure the optimal performance of the WWTP a homogeneous load and constant flowrates of wastewater are required. However, as stated above, in the food and drink industries high variability of loads and amount are presented. Wastewater is therefore collected in equalization or balance tanks before entering to the WWTP. Most of the WWTP integrated on the food and drink industries hold the following steps:

1. Primary treatment: this involves separation of solids and grease in a settling tank (primary clarifier), or by flotation.
2. Secondary treatment: This step consists of biological treatments to reduce organic matter (COD and BOD). In case of high organic loads, anaerobic biological treatment should be applied. On the contrary, if low organic loads are quantified, aerobic biological treatment is recommended.
3. Tertiary treatment: This step includes physicochemical methods such as adsorption, stripping, coagulation or sedimentation to remove nutrients such as phosphorus, sulfide, suspended solids, remaining BOD as well as pathogens. Not all the food industries possess this tertiary treatment. Usually, those industries discharging directly to water bodies, should refine the treated water loads to fulfill with regulations.

Table 1.2 shows main characteristics of the detailed approaches to control and reduce water consumption and wastewater pollution. Indeed, prevention and minimization of wastewater require the adoption of a systematic approach (CIAA, 2002), and to ensure optimal water efficiency, both approaches should be implemented together (Hara et al., 2014).

Finally, best environmental performance is usually achieved procuring best technology or equipment and introducing best environmental practices for the manufacturing of food and drink products. This is recognized by the IPPC Directive definition of "techniques" as "both the technology used and the way in which the installation is designed, built, maintained, operated and decommissioned."

Table 1.2 Approaches for water consumption and wastewater pollution control.

Cleaner production (CP)	End–of–pie treatment (EOP)
Responsibility	
All people through the company are responsible	Solutions are developed by experts often in isolation
Focus	
CP is a continuous evaluation and innovation process	EoP solves only one strict problem
Goal is to meet customers' needs with minimal impact on health and environment	Quality is defined as meeting customers' requirements
Environmental control	
CP is integrated part of product and process development	EoP starts when problems arise
Waste is considered to potential resources	Activities cost the company time and extra money
CP includes technical and nontechnical approaches	Only technology oriented
Handling pollutants	
Pollution prevention through measures at the source	Pollutants controlled by filters and treatment methods
CP is a continuous work to achieve higher development	Only goal is to fulfill standards set by the authority
Involves new practices, attitudes and management techniques and stimulates technical advances	Replies mainly on technical improvements to existing technologies

3.2 Energy efficiency

The agrifood sector is a major consumer of energy, accounting in 2011 for around 30% of the total world energy consumption (FAO, 2011a,b), varying depending on the country or region. Indeed, according to Monforti-Ferrario et al. (2015), in 2013 the energy required to produce food and drink products reached up to the 17% of the EU's gross energy consumption and around 26% of the EU's final energy consumption. Overall, agriculture and livestock management are the most energy intense stages of the food and drink supply chain, accounting for nearly the 33% of the energy consumed in the food production chain. The second most important phase of the food life cycle is industrial manufacturing, which accounts for 28% of the total energy consumed in food and drink life cycle. Moreover, the industrial production together with distribution and packaging are responsible for nearly **half of the total energy** use in the food system.

Different types of energy sources could be found in the food industries. Indeed, actual energy sources on the food industry in the EU-28 mainly came from natural gas (47.8%) and electricity (34%), while fuel oil (7%) and renewable energy sources (3%) are less consumed.

Lower prices make natural gas one of the main energy source in food industries. It is a hydrocarbon gas mixture containing mainly methane, and it is usually used to produce heat for different purposes, such as cooking, room temperature conditioning or steam generation. Regarding the electricity consumption, it is widely used for refrigeration systems or packaging processes, among of the side operations such as motors or compressed air generation. Besides, depending on the country or region different electricity sources could be identified. On average, in Europe, almost half (48.7%) of the net electricity generated in the EU-28 in 2016 came from combustible fuels (such as natural gas, coal and oil), while more than one quarter (25.7%) came from nuclear power stations. Among the renewable energy sources the highest share of net electricity generation in 2016 was from hydropower plants (12.1%), followed by wind turbines (9.7%) and solar power (3.5%)

The energy consumed per unit of food or drink product is very diverse among products and, event for the same product, it is also different depending on the country or the processing procedure. However, on average, energy for cooling and heating, processing machinery and building conditioning are main contributors to the energy consumption:

Regarding the energy required for cooling, refrigerator systems serve for many applications in the food and drink industries and it contributes to around 30% of the total company's energy use (Monforti-Ferrario et al., 2015). Chilling or freezing preservation of foodstuffs, cold storage of raw materials and final products or cooling water for heat exchangers are main activities where cool is used. On the other hand, steam and hot water are also widely used in most of the food subsectors. Defrosting processes, food preserving technologies (pasteurization or dehydration), boiling or cleaning processes consumes large amounts of calories.

Besides, side operations required to support main processing are also significant contributor of energy consumption. Fans, pumps and motors, ventilation system or compressed air generation are one of the major auxiliary activities of this category. Finally, building lighting and heating should be also considered.

Before the global economic crisis, company managers generally looked for cost optimization in other areas before addressing possible savings in energy consumption. However, incredibly tight price-settings of retailer and distribution networks, consumers demanding wide variety of products at low prices and high standard and regulation exert huge pressure on the food and beverage industry. Therefore, nowadays, food and drink companies are looking at energy as key parameter of company's economic and environmental performance.

In spite, the food industry in the EU-28 is largely dominated by SMEs and it makes energy efficiency difficult to be achieved (FoodDrinkEurope, 2018).

Thus, to improve energy efficiency while optimizing production processes, food companies are implementing different approaches. These strategies can be focused on:

(i) the reduction of current energy consumption through strategies to improve the energy efficiency of equipment and industrial facilities,
(ii) energy and heat recovery, or
(ii) the selection of renewable energy sources minimizing the impact of the energy consumed.

3.2.1 Reducing energy consumption at source

Technological and process optimization can decrease energy use in European food industries around 15% by 2020 (Altmann et al., 2013). While certain energy "costs" cannot be avoided, in other activities such as in chilling, freezing, or cooking the way the process is carried out can lead to substantial savings.

Main strategies to reduce energy consumption in food and drink industries could be focused on housekeeping practices, process controls, changes in operating conditions optimization of technological equipment (Navarri and Bedard, 2008). Possible measures span from high-efficiency motors and increased use of combined heat and power (CHP) to intelligent and efficient lighting and appliances' voltage optimization. In 2010, Kamiński and Leduc gathered most suitable energy reduction measures for main energy consumption activities in the food industries. Measures were focused to optimize steam, motor and pumps, compressed air, cooling and refrigeration and lighting systems.

For instance, improving boiler insulation or implementing condensate return systems could save up to 25% and 10% of the total energy consumption respectively, with relatively low or medium implementation costs. Moreover, efficiency of motors and pumps could be easily optimized by installing Variable Speed Drivers. Savings up to 20% of have been reported with this improvement measure, however it has medium implementation costs and relatively high payback period. Similar savings have been reported for pump and motors by properly sizing and improving maintenance. Furthermore, 25% of reduction of energy consumption of the compressed air systems could be also achieved by implementing some managerial actions to support maintenance and monitoring programs, such as leak reporting and repairing or avoiding system pressurizing during nonproductive periods. Higher savings has been reported when installing heat recovery measures, however, it is not always viable solution.

3.2.2 Energy or heat recovery

To identify energy or heat recovery opportunities in the agrifood sector, **Process Integration** appears as a suitable technique. It is a holistic approach to process design which considers the interactions between different unit operations, rather than optimizing them separately (Innhoff et al., 1982). Process integration is an essential tool for the energy efficiency of food and drink industries with the goal of converting conventional processes into an integrated system that optimizes energy efficiency (Maréchal and Muller, 2008). By using process integration, it is possible to identify a process which has the potential to use the heat rejected by another unit, even if the units are not running at optimum conditions on their own.

Process integration studies start by reviewing the main energy consumption activities and measuring energy requirement of each process. However, to meet the goal of maximum energy recovery an appropriate heat exchanger network is required. For that reason, data collection study is followed up by quick assessment of the heat or energy recovering possibilities supported by a traditional **pinch analysis** (Gundersen, 2000). This technique evaluates thermodynamically reachable energy targets for a given process and identifies how to achieve them, allowing an efficient design with minimal energy consumption and optimal heat recovery.

Main energy recovery techniques or measures in the food and drink industries are the integration of combined heat and power production (CHP) and industrial heat exchanger. If surpluses of energy are generated, they could be sold to power companies to obtain an economic return. The use of water as a support media in the process makes also the food and drink manufacturing processes interesting candidates for a combined heat and water recovery.

The advantage of process integration techniques is that there are little changes in processes, and energy savings can be significant. The only required changes are the implementation of heat exchanger network, with pumps, pipes, and heat exchangers. Problems can arise with machines that incorporate heating systems, predesigned for steam or high temperature heat supply. Indeed, ad-hoc studies should be carried out to reduce risk when implementing heat recovery techniques.

3.2.3 Alternative energy sources

Ultimately, no matter how many efforts are made to reduce or recover the energy required in a given company, there will always be an unavoidable energy consumption that will be required to ensure production in the food industries. To minimize the impacts related to the consumption of this energy, the last option is to introduce renewable energy sources.

Despite the variety of renewable energy sources, bioenergy dominates significantly the potential uses in industry. Solar power and electricity from renewables are the only other renewable energy sources offering significant potential, while others, such as geothermal heat, appear to be limited to specific cases where suitable resources are available.

From simple, low-investment technologies or actions, to more complex and demanding integration projects different integration schemes are possible. The selection or prioritization should be made considering the potential to reduce energy consumption and the economic investment of each company or facility. For instance, green power procurement with a third-party power producer or on-site installation of renewable power generation equipment could be the simplest and lowest costs action in food and drink industries. Incorporation of biomass boilers or installation of solar panels are some examples of low-cost integration of renewable energy generation equipment. However, there are other renewable energy production technologies which requires process adaptation such us substituting natural gas-fueled ovens to run on geothermal steam.

Finally, as most of the wastes generated in a food industry are a biomass resource, it can be converted into energy resources (heat and power) for the process (Maréchal and Muller, 2008). Within this approach double benefit could be obtained, from one side obtaining renewable raw materials and energy, and valorization of by-products.

3.3 Raw material efficiency

FAO estimates that globally one third of food produced for human consumption is lost or wasted throughout the entire supply chain, which amounts to about 1.3 billion tons per year (FAO, 2011b). In the EU, around 88 million tonnes of food waste are generated annually with associated costs estimated at 143 billion euros (http://www.eu-fusions.org/phocadownload/Publications/Estimates of European food waste levels.pdf; FUSIONS, 2016).

All the natural resources associated to the cultivation, production, storage and distribution of any food product like fertilizers, fossil fuels, water or energy, are also wasted when food is not used and then discarded. These resources have significant greenhouse gas emissions and environmental impacts (freshwater eutrophication or methane emission from organic waste decomposition). Therefore, the reduction of environmental impact would be significant when decreasing food waste amounts (Peffers et al., 2007).

The amount of food losses and waste varies significantly across the food chain, from the agriculture, food industry, distribution, restaurants to the households. According to a recent report of the European Commission (EC, 2011), food wastes generated at the food industries accounts for the 35% of total food wasted in the food and drink sector.

In order of priority, the objectives of CP in terms of raw material efficiency are, to avoid or reduce at a minimum the generation of food wastes, to reuse food losses in own facilities and to recycle the unavoidable by-products, food losses and waste is often distinguished between unavoidable and avoidable. **The avoidable fraction** is usually the food that is wasted due to processing errors, products with the wrong weight, shape or appearance, or damaged packaging (Quested and Johnson, 2009). In these cases, safety, taste or nutritional value of the food is not affected, however, in a standardized production line these products often end up being discarded (SEPA, 2008; Stuart, 2009). **The unavoidable fraction** defines those wastes or by-products which are not part of the final product. This fraction can be recovered and reused or recycled for different applications being upgraded as raw materials for other products, for food, feed or as a source of chemicals. Sometimes they are just treated as waste and disposed into landfills.

Final disposal or treatment of food losses and waste accounts for large impact when assessing the sustainability of the food industry. The economic consequences are normally focused on the production cost or on the market prices. However, potential economic benefits due to the avoiding externalities or the cost of the resources necessary to produce those wastes should be also taken into account. Thus, when transforming this reduction into monetary savings, minimizing food waste is intended as benefit for supply chain members, households and the entire society (Campoy-Muñoz et al., 2017). In addition, economic benefits due to reduction of waste disposal and treatment costs should be also considered (De Lange and Nahman, 2015).

The environmental impacts of food losses and wastes increases along the supply chain, since all emissions generated on upstream steps are included in the overall impact assessment of the waste material. Scherhaufer et al. (2018) concluded that three quarters of the total global warming potential impact caused by food waste is generated at the primary production step. Moreover, GHG emissions from food processing activities contribute 6%, retail and distribution 7%, food consumption 8%, and food disposal 6% of the total food waste–related global warming potential impact.

Indeed, Tonini et al. (2018) concluded that the global warming potential impact of avoidable food waste ranged from 2000 to 3600 kg CO_2-eq. per ton of waste, depending upon the food waste composition. For instance, in Europe, 186 Mt CO_2-eq can be attributed to food waste (Scherhaufer et al., 2018), which implies 15%–16% of the total impact of the entire food supply chain. In addition, several environmental impacts measured by greenhouse gas (GHG) emissions and footprints concluded that food waste accounts for 23%–24% of the total use of cropland, freshwater resource, and fertilizers related to food production (Cuéllar and Webber, 2010; Thyberg and Tonjes, 2016; Kummu et al., 2012).

Within this framework, The Waste Framework Directive (Directive, 2008/98/EC) laid down a five-step hierarchy of waste management options which must be applied by Member States when developing their national waste policies:

1. Waste prevention
2. Preparing for reuse
3. Recycling
4. Recovery (including energy recovery)
5. Safe disposal (as a last option)

This is aligned with the Reduce, Reuse, and Recycle (3R) strategy, considering that the first level of the European Commission (EC) waste hierarchy deals with waste before it is produced, which includes waste reduction and reuse.

Moreover, global policies are pushing toward the adoption of food losses and waste prevention, valorization, and management strategies. Actually, the United Nations Sustainable Development Goal 12.3 aims to halve per capita food waste, as well as its management costs, while maximizing the value of unavoidable food waste and packaging materials by 2030 (Scherhaufer et al., 2015).

Within the Circular Economy Package (EC (2015) 614 final), the EC introduces similar measures to reduce the amounts of food wastes considering the whole life cycle of a given product. Within this framework, the transition to a more circular economy, where the value of products, materials and resources is maintained in the economy for as long as possible, and the generation of waste minimized, is an essential contribution to the EU's efforts to develop a sustainable, low carbon, resource efficient and competitive economy.

3.3.1 Food waste reduction

First and simplest measures to reduce food waste should always be focused on avoid the generation of them, and thus, taking maximum advantage of the raw material. The approach to avoid generation starts with the quantification of current waste streams and evaluation of main origin and causes.

According to the expertise of the authors of this study, main avoidable food wastes are generated on

(i) raw material reception due to first preparation of foodstuffs like defrosting, cutting, or peeling,
(ii) filling or packaging of the final product, and
(iii) storing due to bad conditions or nonconforming products.

Usually, simple managerial and operational improvement actions are enough to reduce the generation of this avoidable food waste, like stock monitoring, training employees, or establishing agreements with suppliers to purchase more suitable raw material. There are also quite a lot of simple measures like optimization conveyor belts speed and vibration or readjustments on packaging equipment which could avoid large amounts of accidental food spills. Another reduction measure consists on develop markets for "substandard" products. Both commercial and charity organizations could arrange for the collection and sale or use of discarded "substandard" products that are still safe and of good taste and nutritional value (SEPA, 2008).

3.3.2 Reusing food by-products

The most common strategy to reuse food waste or by-products occurs in the processing companies where the recoverability is higher (Perego et al., 2011). The "recoverability" of wasted food is expressed as the opportunity to use it with a minimum additional work. This is one of the main approaches of the "cleaner production" methodology and can be achieved through the use of more efficient processes or better technologies.

Intermediate or final products that do not comply with commercialization criteria can be reintroduced in the production line to be reutilized. Such as cuts of bread, dough in the bakery industry or chocolate crumbs or pieces of chocolate in the confectionery production.

Sometimes, products with lower quality could be manufactured from the recovered materials. A well-known example of this practice occurs in the canned tuna industry, where crumbs generated as side stream cooking and cutting processes of tuna fillets, are also commercialized for lower quality tuna crumbs cans.

3.3.3 Recovery and recycling

Thinking circular instead of linear must be an important step toward a sustainable and efficient raw material management. The hierarchy of waste or by-product management established by the waste framework Directive of the EU parliament (2008) establishes that the reuse for human food or industrial uses must be prioritized. Being the use as

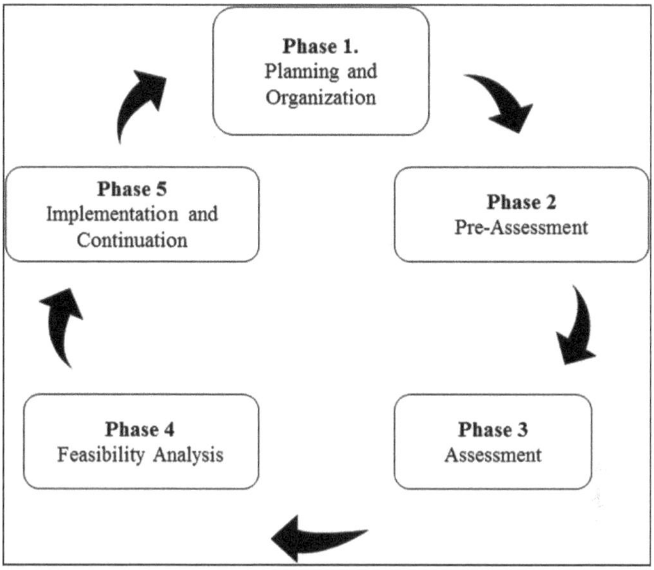

Figure 1.2 Hierarchy of food waste valorization adapted from EU waste framework directive and USA-EPA.
From San Martin, D., Bald, C., Cebrian, M., Iñarra, B., Orive, M., Ramos, S., Zufía, J., 2018. Principles for Developing a Safe and Sustainable Valorisation of Food Waste for Animal Feed: Second Generation Feedstuff. Handbook of Famine, Starvation, and Nutrient Deprivation. From Biology to Policy. Springer International Publishing AG, United Kingdom.

raw material for feed an interesting and very used alternative in many food sectors (Fig. 1.2).

Due to an increasing scarcity of global resources, supply of new raw materials is becoming an important issue to consider for ensuring long-term sustainability of food chains. Hence, food losses and waste are presented as an opportunity for a more sustainable supply of raw materials.

EU waste framework directive promotes the inclusion of biowaste in animal feed, leading to additional benefits for animal feed sector. Availability of environmentally friendly sources of ingredient and decrease of dependence on foreign sources are main benefits that should be considered. In addition, this will reduce the environmental pressure and the overexploitation of soil associated to the agricultural activity by reducing necessary soil destined to produce vegetable raw material for animal feed (Rojas-Downing et al., 2017).

However, the recovery of unavoidable food losses and wastes used to need additional processing steps before being recycled. It is necessary to consider the financial situation of food companies to support the initial investment at the time of implementing any valorization option. Hence, it is essential to obtain suitable and high added-value products to justify the investment and to reduce the costs of obtaining the new valorized product. In addition, type and amount of wastes, the intrinsic potential value for exploitation, contained undesirable substances, the geographical location of producers, intermediaries be involved in the processing and valorization or the potential end-users are main key factors to consider (Galanakis, 2015; San Martin et al., 2017).

In this sense, different studies have demonstrated the viability of the recovery of food losses and waste for other production processes. Indeed, there are plenty of examples of successful valorization of food wastes, like the obtention of new raw materials for animal feed by a simple dehydration process of vegetable by-products produced by food industry and retailers (San Martin et al., 2016).

Nowadays, to maximize the efficiency and added value of the valorization of food wastes, the biorefinery concept has been combined with the traditional valorization schemes. In 2009 the IEA Bioenergy defined **the biorefinery** as the sustainable processing of biomass into a spectrum of bio-based products (food, feed, chemicals, materials) and bioenergy (biofuels, power and/or heat). By producing multiple products, a biorefinery takes advantage of the various components in the biomass of food wastes and their intermediates therefore maximizing the value derived from the biomass feedstock. In 2015 Barkroufa et al. published one the first successful case study of the biorefinery of orange peel wastes. In their study, Barkroufa et al. obtained essential oil, polyphenols, and pectin from orange peel by optimizing microwave and ultrasound technology without adding any solvent.

4. Performance and environmental indicators

To evaluate the achievement of efficiency goals stablished in the CP plan, regular assessment of some sustainability indicators is a key procedure. Two types of indicators could describe the efficiency of processing facility or productive process: performance indicators and environmental impact indicators. On the one hand, performance indicators are associated with production processes and offer a general information about productivity, performance or relative consumption within the industry. On the other hand, environmental impact indicators, convert behavior information into environmental impact according to established models and methods.

4.1 Key Performance Indicators

The use of Key Performance Indicators (KPI) is well known and established throughout all sectors of industry. They provide a simple, focused measure of performance, allowing a concise picture of what is happening in the activity.

One of the major measures for food production is **yield**. This indicator is defined as the number of units coming out of a process divided by the number of units or raw material going into that process over a specified period. This KPI is really important since major costs and environmental impacts of the food industries are originated from the acquisition of raw material for food product and for the packaging. Each food subsector has their own optimal yield and companies should try to achieve that value. If a given company has very low yield, it means that valuable raw material is losing in some point of the production chain. Moreover, even if there are unavoidable raw material loses such as fish or meat bones and peels or fruits stones, this KPI also allows the quantification of those unavoidable parts that are sensitive to valorize.

Apart from yield to ensure and monitor the CP processes in Food industries there are other KPIs that should be also considered. Overall, Water consumption, wastewater generation, energy consumption and food waste generation ratios per kg of final product must be measured and tracked.

4.2 Environmental impact indicator

As mentioned above, the methodology of CP is based on implementing environmental improvement strategies that reduce costs and the use of resources such as water, electricity and raw material in food industry. This methodology helps, on the one hand, to reduce environmental and human health impacts and, on the other hand, to increase the efficiency of the food companies.

Nevertheless, the CP does not include all burdens of the life cycle of a product, process or service in food industry because it is focused on the consumption, emission and waste generation of the manufacturing facility. Moreover, the environmental impacts generated due to the consumption and emissions occurred during the production are not measured. For that purpose, the Life Cycle Assessment (LCA) appears as a suitable internationally standardized method to calculate environmental impact of goods and services (Ramos et al., 2016). Thus, it is useful to combine this procedure with the CP methodology to support a sustainable decision-making (Lopes et al., 2018; Dominguez-Ramos et al., 2007).

The general procedure for conducting an LCA is standardized in ISO 14040:2006a and ISO 14044:2006b. The methodology is divided in four simple steps. Firstly, goal and reference unit of the study should be stablished. Then, all the resources consumed and the emissions generated along the life cycle of a product should be quantified, considering primary production, manufacturing, distribution, use and end-of-life.

Afterward, in the impact characterization step, the inputs and outputs are converted to different environmental impacts. Potential environmental impact on global warming or climate change is one the most common environmental impact considered in most of the studies, however depending on the food product other environmental impacts are also relevant, such es potential impact on freshwater or terrestrial eutrophication, human toxicity or ozone layer depletion. Specific data from food company activity, such as water or energy consumption or wastewater volume discharged, is collected and translated into different environmental impact indicators using specific characterization factors stablished in the official methods. These factors express burdens per unit emission or resource consumed in the context of each indicator (EC-JRC, 2011). For instance, according to the IPCC 2011 methodology (Stocker, 2014), the characterization factor for Climate change impact category of 1 kg of CO_2 emitted is 1 kg CO_2 equivalent, while for 1 kg of methane emitted the characterization factor is 24 kg CO_2 equivalent.

Although different methodologies have been published, in recent years the impact categories and methods recommended EC are the most widely used (EC-JRC, 2011). In Table 1.3, recommended impact categories, units and characterization

Table 1.3 Environmental impact indicators, measurement units, description and methodologies of the impact selected by EC-JRC (updated from 2011).

Impact indicator	Description	Recommended default LCIA method
Acidification (mol H^+ eq)	Accumulated Exceedance (AE) characterizing the change in critical load exceedance of the sensitive area in terrestrial and main freshwater ecosystems, to which acidifying substances deposit	AE (Seppälä et al. 2006, Posch et al. 2008)
Climate change (kg CO_2 eq)	Global Warming Potential calculating the radiative forcing over a time horizon of 100 years	Baseline model of 100 years of the IPCC (based on IPCC 2013) (Stocker, 2014)
Ecotoxicity, freshwater (CTUe)	Comparative Toxic Unit for ecosystems (CTUe) expressing an estimate of the potentially affected fraction of species integrated over time and volume per unit mass of a chemical emitted (PAF m3 year/kg)	USEtox model (Rosenbaum et al. 2008)
Eutrophication, freshwater (kg P eq)	Fraction of nutrients reaching freshwater end compartment (P)	EUTREND model (Struijs et al. 2009)
Eutrophication, marine (kg N eq)	Fraction of nutrients reaching marine end compartment (N)	
Eutrophication, terrestrial (mol Neq)	AE characterizing the change in critical load exceedance of the sensitive area, to which eutrophying substances deposit	AE (Seppälä et al. 2006, Posch et al. 2008)
Human toxicity, cancer (CTUh)	Comparative Toxic Unit for humans (CTUh) expressing the estimated increase in morbidity in the total human population per unit mass of a chemical emitted	USEtox model (Rosenbaum et al. 2008)
Human toxicity, noncancer (CTUh)	Comparative Toxic Unit for humans (CTUh) expressing the estimated increase in morbidity in the total human population per unit mass of a chemical emitted	

Table 1.3 Environmental impact indicators, measurement units, description and methodologies of the impact selected by EC-JRC (updated from 2011).—cont'd

Impact indicator	Description	Recommended default LCIA method
Ionizing radiation, human health (KBq U^{235} eq)	Human exposure efficiency relative to U^{235}	Human health effect model (Frischknecht et al. 2000)
Land use - Several subindicators	Soil quality index	Soil quality index based on LANCA (EC-JRC) (Beck et al. 2010)
Ozone depletion (kg CFC-11 eq)	Ozone Depletion Potential (ODP) calculating the destructive effects on the stratospheric ozone layer over a time horizon of 100 years	Steady-state ODPs 1999 as in WMO assessment
Particulate matter (Disease incidence)	Impact on human health	UNEP recommended model (Fantke et al. 2015)
Photochemical ozone formation, human health (kg NMVOC eq)	Tropospheric ozone concentration increase	LOTOS-EUROS (Van Zelm et al. 2008)
Resource use, fossils (MJ)	Abiotic resource depletion – fossil fuels (ADP-fossil)	CML 2002 (Guinée et al. 2002) and van Oers et al. 2002
Resource use, mineral and metals (kg Sb eq)	Abiotic resource depletion (ADP ultimate reserves)	
Water use (m^3 world eq)	Use deprivation potential (deprivation-weighted water consumption)	Available WAter REmaining (AWARE) (Boulay et al. 2018)

methodologies are detailed. However, when referring to food industries following impact categories needs to be deeply analyzed: acidification, climate change; freshwater, marine and terrestrial eutrophication; cancer and noncancer human toxicity; freshwater ecotoxicity; land use; ozone depletion and resource and water uses.

Finally, impact characterization results should be interpreted. LCA allows the identification of main origin and causes of each of environmental impact, the so-called hot-spot identification. Since the aim of the LCA is to know the specific environmental impact of a given product, this methodology combines perfectly with the CP, focusing the efforts in the main hotspots identified.

All the life cycle stages of the food products contribute to the environmental impacts. However, there are some processes that has major environmental impact. Overall, according to several studies primary production, packaging and use phase are the most relevant processes (Andersson, 2000; Roy et al., 2009). Regarding primary production, fertilization and the use of phytosanitary products for crop cultivation, feed production and manure management of livestock are the hotspots of most of the selected environmental impacts. Extraction of packaging materials and the consumption of water and energy required for its production, is also a major contributor for main environmental impacts such as climate change or resource depletion. On the other hand, the use phase generates environmental impacts mainly due to resources and emissions occurred to produce the electricity required for the cooking, or due to the final disposal of the food wastes and packaging. Finally, in a globalized world, the distribution of food and drinks products is also a relevant step due to GHG and other substances emitted in fossil fuel combustion.

Even if the food manufacturing is not one of the main contributors to the environmental impact, LCA could also evaluate the environmental assessment in a specific equipment or process of the food industry. For instance, the environmental impact generated by the energy consumption of a given equipment or by the raw materials used could be easily measured and tracked. These results allow to select more sustainable techniques (i.e., BAT) or new raw materials or packaging with less environmental impact.

In summary, calculating environmental impact indicators with LCA methodology during CP enables to focus the efforts on main hotspots of the production chain.

5. Future perspectives: food industry 4.0

The term "Industry 4.0" was born in Germany, as part of the "High-Tech Strategy 2020 Action Plan," and it refers to the fourth Industrial Revolution (Kagermann et al., 2013). Industry 4.0 proposes an interconnected production, through the information exchange (smart data) of all the links of the supply chain, connecting workers, machine components, materials, processes, logistics or consumers. The information exchange allows the cooperation with each other and the identification and automatic reaction to unforeseen events downstream and upstream in the supply chain. This means that industries are to become self-organizing and steering, allowing a more efficient production with the highest quality, and thus a better control and management of the resources, which will allow the improvement of the environmental sustainability of the industries.

The Industry 4.0 is related to the concept of "smart" factory of the future, where systems are computer-driven, monitor physical processes and make decisions based on self-organization mechanisms (Smit et al., 2016) Furthermore, it is characterized by two types of integration: (i) horizontal, which refers to the integration of various information technologies (IT) systems across the entire value creation network or chain; (ii) vertical, which refers to the integration of various IT systems at the different hierarchical levels within the company to deliver an end-to-end solution. The key of the concept of industry 4.0 is the use of technologies that allow obtaining, processing and communicating data in an objective, cross-linked and quick way in both dimensions.

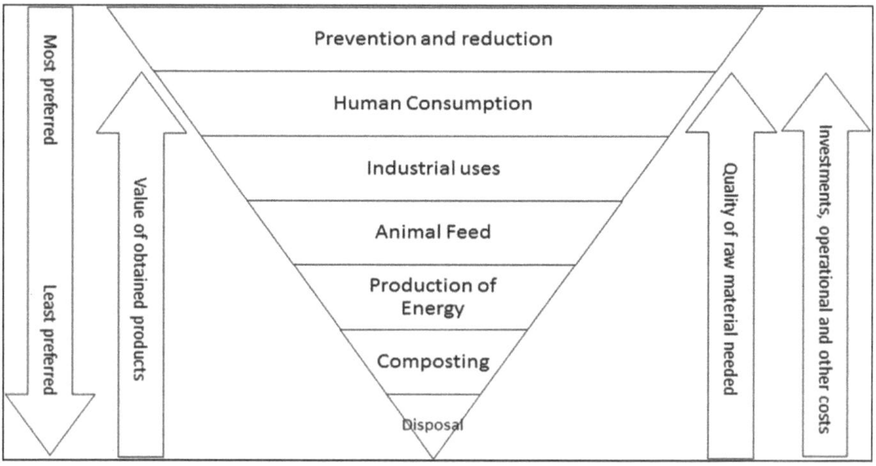

Figure 1.3 Tools of Industry 4.0 that may be applied to the food industry.

In the last years an overwhelming amount of 4.0 technologies have been emerging mostly in the field of manufacturing sector. Those innovations should be selected and adjusted to the characterization and needs of the food industry to reach a successful and effective transformation. Within the food industry, intelligent sensors, autonomous process robots, Internet of Things (IoT), big data, and cloud computing are main tools for this transformation (Fig. 1.3).

All technologies are digitized and connected, which means that the control and prediction capacity of the process management can be significantly increased at any time.

The food industry is at the tail of the implementation of the 4.0 concept due to their traditional component, commonly related to perception of quality, which makes the companies to be reluctant to the implementation of automation or innovative technologies. Anyway, this trend will be changing as long as the companies become aware of the impact of these tools in the sustainability of their business. On the one hand, improving the quality and efficiency of the production (economical sustainability) and, on the other hand, improving the traceability, communication and the consumers' trust along all the value chain (social sustainability); and, least but not last, improving the ecoefficiency of resources guaranteeing a CP strategy (environmental sustainability).

5.1 The use of information in the food industry 4.0: smart data

Having a deep knowledge of the production process and the food product is critical when designing a successful CP strategy. The information and data in the company should be objectively measured, registered and shared to take actions and to decrease the use of inputs and outputs that have an environmental impact.

Food sector is characterized by having a lot of SMEs, many times not only with limited resources but also quite overwhelmed for the high quality of new technologies coming up to the market and without the capacity of assimilating so much information. Moreover, in many cases, the traditional or artisan food processes are perceived or linked to quality, so that makes food industry to be reluctant to innovations too. For this reason, the implementation of intelligent CP strategies used to be difficult due to the lack of objective and reliable data and the limitations in communications systems (Zhang et al., 2017).

Smart sensors are a key tool of the 4.0 transformation which can generate large amount of data in real time. Food product characteristics and critical parameters of the process such as energy, water consumed by each equipment or process should be measured on real. The implementation of these sensors is not a straightforward pace though, due to the fact that food products and processes are complex and subjected to a continuous evolution. Hence, calibrating smart sensors for the food industry is complicated task. The whole variability that may appear when working with so complex matrices and processes needs to be taken into account. Indeed, these calibrations must be developed ad-hoc for each industry or company and they should be maintained for a long time with periodical calibration maintenance.

The type of smart sensors used in the food industry will differ depending on the link and processes of the studied supply chain. Some examples are RFID sensors, which might be used for logistics and product identification; computer vision systems, for external quality and foreign bodies inspection in grading lines; spectroscopic sensors, for physic-chemical quality assessment of the product, waste control, water reuse; spectroscopic imaging, also used for physic-chemical quality assessment and so on.

The huge amount of data generated by the smart sensors, must be converted into useful information. In this sense, big data and data mining techniques are suitable tools to organize and interpret this amount of data supporting the sustainable and smart decision-making. Data need to be cleaned, transformed and analyzed in the shortest time so to take real time decisions and to perform predictions. In the food industry, the type of data acquired will be heterogeneous coming from different sources, such as raw material conditions, stocks, processes, company's management data, water and energy use, suppliers or consumers, and very in different formats: images, video, spectra, text, etc. Data can be processed in real time or can be stored to support on further analysis (Zhang et al., 2017) and to build predictive models, that may anticipate a situation before happening. These techniques are able to process a huge amount of data and to extract the important information (Witkowski, 2017). The use of this information will have an impact on several aspects of the industry, such as processes, technical support, customer service that will help in the development of sustainable production and consumption.

5.2 Food industry 4.0: a change in the cleaner production paradigm

Nowadays companies aim to increase their ecoefficiency, orienting their strategies to improve their economic efficiency, minimizing their ecological impact while trying to

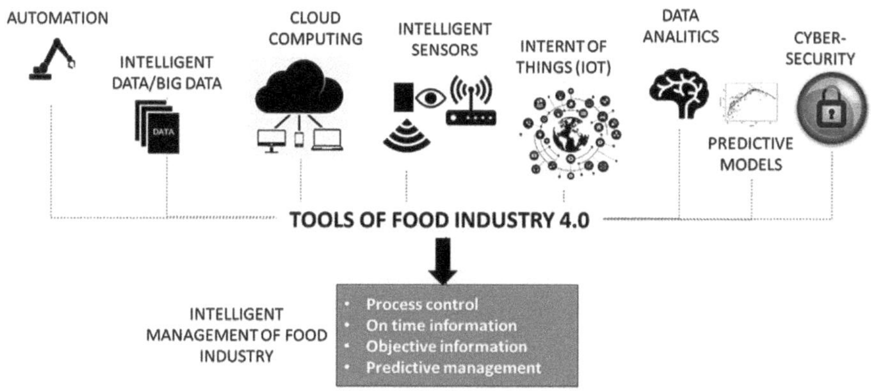

Figure 1.4 Scheme of the paradigm of traditional CP strategies.

accomplish the highest value with the minimum inputs (Henriques and Catarino, 2015). Indeed, CP can be considered as suitable tool to achieve this goal (Henriques and Catarino, 2015) and it has been one of the most important means to accomplish sustainable production for manufacturing enterprises (Zhang et al., 2017). CP complies continuous application of preventive environmental strategies for production, processes and services to increase the ecoefficiency.

Traditionally, the CP strategies in food companies have been focused on prevention, reuse and recycle (Fig. 1.4). Those are proactive actions, based on a limited amount of data, mostly coming from the company.

Besides, Industry 4.0 paradigm states that not only the company is internally connected, but it is also with the rest of the value chain. The company, could provide and receive information in real time about their own processes and the rest of the links of the value chain, adjusting their production to unforeseen events (Fig. 1.5). This leads to optimize the inputs needed for the production (i.e., under demand) and thus minimizing the environmental impact related to the water, energy or raw material efficiency.

This new concept of industry is capable to anticipate to the events before happening, thanks to the use of objective information and historical data which allow to build predictive models, optimizing the environmental and economical sustainability of the processes and the company by means of (Grin et al., 2010; Reid et al., 2010; Michaelis, 2003; Loorbach and Wijsman, 2013; Perrini and Tencati, 2006; Rotmans, 2006):

- Process optimization, through intelligent tools, that allow a better process control by predictive production.
- Raw material optimization, which minimizes the raw material waste.
- Energy optimization.
- Water loss minimization, by means of an efficient use of water.
- Real time recovery-reuse of materials, before they are out of the process.

Therefore, Industry 4.0 technologies offer a wide range of possibilities that imply a change in the paradigm of CP.

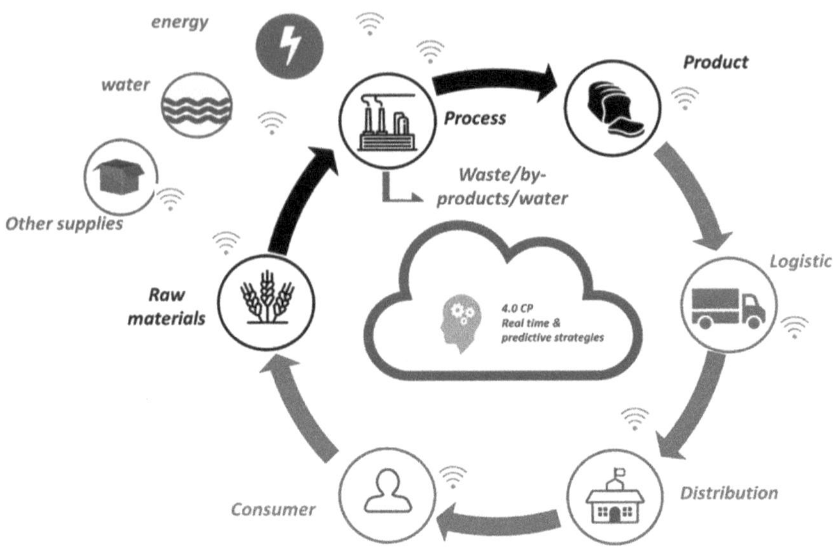

Figure 1.5 New paradigm of value chain based on the Industry 4.0 paradigm, where everything is connected.

References

Altmann, M., Zerhusen, J., Maio, P., Lanoix, J.C., Trucco, P., Egenhofer, C., Fujiwara, N., Behrens, A., Marcu, A., Theises, J., 2013. Energy Efficiency and the ETS. PE 492.468. IP/A/ITRE/ST/2012-11. Brussels, Belgium.

Andersson, K., 2000. LCA of food products and production systems. Int. J. Life Cycle Assess. 5 (4), 239.

Andersson, K., Ohlsson, T., Olsson, P., 1994. Life cycle assessment (LCA) of food products and production systems. Trends Food Sci. Technol. 5 (5), 134.

Beck, T., Bos, U., Wittstock, B., Baitz, M., Fischer, M., Sedlbauer, K., 2010. LANCA—land Use Indicator Value Calculation in Life Cycle Assessment. Fraunhofer, Stuttgart, p. 67.

Boulay, A.M., Bulle, C., Bayart, J.B., Deschênes, L., Margni, M., 2011. Regional characterization of freshwater use in LCA: modeling direct impacts on human health. Environ. Sci. Technol. 45 (20), 8948.

Boulay, A.M., Bare, J., Benini, L., Berger, M., Manzardo, A., Margni, M., Motoshita, M., Núñez, M., Pastor, A.V., Ridoutt, B., 2018. The WULCA consensus characterization model for water scarcity footprints: assessing impacts of water consumption based on available water remaining (AWARE). Int. J. Life Cycle Assess. 23 (2), 368.

Campoy-Muñoz, P., Cardenete, M.A., Delgado, M.C., 2017. Economic impact assessment of food waste reduction on European countries through social accounting matrices. Resour. Conserv. Recycl. 122, 202–209.

CED, 2017. Economic Contribution of the Food and Beverage Industry A Report by the Committee for Economic Development of the Conference Board. The Conference Board Arlington, USA.

CIAA, 2002. Background Document for the Technical Working Group on the "Food and Drink" BAT Reference Document Rev, vol. 7.

Cuéllar, A.D., Webber, M.E., 2010. Wasted food, wasted energy: the embedded energy in food waste in the United States. Environ. Sci. Technol. 44, 646.

Dalin, C., Wada, Y., Kastner, T., Puma, M.J., 2017. Groundwater depletion embedded in international food trade. Nature 543 (7647), 700.

De Lange, W., Nahman, A., 2015. Costs of food waste in South Africa: incorporating inedible food waste. Waste Manag. 40, 167.

Dominguez-Ramos, A., Aldaco, R., Irabien, A., 2007. Life cycle assessment as a tool for cleaner production: application to aluminium trifluoride. IJCRE 5 (1).

EC, 2011. Final report. Preparatory study on food waste across EU 27. by BIO Intelligence Service. Available: http://ec.europa.eu/environment/eussd/pdf/bio_foodwaste_report.pdf.

EC, 2015. Closing the Loop – an EU Action Plan for the Circular Economy COM/2015/0614 Final. http://eur-lex.europa.eu/legal-content/EN/TXT/?uri=CELEX:52015DC0614.

EC-JRC, 2011. ILCD Handbook—Recommendations for Life Cycle Impact Assessment in the European Context. European Commission, Joint Research Centre. Institute for Environment and Sustainability.

EEA, 2018. European Waters. Assessment of Status and Pressures 2018. EEA Report 7/2018. Publications Office of the European Union, Luxembourg.

Fantke, P., Jolliet, O., Evans, J.S., Apte, J.S., Cohen, A.J., Hänninen, O.O., Hurley, F., Jantunen, M.J., Jerrett, M., Levy, J.I., Loh, M.M., Marshall, J.D., Miller, B.-G., Preiss, P., Spadaro, J.V., Tainio, M., Tuomisto, J.T., Weschler, C.J., McKone, T.E., 2015. Health effects of fine particulate matter in life cycle impact assessment: findings from the Basel Guidance Workshop. Int. J. Life Cycle Assess. 20 (2), 276.

Food Drink Europe, 2013. 2013–2014 Data & Trends of the European food and drink industry. Belgium.

FAO, 2011a. 'Energy-Smart' Food for People and Climate, Issue Paper. Food and Agriculture Organization of the United Nations (FAO), Rome, p. 78.

FAO, 2011b. Global Food Losses and Food Waste – Extent, Causes and Prevention. Rome FoodDrinkEurope 2018. Data & Trends of the European Food and Drink Industry 2018, Brussels, Belgium.

FAO, PAR, 2011. Biodiversity for Food and Agriculture Contributing to Food Security and Sustainability in a Changing World Outcomes of an Expert Workshop Held by FAO and the Platform on Agrobiodiversity Research from 14–16 April 2010 in Rome, Italy.

Foster, C., Green, K., Bleda, M., Dewick, P., Evans, B., Mylan, J., 2006. Environmental Impacts of Food Production and Consumption: Final Report to the Department for Environment Food and Rural Affairs. Manchester Business School. Defra, London.

Frischknecht, R., Braunschweig, A., Hofstetter, P., Suter, P., 2000. Human health damages due to ionising radiation in life cycle impact assessment. EIA Rev. 20 (2), 159.

FUSIONS, 2016. Estimates of European Food Waste Levels. IVL Swedish Environmental Research Institute, Stockholm, Sweden.

Galanakis, C.M., 2015. Food Waste Recovery: Processing Technologies and Industrial Techniques. Elsevier Inc., London, U.K.

Grin, J., Rotmans, J., Schot, J., 2010. Transitions to Sustainable Development: New Directions in the Study of Long Term Transformative Change. N.Y.U.O.U. Routledge.

Gundersen, T., 2000. A Process Integration Primer — Implementing Agreement on Process Integration. International Energy Agency, SINTEF Energy Research, Trondheim, Norway.

Guinée, J.B., Gorrée, M., Heijungs, R., Huppes, G., Kleijn, R., van Oers, L., Wegener Sleeswijk, A., Suh, S., Udo de Haes, H.A., de Bruijn, H., van Duin, R., Huijbregts, M.A.J., 2002. Life Cycle Assessment: An Operational Guide to the ISO Standards. Kluwer Academic Publishers, Dordrecht, Netherlands.

Hara, K., Newman, P., Takao, Y., 2014. Sustainable Development Goals: How Can Japanese Local Governments Help? World Sustainability Forum 2014 –Conference Proceedings Paper.

Henriques, J., Catarino, J., 2015. Sustainable value and cleaner production – research and application in 19 Portuguese SME. J. Clean. Prod. 96, 379.

Innhoff, B., Townsend, D.W., Boland, D., Hewitt, G.F., Thomas, B.E.A., Guy, A.R., Marsland, R.H., 1982. A User Guide on Process Integration for the Efficient Use of Energy (IChemE UK).

ISO, 2006a. 14040:2006. Environmental Management-Life Cycle Assessment-Principles and Framework. European Committee for Standardization.

ISO, 2006b. 14044:2006. Environmental Management Life Cycle Assessment Requirements and Guidelines. European Committee for Standardization.

Kagermann, H., Wahlster, W., Helbig, J., 2013. Recommendations for Implementing the Strategic Initiative INDUSTRIE 4.0. Final Report of the Industrie 4.0 Working Group. acatech – National Academy of Science and Engineering, Germany.

Kirby, R., Bartram, J., Richard, C., 2003. Water in food production and processing: quantity and quality concerns. Food Control 14, 283.

Kituyi, M., Thomson, P., 2018. 90% of Fish Stocks Are Used up – Fisheries Subsidies Must Stop. World Economic Forum.

Kummu, M., de Moel, H., Porkka, M., Siebert, S., Varis, O., Ward, P.J., 2012. Lost food, wasted resources: global food supply chain losses and their impacts on freshwater, cropland, and fertiliser use. Sci. Total Environ. 438, 477.

Loorbach, D., Wijsman, K., 2013. Business transition management: exploring a new role for business in sustainability transitions. J. Clean. Prod. 45, 20.

Lopes, J., Medeiros, D.L., Kiperstok, A., 2018. Combining cleaner production and life cycle assessment for reducing the environmental impacts of irrigated carrot production in Brazilian semi-arid region. J. Clean. Prod. 170, 924.

Maréchal, F., Muller, D., 2008. Chapter 8 energy management methods for the food industry. In: Klemes, J., Smith, R., Kim, J.K. (Eds.), Handbook of Water and Energy Management in Food Processing. Elsevier. Woodhead Publishing.

Michaelis, L., 2003. The role of business in sustainable consumption. J. Clean. Prod. 11 (8), 915.

Monforti-Ferrario, F., Dallemand, J.-F., Pinedo Pascua, I., Motola, V., Banja, M., Scarlat, N., Medarac, H., Castellazzi, L., Labanca, N., Bertoldi, P., Pennington, D., Goralczyk, M., Schau, E.M., Saouter, E., Sala, S., Notarnicola, B., Tassielli, G., Renzulli, P., 2015. Energy Use in the EU Food Sector: State of Play and Opportunities for Improvement. EC Joint Research Centre, Institute for Energy and Transport and Institute for Environment and Sustainability, Ispra, Italy.

Navarri, P., Bedard, S., 2008. Chapter 4 Auditing energy and water use in the food industry. In: Klemes, J., Smith, R., Kim, J.K. (Eds.), Handbook of Water and Energy Management in Food Processing. Elsevier. Woodhead Publishing.

Parfitt, J., Barthel, M., Macnaughton, S., 2010. Food waste within food supply chains: quantification and potential for change to 2050. Philos. Trans. R. Soc. Lond. B Biol. Sci. 365 (1554), 3065.

Peffers, K., Tuunanen, T., Rothenberger, M.A., Chatterjee, S., 2007. A design science research methodology for information systems research. JMIS 24 (3), 45.

Perego, A., Perotti, S., Mangiaracina, R., 2011. ICT for logistics and freight transportation: a literature review and research agenda. Int. J. Phys. Distr. Log. 41 (5), 457.

Perrini, F., Tencati, A., 2006. Sustainability and stakeholder management: the need for new corporate performance evaluation and reporting systems. BSE 15 (5), 296.

Posch, M., Seppälä, J., Hettelingh, J.P., Johansson, M., Margni, M., Jolliet, O., 2008. The role of atmospheric dispersion models and ecosystem sensitivity in the determination of characterisation factors for acidifying and eutrophying emissions in LCIA. Int. J. Life Cycle Assess. 13 (6), 477.

Quested, T., Johnson, H., 2009. Household Food and Drink Waste in the UK. Waste & Resources Action Programme (WRAP).

Ramos, S., 2015. Ecodiseño de alimentos mediante análisis de ciclo de vida. Thesis. University of the Basque Country, Bilbao, Spain.

Ramos, S., Larrinaga, L., Albinarrate, U., Jungbluth, N., Ingolfsdottir, G.M., Yngvadottir, E., Landquist, B., Woodhouse, A., Ólafsdóttir, G., Esturo, A., Zufia, J., Pérez-Villarreal, B.J., 2016. SENSE tool: easy-to-use web-based tool to calculate food product environmental impact. Int. J. Life Cycle Assess. 21 (5), 710–721.

Reid, W.V., Chen, D., Goldfarb, L., Hackmann, H., Lee, Y.T., Mokhele, K., Ostrom, E., Raivio, K., Rockstrom, J., Schellnhuber, H.J., Whyte, A., 2010. Earth system science for global sustainability: grand challenges. Science 330, 916.

Rojas-Downing, M.M., Nejadhashemi, A.P., Harrigan, T., Woznicki, S.A., 2017. Climate change and livestock: impacts, adaptation, and mitigation. Clim. Risk Manag 16, 1.

Rotmans, J., 2006. Tools for integrated sustainability assessment: a two-track approach. IAJ 6 (4), 35.

Rosenbaum, R.K., Bachmann, T.M., Gold, L.S., Huijbregts, M.A., Jolliet, O., Juraske, R., Koehler, A., Larsen, H.F., MacLeod, M., Margni, M., McKone, T.E., Payet, J., Schuhmacher, M., Meent, D.V., Hauschild, M.Z., 2008. USEtox—the UNEP-SETAC toxicity model: recommended characterisation factors for human toxicity and freshwater ecotoxicity in life cycle impact assessment. Int. J. Life Cycle Assess. 13 (7), 532.

Roy, P., Nei, D., Orikasa, T., Xu, Q., Okadome, H., Nakamura, N., Shiina, T., 2009. A review of life cycle assessment (LCA) on some food products. J. Food Eng. 90 (1), 1.

San Martin, D., Ramos, S., Zufía, J., 2016. Valorisation of food waste to produce new raw materials for animal feed. Food Chem. 198, 68–74.

San Martin, D., Orive, M., Martínez, E., Iñarra, B., Ramos, S., González, N., Guinea de Salas, A., Vázquez, L., Zufía, J., 2017. Decision making supporting tool combining AHP method with GIS for implementing food waste valorisation strategies. Waste Biomass Valori 8, 1555.

Seppälä, J., Posch, M., Johansson, M., Hettelingh, J.P., 2006. Country-dependent characterisation factors for acidification and terrestrial eutrophication based on accumulated exceedance as an impact category indicator (14 pp). Int. J. Life Cycle Assess. 11 (6), 403.

Scherhaufer, S., Lebersorger, S., Pertl, A., Obersteiner, G., Schneider, F., Falasconi, L., De Menna, F., Vittuari, M., Hartikainen, H., Katajajuuri, J.M., Joensuu, K., 2015. Criteria for and Baseline Assessment of Environmental and Socio-Economic Impacts of Food Waste. FUSIONS Project, Wageningen.

Scherhaufer, P., Höltinger, S., Salak, B., Schauppenlehner, T., Schmidt, T., 2018. A participatory integrated assessment of the social acceptance of wind energy. ERS 45, 164–172.

Scherr, S.J., 1999. Soil Degradation: A Threat to Developing-Country Food Security by 2020? vol. 27. Intl Food Policy Res Inst.

Schug, D., 2016. Reducing water usage in food and beverage processing. Food Eng. https://www.foodengineeringmag.com/.

SEPA, 2008. Svinn I Livsmedelskedjan – Möjligheter till Minskade Mängder. Swedish Environmental Protection Agency, Bromma, Sweden. ISBN 978-91-620-5885-2.

Smit, J., Kerutzer, S., Moeller, C., Carlberg, M., 2016. Industry 4.0, E.U.h.w.e.e.e.s. Policy Department A, Editor 2016.

Smith, V.H., Tilman, G.D., Nekola, J.C., 1999. Eutrophication: impacts of excess nutrient inputs on freshwater, marine, and terrestrial ecosystems. Environ. Pollut. 100 (1–3), 179–196.

Stocker, T. (Ed.), 2014. Climate Change 2013: The Physical Science Basis: Working Group I Contribution to the Fifth Assessment Report of the Intergovernmental Panel on Climate Change. Cambridge University Press.

Struijs, J., Beusen, A., van Jaarsveld, H., Huijbregts, M.A.J., 2009. Chapter 6 aquatic eutrophication. In: Goedkoop, M., Heijungs, R., Huijbregts, M.A.J., De Schryver, A., Struijs, J., Van Zelm, R. (Eds.), ReCiPe 2008 a Life Cycle Impact Assessment Method Which Comprises Harmonisedcategory Indicators at the Midpoint and the Endpoint Level. Report I: Characterisation, vol. 6.

Stuart, T., 2009. Waste: Uncovering the Global Food Scandal. W.W. Norton & Co.

Thyberg, K.L., Tonjes, D.J., 2016. Drivers of food waste and their implications for sustainable policy development. Resour. Conserv. Recycl. 106, 110.

Tilman, D., Clark, M., 2014. Global diets link environmental sustainability and human health. Nature 515 (7528), 518.

Tonini, D., Albizzati, P.F., Astrup, T.F., 2018. Environmental impacts of food waste: learnings and challenges from a case study on UK. Waste Manag. 76, 744.

UN-DESA, 2017. United Nations - Department of Economic and Social Affairs, Population Division. World Population Prospects: The 2017 Revision, Key Findings and Advance Tables. ESA/P/WP/248.

UNEP, 1993. Cleaner Production Worldwide. UNEP Industry & Environment, Paris, France.

UNEP, 1994. Government Strategies and Policies for Cleaner Production. UNEP Industry & environment, Paris.

Valta, K., Kosanovic, T., Malamis, D., Moustakas, K., Loizidou, M., 2013. Water consumption and wastewater generation and treatment in the Food and Beverage Industry. Desalin. Water Treat 53, 12.

Van Oers, L., De Koning, A., Guinée, J.B., Huppes, G., 2002. Abiotic Resource Depletion in LCA. Road and Hydraulic Engineering Institute, Ministry of Transport and Water, Amsterdam.

Van Zelm, R., Huijbregts, M.A.J., Den Hollander, H.A., Van Jaarsveld, H.A., Sauter, F.J., Struijs, J., Van Wijnen, H.J., Van de Meent, D., 2008. European characterisation factors for human health damage of PM10 and ozone in life cycle impact assessment. Atmos. Environ. 42, 441–453.

Waldron, K., 2009. Handbook of waste management and co-product recovery in food processing, vol 1. woodhead publishing series in food science, Technology and Nutrition, pp. 573–596.

Witkowski, K., 2017. Internet of Things, big data, industry 4.0 – innovative solutions in logistics and supply chains management. Procedia Eng. 182, 763–769.

WWAP, 2012. The United Nations World Water Development Report 4: Managing Water under Uncertainty and Risk. UNESCO, Paris.

WWF, 2017. Appetite for Destruction.

Zhang, Y., Ren, S., Liu, Y., Si, S., 2017. A big data analytics architecture for cleaner manufacturing and maintenance processes of complex products. J. Clean. Prod. 142, 626.

Green supply chain

2

Stella Despoudi
Aston Business School, Aston University, Birmingham, United Kingdom

Chapter outline

1. Introduction 35
2. Definition of green food and green food supply chain 39
3. Green farming 40
4. Green food product design 42
4.1 Food product stewardship 42
4.2 Eco-design of food products or design for the environment (DfE) 43
4.3 Eco-labeling of food products' and product certification 44
4.4 Life Cycle Assessment and life cycle costing 47
5. Green food sourcing and procurement 47
6. Green warehousing 49
7. Green food logistics 51
8. Green food supply chains and circular food supply chains 52
8.1 Internal environmental management 52
8.2 Waste management and food waste management 53
8.3 Reverse logistics and circular supply chains 54
9. Conclusion 55
References 57

1. Introduction

Within the conditions imposed by globalization and recession, markets are mainly characterized by high instability and fierce competition (Leonidou and Theodosiou, 2004). In this turbulent environment, continuously changing environmental regulations, for example, are a common occurrence (Cadogan et al., 2006). As a result, it has been posed that traditional supply chains and logistics seem to lack the adequacy to cope with emerging requirements and the continually changing conditions in demand (Emmet and Sood, 2010). In the traditional supply chains there is a linear flow of materials and information that passes from suppliers to end-customers (McKinnon et al., 2010). Beamon (1999) stated that the traditional supply chains are focused on the optimum sourcing of materials and distribution of products to customers. Every activity (supply,

The Interaction of Food Industry and Environment. https://doi.org/10.1016/B978-0-12-816449-5.00002-3

manufacturing, distribution, and retailing) is separated from the others and has its own environmental impact.

Company responsibilities moved from solely shareholder focused to stakeholder. This means that companies have extended responsibilities to everyone that they have an impact on, including suppliers, customers, pressure groups, etc. The larger the company the bigger the environmental impact. This is because a company that produces a large number of products consumes more resources and also more people are buying the products. Over the past decades the concept of Green Supply Chains (GSC) has emerged. There are numerous drivers affecting the GSC management. Walker et al. (2008) distinguish GSC drivers to internal (organizational) and external drivers (stakeholders and interest groups). According to Emmet and Sood (2010) there are five interest groups that drive green initiatives in an organization. These stakeholders are: regulators (e.g., governments), consumers (market pressures), shareholders and employees, community groups (e.g., WWF), and media.

Green evolution started due to an increased occurrence of cases related to environmental pollution such as the BP oil leak in the Gulf of Mexico. Since then, different environmental regulations started to emerge which were posed to companies by governments. Significant environmental degradation caused by human activities led to climatic changes. Climatic changes are causing significant issues in food supply chains as crops and yields are being damaged due to extreme weather events, soil fertility has decreased, and agricultural land is limited. Resource efficiency is more important than even as current use of natural resources is unsustainable. Food companies such as Nestle are trying to identify new resource efficient ways of producing their products by reducing the amount of water usage in production and identifying alternative natural resources (Nestle, 2018a). Nestle also invested in cocoa farms as they have predicted that cocoa production will be extinct in the future. Concerns about the environmental impacts of companies led supply chain management toward stages of improvements. European Union and US authorities are continuously expanding their environmental regulatory compliance schemes. Companies must comply with environmental legislation existing in each country. Robinson and Wilcox (2008) state that environmental laws include regulations describing carbon limits and emission trading requirements, waste (products and packaging), energy efficiency standards, hazardous materials limits and water pollution. Some examples of these regulations are WEEE (Waste of Electrical and Electronic Equipment) Directive, RoHS (Restrictions of Hazardous Substances), Eco-Design of Energy-Using Products (EuP), and EU Emissions Trading System (Koyoto Protocol). The Koyoto protocol was signed in 1997 by the United Nations to tackle global climatic change by reducing greenhouse gas (GHG) emissions (UN, 2018). United Nations also introduced the Sustainable Development Goals that cover 17 goals related social, economic, and environmental issues. All governments should encourage companies to operate according to these goals. Companies started to implement only the required environmental regulations need to operate. However, stakeholder pressures and the need for greening the overall supply chain increased over the years.

Consumers expect companies to be green and they are looking for green products (Boston Consulting Group, 2009). Organizational stakeholders might be drivers in green initiatives as adopting a Green Supply Chain Management (GSCM) approach reduces costs and enhances efficiency. Community groups promote environmental needs and interact negatively with companies that do not operate according to the green regulations/standards. Media can cause reputational risk for companies since they often reveal cases of environmental pollution (e.g., BP's oil spillage in the Gulf of Mexico). The risks of not being green are numerous. Some of these risks are: damaged reputation, noncompliance with regulations and thus penalties, pressure groups protest, impact on sustainability performance. Some of the benefits being green are: improved image, improved sustainability credentials, competitive advantage, and improved business performance due to more efficient use of resources. Although all those factors discussed above influence significantly the green strategies that a business and or supply chain might employ, governments' regulations seem to be the initiator for implementing green initiatives (Lin, 2007; Green et al., 1996).

Companies need to ensure that their overall supply chain is green because if there is an issue at its suppliers the company will be impacted, too. For example, the detection of horsemeat DNA in IKEA's meatballs is a relevant example. Greening in corporations raised the need for a more integrated approach in supply chain management and logistics. GSCM is a new strategic priority for organizations that incorporates an integrated and aligned environmental thinking across the whole supply chain (McKinnon et al., 2010; Mallidis et al., 2019). Adopting a green agenda is an essential element to remain competitive. The green supply chain is a broader view of the supply chain and includes the following activities: product design, procurement, manufacturing, distribution and reverse logistics (Srivastava, 2007; Zhu and Sarkis, 2004). Collaboration and integration of the green supply chain has multiplied benefits for corporations such as energy savings, increased profits, reduce operating costs, comply with new environmental legislation, reputation enhancement, and employee attraction and retention (Despoudi, 2016; Despoudi et al., 2018).

Today's Food Supply Chain (FSC) is facing many pressures due to issues related to fewer natural resources available, limited agricultural land available, population growth, world's food insecurity, climate change, dietary changes, governance of the FSC system, and food waste or Postharvest Food Losses (PHFL) (Defra, 2006). The major natural resources, i.e., food, energy and water are becoming scarce (FAO, 2011). The future scarcity of the natural resources indicates that they need to be preserved and should not intentionally be wasted. The agricultural land is also limited; new ways to grow crops need to be found in places that until now was not possible to farm (Vidal, 2012). The world population has been predicted to reach 9 billion by 2050 and this will require a 70% increase in food production (FAO, 2009). Producing enough food, appropriately distributing it, and minimizing its wastage are some of the challenges that the food industry is facing related to the rising population (Foresight, 2011). According to FAO (2011) food insecurity can be defined as a situation that exists when people do not have consistent and everyday physical, social and economic access to sufficient, safe and nutritious food based on their dietary preferences and needs. Thus, the world's food insecurity issue is becoming a major concern. The rising population, the fewer natural resources available, the possible future insufficient acceleration of technology, and the high levels of food waste rise major concerns about world's food insecurity.

Malthus (1798) and Meadows et al. (1972) highlighted that in a world with finite natural resources, food production is not possible to meet an increasing populations' future needs for food. Criticisms of Malthus' "limits to growth model" stated that this model failed to capture effectively the acceleration of technology until now (Engels and Engels, 1843). Acceleration of technology managed to increase crop yield and create new types of crops (e.g., genetic modified crops). However, the pace of population growth, climate change, income distribution imbalances and the change of consumption patterns are moving faster than technological advancements (Foresight, 2011). This means that in the future technological advances may not be able to keep up with the population growth and the deterioration of the natural environment. Therefore, the issue of future scarcity of natural resources relatively to population increase has been predicted and discussed many years ago. Technology seems to act as a balancing factor of the two aforementioned issues; however, it is not certain for how long technology will keep the balance between the two. Climatic change and future scarcity of natural resources put limits to growth in agriculture and food production, which means that a 70% increase in food production to feed nine billion people is impossible to be achieved (Hodges et al., 2010). Climate change also has and will continue to have in the future severe negative consequences to the FSC (Diaz and Murnane, 2011). Weather changes in the form of extreme weather events, the rise of global temperature and the increase of GHG emissions are the main causes of climate change that will impact significantly the FSC. According to Bennett's Law, increasing wealth pushes people in consumption of higher calories food such as fats, protein, and sugar (Godfray et al., 2010). Those dietary changes affect significantly the FSC as high caloric diets require more natural resources to be spent. The governance of the global FSC at both national and international levels is another challenge that the FSC is facing (FAO, 2002). The globalization of the markets led to changes in power imbalance in the FSC and this creates governance issues in the sector. More precisely, producers are the less powerful in the FSC, while big retailers have dominated the sector (Delloite, 2013).

Another major challenge that the FSC is facing is food losses or PHFL. It has been estimated that between 25% and 50% of food produced is lost or wasted along the supply chain and does not reach consumers, depending on its position in the supply chain (Lundqvist et al., 2008). Reducing PHFL can increase grain supply, food availability and food security without wasting other resources such as land, labor, water and inputs (APO, 2006). According to a recent study conducted by the FAO titled "Global Food Losses and Food Waste" (Gustavsson et al., 2010, p. 4), "food is lost or wasted throughout the supply chain, from the initial agricultural production down to the final household consumption" This means that there are significant amounts of lost food throughout the FSC. The majority of food is lost from the producers to retailers point in the supply chain (Gustavsson et al., 2010). There is a need for development of a sustainable and fair FSC (Driscoll, 2012). Small holder farmers, despite producing more than 70% world's food, represent more than half of the world's hungriest people (Gidney, 2012). Hence, producers need to be supported to enable sustainable food production for now and for the future. Food provisioning in a resource constrained world must be done in a sustainable way to achieve food security for all the people in the world (Krejci and Beamon, 2010). Hence, the greening of the FSCs in continuously changing resource constrained world is essential.

The aim of this chapter is to illustrate how green initiatives are implemented in food supply chains. To fulfill this aim, a number of related topics will be discussed pertaining areas such as for example, green food product, green farming, green food product design, green food sourcing and procurement, green warehousing, green food logistics, and green food supply chains.

2. Definition of green food and green food supply chain

There are different definitions about how to enable green or else sustainable food production and what exactly this involves. Sustain (2015) defines green food as the food that is produced, processed, and traded in ways that:

- Contribute to thriving local economies and sustainable livelihoods;
- Protect the diversity of both plants and animals, and avoid damaging natural resources and contributing to climate change;
- Avoid damaging or wasting natural resources or contributing to climate change;
- Provide social benefits, such as good quality food, safe and healthy products, and educational opportunities.

Beer and Lemmer (2011) stated that environmental sustainability is not enough; food produced must be politically, economically, and socially sustainable. Thus, from an SC perspective, sustainable food production involves adoption of green practices and consideration of other operating environment factors across the supply chain, from production to consumption.

SustainAbility (2011) defined a GSC as a reliable, resilient and transparent, which produces food within ecological limits, empowers food producers, and ensures accessible and nutritious food for all. A sustainable FSC must meet the words need for food and also avoid adverse environmental impacts (Defra, 2006). In the HM government report the "Food 2030" (2010) is stated that green food is food that is produced, processed, and distributed to feed a growing global population in ways which use global natural resources sustainably, enable the continuing provision of the benefits and services, ensure a healthy natural environment provides, promote high standards of animal and welfare, protect food safety, and make significant contribution to rural communities.

In the UK's Strategy for Sustainable Farming and Food, the Government set out the following key principles for a GSC (Defra, 2006, p.9):

- "Produce safe, healthy products in response to market demands, and ensure that all consumers have access to nutritious food, and to accurate information about food products;
- Support the viability and diversity of rural and urban economies and communities;
- Enable viable livelihoods to be made from sustainable land management, both through the market and through payments for public benefits;

- Respect and operate within the biological limits of natural resources (especially soil, water and biodiversity);
- Achieve consistently high standards of environmental performance by reducing energy consumption, minimizing resource inputs, and using renewable energy wherever possible;
- Ensure a safe and hygienic working environment and a high social welfare and training for all employees involved in the food chain; Achieve consistently high standards of animal health and welfare; and
- Sustain the resource available for growing food and supplying other public benefits over time, except where alternative land uses are essential to meet other needs of society."

From all above it could be concluded that greening the FSC is about having the resources and the capabilities in the SC to create sustainable food consistently for now and for the future by balancing all three sustainability elements (i.e., people, planet, profit).

3. Green farming

Green farming or else called organic farming refers to a systems approach to production that is working toward environmentally, socially and economically sustainable production. According to DEFRA, "organic food is the product of a farming system which avoids the use of man-made fertilizers, pesticides; growth regulators and livestock feed additives. Irradiation and the use of genetically modified organisms (GMOs) or products produced from or by GMOs are generally prohibited by organic legislation" (BBC, 2018).

According to EU (2018) organic farming is an agricultural method that aims to produce food using natural substances and processes. This means that organic farming limited environmental impact as it promotes (a) responsible use of energy and natural resources, (b) maintenance of biodiversity, (c) preservation of regional ecological balances, (d) enhancement of soil fertility, (e) maintenance of water quality, and (d) high standard of animal welfare (EU, 2018). According to the UK Government organic farming involves the following: (a) avoiding artificial fertilizers and pesticides, (b) using crop rotation and other forms of husbandry to maintain soil fertility, (c) controlling weeds, pesticides and diseases using husbandry techniques and where necessary approved materials to control pests and diseases (DEFRA, 2016).

Some examples of natural resources conservation as part of organic farming are listed below (Ball et al., 2018).

- Animal Welfare: Elimination of cruelty; safe handling, housing, slaughter and transport
- Soil conservation: Conservation forest, species; prevent soil erosion and pollution, prevent loss of arable land and biodiversity, responsible farming methods, elimination of contaminant and pollutant agents
- Water conservation: reducing water consumption, efficient water use, wastewater reuse and recovery

There are incentives for UK farmers to engage in organic farming. An example of this is the Countryside Stewardship (CS) scheme which provides financial incentives for farmers and land managers to look after their environment by: (a) conserving and restoring wildlife habitats, (b) flood risk management, (c) woodland creation and management, (d) reducing widespread water pollution from agriculture, (e) keeping the character of the countryside, (f) preserving historical features in the landscape, and (g) encouraging educational access (UK Government, 2019).

Organic farming has several advantages for the environment, farmers and consumers. However, it comes with certain disadvantages (Howe, 2017). The main benefit for consumers is the nutritional value of organic food as it has high mineral content compared to nonorganic food. Organic farming fosters the life of the soil and thus it captures its benefits i.e., nutrients. In that way plants are healthier and people and animals too. Additionally, a major benefit to consumers of organic food is that it is free of contamination with health harming chemicals such as pesticides, fungicides and herbicides. Thus, populations' exposure to chemicals and its associated implications is reduced. Organic food is found to taste better due to its sugar content which is related to the quality of the healthy grown food. There are numerous advantages for farmers too. Organic farming can reduce production costs as farmers do not need to buy chemicals e.g., pesticides and fertilizers as they use natural ones. The negative environmental implications of farming as explained above are significantly reduced in organic farming. By doing so, farmers can save energy and resources. It has a direct impact on the global warming, ground water cleanness, and biodiversity. However, organic farming has higher production costs as farmers need more workers. In the short-term, it is not very productive compared to the industrialized agriculture. The crops/harvest are more sensitive to infestations as some pesticides and genetically modified crops protect plants from illnesses. Some of the disadvantages of organic farming can be eliminated through smart farming or else called Agriculture 4.0. Smart Farming is a farming management concept using modern technology to increase the quantity and quality of agricultural products (Schuttelaar, 2017). This can be achieved by having access to GPS, soil scanning, data management, and Internet of Things technologies. In this way, farmers can precisely measure variations within a field and adapt their strategies accordingly, this can greatly increase the effectiveness of pesticides and fertilizers or even use no chemicals. Smart farming can also be applied to animals by adjusting their nutrition and thus preventing disease and enhancing herd health. Some examples of technologies that can be used as part of smart farming are (Schiforce, 2018):

- Sensing technologies, including soil scanning, water, light, humidity, temperature management;
- Software applications -specialized software solutions that target specific farm types;
- Communication technologies, such as cellular communication;
- Positioning technologies, including GPS;
- Hardware and software systems that enable IoT-based solutions, robotics and automation;
- Data analytics, that underlies the decision-making and prediction processes.

According to Beecham Research (2014) Smart farming can be used in different aspects of farming. Some of these are: (a) fleet management: tracking of farming vehicles, (b) arable farming, large and small field farming, (c) livestock monitoring, (d) indoor farming: greenhouses and stables, (e) fish farming, (f) forestry, (g) storage monitoring of water and fuel tanks.

4. Green food product design

The supply chain starts with the design of the product. Thus, to achieve a green FSC, product design should be considered first. Green food products need to have zero emissions, reduced material usage, purchased in an ethical way, reduced chemical usage i.e., organic, green packaging in terms of design and in terms of keeping it fresh longer to reduce wastage, and be easily recycled. New practices and models are required for a successful green food product, such as Design for the Environment (DfE), Product Stewardship, eco-labeling, Life Cycle Assessment and Costing, and Green Packaging.

4.1 Food product stewardship

Companies' responsibilities have been extended to stakeholders from shareholders. Similarly, a green product's responsibilities were extended to all the parties involved in its lifecycle. Product stewardship can be defined as the shared responsibilities that all the participants in a product's lifecycle have for minimizing its environmental and health impacts (Product Stewardship Institute, 2011). Producers' responsibilities are now extended from only downstream in the supply chain (customer end of the supply chain) to upstream supply chain (supplier end of the supply chain). Manufacturers and retailers also are responsible for products impacts. The responsibilities in a supply chain do not stop when the product is delivered to consumers. All supply chain entities are responsible for the postconsumption stage or else called end-of-life of a product. Collaboratively they should aim to identify ways to enable the end-of-life process as much as possible. Companies that produce goods are responsible for the whole lifecycle of the product; from raw material extraction to use and disposal. Fig. 2.1 below shows the green product stewardship concept and the different supply chain entities that are responsible for a food product which are producers, manufacturers, retailers, con.

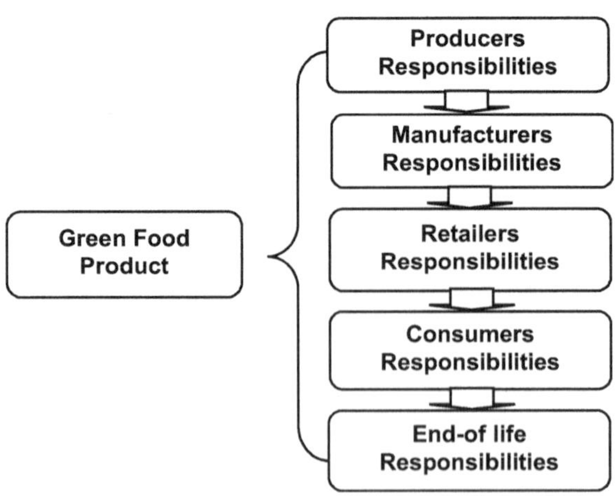

Figure 2.1 Food Product stewardship and the different shared responsibilities of GSC entities (authors own creation).

Nestle provides a good example of product stewardship. Nestle's vision is a water-secure world that enables people, cultures, business and nature to prosper, now and in the future. To achieve this, it aims to ignite and nurture global and local leadership in credible water stewardship that recognizes and secures the social, cultural, environmental and economic value of freshwater (Nestle, 2018b). Its approach to water stewardship is based on the Alliance for Water Stewardship (AWS) Standard which is a publicly recognized and credible model to ensure the use of water is socially equitable, environmentally sustainable and economically beneficial (AWS, 2018). To achieve water stewardship, Nestle focuses its efforts and resources on four key areas: in our factories, continuously improving water use efficiency; in watersheds, working with partners to protect shared water resources; across our agricultural supply chain, where water challenges are putting the sourcing of our raw materials at risk; in the communities where we are present, to widen access to clean water and sanitation (Nestle, 2018c).

4.2 Eco-design of food products or design for the environment (DfE)

Efficient and environmentally friendly GSCs require products that will be easily disposed, reused and remanufactured. Eco-design is way of improving existing products' or brand-new products' design by ensuring that their whole lifecycle has minimum impact to the environment (Charter and Tischner, 2001). Minimum impact on the environment can be translated into minimum usage of resources as well as minimum waste and pollution during production, thus enhancing products' recyclability, reuse and recovery. Crane et al. (2008) and Zhu et al. (2006) state that "design for the environment" is a strategic tool for designing products that emphasizes the 3 R's (recover, reuse and recycle). Those three products' environmental aspects have to be examined in the initial stage. DfE might also consider the energy usage in manufacturing the product, use of abundant materials, toxicity of materials and waste of material production (Hitchcock and Willard, 2007; Slack et al., 2010). Imbedding green initiatives in the product design stage can significantly reduce current product impacts, and also provide a platform for new-to-the-world ideas and New Product Development (NPD) which may lead to competitive advantage.

Food products should be produced in a way that minimizes their environmental impact. This will start from the product phase were minimum inputs are used as part of the smart farming. The machines that will be used for harvesting and for helping with the growing of the produce should not produce a lot of emissions. Also, reduced chemical or chemical free production is encouraged as part of the organic farming. In the food manufacturing the production itself and the resources used should be considered. Also, at this stage the packaging of the food product should be "smart" packaging meaning that is made from sustainable resources, it was manufactured in a sustainable way, it preserves the product as much as possible to limit waste, it can be easily opened and sealed by consumers, and easily recycled.

To achieve a greener product packaging companies should transform their products' into being lighter in weight, easier to recycle and using recycled materials in manufacturing. Disposable single-trip packaging is changed into reusable multitrip packaging (Wrap, 2011). Decisions about products' package need to balance the relative costs of materials and possible environmental impacts (Deftra, 2011). Many countries follow specific environmental policies about product packaging. Some examples on these policies in the UK are: the European Packaging Waste Directive 94/62/EC, the Producer Responsibility Obligations (Packaging Waste) Regulations and the Packaging (Essential Requirements) Regulations (Industry Council for Packaging and Environment, 2011).

Asda embedded green initiatives in NPD in collaboration with WRAP (WRAP, 2018a). Asda has a formal, well-structured process for NPD which helps to create new and improved products, which meet changing consumer expectations. This pilot project was done in collaboration with three key suppliers of Asda in an effort to embed green criteria into the NPD process. By doing so is helped to consider sustainability from the very beginning of the process, to map the main sustainability impacts of selected categories defining how suppliers can tackle these, and to deliver a series of inspiring and imaginative NPD ideas that are sustainability-led. One these suppliers was Yeo Valley which considered single-serve desserts and explored ideas around sustainable packaging innovation, food waste reduction, sustainable sourcing, and product reformulation. The new idea that the company had was called dual purpose product concept which can be eaten as a soft mousse or as an ice cream. Comes with a stick that can be used as a spoon or ice cream stick – to aid decanting to freezer.

Greencore was another company participated in this project and it considered a wider product range and explored a number of ideas from new packaging formats, more efficient ways to cook, formats to reduce product waste, to various sustainability communications and behavior change opportunities. In particular they developed and innovative pack that allow microwaving to get finish on top instead of oven. Consumers want the convenience of microwave and quality of oven.

For all the different food products, opportunities for recycling as well as the environmental impact of the product should be highlighted in the packaging. Fig. 2.2 presents the different aspects of green food product design across the different stages of the GSC which are: production, manufacturing, retail, consumers, postconsumption and the respective elements of developing green food products are each stage. It is important that all these different aspects are connected as they impact each other.

4.3 Eco-labeling of food products' and product certification

Eco-labeling of food products is another way of signaling an eco-designed product. Products' environmental labeling might refer to the materials used to manufacture it, the recyclability of the product or the packaging, and its carbon footprint. Eco-labeling of products might also include certification of superior environmental performance such as the EU Flower ecolabel (Industry Council for Packaging and Environment, 2011).

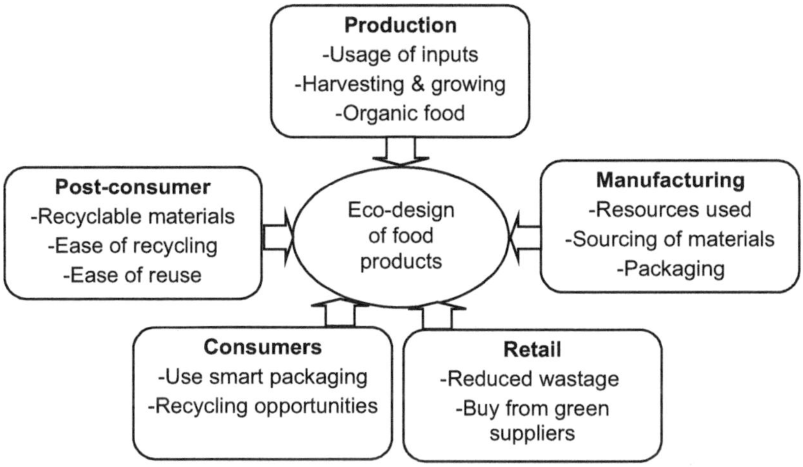

Figure 2.2 The different aspects of green food product design (authors own creation).

In the food industry products are certified as organic. Organic standards consider at all the aspects of organic food such as manufacturing and production, storage and sales. They consider everything from packaging to animal welfare and wildlife conservation and ban unnecessary and harmful food additives in organic processed foods. Any product sold as "organic" in the EU needs to comply with a set of standards. Organic farming and production started to be regulated at EU level since 1991 (IFOAM, 2018). Today the European requirements for organic production are set by Council Regulation (EC) No 834/2007 defining the aims, objectives and principles of organic farming and production, and by two implementing regulations (No 889/2008 and No 1235/2008) detailing the organic production, labeling, control and import rules. All products labeled as organic and sold in the EU must be produced in accordance with these regulations.

Products certified to these standards can use the EU Green Leaf organic symbol or the Soil Association organic symbol on their packaging which is an assurance to customers about what they are buying is genuine and fully traceable back to the farm (Soil Association, 2018a). These logos give a coherent visual identity to European Union produced organic products sold in the EU. In this way, it is easier for EU based consumers to identify organic products and helps farmers to market them across all EU countries (EU, 2018). In order for companies to obtain the organic logo their products need to be certified by an authorized control agency or body. This means that they have fulfilled strict conditions on how they are produced, transported and stored.

An example of an organic certified business from Soil Association is Green Gourmet which is a dynamic food innovation company, based in the South West of the UK (Soil Association, 2018b). Their certified product range includes organic sausage, organic pork meatballs and organic mince beef which have numerous listings with school caterers, helping caterers serve more organic foods in the UK. Organic product certification helped them to grow their business in school meals, having increased sales, and it opened up new channels. Organic certification can be obtained for different food products which are as follows: vegetables and fruits, meat, dairy products, and fish (Organic Food, 2018).

Organic vegetables and fruits mean that plants come from fields that were free from chemical application of fertilizers and pesticides. Organic meat means that the meat does not contain any chemicals as the animals were fed with natural food which did not contain chemicals. Specifically, this requires that animals should not be raised with antibiotics or growth stimulants. Animals should have humane treatment, have access out-of-door, e.g., not caged chickens, be fed 100% certified organic feed and must graze in certified organic pastures. Organic dairy products include milk from all dairy animals, i.e., cows, goats, and sheep which were only fed organic feed and they graze in organic pastures. Organic fish are fishes that are grown in fish farms such as salmon, trout, cod, halibut and sea bass and they are fed with organic feed.

At US organic products are regulated by federal standards, meaning product labeling is strictly controlled. Certification needs to be according to the requirements set up by the United State Department of Agriculture (USDA) and monitored by the National Organic Program (NOP). According to EFCR (2018) here are four different categories of organic food products that can be sold to the market which are:

(a) Products sold, labeled, or represented as '100% organic.' A raw or processed agricultural product sold, labeled, or represented as "100% organic" must contain (by weight or fluid volume, excluding water and salt) 100% organically produced ingredients. If labeled as organically produced, such product must be labeled pursuant to §205.303.

(b) Products sold, labeled, or represented as 'organic.' A raw or processed agricultural product sold, labeled, or represented as 'organic' must contain (by weight or fluid volume, excluding water and salt) not less than 95% organically produced raw or processed agricultural products. Any remaining product ingredients must be organically produced, unless not commercially available in organic form, or must be nonagricultural substances or nonorganically produced agricultural products produced consistent with the National List in subpart G of this part. If labeled as organically produced, such product must be labeled pursuant to §205.303.

(c) Products sold, labeled, or represented as 'made with organic (specified ingredients or food group(s)).' Multiingredient agricultural product sold, labeled, or represented as 'made with organic (specified ingredients or food group(s))' must contain (by weight or fluid volume, excluding water and salt) at least 70% organically produced ingredients which are produced and handled pursuant to requirements in subpart C of this part. No ingredients may be produced using prohibited practices specified in paragraphs (f) (1), (2), and (3) of §205.301. Nonorganic ingredients may be produced without regard to paragraphs (f) (4), (5), (6), and (7) of §205.301. If labeled as containing organically produced ingredients or food groups, such product must be labeled pursuant to §205.304.

(d) Products with less than 70% organically produced ingredients. The organic ingredients in multiingredient agricultural product containing less than 70% organically produced ingredients (by weight or fluid volume, excluding water and salt) must be produced and handled pursuant to requirements in subpart C of this part. The nonorganic ingredients may be produced and handled without regard to the requirements of this part. Multigradient agricultural product containing less than 70% organically produced ingredients may represent the organic nature of the product only as provided in §205.305.

(e) Livestock feed. (1) A raw or processed livestock feed product sold, labeled, or represented as '100% organic' must contain (by weight or fluid volume, excluding water and salt) not less than 100% organically produced raw or processed agricultural product. (2) A raw or processed livestock feed product sold, labeled, or represented as 'organic' must be produced in conformance with §205.237."

4.4 Life Cycle Assessment and life cycle costing

Life Cycle Assessment (LCA) is the process of examining the impacts of a product over its entire lifetime (Hitchcock and Willard, 2007). LCA measures all the product's possible environmental impacts at each stage of its lifecycle. By implementing LCA, companies can achieve cost and energy savings, reduction in materials usage, and enhanced environmental sustainability. This method of assessment is used for a company's own products and its suppliers' raw materials/products. Life cycle costing examines the costs related to a product's lifecycle (materials needed, operating costs and disposal costs). This procedure reveals the real product's costs before manufacturing it. Specific characteristics of the products are being tested, such as longevity, safety risks and recyclability costs (Hitchcock and Willard, 2007; Charter and Tischner, 2001).

Nestle is committed in assessing and optimizing the environmental impact of its products and they are using the LCA. By using LCA they assess and improve their products from development to disposal and enable them to focus on minimizing our environmental impact across our products' lives (Nestle, 2018b). LCAs is core part of their decision-making, and it helps them to improve their environmental performance while responding to stakeholder concerns and providing them with credible evidence to support specific marketing claims. Some examples of areas where they use LCA for developing products are: water use in agriculture, CO2 emissions in manufacture, or the amount of energy used in distribution (Nestle, 2018d).

5. Green food sourcing and procurement

Material Sourcing is strategic function of a business and contributes significantly in managing waste, the protection of product claims and the organization's image (Charter and Tischner, 2001). Green purchasing is an additional element of the purchasing managers sourcing criteria in addition to quality, cost and delivery time. It is not enough if a company operates according to environmental standards. The company's suppliers need to follow the same standards as well and offer green products. Thus, the extended supply chain needs to be green. There are several risks associated with sourcing the right materials from the right suppliers. Supply chain risks are increasing, and global supply chains are facing even more risks than the local ones. Purchasing can add value to the entire supply chain, but simultaneously can create several reputational risks. Different ways are used by companies to map theirs supply chain risks. Supplier risk is one of the major risks that if it is ignored it could be proved disastrous for a company. Procurement can be used as a way to map the sourcing risks per supplier, country, and type of material. Understanding the sourcing risks, a company can decide whether to accept, reduce, hedge or avoid the risk. Sourcing from a supplier who is located in a region with political unrest or with frequent extreme climatic events is a risky option. Companies can make informed decisions about their supplier base through the procurement process. The procurement process can act as insurance to the company.

Governments publish Environmental Preferable Purchase policies, and gray and black chemicals lists that companies need to comply with (Hitchcock and Willard, 2007). Procurement managers need to ensure that their suppliers operate according to governments' policies. Procurement departments need to have specific purchasing practices that enhance the environmental aspects of their supply chains (Business Link UK, 2011a,b,c,d). Overall, green sourcing in a supply chain can be achieved by adopting Environmentally Preferable Purchasing (EPP) policies, by using service contracts to align suppliers' interests with the business and by implementing frequent and proactive supplier audit programs (Hitchcock and Willard, 2007). Suppliers audits must be produced according to specific criteria and their improvement should be assessed systematically (Min and Galle, 1997). A more integrated approach toward purchasing is the implementation of supply chain environmental management systems. Companies can implement SCEMS to choose the appropriate environmental policy, prioritize activities, forward planning, monitoring and evaluate the outcome (Hitchcock and Willard, 2007). Identifying and building relationships with suppliers that share the same green values as with the core company are critical to a company's green credentials. Complexity of the future supply chains implies transparency of information and visibility across the whole chain. Collaboration of all members in a supply chain is the fundamental principle. Shared information makes supply chains more agile; they can respond in demand fluctuations easier. In Capgeminy's Report (2011) it is mentioned that "Joint Scorecard and Business Plan" can measure how effective the collaboration between trading partners is. Companies can use partnerships and good supplier relationships to handle and align the suppliers' interests in their way. Suppliers need to comply with governmental legislations and companies' environmental policies (Business Link UK, 2011a). Frequent suppliers' audits are used to ensure their environmental performance meets the company's standards. ISO 14001 certification and Audit Scheme (EMAS) can be used by companies to assess their suppliers' environmental performance continuously.

Recent green scandals indicate that there is a need to increased transparency in the supply chain with a particular focus on the sourcing part. In order companies to achieve sustainability in sourcing they need to achieve sustainability in terms of the environmental, social, and economic aspects. In the case of IKEA, the risk came from its suppliers who without informing the core company combined horsemeat with beef meat and then a sample was taken from IKEA and they found out about it. Horsemeat is not bad for humans to consume, but the company as well as the consumers need to know exactly what they are buying. Making sure that the company's suppliers operate according to the company's green standards and conducting frequent audits to suppliers are the first steps toward ensuring a green food sourcing. To achieve a green food sourcing and procurement function as part of the overall GSC, the environmental, economic, and social performance of these activities need to be balanced.

The environmental aspects can be achieved through resource depletion, waste minimization, reduction of GHG emissions and other polluting factors. Green procurement also implies green power sourcing. GSCs should use a significant amount of renewable energy to decrease pollution rates and costs. The social aspects can be improved through fair trade, respect of labor standards, equality and diversity. While, the economic through value addition, cost reduction, and resource efficiency. All the aforementioned sustainability examples here are considered from the sourcing perspective only and not whole organizational perspective. Fig. 2.3 below presents the three aspects of green sourcing and procurement.

Unilever is committed to green procurement though sustainable farming methods that will help to increase yields, mitigate the effects of climate change and provide economic and social benefits to farmers, their families, and the surrounding communities (Unilever, 2018). They believe that sustainable agriculture will play a vital role in achieving the UN's Global Goals for Sustainable Development, particularly those on eradicating hunger and poverty. Sourcing sustainably helps them to secure their supplies and reduces risk and volatility in their raw material supply chains. It also opens up opportunities for innovation: by focusing on people's sustainable living needs and consumer preference, we build stronger brands. Sustainable farming methods can also improve the quality of our products, such as our sauces, soups, dressings or ice creams.

6. Green warehousing

Eco-designed buildings are using energy more efficiently, in addition to reduced operating costs. Green design of buildings is about efficient usage of electricity, use of recycled materials or use of renewable energy systems which decreases the carbon footprint (Mangan et al., 2008; Esty et al., 2006; Guardian, 2011a,b). Greening building design can be achieved by implementing specific environmental standards and by

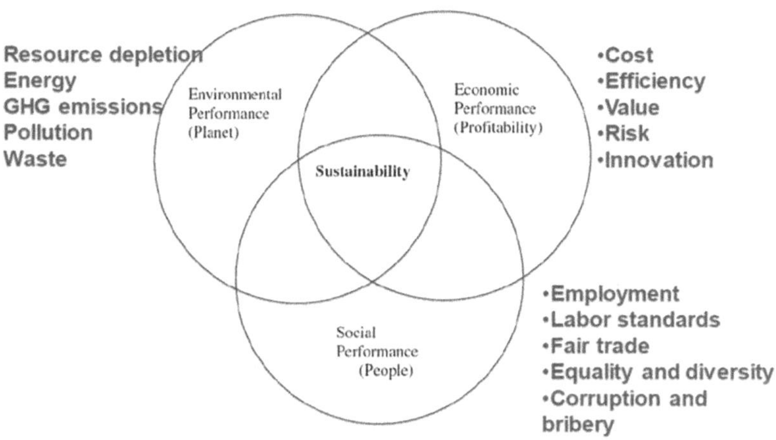

Figure 2.3 Green Sourcing and procurement (authors own creation).

adopting various certifications. Leadership in Energy and Environmental Design (LEED) is a green building certification system provided by the Green Building Certification Institute (GBCI). Buildings can be certified as gold, silver or platinum according to their contribution in energy saving, footprint, water usage and recovery of resources (Green Building Certification Institute, 2011; World Green Building Council, 2011; UK Green Building Council, 2011). Traditional warehouses changed into "collaborative warehouses" and "strategic warehouses" (Capgemini Report, 2011; UPS Report, 2011). Manufacturers, retailers and Logistics Service Providers (LSP) need to collaborate, share information and storage space for improving the environmental aspects of their supply chain. By adopting the collaborative warehousing concept companies ensure high warehouse utilization, footprint reduction and logistics optimization through more efficient transportation modes.

The location of the warehouse will determine its green credentials (Napolitano, 2013). Companies investigate moving their distribution facilities closer to inbound ports or closer to customers to save local transportation energy consumption and thus reduced environmental emissions. This will have a direct impact on company's carbon footprint which is about total emissions caused by an individual, event, organization, or product, expressed as carbon dioxide equivalent. The location of the warehouses will be primarily decided based on the logistics strategy. Modernized lighting is another way of warehouse greening which focuses on providing a true illumination option at a resourceful rate and it is environmentally friendly as it reduces usage and consumption rates. Recycling should be a core part of every green warehouse. Energy management systems that autonomously oversee all the utilities one facility uses without much human intervention could be used as well. Some examples of energy management systems are: integration of timers, thermostats, and gauges for all forms of electricity, gas, heat, and water, energy. Current warehousing equipment and machinery that operates using gas and oil could be replaced by electric ones. For example, electric forklifts could be use that they will simply need to be charged. All the aforementioned examples of green warehouses could be also used by food companies that aim to achieve GSC. Fig. 2.4 below shows the different options available for green warehousing. An example of green warehouse in the food and drinks sector is Coca Cola's Brampton warehouse which was converted to an energy efficient lighting

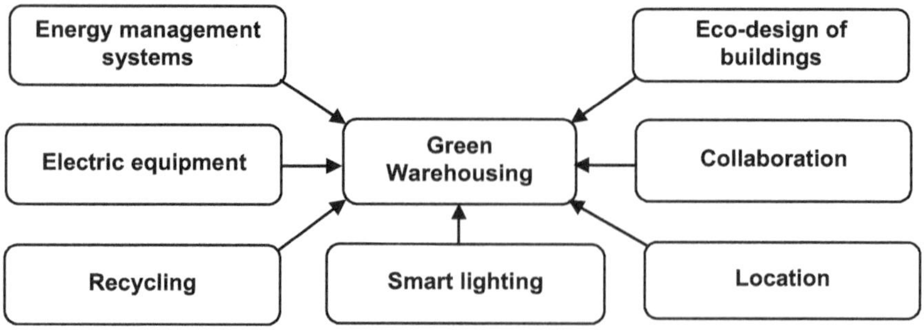

Figure 2.4 Green Warehousing Options (author's own creation).

system that uses 50% less energy and provides 50% more light (Ridler, 2017). These new fixtures also operate on motion sensors for even greater savings. In the distribution channels, the company has installed 2000 EMS-55 energy management devices in vending machines. These devices activate lights and adjust cooling based on use, leading to improved energy efficiency by up to 35%.

7. Green food logistics

As companies expand globally the need for green logistics becomes even more important. Global expansion of companies results in increased transportation and distribution of goods. Green logistics means less unnecessary freight movement, less energy consumption, better utilization of the transport vehicles, and overall having reduced environmental impact (McKinnon et al., 2010). Environmental footprint reduction in logistics can be achieved by using "scale" in the transportation of goods (e.g., using larger containers or even larger ships) and creating more efficient ways of transporting and storing of goods (e.g., centralized warehouses; Mangan et al., 2008). Redesigning supply chains to improve the environmental footprint of logistics network can be done by avoiding the creation of "bulks" in the movement of freights and having more centralized deliveries directly to customers. Sourcing locally is another way of reducing the negative impacts of freight movements, by minimizing all unnecessary movements. However, local sourcing might be more expensive (Capgemini, 2011). As transportation increases there is a need for using bigger means of transport i.e., containers, vessels or creating "hubs" in large ports/central locations as intermediaries to facilitate the logistics function (Mangan et al., 2008). Sharing deliveries with suppliers or other partners (i.e., optimizing trucks load capacity) and improving transportation modes efficiency (using alternative fueling, ensuring vehicles' eco-operation), is another way of reducing the impact of transportation (UPS Report, 2011). Environmental sustainability of logistics can also be enhanced by relocating distribution centers in more strategic places. "Port Centric Logistics" is a new way of reducing the unnecessary movement of freights in sea (Mangan et al., 2008). Many companies nowadays use ports as warehouses and minimize the mileage of freights as well as the "less than full container load" (LCL). An essential requirement in greening logistics is collaboration (Capgemini, 2011). Manufacturers, retailers and third-party logistics move toward sharing information, transportation and warehouses to reduce their costs and carbon footprint as well.

Many food logistics companies moved toward green logistics. An example of such company is the Alliance Shippers Inc that commitment to reduce negative impacts on the environment and its own carbon footprint through the EverGreen refrigeration units to its fleet (Food Logistics, 2018). The EverGreen refrigerator is refrigerated trailers which are equipped with two-way cellular tracking devices that draw power from the battery source within the refrigeration units, allowing Alliance to always have a GPS location and the ability to turn the refrigeration unit on and off at any time. The company has also invested in solar panels to charge batteries in its trailers, lower fuel consumption and reduce emissions to benefit customers and the planet. By building its fleet of refrigerated trailers with the most current technology, Alliance Shippers is able to move refrigerated goods through the nation's railroad system, reducing CO_2 output by 67%. Another example of green logistics is Nature's Frequencies' Food Freshness Card which is helping to reduce shrink and spoilage in the world's supermarkets by keeping food fresher for longer periods of time (Food Logistics, 2018). By implementing these cards supermarkets have decreased shrink by upwards of 30%. The company's patented technology was named the "Best New Food Safety Solution in 2017" at the United Fresh Produce Association and continues to see growth in its sustainable practices. DHL also in collaboration with a retailer customer they improved their green credentials in logistics (DHL, 2018). Specifically, one of their customers in the retail sector wanted to reduce carbon emissions in its transport operations between suppliers and Distribution Centers in the UK. DHL worked together with the customer to develop a solution with lasting benefits. The teardrop trailer was chosen as the green solution to fulfill requirements in terms of reducing fuel usage and carbon emissions, as well as allowing the customer to carry more stock per trailer, cutting the number of journeys they need to make. The unique aerodynamic shape of these trailers generates fuel savings of approx. 10% and a 10% increase in cubic capacity. We deployed 399 aerodynamically optimized teardrop trailer trucks that reduce customer's carbon footprint by more than 2000 tones every year.

8. Green food supply chains and circular food supply chains

Supply chain design decisions are crucial decisions that influence all the activities related to them. Supply chain designers have to find out the right place of having warehouses, distribution centers and eco-friendly transportation of products/materials. To have a green supply chain design, redesigning of the whole supply chain is required. Mangan et al. (2008) stated that to achieve a green supply chain, products' and supply chains design has to be synchronized. The overall greening of the supply chains requires green internal mechanisms such as internal environmental management and waste management principles. Also, to achieve greening of the supply chain feedback mechanisms in the end-of-use stage of the GSC or else postconsumer stage need to be implemented. These mechanisms refer to the reverse logistics or else a newly introduced concept of circular supply chains.

8.1 Internal environmental management

Companies that want to implement green initiatives should first make sure that they comply with environmental regulations and key environmental performance indicators. Companies are required to produce and publish Environmental reports against specific Key Performance Indicators (KPIs/Business Link UK, 2011a,b,c,d). For the future supply chains, the KPIs are developed to become Sustainable and incorporate the measurement of energy consumption, CO_2 emissions, traffic congestion and infrastructure simplification (Capgemini Report, 2011). Common practices that companies adopt for greening their businesses are the Environmental Management Systems and ISO 14000 certification. The ISO 14000 family of standards provides practical tools for companies and organizations of all kinds looking to manage their environmental responsibilities (ISO, 2018). ISO 14001:2015 and its supporting standards such as ISO 14006:2011 focus on environmental systems to achieve this. The other standards in the family focus on specific approaches such as audits, communications, labeling and life cycle analysis, as well as environmental challenges such as climate change. ISO 14001:2015 sets out the criteria for an environmental management system and can be certified to. It maps out a framework that a company or organization can follow to set up an effective environmental management system. Moreover, "top management" commitment and support are critical for businesses environmental success (Rice, 2003; Lippmann, 1999; Deloitte Report, 2011). Implementing a GSC management approach requires strategic alignment with the core strategy of the business. Top management people need to promote and integrate the green strategy internally and externally in the whole the supply chain (Zailani and Wooi, 2010). Internally it can be achieved by prioritizing the green organizational goals and by motivating all employees to promote the new strategies. Externally this can be done by encouraging and influencing suppliers and customers to participate in the promotion of green initiatives. Integration and cross-functional collaboration are the key success factors for achieving a long-term benefit in a green supply chain.

8.2 *Waste management and food waste management*

Greening a supply chain requires efficient waste management. There is a high need for companies not just to minimize waste but eliminate it. Waste can be used packaging, raw materials used, energy used, solid and hazardous waste or even waste in manufacturing (e.g., Toyota's seven types of waste). There are different waste management options for each type of waste produced. The waste hierarchy must be taken into account when choosing a waste management option. In the Waste management hierarchy as seen in Fig. 2.5, activities are listed from environmentally friendly to least

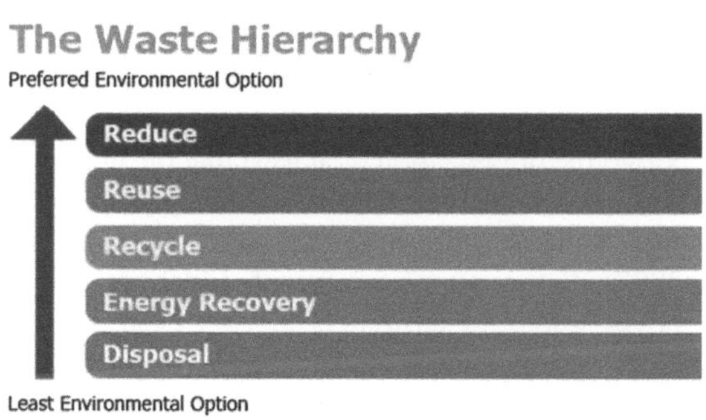

Figure 2.5 Waste hierarchy

environmentally friendly. According to Business Link UK (2011a,b,c,d) these activities include reduction, reuse, recycle and compost, energy recovery and disposal. Reduction of waste is the most effective way for reducing cost and having a less impact on the environment. Disposal of waste should be the last choice as it does not offer any substantial benefits to the environment. No matter what waste management option a company might choose, compliance with waste legislation is required (e.g., European Waste Framework Directive). "Zero waste" is a more integrated approach of eliminating waste across the supply chain. Zero Waste considers the whole life cycle of products, processes and systems (Zerowaste Alliance, 2011). By adopting Zero Waste strategies, waste can be prevented from the design stage.

Food is lost or wasted throughout the supply chain from initial agricultural production down to final household consumption (Gustavsson et al., 2010). "Producing and appropriately distributing enough food to feed a rising population presents many challenges for the industry, reducing food waste across the supply chain is a primary target to ensure global food security" (Mena et al., 2011, p.649). Interventions to reduce food waste are seen as important efforts to reduce food insecurity and to realize agriculture's potential to meet the world's need for food (World Bank, 2011). Environmental and human priorities lie in addressing food waste reduction rather than finding better ways to treat food that might be lost in the supply chain such as value adding activities (Foresight, 2011). Food waste needs to be significantly reduced or even achieve zero PHFL in the face of a sustainable future (World Economic Forum, 2011). Reducing food waste can increase grain supply and food security without wasting other resources such as land, labor, water and inputs (Kader, 2005; FAO, 2006; Hodges et al., 2010; The World Bank, 2011; Foresight, 2011). By reducing food waste both the profitability and the operational performance of all supply chain partners will be increased (Chapman, 2010). Food waste does not only have environmental and economic impacts, but also social impacts; it is a cause of poor nutrition and has significant effects on health and life expectancy (FAO, 2006). All food companies should have food waste avoidance policies in the first place and then also have food waste reduction policies. These policies should be about food waste prevention at operational stage such as in product processing, but also at postoperational stages such as sales in the supermarkets. It is essential that product labeling provides the correct information to customers, so they will know when until when the food product should be consumed.

8.3 Reverse logistics and circular supply chains

In terms of becoming more efficient, supply chains need to have activities that facilitate products return back into the organization or at least that the materials could be reused. Reverse Logistics ensure that there is a flow of taking back of used, defective or returned items as well as reuse, remanufacture or recycling (Carbone and Moatti, 2008; Srivastava, 2007). Future supply chains should have two flows of goods; a downstream that moves goods to customers and an upstream that takes products back. Reverse Logistics optimize the flow of taking back products and moving them forward in the supply network again. The European Recycling Platform (ERP) helps companies to implement effectively and efficiently WEEE Directive, which all companies are required to comply with (European Recycling Platform, 2011). It seems that the main driver of reverse logistics is governmental regulations. In the food sector reverse food logistics can be used for recycling of the packing and the wasted product as well. Most of the food packaging is plastic wrappers or containers. Plastic itself is harmful for the environment due to its quality for being long lasting and nondecomposition properties. Recently there have been many efforts to reduce the plastic packaging of food as it is found to have huge implications for the environment, but also about the humans that consume these food products. The idea is not to demotivate people from buying packaged food from stores, but instead, use food that comes with either biodegradable or with nature-friendly packaging. This way it ensures that the packaging content does not impact anything natural in a negative manner. Thus, companies should ensure that all the food packing that they use it could be recycled and identify new innovative ways of reducing packaging or identifying more environmentally friendly packaging options. Also, product take-back schemes are part of the reverse food logistics. For example, in many countries they use the beer bottle take-back scheme which is seen as an effort to recycle and be able to reuse these bottles.

Circular supply chains are similar to reverse logistics and they are based on the circular economy principles. A circular economy is an alternative to a traditional linear economy (make, use, dispose) in which we keep resources in use for as long as possible, extract the maximum value from them while in use, then recover and regenerate products and materials at the end of each service life (WRAP, 2018b). Circular economy is about looking beyond the current take-make-waste extractive industrial model, a circular economy aims to redefine growth, focusing on positive society-wide benefits (Ellen-McArthur, 2018a). It entails gradually decoupling economic activity from the consumption of finite resources and designing waste out of the system. It involves transition to renewable energy sources, and it builds on economic, natural, and social capital. The need to develop circular food supply chains is highlighted by the EllenMcArthur foundation (2018b) whose primarily purpose is to enable the transition of companies toward circular economy. The main driver for this change of the FSC is that the provision of healthy food for nine billion people by 2050 under today's degenerative food and agriculture system will lead to substantial further environmental and social damage across rural and urban communities globally (EllenMcArthur, 2018b).

9. Conclusion

Companies need to understand how food future supply chains will be affected by climate change and adapt their supply chain strategies accordingly; to both benefit and mitigate risks. "Greening the food supply chain" does not only have environmental benefits, it also reduces operational costs and increases flexibility of the supply chain. GSC is focused on energy reduction, scarce resources usage reduction, waste disposal minimization-elimination and decrease of carbon footprint. However, all these cannot be achieved unless fundamental organizational changes take place. Achieving a green FSC requires strategic business transformation. This transformation will take place only if companies align and implement the green initiatives into their core strategy continuously; maintaining a long-term perspective. Companies need firstly promote and support this strategy internally, improving the strategy applied within the company. Once this action has been conducted, the next stage is to influence the way that the entire supply chain (all suppliers and customers) operate.

The first stage of designing greener FSCs includes placing a focus on product redesign. After the redesign of products, all the activities that surround their supply i.e., the entire supply chain will be reinvented as well. Products and their packaging need to be manufactured in an environmental way to facilitate the closed-loop of the green supply chain. Product Stewardship is the main characteristic that all future GSCs must incorporate into their product's design. To achieve this, products must have the appropriate design (i.e., the eco-design). Governmental regulations also require companies to design and manufacture their products in a sustainable way. GSC design increases all parties' responsibilities for the products impacts to the environment. All the members in a supply chain must follow environmental policies, publish environmental reports and implement environmental management systems. Food companies must ensure that their suppliers manufacture and provide them green products to avoid reputational risks. Frequent audits and assessments in every stage of the supply chain are two essential characteristics of the future GSCs (from product audits and assessment to suppliers' audits). Greening supply chains require a life cycle thinking of the whole supply chain that all the processes are linked to each other with common goals, regulation compliance and ensuring the products' taken back (reverse logistics). Reverse food logistics require the products to be reused, recycled, recovered and disposed. Redesigning greener supply chains requires not only reverse logistics but also the appropriate waste management mechanisms. Every organization has to indentify the waste it produces and try to take advantage of it; ideally creating more desirable ways to eliminate waste.

The GSC design requires energy reduction across the chain, from using energy efficient ways to produce products, to energy efficient storage buildings. Future green buildings will be required to use alternative energy resources, built from recycled materials and to have environmental certification (e.g., LEED certification). Integration and collaboration in the future supply chain will make transportation more efficient. Containers optimization, efficient freight movement (full loads), sharing warehouses and transportation are some of the main activities of green logistics. All parties in the supply chains of the future will have full visibility of all the activities. However, it seems that there is a discrepancy between the green framework and what actually happens in the implementation of green design of the supply chains (Delloite, 2011; Preuss, 2005). In reality it is quite complex, as policies and regulations change frequently. Also, top management support is not always viable. Organizations, before starting transformation of their supply chains, have to ensure that the company's culture is open and that senior staff can support this new strategy. A limitation of the supply chain greening is that when implemented for the first time it requires redesign of all activities and some companies might not be able to afford it. It is difficult to provide a definitive answer to the question of how the design of supply chains, and logistics will change if green initiatives are implemented. This is because every supply chain has its unique characteristics and needs. From the above, it is suggested that to achieve a complete answer to how the GSCs will affect the design of future supply chains, a plethora of variables should be considered (such as industry characteristics and company strategy).

Overall, FSC greening requires high collaboration and visibility across the supply chain. Future FSC need to follow more integrated strategies and methods in sourcing, product design, transportations, storage, supplier relationship management, and reverse logistics. However, every business must find its own unique way for redesigning its supply chain and logistics. Food companies need to identify ways of how to best design, engineer and benefit from their supply chains while simultaneously reducing their carbon footprint. Green supply chain strategies add value to FSC by extending their role from end-customer delivery to product recovery and reuse. FSC managers must identify which designs fit best to their core business strategy, producing optimum benefits for all.

References

APO, 2006. Postharvest Management of Fruit and Vegetables in the Asia-Pacific Region. http://www.apo-tokyo.org/00e-books/AG-18_PostHarvest/AG-18_PostHarvest.pdf.

AWS, 2018. Alliance for Water Stewardship. https://a4ws.org/about/.

Ball, B.C., Hargreaves, P.R., Watson, C.A., 2018. A framework of connections between soil and people can help improve sustainability of the food system and soil functions. Ambio 47 (3), 269–283. https://doi.org/10.1007/s13280-017-0965-z.

BBC, 2018. What Does Organic Mean? https://www.bbcgoodfood.com/howto/guide/organic.

Beamon, M.B., 1999. Designing the green supply chain. J. Logist. Inf. Manag. 12 (4), 332–342.

Beecham Research, 2014. Towards Smart Farming. https://www.beechamresearch.com/files/BRL%20Smart%20Farming%20Executive%20Summary.pdf.

Beer, S., Lemmer, C., 2011. A critical review of 'green' procurement: life cycle analysis of food products within the supply chain. Worldw. Hosp. Tour. Themes 3 (3), 229–244.

Boston Consulting Group, 2009. Capturing the Green Advantage for Consumer Companies. https://www.bcgperspectives.com/content/articles/capturing_green_advantage_for_consumer_companies/.

Business Link UK, 2011a. Selecting a Sustainable Supplier. http://www.businesslink.gov.uk/bdotg/action/detail?itemId=1086763571&type=RESOURCES.

Business Link UK, 2011b. Integrate Environmental Policy into Your Business. http://www.businesslink.gov.uk/bdotg/action/detail?itemId=1083350600&type=RESOURCES.

Business Link UK, 2011c. Being a Supplier with Good Environmental Credentials. http://www.businesslink.gov.uk/bdotg/action/detail?itemId=1082899670&type=RESOURCES.

Business Link UK, 2011d. The Waste Hierarchy. http://www.businesslink.gov.uk/bdotg/action/detail?itemId=1079432000&r.l1=1079068363&r.l2=1086048413&r.l3=1079431917&r.s=sc&type=RESOURCES.

Cadogan, J.W., Cui, C.C., Morgan, R.E., Story, V.M., 2006. Factors facilitating and impeding the development of export market-oriented behaviour: a study of Hong Kong manufacturing exporters. Ind. Mark. Manag. 35 (5), 634–647.

Capgemini, 2011. Future Supply Chain 2016: Serving Customers in a Sustainable Way. http://www.capgemini.com/insights-and-resources/by-publication/future_supply_chain_2016/.

Carbone, V., Moatti, V., 2008. Greening the supply chain: preliminary results of a global survey. Supply Chain Forum Int. J. 9 (2), 66–76.

Chapman, P., 2010. Reducing product losses across the food supply chain. In: Mena, C., Graham, S. (Eds.), Delivering Performance in Food Supply Chains. Woodhead Publishing.

Charter, M., Tischner, U., 2001. Sustainable Solutions: Developing Products and Services for the Future. Greanleaf Publishing Limited, Sheffield, UK.

Crane, A., Matten, D., Spence, L.,J., 2008. Corporate Social Responsibility: Reading and Cases in a Global Context. Routledge, New York, USA.

Defra, 2006. Food Industry Sustainability Strategy. http://www.defra.gov.uk/publications/files/pb11649-fiss2006-060411.pdf.

DEFRA, 2016. What Counts as Organic Farming. https://www.gov.uk/guidance/organic-farming-how-to-get-certification-and-apply-for-funding#what-counts-as-organic-farming.

Deftra, 2011. Resource Efficiency. http://www.defra.gov.uk/news/2011/03/07/resource-efficiency/.

Delloite, 2011. The Food Value Chain: A Challenge for the Next Century. http://www2.deloitte.com/content/dam/Deloitte/ie/Documents/ConsumerBusiness/2015-Deloitte-Ireland-Food_Value_Chain.pdf.

Delloite, 2013. The Food Value Chain: A Challenge for the Next Century. http://www2.deloitte.com/content/dam/Deloitte/ie/Documents/ConsumerBusiness/2015-Deloitte-Ireland-Food_Value_Chain.pdf.

Deloitte Report, 2011. Green Means Go, Green Supply Chain Management: Opportunity Today, Imperative Tomorrow. http://www.deloitte.com/assets/Dcom-Canada/Local%20Assets/Documents/ca_en_consulting_greenmeansgo_aug08.pdf.

Despoudi, S., 2016. An Investigation of the Collaboration – Postharvest Food Loss Relationship and the Effect of the Environmental Turbulence Factors. Loughborough University. https://dspace.lboro.ac.uk/dspace-jspui/bitstream/2134/21785/1/Thesis-2016-%20Despoudi.pdf.

Despoudi, S., Papaioannou, G., Saridakis, G., Dani, S., 2018. Does collaboration pay in agricultural supply chain? An empirical approach. Int. J. Prod. Res. https://doi.org/10.1080/00207543.2018.1440654.

DHL, 2018. Green Optimisation in Transport. https://www.logistics.dhl/om-en/home/all-products-and-solutions/green-logistics.html.

Diaz, H.F., Murnane, 2011. Cilmate Extremes and Society. Cambridge University Press. http://www.cambridge.org/ar/academic/subjects/earth-and-environmental-science/climatology-and-climate-change/climate-extremes-and-society.

Driscoll, M., 2012. How Can We Build a Sustainable Farming System for All? http://www.theguardian.com/sustainable-business/blog/farming-system-principles-based-sustainable.

EFCR, 2018. 205.301 Product Composition. https://www.ecfr.gov/cgi-bin/text-idx?SID=786c250372e4a0197dcd3602162e42b4&mc=true&node=sp7.3.205.d&rgn=div6#se7.3.205_1300.

EllenMcArthur, 2018a. What is Circular Economy. https://www.ellenmacarthurfoundation.org/circular-economy/concept.

EllenMcArthur, 2018b. Cities and the Circular Economy of Food. https://www.ellenmacarthurfoundation.org/assets/downloads/Cities-and-the-circular-economy-for-food-1.pdf.

Emmet, S., Sood, V., 2010. Green Supply Chains: An Action Manifesto. John Wiley & Sons Ltd, Cornwall, UK.

Engels, F., 1843. Outlines of a critique of political economy. In: Engels, F. (Ed.), 1956, on Marx's Capital. Progress Publishers, Moscow.

Esty, D.C., Wiston, A.C., 2006. Green to Gold: How Smart Companies Use Environmental Strategy to Innovate, Create Value, and Build Competitive Advantage. Yales University, New Heaven and London, UK.

EU, 2018. Organic Farming. https://ec.europa.eu/info/food-farming-fisheries/farming/organic-farming/organics-glance_en.

European Recycling Platform, 2011. http://erprecycling.prodigentia.net/index.php?content=12.
FAO, 2002. World Agriculture: Towards 2015/2030 Summary Report. ftp://ftp.fao.org/docrep/fao/004/y3557e/y3557e.pdf.
FAO, 2006. Postharvest Management of Fruit and Vegetables in the Asia-Pacific Region. http://www.apo-tokyo.org/00e-books/AG-18_PostHarvest/AG-18_PostHarvest.pdf.
FAO, 2009. The State of Food Insecurity in the World. ftp://ftp.fao.org/docrep/fao/012/i0876e/i0876e.pdf.
FAO, 2011. Global Food Losses and Waste. http://www.fao.org/fileadmin/user_upload/ags/publications/GFL_web.pdf.
Food Logistics, 2018. Food Logistics Names 2018 Top Green Providers. https://www.foodlogistics.com/awards/article/21006517/food-logistics-top-green-providers-2018.
Foresight, 2011. The Future of Food and Farming: Final Project Report. The Government Office for Science, London.
Gidney, M., 2012. Why a Fair Supply Chain Is Key to Achieving a Sustainable Food System. http://www.theguardian.com/sustainable-business/fairtrade-partner-zone/fair-supply-chain-sustainable-food-system.
Godfray, H.C., Crute, I.R., Haddad, L., Lawrence, D., Muir, J.F., Nisbett, N., Pretty, J., Robinson, S., Toulmin, C., Whiteley, R., 2010. The future of global food system. Philos. Trans. R. Soc. B 365, 2769–2777.
Green Building Certification Institute, 2011. http://www.gbci.org/Homepage.aspx.
Green, K., Morton, B., New, S., 1996. Purchasing environmental management: interactions, policies and opportunities. Bus. Strateg. Environ. 5 (5), 1188–1197.
Guardian, 2011a. Energy Efficient Buildings Are Vital to Sustainability. http://www.guardian.co.uk/sustainable-business/energy-efficient-buildings1.
Guardian, 2011b. Body Shop's Ethics Under Fire after Colombian Peasant Evictions. http://www.guardian.co.uk/world/2009/sep/13/body-shop-colombia-evictions?INTCMP=ILCNETTXT3487.
Gustavsson, J., Cederberg, C., Sonesson, U., Otterdijk, R., Meyberg, A., 2011. Global Food Losses and Waste. http://www.fao.org/fileadmin/user_upload/ags/publications/GFL_web.pdf.
Hitchcock, D., Willard, M., 2007. The Business Guide to Sustainability: Practical Strategies and Tools for Organizations, second ed. Earthscan Ltd, London, UK.
Hodges, R., Buzby, J.C., Benett, B., 2010. Postharvest Losses and Waste in Developed and Less Developed Countries: Opportunities to Improve Resource Use, Journal of Agricultural Science. Cambridge University.
Howe, M., 2017. Advantages and Disadvantages Organic Farming. https://small-farm-permaculture-and-sustainable-living.com/advantages_and_disadvantages_organic_farming/.
IFOAM, 2018. Organic Regulations. https://www.ifoam-eu.org/en/what-we-do/organic-regulations.
Industry Council for Packaging and Environment, 2011. http://www.incpen.org/displayarticle.asp?a=11&c=2.
ISO, 2018. ISO 1400 Familiy. https://www.iso.org/iso-14001-environmental-management.html.
Kader, A.A., 2005. Increasing food availability by reducing postharvest losses of fresh produce. In: Proceedings. 5th International Postharvest Symposium.
Krejci, C., Beamon, B., 2010. Environmentally-conscious supply chain design in support of food security. Oper. Supply Chain Manag. 3 (1), 14–29.
Leonidou, L.C., Theodosiou, M., 2004. The export marketing information system: an integration of the extant knowledge. J. World Bus. 39 (1), 12–36.

Lin, C.Y., 2007. Adoption of green supply chain practices in taiwan's logistics industry. J. Int. Manag. Stud. 2 (2), 90–98.
Lippmann, S., 1999. Supply chain environmental management: elements of success. Corp. Environ. Strat. 6 (2), 175–182.
Lundqvist, J., De Fraiture, C., Molden, D., 2008. Saving Water: From Field to Fork - Curbing Losses and Wastage in the Food Chain. SIWI Policy Brief. SIWI.
Mallidis, I., Despoudi, S., Vlachos, D., Dekker, R., Iakovou, E., 2018. An assessment of the impact of sulphur limit fuel regulations on supply chain network design decisions. Ann. Oper. Res. https://doi.org/10.1007/s10479-018-2999-4.
Malthus, T., 1798. An Essay on the Principle of Population. http://129.237.201.53/books/malthus/population/malthus.pdf.
Mangan, J., Lalwani, C., Butcher, T., 2008. Global Logistics and Supply Chain Management. John Wiley & Sons Ltd, West Sussex: England.
McKinnon, A., Cullinane, S., Browne, M., Whiteing, A., 2010. Green Logistics: Improving the Environmental Sustainability of Logistics. Kogan Page, London.
Meadows, D.H., Meadows, D.L., Randers, J., Behrens, W.W., 1972. The Limits to Growth: A Report for the Club of Rome's Project on the Predicament of Mankind. Universe Books, New York.
Mena, C., Adenso-Diaz, B., Yurt, O., 2011. The causes of food waste in the supplier-retailer interface: evidences from UK and Spain. Resour. Conserv. Recycl. 55, 648–658.
Min, H., Galle, W.P., 1997. Green purchasing practices: trends and implications. Int. J. Purch. Mater. Manag. 33 (3), 10–17.
Napolitano, M., 2013. Seven Trends in Sustainable Warehouse Design. http://www.supplychain247.com/article/7_trends_in_sustainable_warehouse_design/green.
Nestle, 2018a. Resource Efficiency. https://www.nestle.co.uk/csv2015/environmental-sustainability/resource-efficiency.
Nestle, 2018b. Advocate for Effective Water Policies and Stewardship. https://www.nestle.com/csv/impact/water/policy-stewardship.
Nestle, 2018c. Nestlé Joins Research Effort to Improve Products' Environmental Performance. https://www.corporate.nestle.ca/en/media/newsandfeatures/lifecycleassessment.
Nestle, 2018d. Product Life Cycle. https://www.nestle-ea.com/en/csv/planet/product%20-life%20cycle/home.
Organic Food, 2018. Types of Organic Food. http://www.organicfood.sg/types_of_organic_food.html.
Preuss, L., 2005. Rhetoric and Reality if Corporate Greening: A View from the Supply Chain Management Function, vol. 14. Business Strategy and Environment, pp. 123–139.
Product Stewardship Institute, 2011. http://productstewardship.us/displaycommon.cfm?an=1&subarticlenbr=55.
Rice, S., 2003. Commitment to excellence: practical approaches to environmental leadership. Environ. Qual. Manag. 12 (4), 9–22.
Ridler, J., 2017. Coca-Cola-Makes-Multi-Million-Pound-Warehouse-investment. https://www.foodmanufacture.co.uk/Article/2017/06/23/Coca-Cola-makes-multi-million-pound-warehouse-investment.
Robinson, D.R., Wilcox, A., 2008. The Greening of Supply Chains: A Special Supplement to Supply Chain Management Review. Supply Chain Management Review, pp. 61–66. October.
Schiforce, 2018. Smart Farming, or the Future of Agriculture. https://medium.com/sciforce/smart-farming-or-the-future-of-agriculture-359f0089df69.

Schuttelaar, 2017. Smart Farming Is Key for the Future of Agriculture. https://www.schuttelaar-partners.com/news/2017/smart-farming-is-key-for-the-future-of-agriculture.

Slack, N., Chambers, S., Johnston, R., 2010. Operations Management. Pearson Education Limited, Essex, UK.

Soil Association, 2018a. Organic Standards. https://www.soilassociation.org/certification/food-drink/what-is-organic-certification/.

Soil Association, 2018b. Green Gourmet. https://www.soilassociation.org/certification/food-drink/meet-our-licensees/green-gourmet/.

Srivastava, S.K., 2007. Green supply chain management: a state-of-the art literature review. Int. J. Manag. Rev. 9 (1), 53−80.

Sustain, 2015. Sustainable Food. http://www.sustainweb.org/publications/the_sustain_guide_to_good_food/ [Accessed 10.02.15].

SustainAbility, 2011. Appetite for Change. http://www.sustainability.com/library/appetite-for-change#.Tzzj5Vy15WU [Accessed 03.02.12].

The World Bank, 2011. Missing Food: the Case of Postharvest Grain Losses in Sub-Saharan Africa. http://siteresources.worldbank.org/INTARD/Resources/MissingFoods10_web.pdf.

UK Government, 2019. Countryside Stewardship. https://www.gov.uk/government/collections/countryside-stewardship-get-paid-for-environmental-land-management.

UK Green Building Council, 2011. http://www.ukgbc.org/site/home.

UN, 2018. Koyoto Protocol. In: https://unfccc.int/process/the-kyoto-protocol.

Unilever, 2018. Sustainable Sourcing. https://www.unilever.com/sustainable-living/reducing-environmental-impact/sustainable-sourcing/.

UPS, 2011. How Green Is Your Business. http://www.ups.com/bussol?loc=en_US&viewID=categoryView&contentID=ct1_solg_cat_decision_green.

Vidal, J., 2012. The Future of Food. Guardian. http://www.theguardian.com/global-development/2012/jan/22/future-of-food-john-vidal.

Walker, H., Di Sisto, L., McBain, D., 2008. Drivers and barriers to environmental supply chain management practices: lessons from the public and private sectors. J. Purch. Supply Chain Manag. 14 (1), 69−85.

World Economic Forum, 2011. Driving Sustainable Consumption: Value Chain Waste. https://members.weforum.org/pdf/sustainableconsumption/DSC%20Overview%20Briefing%20-%20Value%20Chain%20Waste.pdf.

World Green Building Council, 2011. http://www.worldgbc.org/about-worldgbc/who-we-are.

Wrap, 2011. http://www.wrap.org.uk/downloads/15203-05_Overall_CS_LoRes.08f25a2f.3740.pdf.

WRAP, 2018a. Designing in Sustainability with Asda and Three Key Suppliers. http://www.wrap.org.uk/sites/files/wrap/Asda_Sustainable_New_Product_Development_case_study.pdf.

WRAP, 2018b. WRAP and the Circular Economy. http://www.wrap.org.uk/about-us/about/wrap-and-circular-economy.

Zailani, S., Wooi, G.C., 2010. Green supply chain initiatives: investigation on the barriers in the context of SMEs in Malaysia. Int. Bus. Manag. 4 (1), 20−27. Medwell Journals.

Zerowaste Alliance, 2011. http://www.zerowaste.org/.

Zhu, Q., Sarkis, J., 2004. Relationships between Operational Practices and Performance among early adopters of green supply chain management practices in Chinese manufacturing enterprises. J. Oper. Manag. 22, 265−289.

Zhu, Q., Sarkis, J., Lai, K., 2006. Green Supply Chain Management Implications for 'Closing the Loop'. Logist. Transport. Rev. 44 (1), 1−18.

Life cycle assessment in the food industry

3

Adriana Del Borghi, L. Moreschi, M. Gallo
Department of Civil, Chemical and Environmental Engineering (DICCA), University of Genoa, Genova, Italy

Chapter outline

1. Introduction 63
2. Life cycle assessment (LCA) 65
2.1 The application of LCA to food sector 68
2.1.1 Functional unit 68
2.1.2 System boundary 69
2.1.3 Allocation 71
2.1.4 Crop cultivation 75
2.1.5 Manure management 78
2.1.6 Wastewater management 78
2.1.7 Packaging 79
2.2 LCA databases 81
2.2.1 LCA food database 81
2.2.2 World food life cycle assessment database (WFLDB) 81
2.2.3 Ecoinvent 82
2.2.4 Agri-footprint 89
3. Labeling schemes 90
4. EU organic logo and other standards on organic farming 98
5. Product environmental footprint (PEF) 99
6. Environmental product declaration (EPD) 100
7. Examples of product category rules (PCR) related to food 107
8. Conclusions 111
References 112

1. Introduction

In the next years, the main goal of the research in environmental sustainability will necessarily be the protection of the planet in order to guarantee to the future generations the possibility of living healthy and carrying out their lives (Klugman, 2011). Indeed, one of the most struggling challenges that the scientific community and the

The Interaction of Food Industry and Environment. https://doi.org/10.1016/B978-0-12-816449-5.00003-5

agricultural sector will face is the requirement of feeding the more and more numerous world population without overexploiting the planet and its resources. This issue was deeply analyzed during the Milan Expo 2015 "*Feeding the Planet, Energy for Life*", which established as its inheritance a chart enclosing all the goals to achieve in order to understand the connections between environmental sustainability and food equity, safeguarding biodiversity and ecosystems by means of a smart agriculture (EXPO 2015). Analogous goals were presented by the Roadmap to a Resource Efficient Europe (COM(2011) 571) and afterward also by the European Common Agricultural Policy (CAP) (European Commission, 2012) with the statement of the three high priority fields: the improvement of the sustainability in the agriculture; the reinforcement of the conservation of biodiversity and of ecological farming and forestry systems; and the handling of the water management and use in accordance with the possible climate change scenarios that could occur in the near future. In 2013, the reorganization of the CAP with a long-term plan named "CAP Reform 2014−20" was carried out with the aim of facing the urgency to create a higher level of safe and quality food, yet pursuing at the same time a continuous protection of those natural resources that the agricultural sector needs to maintain its productivity standards. In order to achieve this goal, the food sector should go through a deep renewal that allows it to become more competitive in the market, above all with the implementation of a well performing supply chain, which guarantee the maintenance of a blooming rural economy (European Commission, 2013; European Parliament and Council, 2013). Analogously to all industrial processes, also food production significantly contributes to consumption of resources and presents remarkable environmental impacts to be evaluated through a life cycle approach. Life Cycle Assessment (LCA) (ISO, 2006a) is a tool used for the quantification of the environmental burdens along the life cycle of a product or a process, whose contributions to several impact categories and use of resources can be evaluated. From this perspective, LCA studies can represent a necessary decision tool in order to plan the appropriate strategies for life cycle management and the improvement of the environmental performance of products, aiming to the promotion of a shifting toward sustainable agriculture, farming and food production systems and leading toward more sustainable food consumption pattern. Over the last years, the food industry has seen the birth of several LCA studies defining the impact assessment of different products and consumer goods (Roy et al., 2009; Al-Ansari et al., 2015), such as bread (Espinoza-Orias et al., 2011), beer (Cordella et al., 2007), wine (Pattara et al., 2012; Martins et al., 2018; Ponstein et al., 2019) and tomato products (Andersson et al., 1998; Del Borghi et al., 2014); milk (Eide, 2002) and butter (Flysjö, 2011); meat (Nunez et al., 2005; Gallo et al., 2015a; Cesari et al., 2018; Vitali et al., 2018); fishery (Hospido et al., 2006; Biermann and Geist, 2019); apple (Milà i Canals et al., 2006; Mouron et al., 2012), orange (Coltro et al., 2009), melon (de Figueirêdo et al., 2014), peach (Ingrao et al., 2015; Nikkhah et al., 2017), raspberry (Tecco et al., 2016), sugar (Yuttitham et al., 2011), olive (Avraamides and Fatta, 2008; Proietti et al., 2014), lettuce (Foteinis and Chatzisymeon, 2016), beans (Romero-Gámez et al., 2012; Abeliotis et al., 2013), peas (Naudin et al., 2014), broccoli (Milà i Canals et al., 2010; Atallah et al., 2014); breakfast cereals and snacks (Jeswani et al., 2015); and animal feed (Strazza et al., 2015).

LCA, originally developed to be used as decision support tool for environmental management, has now several related applications such as external communication through environmental labels and declarations. New consumer awareness and behavior favoring greener products and services is shifting industry toward more environmentally sustainable production systems (Miranda-Ackerman and Azzaro-Pantel, 2017). Therefore, a further step in applying LCA to food industry would be the spread of environmental certification and labeling schemes utilization for Business to Business (B2B) and Business to Consumer (B2C) communication. The International Standards Organisation (ISO) has classified the existing environmental labels into three typologies—Type I, II, and III—and has specified the preferential principles and procedures for each one of them (ISO 14021, ISO 14024 and ISO 14025) (ISO, 2006a; ISO, 2016; ISO, 2018; UNOPS, 2009). Despite Type I Ecolabels do not apply to food and feed products, Type III environmental declarations such as the Environmental Product Declaration (EPD) show a wide application of environmental labels in the food and beverages sector. At European level, the European Commission's Product Environmental Footprint (PEF), a multi-criteria measure for measuring and communicating the environmental performance of products, includes 11 food categories (COM 196 and European Commission, 2013; Galatola and Pant, 2014). This chapter presents and discusses the implementation of LCA in different food industries. Specific methodological issues related to the food sector will be discussed with particular attention to functional unit, system boundary, allocation rules, crop cultivation, manure management and packaging. Moreover, several applications of environmental labels to food products are presented.

2. Life cycle assessment (LCA)

LCA is a tool to evaluate the potential environmental impacts deriving from products, processes or services along their life cycle. The standardized methodology of LCA is described in the standards ISO 14040-44 (ISO, 2006b; ISO, 2006c). The scope of an LCA study considers not only the actual processing stage but also upstream (e.g., raw material production, agriculture, livestock, fisheries and aquaculture and packaging production) and downstream (e.g., product distribution, consumption or use phase, and waste disposal) processes including the transport activities needed in all the stages. The life cycle impact assessment results quantify the multiple environmental impacts by means of several characterization models—each with its own equivalent unit of measurement—and facilitate the identification of the hot spots (i.e., the main life cycle stages and activities associated with the most relevant impacts). According to ISO 14040, the principles and framework for Life Cycle Assessment include (Fig. 3.1): the goal and scope definition of the LCA; the life cycle inventory analysis (LCI) phase; the life cycle impact assessment (LCIA) phase; the life cycle interpretation phase; reporting and critical review of the LCA; limitations of the LCA; relationship between the LCA phases; and conditions for use of value choices and optional elements.

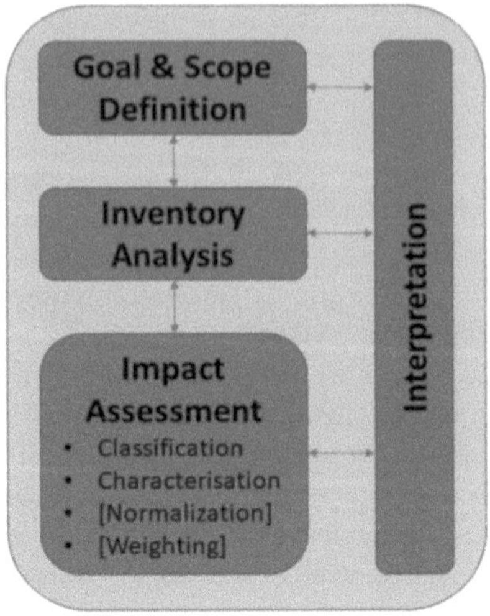

Figure 3.1 Phases of the life cycle assessment.

Among the above-said principles, only the first four constitute the real work phases for an LCA study and are explained in the following of the paragraph.

The "goal and scope definition" phase is generally meant to clearly identify the intention of the application and may be improved during the study. On the one hand, the goal of the study should state the intended application, the reasons that led to the start of the study, the final audience interested in the results of the study and if the results are meant to be used for comparative evaluations. On the other hand, the scope of the study should describe the product system involved in the study and its functions, the system boundaries, the allocation procedures utilized, the impact categories chosen to be characteristic and representative for the study, the data quality requirements and, in general, all the assumptions and the choices made for the realization of the study. In addition, the scope of the study has to define the "functional unit", providing a reference to which the input and output data are normalized and the results are referred and allowing the comparison with other systems—and then LCA studies—with the same functional unit. The "Life Cycle Inventory analysis (LCI)" phase includes data collection and calculation procedures that allow quantifying the input and output flows of a product system. These incoming and outgoing flows have to consider all the energy, raw material and ancillary inputs and can include the use of resources and releases in the air, water and soil associated with the system. In order to allow the normalization with the functional unit, also data on products, coproducts and waste quantities need to be collected. Since data may derive not only from actual measurements but also from calculations and estimations, a check on data validity needs to be conducted during the process of data collection to confirm and provide evidence that the data quality requirements for the intended application have been fulfilled.

The inputs and outputs data can then be allocated to the different products according to clearly defined procedures that have to be stated and explained together with the allocation procedure.

Starting from these data, the study can derive some interpretations, in relation to the goal and the field of application of the LCA study. These data also form the basis for the assessment of the impact of the life cycle. The "Life Cycle Impact Assessment (LCIA)" phase consists in the evaluation of the environmental performance of the system analyzed and includes the collection of indicator results for the different impact categories, which together represent the LCIA profile for the product system. The choice of the impact categories has to be coherent with the goal and scope definition in order to define the most representative ones. Each impact category is associated with at least one characterization model which, by means of a set of characterization or emission factors, allows converting all the data collected during the LCI phase into well defined environmental impacts with their own units of measurements. The most common impact categories refer to the following environmental impacts:

- the Acidification Potential (AP)—measured in sulfur dioxide equivalents (SO_2 eq.)—consisting in a drop in pH of soils, lakes, forests, due to air emissions of acidifying compounds, with harmful effects on living organisms, e.g., "acid rains";
- the Global Warming Potential (GWP)—measured in carbon dioxide equivalents (CO_2 eq.) – consisting in a change in the greenhouse effect, usually calculated for 100 years, due to emissions and absorptions attributable to humans, such as carbon dioxide (CO_2), methane (CH_4), nitrous oxide (N_2O), and other greenhouse gases;
- the Eutrophication Potential (EP)—measured in phosphate equivalents (PO_4^{3-})—consisting in a reduction in dissolved oxygen levels in water media with collapse of fish and other aquatic species due to excess addition of large quantities of mineral nutrients such as nitrogen and phosphorous and subsequent dramatic increase in flora that feed on these nutrients;
- the Photochemical Ozone Creation Potential (POCP)—generally measured in ethylene equivalents (C_2H_4 eq.) or in Nonmethane volatile organic compound equivalents (NMVOC eq.)—consisting in the formation of ozone at ground level due to air emissions of unburnt hydrocarbons and nitrogen oxides in presence of solar radiation. This phenomenon is harmful for living organisms and often present in large urban centers.

Other possible impact categories or indicators useful for the food sector may be:

- the Land Use—measured in square meters per year (m^2a)—consisting in an impact on biodiversity. Biodiversity depends on the type of use of soil and dimensions of area. In this impact category both regional and local impacts are taken into consideration and the damage related to land use results from both conversion and occupation of soil;
- the Abiotic Depletion Potential—measured in antimony equivalents (Sb eq.) for elements or in terms of energy (MJ, net calorific value) for fossil fuels—consisting in depletion of nonrenewable resources, such as fossil fuels, metals and minerals;
- generic indicators on the water use and consumption being water scarcity indicators (e.g. Boulay, 2018) or Water Footprint indicator in accordance to the ISO 14046 standard (ISO, 2017);
- indicators on the waste production;
- indicators on the use of resources.

The "Life cycle interpretation" phase runs through the other phases and consists in the identification of the significant issues based on the results of the LCI and LCIA phases, a continuous check on the completeness and the consistency of the data collected and on the sensitivity of the final results with reference to the potential uncertainties linked with the data and the definition of the conclusions, limitations and recommendations of the study.

2.1 The application of LCA to food sector

Despite LCA has widely been applied for products and services, there are still several methodological issues which come into focus when LCAs are applied to food sector. Functional unit, system boundary, allocation rules, crop cultivation and packaging are issues that need to be deeply discussed when performing an LCA study in the food industry. The following main product categories can be considered:

a. Products of agriculture
b. Live animals and animal products (e.g., dairy products, egg products, meat products)
c. Fish and other fishing products
d. Beverages

2.1.1 Functional unit

According to ISO 14044, the functional unit has to be consistent with the goal and scope of the study and it needs to be clearly defined and measureable. The functional unit consists of a qualitatively defined function or property and its quantification via a unit. It is defined as a quantified performance of the product for use as a reference unit. For LCA not covering a full life cycle, the concept of functional unit is transferred into a declared unit, related to the typical applications of products. The functional unit can be defined answering questions such as: 'what', 'how well', 'how much', and 'how long'. Agricultural products (i.e., crop products) are usually based on a mass reference of 1 kg of output fresh product (Nemecek et al., 2014). For vegetable products, the declared unit are usually 1 kg of product, including its packaging (the weight of the packaging is usually not included in this 1 kg). The reference flow in the LCA is defined at the point where the product arrives at the customer gate, i.e., any losses occurring before then should be taken into account. For live animal production, the reference flow can be defined as 1 kg animal, 1 kg egg, 1 kg milk. For transformed food items, the reference flow can be 1 kg animal product with different coproducts (meat of different grades, skin, fat, bones, etc.) allocated to the total weight of the animal. For example, for meat of mammals, the declared unit is 1 kg of pure meat (chilled or frozen) and it does not include bones or other not edible parts. For meat of poultry, the declared unit is just 1 kg of poultry meat (fresh, chilled or frozen). The animal live weight, the carcass yield and the boning yield considered to produce edible meat are important issues for the declared unit definition. For raw milk, the declared unit can be 1 L of liquid raw milk, excluding packaging, entering the processing plant or being dispensed at a raw milk distributor. Despite their spread in the LCA studies and especially in the labeling schemes, the utilization of this type of functional units defining mass or volume is getting questioned in the scientific community as not the most fit for the food sector. The criticism rises from the consideration that the scope of nutrition is not as much to fulfill a required target of weight or volume but rather to provide to human beings the energy and the nutrients needed to live. The necessity of define some functional units more suitable for the food sector is also strictly link with the opportunity of performing comparison between different foods by means of functional units more coherent with the scope of their product systems.

Different functional units have been proposed among the scientific community analyzing the effects of the comparison with the traditional ones (Xu et al., 2018): compares the carbon footprints of several carbohydrate-rich foods obtained by using different functional units based on weight, energy content, protein content, carbohydrate content, and two comprehensive trophic indexes based on nutritional profiling models; a Nutrient Density Unit (NDU) reflecting total protein, essentially fatty acids and dietary fiber per energy density has been investigated (van Dooren, 2016); also a further step comparing different a functional units based on protein content in foods and also on their digestibility is pursued highlighting a small improvement and change in the results compared to the significantly bigger amount of data required (Sonesson et al., 2017). A completely different approach can be made considering a functional unit based on economic values of foods instead of their nutritional quality: the use of this type of functional unit may be more effective in terms of business to consumer communication (B2C) leading the consumers toward the decrease of their environmental impacts in line with their economic capacity and purchasing power (van der Werf and Salou, 2015).

On the one hand, from a labeling schemes point of view, these potential improvements in the functional units based on nutritional properties may represent an obstacle for the diffusion of the environmental labels in the food sector leading to a too higher amount of technical skills and scientific knowledge required to the consultants for the LCA studies in comparison with the current functional units based on mass or volume. On the other hand, from a scientific research point of view, the identification and the creation of new functional units more fit with the actual scope of nutrition may enhance the debate and the research on more sustainable product systems in the food sector. The first step of this approach could be the identification of the most suitable functional units for the detection of sustainable foods in correlation with their nutrient values (Masset et al., 2015). Moreover, in continuity with this path, the further step could be the improvement of the concept of functional units toward a switch with the concept and the comparison of complete dietary patterns (Pairotti et al., 2015) comprehending the daily energy intake and the amount of all the nutrients required by an healthy diet, e.g., according to the standards set up by the European Food Safety Authority (EFSA).

2.1.2 System boundary

The system boundaries include all processes linked to the product supply chain relative to the unit of analysis. The inclusion of all attributional processes from Cradle-to-grave is a default approach. System boundaries can be divided into foreground processes and background processes. Foreground processes are core processes for which direct access to information is available. Moreover, the life cycle of products can be separated into different life cycle stages: upstream processes (from cradle-to-gate), core processes (from gate-to-gate) and downstream processes (from gate-to-grave). All elementary flows at resource extraction need to be included.

For products of agriculture, the following attributional processes are usually part of the product system and classified as upstream processes:

- Production of seeds, cuttings or plants for the cultivation
- Production of fertilizers used in the agriculture
- Production of auxiliary products used such as detergents for cleaning, etc.
- Production of semiproducts used in the core process, if applicable
- Production of primary and secondary packaging

The following attributional processes are part of the product system and classified as core processes:

- External transportation to the core processes
- Agriculture including e.g., operations at the farm(s), emissions of nitrous gases. The cradle for the agriculture is soil preparation and cultivation
- Maintenance (e.g., of the machines)
- Preparation of the final product
- Waste treatment of waste generated during manufacturing

Usually, the technical system do not include:

- Manufacturing of production equipment, buildings and other capital goods
- Business travel of personnel
- Travel to and from work by personnel
- Research and development activities

The following attributional processes can be part of the product system and classified as downstream processes:

- Transportation of the product to an average customer or consumer
- Customer or consumer use of the product
- End-of-life processes of any wasted part of the product (e.g., peel of oranges)
- End-of-life processes of packaging waste

For live animals and animal products, the following additional attributional processes should be part of the product system and classified as upstream processes:

- Feed production, e.g., cultivation, harvest and refining
- Animal breeding (including enteric fermentation) (e.g., mammals, poultry rearing, laying hen farming)
- Farm management
- Slaughterhouse activities
- Production of the product, e.g., milking of cows
- Preparation of the final product (e.g., slaughter activities, meat processing, packaging of the final product)
- Manure management
- Wastewater treatment generated during slaughterhouse process

For fish and other fishing products, the system boundaries should include the following additional processes:

- Preparation before catching in ocean
- Production of fish feed
- The fishery, farming or catching in the ocean. This includes e.g., air and water emissions and emissions from energy wares used in the fishery
- Fish processing near the coast
- Maintenance of the ships

Boundaries to nature are defined as flows of material and energy resources from nature into the system. Emissions to air, water and soil cross the system boundary when they are emitted from or leaving the product system.

2.1.3 Allocation

Allocation can be defined as the partitioning of input or output flows of a process or a product system between to the product system under study and other product systems. In case of multi-functionality, the following decision hierarchy can be set:

(1) subdivision or system expansion;
(2) allocation based on a relevant underlying physical relationship;
(3) allocation based on some other relationship, such as economic value.

A critical difference between different approaches is that the method of avoiding allocation by expanding the system boundary is considered not applicable within in attributional LCA used and the concept of modularity. For example, in the context of ISO 14025 (ISO, 2006a), the following step-wise procedure are usually applied for multifunctional products and multiproduct processes:

1. Allocation is preferable to be avoided, if possible, by dividing the unit process into two or more sub-processes and collecting the environmental data related to these sub-processes.
2. If allocation cannot be avoided, the inputs and outputs of the system may be partitioned between its different products or functions in a way that reflects the underlying physical relationships between them; i.e., they should reflect the way in which the inputs and outputs are changed by quantitative changes in the products or functions delivered by the system.
3. Where physical relationships alone cannot be established or used as the basis for allocation, other allocation method can be defined for each process, including economic allocation.

While for agricultural products allocation is usually not a key issue and generally solved through economic allocation (Nemecek et al., 2014), allocation for multifunctional products represents a priority to be considered for meat and dairy products.

The example of meat of mammals is presented in the following. Mammal are defined as nonreproducing mammal destined to meat production (ex. calf in cow meat farming), while reproductive mammal are mammal of female gender that has reproduced and that at end of career (when no longer destined to reproduction and/or milk production) is destined to meat production (ex. reproductive cow in cow meat farming).

In the process of meat production, the following life cycle stages involve dealing with multi-functional processes:

1. production of feed;
2. production of meat from reproductive mammals (e.g., meat from dairy cows at end of career, etc.);
3. slaughterhouse activities

Given that both Mammals and Reproductive Mammals can be destined for slaughter, meat chain coproducts may vary in relation to the type of mammal considered. For example, possible coproducts of cows are meat, milk and leather, while sheep coproducts are meat, milk and wool. Only meat can be produced from pigs and meat or milk from goats. In all cases, impact allocation for all products and coproducts can be done considering the allocation method proposed by the International Dairy Federation in 479/2015 Bulletin (FIL-IDF, 2015). According to this biophysical allocation method, the allocation factor for milk is calculated as follows:

$$AF = 1 - 6.04 * BMR$$

Where BMR = M_{meat}/M_{milk}; M_{meat} is the mass of live weight of all animals sold including bull calves and culled mature animals per year; and M_{milk} is the mass of fat and protein corrected.

milk (FPCM) sold per year (corrected to 4% at and 3.3% protein), calculated with the following equation:

$$FPCM(kg/yr) = Production(kg/yr) * [0.1226 * fat\% + 0.0776 * protein\% + 0.2534]$$

All cases contemplate just the impacts of the female breeding mammal's lifecycle.

For the dairy farm system where the focus is milk production, the meat generated from surplus calves and cull dairy cows is an important coproduct. When the object of the analysis is meat from cull dairy cow, the impacts that should be considered are the following:

- impacts of the reproductive mammal's lifecycle before entering in the reproduction phase; in this case, the impact can be allocated on the biophysical basis to all the products (i.e., reproductive mammals meat at the end of career - milk, surplus calves) generated during the entire life cycle of the reproductive mammal. The impacts to be taken into consideration are those related to (e.g., dairy cow) meat;
- the impacts of the reproductive mammal's lifecycle starting from the last breeding phase up to its slaughter.

When the object of the analysis is, instead, meat coming from veal calves, the impacts that should be considered are:

- impacts of the reproductive mammal's lifecycle before entering in the reproduction phase; in this case, the impact should be allocated on the biophysical basis to all the products (i.e., reproductive mammals meat at the end of career - milk, surplus calves) generated during the entire life cycle of the reproductive mammal. The impacts to be taken into consideration are those related to (e.g., surplus calves) meat;

- mammal (e.g., surplus calf) lifecycle environmental impact.

For products deriving from dairy farm systems it is therefore necessary to determine total emissions and to allocate them between milk and meat.

The allocation factor for milk (AF_{milk}) can be calculated as follows (FIL-IDF, 2015):

$$AF_{milk} = 1 - 6.04 * \frac{M_{meat}}{M_{milk}}$$

Where M_{meat} is the mass of live weight of all animals sold including bull calves and culled mature animals per year, and M_{milk} is the mass of fat and protein corrected milk sold per year (corrected to 4% at and 3.3% protein).

M_{milk} is calculated by multiplying milk production (per farm) by the ratio of the energy content to the energy content of standard milk 4% fat and 3.3% true protein content, using the following formula:

$$M_{milk} = \text{Milk production} * [0.1226 * \text{fat}\% + 0.0776 * \text{true protein}\% + 0.2534]$$

Then it is possible to calculate the allocation factor for meat AF_{meat}:

$$AF_{meat} = 1 - AF_{milk}$$

In the case of meat from mammals coming from systems devoted to meat production only (e.g., cattle meat), the mammal is the object of meat transformation processing and the impacts that need to be considered should be:

- part of impacts of the reproductive mammal's lifecycle before entering in the reproduction phase; in this case, the impact should be allocated on the biophysical basis to the meat products generated during the entire life cycle of the reproductive mammal products (i.e., generated mammals' meat and reproductive mammal's meat);
- impact of the reproductive mammal lifecycle phase dedicated to the reproduction and nurturing of mammals, considering the total amount of produced meat (from the generated mammals and the reproductive mammal itself) following the biophysical allocation approach;
- mammal lifecycle environmental impact.

The possible coproducts of slaughterhouse activities are quite different in meat production systems (e.g., meat, leather, fat, etc.) and may vary from one considered specie to another; then mass allocation is controversial, as the outputs (in terms of mass) are not linearly related to the mass of inputs. Consequently, the allocation method suggested by the Cattle model working group (CMWG) (EC, 2013) is the economic allocation. The impact for each meat product and sub-product that exit the slaughterhouse can be evaluated considering their average historical economic value. The following breakdown list represents the main considered products and by-products exiting the slaughterhouse plant:

a. Fresh meat and products suitable for human consumption
b. Hides and skins, sold to leather industry
c. Products sold for rendering (e.g., fat)
d. Specified risk material (e.g., skull, spinal cord, etc. considered category I by-products, sent to incineration)
e. Products not suitable for human consumption (category II by-products sent to recovery/energy plants)
f. Products for animal feed applications, such as pet food or feed or to compost (category III by-products sent to recovery)

In the case of meat of poultry, allocation between different poultry products and co-products suitable for human consumption should be based on economic allocation as well.

The management of animal by-products not intended for human consumption in European slaughterhouses is regulated by the EU laws (EC, 2002). It could be considered as representative as the global average management situation: high-risk wastes potentially infected with BSE require high-temperature incineration, while other low-risk materials can be rendered to produce fuel or recycled as animal feeds and pet food (World Bank, 2009). In the case of eggs production, main product and co-products are eggs, spent hens for slaughter and manure. In this case, the suggested allocation method, following the Livestock Environmental Assessment and Performance (LEAP) guidelines (FAO, 2014), is the economic allocation. The impact for each product and sub-product that exit the farm should be evaluated considering its average economic value. The latter can be calculated considering the historical average economic values. The LEAP guidance strongly recommends setting as default approach that manure should be to consider as a residual at the farm gate. This results in a clean separation of the system where all postfarm emissions from use of the manure are assigned to that use, while all on-farm management is assigned to the main product from the farm (birds, eggs). Excluding the system expansion approach, there are two possibilities regarding manure/litter allocation, depending on if it considered being a coproduct or a waste:

1. Coproduct: when manure is a valuable output of the farm (e.g., it is sold to be used as fertilizer or as fuel in a nearby power plant), the system of manure production cannot be separated from the system of animal production and the full supply chain emissions to the farm gate should be shared by these coproducts.
2. Waste: when it is disposed to landfill, incineration without energy recovery or sent to a treatment facility, all of the emissions associated with on-farm manure management are assigned to the birds and eggs at farm gate. Emissions associated with the final disposition of litter are within the system boundary too and should be accounted and assigned to the animal products.

2.1.4 Crop cultivation

The most important issue in the crop cultivation is data quality. Accurate data about cultivation, harvesting and transformation are fundamentals. In particular, crop yield data directly affects the functional unit, while water, fertilizers, pesticides and diesel consumption data are relevant production inputs. As a general rule, yield and consumption data should derive from primary data, while emissions due to fertilizer use can be calculated starting from type and amount of fertilizers application. Crop yield can vary because of conditions such as weather, soil, location, input intensity, irrigation, and rotation. Water use varies with species, climatic and soil conditions, and with the growth period. Systemic and selective herbicides are usually used for the pre- and postemergence control of grass and broad-leaved weeds. If there are no site or region-specific data available, emissions due to fertilizer use can be calculated according to specific emission factor. The following air emissions and water emissions should be included in the inventory: ammonia, N_2O and NO direct emissions and N_2O indirect emissions, nitrates and phosphorous. Ammonia volatilized can be estimated using the EMEP/CORINAIR emission factors (EMEP/EEA, 2013). Table 3.1 reports the total NH_3 emission from cultures due to fertilizer volatilization.

N_2O and NO direct emissions can be estimated using the emission factors calculated by (Bouwman et al., 2002). Table 3.2 reports the direct N_2O and NO emissions from cultures due to fertilizer use.

Table 3.1 Total NH3 emissions from cultures due to fertilizer volatilization.

	Emission factor [kg NH_3–N volatilized per kg of N in fertilizers used]	
Type of fertilizers	**Low-pH soil**	**High-pH soil**
Ammonium nitrate	0.037	
Ammonium phosphate	0.113	0.293
Ammonium solutions	0.037	
Ammonium solutions	0.125	
Ammonium sulfate	0.013	0.27
Anhydrous ammonia	0.011	
Animal manure	0.2	
Calcium ammonium nitrate	0.022	
Calcium nitrate	0.009	
Other NK and NPK	0.037	
Urea	0.243	
Urea ammonium sulfate	0.195	

Table 3.2 Total N_2O and NO emissions from cultures due to fertilizer use.

Type of fertilizers	Emission factor [kg N_2O per kg of N in fertilizers used]	Emission factor [kg NO per kg of N in fertilizers used]
Ammonia, direct application	0.009	0.005
Ammonium nitrate	0.008	0.006
Ammonium phosphates	0.009	0.007
Ammonium sulfate	0.01	0.007
Animal manure	0.008	0.005
Calcium ammonium nitrate	0.007	0.006
Compound NK-N	0.009	0.008
Compound NPK-N	0.008	0.006
Generic mineral N fertilizers	0.01	0.007
Nitrogen solutions	0.01	0.007
Other compound NP-N	0.009	0.006
Other straight N	0.012	0.01
Urea	0.011	0.007

Indirect emissions of N_2O can be estimated using the emission factors proposed by the IPCC (IPCC, 2006a). Table 3.3 reports the indirect N_2O emissions from cultures due to fertilizer use.

Nitrates leaching and run-off should be estimated using the most accurate methodology available. If more accurate methodologies cannot be used, they can be estimated using the emission factor proposed by the IPCC (IPCC, 2006b). Table 3.4 reports the total NO_3^- emissions due to leaching and run-off.

Three different kinds of phosphorus emission to water can be distinguished:

- leaching of soluble phosphate to ground water (inventoried as “phosphate, to ground water”),

Table 3.3 Total N2O emissions from cultures due to fertilizer volatilization.

	Emission factor N_2O indirect emission [kg N_2O-N emitted]
Emissions per kg of NH_3-N volatilized from fertilizers used	0.01
Emissions per kg of $NO_3\cdots N$ lost by leaching/run-off	0.0075

Table 3.4 Total NO3- emissions due to leaching and run-off.

	Emission factor NO_3^- - indirect emission [kg NO_3^- - N emitted]
Emissions per kg of N in fertilizers used	0.3

- run-off of soluble phosphate to surface water (inventoried as "phosphate, to river"),
- erosion of soil particles containing phosphorus (inventoried as "phosphorus, to river")

According to (Prasuhn, 2006) phosphorus leaching to the ground water can be estimated as an average leaching, corrected by phosphorus fertilization:

$$P_{gw} = P_{gwl} * F_{gw}$$

P_{gw} = quantity of phosphorus leached to ground water (kg/ha)
P_{gwl} = average quantity of P leached to ground water for a land use category (0.07 kg P/ha for arable land)
F_{gw} = correction factor for fertilization with slurry

$$F_{gw} = 1 + 0.2/80 * P_2O_{5sl}$$

P_2O_{5sl} = quantity of P_2O_5 applied.

Run-off to surface waters can be calculated in a similar way to leaching to ground water:

$$P_{ro} = P_{rol} * F_{ro}$$

P_{ro} = quantity of phosphorus lost through run-off to rivers (kg/ha)
P_{rol} = average quantity of P lost through run-off to rivers for a land use category (0.175 kg P/ha for arable land)
F_{ro} = correction factor for fertilization with phosphorus, calculated as:

$$F_{ro} = 1 + 0.2/80 * P_2O_{5min} + 0.7/80 * P_2O_{5sl} + 0.4/80 * P_2O_{5man}$$

P_2O_{5min} = quantity of P_2O_5 applied with mineral fertilizers (kg/ha)
P_2O_{5s} = quantity of P_2O_5 applied with slurry (kg/ha)
P_2O_{5man} = quantity of P_2O_5 applied with solid manure (kg/ha)

Phosphorus emissions through erosion to surface waters, if there is no more accurate information available, could be estimated using the default value 0.53 kg P_2O_5/ha, derived from an elaboration made using the SALCA-P model (Prasuhn, 2006) (considering 1.5 $t \cdot ha\text{-}1 \cdot yr^{-1}$ of eroded soil). To calculate the impact of pesticides production the content in active substance of the specific products should be considered. If no specific data are available, the total amount of active ingredients used per hectare for a specific crop in a specific country can be estimated applying the MEXALCA approach (Roches et al., 2010; Nemecek et al., 2011) that uses crop yield and intensity indices per country for pesticide use based on FAOSTAT data. The modeling of pesticide emissions and the definition of the interface between inventory and impact assessment is matter of a current scientific debate (SETAC, 2013; SETAC, 2014).

2.1.5 Manure management

At farm level, the manure management represents a key issue that must to be deeply analyzed in order to evaluate the environmental impact of meat production system. It is usually associated with the emission of greenhouse gases, with particular regard to methane (CH_4) and nitrous oxide (N_2O). One of the following scenarios can be taken into account for manure management:

- If it is used within the farm, both emissions related to manure management and emissions due to manure field spreading have to be accounted within the system boundaries;
- if it is used outside the farm, only manure management emissions inside the farm needs to be considered;
- if it is considered as waste, impacts related to landfill (and possibly incineration) need to be accounted for.

If specific data are not available, data could be evaluated considering the methodology reported in the IPCC Guidelines for National Greenhouse Gas Inventories (IPCC, 2006b) or in literature (Gallo et al., 2015b). The Guidelines provides guidance on estimating the quantities of greenhouse gases which are emitted in function of the specific manure management system. As regard the emission of ammonia (NH_3) if specific inventory national data are not available, data could be evaluated considering the methodology reported in the EMEP/EEA air pollutant emission inventory Guidebook (EMEP/EEA, 2013). The Guidelines provides guidance on estimating the quantities of ammonia, which are emitted for animal husbandry and manure management.

2.1.6 Wastewater management

Among the core processes in the different sectors of the food industry, the wastewater management represents a relevant process in the production of meat of mammals and poultry—respectively during the slaughterhouse process and during the manufacturing phases—but also in the agri-food sector it should be included in the system boundaries, except for the reuse of wastewater sludge. The waste water treatment is a process phase aimed at the conversion of wastewater into an effluent suitable to be returned to the water cycle with minimum impact on the environment—in accordance with the impurity concentrations defined by the international and national laws—or directly reused. The treatment process is operated in a wastewater treatment plant (WWTP), generally referred to as a Water Resource Recovery Facility (WRRF) or a sewage treatment plant. The treatment of wastewater is part of the overarching field of sanitation. Sanitation may also include the management of human waste and solid waste as well as storm water (drainage) management. By-products from wastewater treatment plants, such as screenings, grit and sewage sludge may also be treated in a wastewater treatment plant. The direct and indirect environmental impacts deriving from the waste water management may vary depending on the type of plant chosen, i.e., its construction and operation, and on the related different amount of greenhouse gases emissions expected, such as carbon dioxide (CO_2), methane (CH_4) and nitrous oxide (N_2O). The choice of the most suitable waste water treatment plant may be conducted evaluating the environmental impacts deriving from the process in accordance with predefined guidelines and rules, such as PCRs (Del Borghi et al., 2008), and considering all the existing energy recovery systems from waste water utilities that may help reducing the overall environmental impacts.

2.1.7 Packaging

Protecting food from the outside contaminations, damage and influences is the principal role of food packaging, in particular for beverage. The perfect packaging keep the ingredient and nutritional factors unaltered from the producer to the consumer. Moreover, traceability, convenience, and tamper indication are packaging functions of increasing importance. Food packaging must satisfy industry necessities and consumer needs, containing food in a cost-effective way that guarantee food safety, and minimizes environmental impact (Marsh and Bugusu, 2007). In the last years, growing concerns about environment and sustainability have raised the interest of consumers about the environmental score of items, food and every aspect of their life. Packaging is fundamental for the conservation and transportation of food, but usually generates huge environmental burdens related to its production and end-of-life. No matter how waste packaging materials are treated, the increasing the consumption of materials for the packaging production and disposal increases the environmental burden (Huang and Ma, 2004). LCA of packaging can be performed to communicate to stakeholders, to choose among different materials, to perform eco-design studies, to evaluate end-of-life burdens (Del Borghi et al., 2016). Many scientific papers, researches and studies focused on the importance of stakeholders packaging perceptions in particular about end-of-life of packaging results (Meylan et al., 2014; Bertoluci et al., 2014; Bjelkemyr et al., 2015). Concerning packaging material, many studies present a comparison among different materials, giving information about the best material choice. In particular, studies on beverage compare glass, metal and plastic to assess which is the best solution in terms of sustainability and environmental scores, including end-of-life scenarios (Cordella et al., 2007; Humbert et al., 2009; Detzel and

Mönckert, 2009; Amienyo et al., 2012; Amienyo and Azapagic, 2016; Del Borghi et al., 2018). In product–package systems, the production of packaging material represents the main hotspot for most environmental impacts, contributing between 59% and 77% of total impacts (Amienyo et al., 2012). In particular, in the preserved food industry, the production of packaging is the major hotspot accounting for most of the total global impacts, due to the high amount of nonrenewable energy used for the production of materials (glass and steel) (Del Borghi et al., 2018).

On the contrary, carton-based containers production presents lower energy consumption than glass and steel materials manufacturing leading to lower CO_2 emissions in the packaging subsystem (Del Borghi et al., 2014). If compared to glass and metal, from an LCA point of view plastic packaging shows better results thanks to its lighter weight. Glass can be a better choice compared to plastic, but only if glass have a reuse factor higher than 80% (Vellini and Savioli, 2009). If the reuse of glass bottles is included in the model, the carbon footprint of the drink in glass bottles resulted comparable to that in aluminum cans and 0.5 PET bottles (Amienyo et al., 2012). The improvement of packaging category is one of the most prolific in terms of scientific papers production. Several eco-design studies have been performed on wine and olive oil packaging (Point et al., 2012; Neto et al., 2012; Benedetto, 2013; Cleary, 2013; Amienyo et al., 2014; Rinaldi et al., 2014; Fusi et al., 2014; Accorsi et al., 2015; Eriksson et al., 2016). All these studies agreed that the use of lighter bottles is the best alternative to reduce the packaging environmental impacts in wine and olive oil bottles. A 10% decrease in bottle weight would reduce the GWP, up to 4%, while wine packaged in cartons instead of glass bottles would cause a GWP saving around 5%. Moreover, refillable bottles can reduce the packaging impacts up to 87% if compared to the results of the traditional disposable bottle. In the vegetable sector, several LCA studies have been performed on preserved products identifying the environmental hotspots of the whole supply chains and some improvements options. As packaging production resulted to be the life cycle stages with the highest impact in all the considered categories, some improvement options related to packaging subsystem has been identified: the reduction of weight and the switch to different packaging materials, such as plastic packaging has been has been proposed with the aim of reducing GWP (Del Borghi et al., 2014; Manfredi and Vignali, 2014). As reported by many research, one of the most significant phase along the life cycle of packaging is the end-of-life stage. In particular, a research about the end-of-life of packaging taking into account many different fractions such as newspaper, cardboard, mixed paper, glass bottles and jars, aluminum cans, tin-plated steel cans, plastic bottles and other conventionally recoverable materials, highlighted the fact that the recycling option allowed saving large amount greenhouse gas emissions. In particular, glass coming from packaging, if recycled and reused for the production of new materials generate new glass that require only the 35% of energy less than virgin glass (Morris, 2004). Similar results where reached by other authors (Mata and Costa, 2001; Simon et al., 2016).

2.2 LCA databases

One of the critical issues of LCA is retrieving data to be used in LCI phase. As a general rule, data quality is assessed against the following criteria: technological representativeness, geographical representativeness, time-related representativeness, completeness, parameter uncertainty, methodological appropriateness and consistency. In the following, the most reliable databases are presented.

2.2.1 LCA food database

LCA Food Database is a result of the project "Lifecycle Assessment of Basic Food" (2000–2003) (Nielsen et al., 2003). The database provides environmental data on processes in food products' product chains and on food products at different stages of their product chains. The databases included several processes (agriculture, aquaculture, fishery, industrial processing, trade, cooking, transport, energy, water supply, and waste treatment) and food products (crops and crop based products, milk and milk based products, vegetables, meat, fish, packaging). Input/output data on processes in the food sector are derived from a variety of sources: data on production in agriculture and fishery have been determined by a "top-down" approach where statistical data on a national level have been broken down to represent specific processes, while data on other processes than agriculture and fishery have been determined by a "bottom-up" approach where data from a limited number of sources have been used to represent the national level. Data include main inputs (resources, raw materials, water and energy) and outputs (products and waste as well as emissions to air and water). Data do in most cases not include:

- Capital goods such as buildings and machinery (except for traction).
- Cleaning agents and most other chemicals (although it could be relevant in terms of toxicity).
- Veterinary medicine etc. (Sørensen et al., 2002).
- Pesticides, heavy metals etc.

Product models are market based (Nielsen, 2003) and models include processes, which are influenced by a marginal change in demand for a considered food product (Jensen and Andersen, 2003).

2.2.2 World food life cycle assessment database (WFLDB)

The World Food Life Cycle Assessment Database (WFLDB) is a global initiative of leaders in the private food sector launched in 2012. The purpose is to drive sustainable change in the agrofood sector by delivering sound environmental sustainability data (Nemecek et al., 2014). The main aim of the WFLDB is to create a database that represents agricultural primary products and processed food products. The geographical focus is global, i.e., products that are dealt on the global market are represented. The WFLDB covers more than 1600 LCI related to agriculture, food processing and consumption activities. The data are documented in the electronic EcoSpold v1 format.

The documentation of unit processes includes an assessment of uncertainties. Data include information on food waste and water use. The database, covers among others, the following datasets:

- Agricultural production services: application of fertilizers, machinery hours
- Vegetable production: spinach, salad, tomatoes, lettuce, potatoes, onions, asparagus, etc.
- Fruits: apples, strawberries, cherries, grapes, oranges, bananas, vine, melons
- Animal products: pork, veal, beef, lamb, poultry, eggs
- Fish: codfish, herring, mackerel, salmon
- Dairy products: butter, milk, milk powder, yogurt, cheese
- Meat alternatives: tofu, falafel, quorn, soy vegetarian mince, etc.
- Staple food: noodles, pasta, bread, wheat flour
- Drinks: apple and orange juice, mineral water, tap water, beer, wine, milk, coffee, tea, soy milk
- Sweets: chocolate, ice cream, cakes
- Meals: roast, lasagne, soups, canteen meals, recipes
- Household appliances: cooking stoves and ovens, microwaves, refrigerators, carbonization devices, coffee machine
- Food consumption: packages, transports, cooking, consumption patterns
- Pet food: cat food

2.2.3 Ecoinvent

Ecoinvent is a not-for-profit association. The Ecoinvent database provides well-documented process data for thousands of products that can be used in a broad range of environmental studies including LCA, EPD, Design for Environmental or Carbon Footprinting. It allows conducting studies on different levels of detail: from screenings for basic, initial answers to extensive studies such as peer-reviewed, ISO complaint studies. Ecoinvent includes over 14,700 LCI datasets in many areas such as energy supply, agriculture, transport, biofuels and biomaterials, bulk and specialty chemicals, construction materials, wood, and waste treatment (Ecoinvent Center, 2018).

Table 3.5 reports all the market activity processes related to the food sector present in the Ecoinvent database, associated with their ISIC Code (International Standard Industrial Classification of All Economic Activities).

Version 3.3 of the Ecoinvent database included a first part of the WFLDB covering several hundreds of agricultural datasets, among others in the fields of:

- vegetables (such as asparagus, carrots, onions, potatoes)
- fruits (such as lemons, mandarins, oranges, peaches)
- cereal and grain products (such as oat, rice, sweetcorn, wheat)
- Tea
- coffee (such as Robusta and Arabica)
- sugar and confectionery (such as cocoa and vanilla)

For many products, it has been possible to choose production in different countries and using different technologies, such as open field, nonheated greenhouse or even in heated greenhouse. Furthermore, a distinction is made between fresh grade and processing grade. In Phase Two, the WFLDB will build on Phase One achievements

Table 3.5 "Market activity" processes in the food sector (Ecoinvent 3.5).

Process name	ISIC classification no.
Market for alfalfa-grass mixture, Swiss integrated production	129
Market for alfalfa-grass silage	129
Market for almond	125
Market for apple	124
Market for apricot	124
Market for aubergine	113
Market for avocado	122
Market for banana	122
Market for barley grain	111
Market for barley grain, feed	111
Market for barley grain, feed, organic	111
Market for barley grain, feed, Swiss integrated production	111
Market for barley grain, organic	111
Market for barley grain, Swiss integrated production	111
Market for bell pepper	128
Market for broccoli	113
Market for butter, from cow milk	1050
Market for buttermilk, from cow milk	1050
Market for cabbage red	113
Market for cabbage white	113
Market for carrot	113
Market for carrot seed, Swiss integrated production, at farm	113
Market for cattle for slaughtering, live weight	141
Market for cauliflower	113
Market for celery	113
Market for cheese, from cow milk, fresh, unripened	1050
Market for chicken for slaughtering, live weight	14
Market for clover seed, Swiss integrated production, at farm	119
Market for cocoa bean	127
Market for coconut husk	161

Continued

Table 3.5 "Market activity" processes in the food sector (Ecoinvent 3.5).—cont'd

Process name	ISIC classification no.
Market for coconut oil, crude	1040
Market for coconut, dehusked	126
Market for coffee, green bean	127
Market for compost	161
Market for cotton fiber	116
Market for cotton seed	116
Market for cottonseed oil, crude	1040
Market for cottonseed oil, refined	1040
Market for cow milk	141
Market for cream, from cow milk	1050
Market for cucumber	113
Market for evaporation of milk	1050
Market for fava bean, feed, Swiss integrated production	111
Market for fava bean, organic	111
Market for fava bean, Swiss integrated production	111
Market for fennel	128
Market for fish oil	1040
Market for fishmeal, 63%–65% protein, from anchovy	3
Market for fishmeal, 65%–67% protein, from anchovy	3
Market for fodder beet, Swiss integrated production	119
Market for glucose	1062
Market for grape	121
Market for grass seed, Swiss integrated production, at farm	119
Market for grass silage, organic	119
Market for grass silage, Swiss integrated production	119
Market for grass, organic	119
Market for grass, Swiss integrated production	119
Market for green asparagus	113
Market for hay	119
Market for hay, organic, intensive	119

Table 3.5 "Market activity" processes in the food sector (Ecoinvent 3.5).—cont'd

Process name	ISIC classification no.
Market for hay, Swiss integrated production, extensive	119
Market for hay, Swiss integrated production, intensive	119
Market for iceberg lettuce	113
Market for jatropha seed	12
Market for jute fiber	116
Market for kenaf fiber	116
Market for kiwi	125
Market for lemon	123
Market for lettuce	113
Market for linseed	111
Market for linseed seed, at farm	111
Market for maize grain	111
Market for maize grain, feed	111
Market for maize grain, feed, organic	111
Market for maize grain, feed, Swiss integrated production	111
Market for maize grain, organic	111
Market for maize grain, Swiss integrated production	111
Market for maize seed, at farm	111
Market for maize seed, organic, at farm	111
Market for maize seed, Swiss integrated production, at farm	111
Market for maize silage	119
Market for maize silage, organic	119
Market for maize silage, Swiss integrated production	119
Market for maize starch	1062
Market for Mandarin	123
Market for manure, liquid, cattle	161
Market for manure, liquid, swine	161
Market for manure, solid, cattle	161
Market for meat and bone meal	1040
Market for melon	113

Continued

Table 3.5 "Market activity" processes in the food sector (Ecoinvent 3.5).—cont'd

Process name	ISIC classification no.
Market for milking	162
Market for mint	128
Market for miscanthus, chopped	129
Market for molasses, from sugar beet	1072
Market for oat grain	111
Market for oat grain, feed	111
Market for oat seed, Swiss integrated production, at farm	111
Market for olive	126
Market for onion	113
Market for operation, dried roughage store, air dried, solar	163
Market for operation, dried roughage store, cold-air dried, conventional	163
Market for operation, dried roughage store, nonventilated	163
Market for orange, fresh grade	123
Market for orange, processing grade	123
Market for palm date	122
Market for palm date, organic	122
Market for palm fruit bunch	126
Market for palm kernel oil, crude	1040
Market for palm oil, crude	1040
Market for palm oil, refined	1040
Market for papaya	122
Market for paris market carrot	113
Market for peach	124
Market for peanut	111
Market for peanut seed, at farm	111
Market for pear	124
Market for phacelia silage	161
Market for pineapple	122
Market for polyester-complexed starch biopolymer	1062

Table 3.5 "Market activity" processes in the food sector (Ecoinvent 3.5).—cont'd

Process name	ISIC classification no.
Market for potato	113
Market for potato grading	163
Market for potato seed, at farm	113
Market for potato seed, organic, at farm	113
Market for potato seed, Swiss integrated production, at farm	113
Market for potato starch	1062
Market for potato, organic	113
Market for potato, Swiss integrated production	113
Market for poultry manure, dried	161
Market for protein pea	111
Market for protein pea, feed, Swiss integrated production	111
Market for protein pea, organic	111
Market for protein pea, Swiss integrated production	111
Market for radish	113
Market for rape oil, crude	1040
Market for rape oil, crude	1040
Market for rape seed	111
Market for rape seed, organic	111
Market for rape seed, Swiss integrated production	111
Market for red meat, live weight	141
Market for rice	112
Market for rye grain	111
Market for rye grain, feed, organic	111
Market for rye grain, feed, Swiss integrated production	111
Market for rye grain, organic	111
Market for rye grain, Swiss integrated production	111
Market for ryegrass silage	161
Market for ryegrass—Egyptian and Persian clover-mixture silage	161
Market for ryegrass—red and Egyptian clover-mixture silage	161
Market for sheep fleece in the grease	144

Continued

Table 3.5 "Market activity" processes in the food sector (Ecoinvent 3.5).—cont'd

Process name	ISIC classification no.
Market for sheep for slaughtering, live weight	144
Market for skimmed milk, from cow milk	1050
Market for solid manure loading and spreading, by hydraulic loader and spreader	161
Market for soybean	111
Market for soybean meal	1040
Market for soybean oil, crude	1040
Market for soybean oil, refined	1040
Market for soybean, feed	111
Market for soybean, organic	111
Market for soybean, Swiss integrated production	111
Market for spinach	113
Market for spray-drying of milk	1050
Market for straw	111
Market for straw, organic	111
Market for straw, stand-alone production	111
Market for strawberry	125
Market for sugar beet	111
Market for sugar, from sugar beet	1072
Market for sugar, from sugarcane	1072
Market for sugarcane	114
Market for sunflower seed	111
Market for sunflower seed, Swiss integrated production	111
Market for sunflower silage	161
Market for sweet corn	111
Market for sweet sorghum stem	111
Market for swine for slaughtering, live weight	145
Market for tea, dried	127
Market for tilapia	322
Market for tomato, fresh grade	113

Table 3.5 "Market activity" processes in the food sector (Ecoinvent 3.5).—cont'd

Process name	ISIC classification no.
Market for tomato, processing grade	113
Market for trout, from aquaculture	322
Market for vanilla	128
Market for vegetable oil, refined	1040
Market for vinasse, from fermentation of sugar beet	161
Market for vinasse, from fermentation of sugar beet molasses	161
Market for vinasse, from fermentation of sugarcane	161
Market for vinasse, from fermentation of sugarcane molasses	161
Market for vinasse, from fermentation of sweet sorghum	161
Market for wheat grain	111
Market for wheat grain, feed	111
Market for wheat grain, feed, organic	111
Market for wheat grain, feed, Swiss integrated production	111
Market for wheat grain, organic	111
Market for wheat grain, Swiss integrated production	111
Market for white asparagus	113
Market for wood chips and particles, willow	129
Market for yogurt, from cow milk	1050
Market for zucchini	113

of more than 900 environmental footprint datasets and sub-datasets for plant, animal and final food products. Over the next 3 years, more food data will be developed for a broader set of products, including dairy, fruits, oils and fats, sweeteners and flavors, as well as for agricultural best practices. The fifth update of Ecoinvent version three includes over 2000 new and updated datasets including processes related to aquaculture and fish capture.

2.2.4 Agri-footprint

Agri-footprint is a comprehensive database created by Blonk Consultants—a sustainability consulting firm specializing in the complexities of agricultural LCA—in 2014 to meet the increasing demand for agricultural LCAs: methodologies are now reaching maturity and can support governments and international bodies in the regulation of sustainable agriculture, and production of bio-based chemicals and energy.

The Agri-footprint database allows the users to access to a wide range of agriculture-specific impact categories such as water and land use, land use change, fertilizers, and soil carbon content and to comply with already existing initiatives and methodologies, such as the ISO standards or the PEF initiative of the European Commission. The database is completely focused on the agricultural and farming products and, according to the intrinsic characteristics of this industrial sector, allows the users to choose among three predefined allocation options: mass, energy and economic. The methodologies and data quality in Agri-footprint 4.0 are compliant with the PEF initiative of the European Commission, ILCD and ReCiPe, and have been reviewed by RIVM (Dutch national institute for public health and the environment). Agri-footprint 4.0 contains approximately 3500 products and processes specific to agricultural LCA: crops, products and intermediate products, feed compounds, food products, animal production systems and background processes such as transport, auxiliary inputs for processing, and fertilizers. Table 3.6 reports an exhaustive summary of the processes furnished by the Agri-footprint database. Intermediate products, seeds production and animal feed production are excluded from the list.

3. Labeling schemes

Environmental labels serve different communication purposes and consist of different formats. The International Organization for Standardization (ISO) provides general definitions and principles for the establishment of voluntary ecolabels through the standard ISO 14020 (ISO, 2000). Furthermore, three broad types of voluntary labels have been defined through the following standards:

(1) ISO 14024 (ISO, 2018) on Type I environmental labels or ecolabels—these are multi-criteria based, third-party verified labels awarded to products that fulfill certain product environmental criteria based on life cycle considerations;
(2) ISO 14021 (ISO, 2016) on Type II environmental labels, known as self-declared environmental claims, issued in the form of a claim, stamp, label, or declaration;
(3) ISO 14025 (ISO, 2006a) on Type III environmental declarations, known also as EPD. These are third-party verified, quantitative declarations based on a LCA of the product, according to ISO 14040 (ISO, 2006b) and apply Product Category Rules (PCR) that are designed specifically for the particular product group in focus.

While Ecolabels (under the EU definition) can be applied to different product groups (e.g., cosmetics, hygiene, cleaning, clothing, paints, electronics equipment, building materials, household appliances, etc.), they do not apply to food and feed products. Despite not intending to develop Ecolabel criteria for food and feed products at this time, the Commission declares on the EU Ecolabel website that this question could be revisited in the future considering the possible role of the EU Ecolabel within the framework of the development of any wider EU food strategy, in particular in light of developments in methodologies, and other tools, for measuring the environmental impact (including by, for example, environmental footprinting) of products (Hollins, 2011).

Table 3.6 Food processes (Agrifootprint).

Product	Countries/Regions covered
Barley grain, dried, at farm	AR, AT, AU, BE, BG, CA, CH, CZ, DE, DK, EE, ES, FI, FR, GR, HU, IE, IT, LT, LV, NL, PL, PT, RO, SE, SK, UK
Barley straw, at farm	AR, AT, AU, BE, BG, CA, CH, CZ, DE, DK, EE, ES, FI, FR, GR, HU, IE, IT, LT, LV, NL, PL, PT, RO, SE, SK, UK
Beans, dry, canned, at plant	NL
Beef cattle for slaughter, at beef farm	IE
Beef coproduct, food grade bones, from beef cattle, at slaughterhouse, PEF compliant	IE
Beef coproduct, food grade bones, from dairy cattle, at slaughterhouse, PEF compliant	NL
Beef coproduct, food grade fat, from beef cattle, at slaughterhouse, PEF compliant	IE
Beef coproduct, food grade fat, from dairy cattle, at slaughterhouse, PEF compliant	NL
Beef coproduct, food grade, from beef cattle, at slaughterhouse	IE
Beef coproduct, food grade, from dairy cattle, at slaughterhouse	NL
Beef meat, fresh, from beef cattle, at slaughterhouse	IE
Beef meat, fresh, from dairy cattle, at slaughterhouse	NL
Bleaching earth, at plant	RER
Brewer's grains, wet, at plant	NL
Broad bean, hulls, at plant	NL
Broad bean, meal, at plant	NL
Broccoli, at farm	FR, NL
Broiler parents <20 weeks, breeding, at farm	NL
Broiler parents >20 weeks, for slaughter, at farm	NL
Broilers, for slaughter, at farm	NL
Calves, at dairy farm	NL
Carrot, at farm	BE, NL

Continued

Table 3.6 Food processes (Agrifootprint).—cont'd

Product	Countries/Regions covered
Cassava peels, fresh, from processing with use of coproducts, at plant	TH
Cassava pomace (fibrous residue), fresh, from processing with use of coproducts, at plant	TH
Cassava root dried, from tapioca processing, at plant	TH
Cauliflower, at farm	ES, NL
Cheese, from cheese production, at plant	NL
Chicken coproduct, food grade, at slaughterhouse	NL
Chicken meat, fresh, at slaughterhouse	NL
Chickpea, at farm	AU, IN, US
Chickpea, canned, at plant	NL
Chicory root, at farm	BE, NL
Citrus pulp dried, from drying, at plant	BR, US
Coconut, at farm	ID, IN, PH
Consumption eggs, broiler parents >20 weeks, at farm	NL
Consumption eggs, laying hens >17 weeks, at farm	NL
Consumption potato, at farm	NL
Cream, full, from processing, at plant	NL
Cream, skimmed, from processing, at plant	NL
Curly kale, at farm	ES, NL
Dicamba, at plant	RER
Fodder beets cleaned, from cleaning, at plant	NL
Green bean, at farm	KE, MA, NL
Groundnuts, with shell, at farm	AR, CN, ID, IN, MX, SD, SN, UG, US-AL, US-FL, US-GA, US-NC, US-TX, US
Hatching eggs, broiler parents >20 weeks, at farm	NL
Landed fish, from fishery, at plant	CL, DE, DK, NO, PE, UK
Laying hens <17 weeks, breeding, at farm	NL

Table 3.6 Food processes (Agrifootprint).—cont'd

Product	Countries/Regions covered
Laying hens >17 weeks, for slaughter, at farm	NL
Lentil, at farm	AU, CA
Lentil, canned, at plant	NL
Lindane, at plant	RER
Linuron, at plant	RER
Lucerne, dried, at farm	ES, IT
Lupine, at farm	AU, DE, PL
Lupine, hulls, at plant	NL
Lupine, meal, at plant	NL
Maize bran, from wet milling (drying), at plant	DE, FR, NL, US
Maize flour, from dry milling, at plant	DE, FR, IT, NL, PL, US
Maize germ, dried, from wet milling (germ drying), at plant	DE, FR, NL, US
Maize middlings, from dry milling, at plant	DE, FR, IT, NL, PL, US
Maize silage, at dairy farm	NL
Maize solubles, from wet milling (steepwater dewatering), at plant	DE, FR, NL, US
Maize starch, from wet milling (starch drying), at plant	DE, FR, NL, US
Maize, at farm	AR, BE, BG, BR, CA, CN, CZ, DE, ES, FR, GR, HU, ID, IN, IT, MX, PH, PK, PL, PT, RO, SK, TH, UA, US-CO, US-GA, US-IA, US-IL, US-IN, USKS, US-KY, US-MI, US-MN, US-MO, US-NC, US-ND, US-NE, US-NY, US-OH, US-PA, US-SC, US-SD, US-TX, US-WI, US, VN, ZA
Maize, market mix, at regional storage	DE, FR, IT, NL, PL, US
Meatless hybrid, dehydrated (dry) rice	NL
Meatless hybrid, hydrated (wet) rice	NL
Meatless hybrid, hydrated (wet) tapioca	NL
Meatless hybrid, hydrated (wet) wheat	NL
Milk powder skimmed, from drying, at plant	NL
Milk powder whole, from drying, at plant	NL

Continued

Table 3.6 Food processes (Agrifootprint).—cont'd

Product	Countries/Regions covered
Oat grain peeled, from dry milling, at plant	BE, NL
Oat grain, at farm	AT, BE, BG, CA, CZ, DE, DK, EE, ES, FI, FR, GR, HU, IT, LT, LV, NL, NO, PL, PT, RO, SE, SK, UK, US-KS, US-MI, US-MN, US-ND, US-NE, US-NY, US-PA, US-SD, US-WI, US
Oat grain, dried, at farm	AT, BE, BG, CA, CZ, DE, DK, EE, ES, FI, FR, GR, HU, IT, LT, LV, NL, NO, PL, PT, RO, SE, SK, UK, US-KS, US-MI, US-MN, US-ND, US-NE, US-NY, US-PA, US-SD, US-WI, US
Oat grain, market mix, at regional storage	BE, NL
Oat straw, at farm	AT, BE, BG, CA, CZ, DE, DK, EE, ES, FI, FR, GR, HU, IT, LT, LV, NL, NO, PL, PT, RO, SE, SK, UK, US-KS, US-MI, US-MN, US-ND, US-NE, US-NY, US-PA, US-SD, US-WI, US
Offal, from fishery, at plant	CL, DE, DK, NO, UK
Oil palm fruit bunch, at farm	ID, MY
One-day-chickens, at hatchery	NL
Onion, at farm	FR
Pea, at farm	AU, CN, CZ, DE, ES, FR, IT, RO, SE, UK
Pea, canned, at plant	RER
Pea, fibers, at plant	RER
Pea, meal, at plant	RER
Pea, protein concentrate, at plant	RER
Pea, protein-isolate, at plant	RER
Pig meat, fresh, at slaughterhouse	NL
Pigeon pea, at farm	IN
Piglets, sow-piglet system, at farm	NL
Pigs to slaughter, pig fattening, at farm	NL
Potato juice concentrated, from wet milling, at plant	DE, NL
Potato protein, from wet milling, at plant	DE, NL
Potato pulp pressed fresh+silage, from wet milling, at plant	DE, NL
Potato pulp, from drying, at plant	DE, NL

Table 3.6 Food processes (Agrifootprint).—cont'd

Product	Countries/Regions covered
Potato starch dried, from wet milling, at plant	DE, NL
Potato, at farm	AT, BE, BG, CH, CN, CZ, DE, DK, EE, ES, FI, FR, GR, HU, IN, IT, LT, LV, NL, NO, PL, PT, RO, RU, SE
Raw milk, at dairy farm	NL
Refined coconut oil, at plant	ID, IN, PH
Refined maize germ oil, (pressing), at plant	DE, FR, NL, US
Refined maize germ oil, (solvent), at plant	DE, FR, NL, US
Refined palm kernel oil, at plant	ID, MY
Refined palm oil, at plant	NL
Refined rice bran oil, at plant	CN
Refined soybean oil, from crushing (pressing), at plant	AR, BR, NL
Refined soybean oil, from crushing (solvent), at plant	AR, BR, NL
Refined soybean oil, from crushing (solvent, for protein concentrate), at plant	AR, BR, DE, ES, IT, NL, PT, UK
Refined sunflower oil, from crushing (pressing) at plant	AR, CN, UA
Refined sunflower oil, from crushing (solvent) at plant	AR, CN, UA
Rice brokens, from dry milling, at plant	CN
Rice brokens, from dry milling, parboiled, at plant	CN
Rice brokens, from dry milling, raw, at plant	CN
Rice fiber, protein extraction, at plant	GLO
Rice protein, protein extraction, at plant	GLO
Rice starch, protein extraction, at plant	GLO
Rice without husks, from dry milling, at plant	CN
Rice without husks, from dry milling, parboiled at plant	CN
Rice without husks, from dry milling, raw, at plant	CN

Continued

Table 3.6 Food processes (Agrifootprint).—cont'd

Product	Countries/Regions covered
Rice, at farm	CN, IN, US-AR, US-CA, US-LA, US-MO, US-MS, US-TX
Rice, early, alternate wet and dry, at farm	CN
Rice, early, continuous flooding, at farm	CN
Rice, late, alternate wet and dry, at farm	CN
Rice, late, continuous flooding, at farm	CN
Rice, production mix, at farm	US
Rice, single, alternate wet and dry, at farm	CN
Rice, single, continuous flooding, at farm	CN
Rye flour, from dry milling, at plant	BE, DE, NL
Rye grain, dried, at farm	AT, BE, BG, CZ, DE, DK, EE, ES, FI, FR, GR, HU, IT, LT, LV, NO, PL, PT, RO, SE, SK
Rye grain, market mix, at regional storage	BE, DE, NL
Rye middlings, from dry milling, at plant	BE, DE, NL
Rye straw, at farm	AT, BE, BG, CZ, DE, DK, EE, ES, FI, FR, GR, HU, IT, LT, LV, NO, PL, PT, RO, SE, SK
Sorghum, at farm	AR, IN, MX, NG, US, ZA
Sow to slaughter, sow-piglet system, at farm	NL
Soybean lecithin, from crushing (solvent), at plant	DE, ES, FR, IT, NL, PT, UK
Soybean protein concentrate, from crushing (solvent, for protein concentrate), at plant	AR, BR, DE, ES, FR, IT, NL, PT, UK
Soybean, at farm	AR, BR, CA, CN, FR, HU, IN, IT, MX, PL, RO, US-AR, US-IA, US-IL, USIN, US-KS, US-KY, US-LA, US-MD, US-MI, US-MN, US-MO, US-MS, US-NC, US-ND, US-NE, US-OH, US-PA, US-SD, US-TN, US-VA, US-WI, US
Soybean, heat treated, from heat treating, at plant	NL
Spinach, at farm	BE, NL
Standardized milk, full, from processing, at plant	NL
Standardized milk, skimmed, from processing, at plant	NL

Table 3.6 Food processes (Agrifootprint).—cont'd

Product	Countries/Regions covered
Sugar beet pulp, dried, from pulp drying, at plant	DE, NL
Sugar beet pulp, pressed, from wet pulp pressing, at Suiker Unie plants	NL
Sugar beet pulp, wet, from sugar production, at plant	BE, DE, ES, FR, IT, NL, PL, UK
Sugar beet, at farm	AT, BE, CH, CZ, DE, DK, ES, FI, FR, HU, IT, LT, NL, PL, RO, SE, UA, UK
Sugar beet, market mix, at regional storage	BE, DE, ES, FR, IT, NL, PL, UK
Sugar cane, at farm	AR, AU, BR, CN, CO, ID, IN, PK, SD, TH, US, VE
Sugar, from sugar beet, from sugar production, at plant	BE, DE, ES, FR, IT, NL, PL, UK
Sugar, from sugar cane, from sugar production, at plant	AU, BR, IN, PK, SD, US
Tapioca starch, from processing with use of coproducts, at plant	TH
Tapioca starch, from processing without use of coproducts, at plant	TH
Triticale straw, at farm	AT, CH, CZ, DE, ES, FR, HU, LT, LV, NL, PL, PT, RO, SE, SK
Triticale, dried, at farm	AT, CH, CZ, DE, ES, FR, HU, LT, LV, NL, PL, PT, RO, SE, SK
Wheat bran, from dry milling, at plant	BE, CZ, DE, ES, FR, HU, IT, NL, PL, PT, RO, UK
Wheat bran, from wet milling, at plant	BE, DE, ES, FR, IT, NL, UK
Wheat flour, from dry milling, at plant	BE, CZ, DE, ES, FR, HU, IT, NL, PL, PT, RO, UK
Wheat germ, from dry milling, at plant	BE, CZ, DE, ES, FR, HU, IT, NL, PL, PT, RO, UK
Wheat grain, dried, at farm	AR, AT, BE, BG, BR, CA, CH, CN, CZ, DE, DK, EE, ES, FI, FR, GR, HU, IE, IN, IT, LT, LV, MX, NL, NO, PK, PL, PT, RO, SE, UA, UK
Wheat grain, durum, dried, at farm	US-MT, US-ND
Wheat grain, market mix, at regional storage	BE, CZ, DE, ES, FR, HU, IT, NL, PL, PT, RO, UK

Continued

Table 3.6 Food processes (Agrifootprint).—cont'd

Product	Countries/Regions covered
Wheat grain, production mix, at farm	US
Wheat grain, spring, dried, at farm	US-ID, US-MN, US-MT, US-ND, US-OR, US-SD, US-WA
Wheat grain, winter, dried, at farm	US-AR, US-CO, US-DE, US-GA, US-ID, US-IL, US-KS, US-KY, US-MI, USMN, US-MO, US-MS, US-MT, US-NC, US-ND, US-NE, US-OH, US-OK, US-OR, US-PA, US-SD, US-TX, US-WA
Wheat starch, from wet milling, at plant	BE, DE, ES, FR, IT, NL, UK
Wheat straw, at farm	AR, AT, BE, BG, BR, CA, CH, CN, CZ, DE, DK, EE, ES, FI, FR, GR, HU, IE, IN, IT, LT, LV, MX, NL, NO, PK, PL, PT, RO, SE, UA, UK
Wheat straw, durum, at farm	US-MT, US-ND
Wheat straw, spring, at farm	US-ID, US-MN, US-MT, US-ND, US-OR, US-SD, US-WA
Wheat straw, winter, at farm	US-AR, US-CO, US-DE, US-GA, US-ID, US-IL, US-KS, US-KY, US-MI, USMN, US-MO, US-MS, US-MT, US-NC, US-ND, US-NE, US-OH, US-OK, US-OR, US-PA, US-SD, US-TX, US-WA
Whey powder, from drying, at plant	NL
White rice, from dry milling, at plant	CN
White rice, from dry milling, parboiled, at plant	CN
White rice, from dry milling, raw, at plant	CN

Even though the application of the EU Ecolabel to food and feed products is currently not feasible from a methodological and technical point of view, the current Commission aim is also that of deeply analyzing the possible synergies with other existing labels (e.g., EU Organic logo).

4. EU organic logo and other standards on organic farming

Currently, no specific labels focused on the food sector for the evaluation of the environmental impacts—especially LCA-based—of products exist in the market, whereas a wide set of labels can be found for the certification of organic farming.

Organic farming is a method of crop and livestock production that neglects the use of pesticides, fertilizers, genetically modified organisms, antibiotics and growth hormones. Organic production is a holistic system designed to optimize the productivity and fitness of diverse communities within the agro-ecosystem, including soil organisms, plants, livestock and people, whose principal goal is to develop enterprises and production methods that are sustainable and harmonious with the environment. Organic farming promotes the use of crop rotations and cover crops, and encourages balanced host/predator relationships; recycles organic residues and nutrients produced on the farm back to the soil; practices preventative insect and disease control methods; only allows the use of organically approved pesticides identified in the Permitted Substances List (PSL) of the organic standards (OMAFRA, 2009).

The EU Organic logo (EC, 2007; EC, 2008) can be applied only on products that have been certified as organic by an authorized control agency or body since fulfilling strict conditions on how they are produced, transported and stored. The EU Organic logo gives a coherent visual identity to European Union produced organic products sold in the EU. This makes it easier for EU based consumers to identify organic products and helps farmers to market them across all EU countries. The European Union organic farming rules cover agricultural products, including aquaculture and yeast. They include every phase of the production process, from seeds to the final processed food on the market. This means that there are specific provisions covering a large variety of products, such as:

- seeds and propagating material such as cuttings, rhizome etc. from which plants or crops are grown;
- living products or products which do not need further processing;
- feed;
- products with multiple ingredients or processed agricultural products for use as food.

EU regulations on organic production exclude products from fishing and hunting of wild animals but include harvest of wild plants when certain natural habitat conditions are respected. Specific rules are present for wine and aquaculture. However, the EU has passed new legislation that will come into force on 1st January 2021.

Other known labels and standards for the organic farming may refer to the Bud Labels provided by Bio Suisse for food products fully organic—the "Bio Suisse Bud" if produced in Switzerland, the "Bio-Bud" if more than 10% of raw materials is imported but subject to equivalent organic standards; to the JAS standards provided by the Agricultural Organic Standard (JAS) on the basis with the Guidelines for the Production, Processing, Labeling and Marketing of Organically Produced Foods which were adopted by the Codex Alimentarius Commission; or to the existing national standards provided by the European countries that have adopted their own labeling schemes for organic farming.

5. Product environmental footprint (PEF)

The European Commission's Product Environmental Footprint is a proposed European methodology for calculating the life cycle environmental impact of products, developed in the context of one of the building blocks of the Flagship initiative of the Europe 2020 Strategy—"A Resource-Efficient Europe" (COM 21, 2011). In fact, one of the objectives of the European Commission's "Roadmap to a Resource Efficient Europe" is to establish a common methodological approach to enable Member States and the private sector to assess, display and benchmark the environmental performance of products, services and companies based on a comprehensive assessment of environmental impacts over the life cycle ('environmental footprint') (COM 571, 2011). The PEF Guide, published in 2013, provides a method for modeling the environmental impacts of the flows of material/energy and the emissions and waste streams associated with a product throughout its life cycle. It provides guidance on how to calculate a PEF, as well as how to develop product category-specific methodological requirements for use in Product Environmental Footprint Category Rules (PEFCRs). The PEF was in a pilot phase between 2013 and 2016, where PEFCR documents and other aspects of the methodology and format of communication were developed. The first wave of pilots started with 14 nonfood products and was complemented in 2014 with the second wave of 11 food products (beer, coffee, dairy, feed for food producing animals, marine fish, meat -bovine, pigs and sheep, pasta, packed water, pet food [cats and dogs], olive oil, wine). As of September 2018, PEFCRs for 11 categories were developed: Beer, Dairy, Decorative paints, Feed for food producing animals, IT equipment, Leather, Packed water, Pasta, Pet Food, Rechargeable batteries, Wine. In the period between the end of the Environmental Footprint pilot phase and the possible adoption of policies implementing the PEF methods, a transition phase is established. Table 3.7 reports the food-related PEFCRs and the composition of Technical Secretariats responsible of their writings.

6. Environmental product declaration (EPD)

An EPD, also referred to as Type III environmental declaration, is a standardized (ISO 14025) and LCA-based tool to communicate the environmental performance of a product (Grahl and Schmincke, 2007). Despite the development and use of EPDs is a voluntary act, their demand in recent years has increased (Ibanez-Fores et al., 2016; Ingwersen and Stevenson, 2012; Strazza et al., 2010) as well as the number of program operators, e.g., the bodies supervising and administrating the Type III Environmental Declaration Program, the development of Product Category Rules (PCR) and EPDs and the verification process (Del Borghi et al., 2008; Strazza et al., 2013). The International EPD System (IES), a global program for environmental declarations having its origin in Sweden, has an online database of over 900 EPDs covering a wide range of product categories in 40 countries. The IES is a global program for Type III environmental declarations operating in accordance with ISO 14025 with the aim of enable and support organizations in any country to communicate quantified environmental information on the life cycle of their products. The scope of the program includes any type of product from any organization in any country where there is a market demand to communicate its life cycle-based environmental information. EPDs are

Table 3.7 PEFCRs in the food sector.

Food product	Members of the technical secretariat
Beer	The Brewers of Europe, AB-InBev, Carlsberg Group, HEINEKEN, SABMiller, European Aluminum, The European Container Glass Federation – FEVE, Beverage Industry Environmental Roundtable – BIER, Blonk Consultants, Bocconi University
Dairy	European Dairy Association (EDA), Alliance for Beverage Carton and the Environment (ACE), ACTALIA, BEL Group, French Environment and Energy Management Agency (CGDD), Coopérative Laitière de la Sèvre (CLS), Center National Interprofessionnel de l'Économie Laitière (CNIEL) and Association de la Transformation Laitière Française (ATLA), Constantia Flexibles, Danone, DMK GROUP, European Container Glass Federation (FEVE), Fonterra, FrieslandCampina
Packed water	European Federation of Bottled Waters (EFBW), European Container Glass Federation (FEVE), PETCORE Europe, Union Européenne des Transporteurs Routiers (UETR), Danone Waters, Ferrarelle, Nestlé Waters, San Benedetto, Spadel, Quantis
Pasta	UN.A.F.P.A., Barilla G. e R. Fratelli S.p.A., Pasta Zara S.p.A., Pastificio Lucio Garofalo S.p.A., Life Cycle Engineering
Wine	Comité Européen des Entreprises Vins (CEEV), Pernod Ricard Winemakers Spain, Comité Interprofessionnel du Vin de Champagne (CIVC) and three Champagne producers represented by CIVC, Unione Italiana Vini (UIV), Soc. Agr. Salcheto, The European Container Glass Federation (FEVE), Amcor, Nomacorc, C.E. Liège, IHOBE – Public agency of environment of the Basque Government, Institut Français de la Vigne et du Vin (IFV), Lavola, Ecole supérieure d'agricultures (ESA) - Angers

based on Product Category Rules (PCR) providing rules, requirements, and guidelines for a defined product category (IES, 2017). The IES includes an extensive PCR library covering the following 12 categories with the related sub-categories: chemical products; construction products; electricity, steam and fuels; food and beverages (bakery products; bottled water, wine and beer; dairy and egg products; fruits, vegetables and juice; grain mill products; olive oil; other; pasta); furniture and other goods; infrastructure and buildings; machinery and equipment; metal, plastic and glass products; paper products (absorbent hygiene products; packaging and materials; tissue products); services; textiles, footwear and apparel; vehicles and transport equipment (aircraft; carts and trolleys; rolling stock—trains and trams). Table 3.8 reports all the published EPDs of food and beverage category found on the International EPD System website. At the time of writing, 149 EPDs have been found: 53 bakery products, 32 dairy and egg products, 16 Fruits, vegetables and juice, 15 pasta, 8 bottled water, wine, and beer, eight olive oil, five Grain mill products and 12 Others (including sauces, meat, sugar).

Table 3.8 Published EPD for the "Food and Beverages" category.

No.	Product/Service	Company
1	Accadì Fresh lactose-free milk	Granarolo S.p.A.
2	Accadì Lactose-free mascarpone cheese	Granarolo S.p.A.
3	Accadì Lactose-free ricotta cheese	Granarolo S.p.A.
4	Accadì Lactose-free stracchino cheese	Granarolo S.p.A.
5	Accadì lactose-free semiskimmed milk	Granarolo S.p.A.
6	Accadì lactose-free skimmed milk	Granarolo S.p.A.
7	Accadì lactose-free whole milk	Granarolo S.p.A.
8	Alberta Peas	Alberta Pulse Growers
9	Barilla Arrabbiata sauce	Barilla
10	Barilla Basilico Sauce 2015 Limited Edition	Barilla
11	Barilla Basilico sauce	Barilla
12	Barilla Dry Semolina Pasta Selezione Oro Chef	Barilla
13	Barilla Durum wheat semolina pasta	Barilla
14	Barilla Durum wheat semolina pasta 5 kg for FoodService	Barilla
15	Barilla Pestati sauce	Barilla
16	Barilla Pesto al basilico 100% vegetale sauce	Barilla
17	Barilla Pesto alla Genovese sauce	Barilla
18	Birrificio Angelo Poretti 4 Luppoli Originale beer	Carlsberg Italia SpA
19	Birrificio Angelo Poretti 5 and 6 Luppoli Bock beer	Carlsberg Italia SpA
20	Borges Extra virgin olive oil	Borges Agricultural and Industrial Edible oils, SAU
21	CJ Glass Noodle (400g)	CJ Cheiljedang
22	COOP Beef Meat	COOP Italia
23	COOP Veal Meat	COOP Italia
24	Carlsberg beer	Carlsberg Italia SpA
25	Cerelia Mineral water	Cerelia Srl
26	Classic Potato Selenella®	Consorzio Patata Italiana di Qualità Soc. Cons. R.L.
27	Cuor di Lino - Italian soft bread	Barilla
28	De Cecco Durum wheat semolina pasta	De Cecco

Table 3.8 Published EPD for the "Food and Beverages" category.—cont'd

No.	Product/Service	Company
29	De Cecco Egg pasta	De Cecco
30	De Cecco Extra virgin olive oil	De Cecco
31	Dried durum wheat semolina pasta Patrimoni d'Italia	Consorzio Agrario del Nordest
32	Dry rice Terra e Vita	Terra e Vita S.n.C.
33	Farina Bio Molino Grassi - organic soft wheat flour	Molino Grassi S.p.A.
34	Ferrarelle Mineral water	Ferrarelle S.p.A.
35	Filiz Dry Semolina Pasta	Barilla
36	Food service organic Sgambaro Pasta (5 kg package)	Sgambaro S.p.A.
37	Food service Sgambaro Pasta (5 kg)	Sgambaro S.p.A.
38	Fresh semiskimmed "Selezione Mugello" milk	Centrale del Latte della Toscana S.p.A.
39	Fresh whole high quality "Selezione Mugello" milk	Centrale del Latte della Toscana S.p.A.
40	Gran Cereale Biscotto classico e alla frutta - wholegrain biscuits	Barilla
41	Granarolo Accadì Cream	Granarolo S.p.A.
42	Granarolo Accadì Mozzarella	Granarolo S.p.A.
43	Granarolo Accadì Ricotta cheese	Granarolo S.p.A.
44	Granarolo Accadì organic semiskimmed lactose-free ESL milk	Granarolo S.p.A.
45	Granarolo Alta Qualità flavored yogurt	Granarolo S.p.A.
46	Granarolo Alta Qualità white yogurt	Granarolo S.p.A.
47	Granarolo Creamy mascarpone cheese	Granarolo S.p.A.
48	Granarolo Egg Pasta	Granarolo S.p.A.
49	Granarolo Fresh milk "Piacere leggero" in PET bottles	Granarolo S.p.A.
50	Granarolo Fresh organic eggs	Granarolo S.p.A.
51	Granarolo G+ drink made with semiskimmed and without fat milk	Granarolo S.p.A.
52	Granarolo high quality pasteurized milk	Granarolo S.p.A.
53	Granarolo Italian UHT milk (whole, partially skimmed and skimmed)	Granarolo S.p.A.

Continued

Table 3.8 Published EPD for the "Food and Beverages" category.—cont'd

No.	Product/Service	Company
54	Granarolo Lactose-free UHT cream	Granarolo S.p.A.
55	Granarolo Mozzarella made from high quality fresh milk	Granarolo S.p.A.
56	Granarolo organic pasteurized semiskimmed milk	Granarolo S.p.A.
57	Granarolo organic pasteurized whole milk	Granarolo S.p.A.
58	Granarolo organic rice drink	Granarolo S.p.A.
59	Granarolo organic soy drink	Granarolo S.p.A.
60	Granarolo Prima Natura Bio yogurts	Granarolo S.p.A.
61	Granarolo soy drink	Granarolo S.p.A.
62	Granarolo soy drink ESL	Granarolo S.p.A.
63	Granarolo soy yogurt	Granarolo S.p.A.
64	Granarolo soy-based dressing UHT	Granarolo S.p.A.
65	Granarolo Yomo organic Yogurt	Granarolo S.p.A.
66	Granarolo Yomo lactose-free yogurt	Granarolo S.p.A.
67	Granarolo durum wheat semolina pasta	Granarolo S.p.A.
68	Harrys 100% Mie Nature	Barilla
69	Harrys American Sandwich Complet	Barilla
70	Harrys American Sandwich Nature	Barilla
71	Harrys Brioche Tranchée	Barilla
72	Harrys Extra Moelleux	Barilla
73	Italia zuccheri sugar from sugar beet	CO.PRO.B. S.C.A. - Cooperativa produttori Bieticoli Soc. Coop. Agricola
74	Italian apples	Assomela - Società Cooperativa
75	Kronenbourg 1664 beer	Carlsberg Italia SpA
76	Long rice (SOS)	Fallera Nutricio SL
77	Mineral water Lete and Sorgesana	Lete S.p.A.
78	Misko Dry Semolina Pasta	Barilla
79	Monini Bios organic Extra Virgin oil	Monini S.p.A.
80	Monini D.O.P. Umbria extra virgin olive oil	Monini S.p.A.
81	Monini Extra Virgin olive oil Classico	Monini S.p.A.
82	Monini Extra Virgin olive oil Delicato	Monini S.p.A.

Table 3.8 Published EPD for the "Food and Beverages" category.—cont'd

No.	Product/Service	Company
83	Monini Extra Virgin olive oil Gran Fruttato	Monini S.p.A.
84	Montana Canned beef meat	INALCA S.p.A.
85	Montana Frozen Hamburger	INALCA S.p.A.
86	Mulino Bianco Abbracci	Barilla
87	Mulino Bianco Batticuori - Italian biscuits with chocolate	Barilla
88	Mulino Bianco Camille	Barilla
89	Mulino Bianco Campagnole	Barilla
90	Mulino Bianco Chicchi di Cioccolato biscuits	Barilla
91	Mulino Bianco Cornetti	Barilla
92	Mulino Bianco Cracker	Barilla
93	Mulino Bianco Fette Biscottate (Dorate, Integrali, Malto d'Orzo, Rustiche e Dolcifette)	Barilla
94	Mulino Bianco Fiori d'Acqua crackers	Barilla
95	Mulino Bianco Fiori di latte biscuits	Barilla
96	Mulino Bianco Flauti	Barilla
97	Mulino Bianco Galletti biscuits	Barilla
98	Mulino Bianco Gemme Integrali	Barilla
99	Mulino Bianco Girotondi biscuits	Barilla
100	Mulino Bianco Gran Bauletto - Italian soft bread	Barilla
101	Mulino Bianco Granetti	Barilla
102	Mulino Bianco Macine	Barilla
103	Mulino Bianco Magie di Nocciola biscuits	Barilla
104	Mulino Bianco Michetti	Barilla
105	Mulino Bianco Pagnotta di Gran Duro e Integrale - Italian soft bread	Barilla
106	Mulino Bianco Pan Bauletto - Italian soft bread	Barilla
107	Mulino Bianco Pan Briochè - soft bread	Barilla
108	Mulino Bianco Pan Carrè	Barilla
109	Mulino Bianco Pan Goccioli	Barilla

Continued

Table 3.8 Published EPD for the "Food and Beverages" category.—cont'd

No.	Product/Service	Company
110	Mulino Bianco Pangrì - breadsticks	Barilla
111	Mulino Bianco Plumcake	Barilla
112	Mulino Bianco Saccottini	Barilla
113	Mulino Bianco Segreti di Bosco	Barilla
114	Mulino Bianco Tarallucci	Barilla
115	Mulino Bianco Torta al Limone	Barilla
116	OVOMET® Eggshell membrane powder and OVOCET® Calcium for human consumption	EGGNOVO S.L.
117	OVOPET® Eggshell membrane powder	EGGNOVO S.L.
118	Organic extra virgin olive oil produced by LUQUE ECOLOGICO, SL	LUQUE ECOLOGICO, S.L.
119	Pan di Stelle Biscotto	Barilla
120	Pan di Stelle La Merenda	Barilla
121	Pan di Stelle Mooncake	Barilla
122	Pan di Stelle Torta - Italian cake	Barilla
123	Pavesi Gocciole biscuits with chocolate chips	Barilla
124	Pavesi Gran cracker	Barilla
125	Pavesi Pavesini biscuits	Barilla
126	Pavesi Petit dried biscuit	Barilla
127	Pavesi Ringo	Barilla
128	Plant-based rice drink with vitamin and calcium	Unigrà Srl
129	Plant-based soy drink with vitamin and calcium	Unigrà Srl
130	Round rice (SOS)	Fallera Nutricio SL
131	SCUDO Complete rabbit feed	Mignini and Petrini SpA
132	Semola Bio Molino Grassi - organic durum wheat semolina	Molino Grassi S.p.A.
133	Semola Kronos® Molino Grassi - semolina	Molino Grassi S.p.A.
134	Stracchino Alta Qualità fresh cheese	Granarolo S.p.A.
135	Tuborg beer	Carlsberg Italia SpA
136	Valfrutta Borlotti beans	Conserve Italia soc. coop agricola

Table 3.8 Published EPD for the "Food and Beverages" category.—cont'd

No.	Product/Service	Company
137	Valfrutta Chopped tomatoes	Conserve Italia soc. coop agricola
138	Valfrutta Pear nectar	Conserve Italia soc. coop agricola
139	Voiello durum wheat dried semolina pasta	Barilla
140	Wasa Crisp'n Light 7 grains	Barilla
141	Wasa Frukost	Barilla
142	Wasa Havre and Vitalité	Barilla
143	Wasa Husman	Barilla
144	Wasa Light Rye, Integrale and Delikatess	Barilla
145	Wasa Multigrain, Mehrkorn and Surdeg Flerkorn	Barilla
146	Wasa Rågi	Barilla
147	Yellow label Sgambaro pasta	Sgambaro S.p.A.
148	Yomo Food service Yogurt	Granarolo S.p.A.
149	Zeus Kiwi fruit	Zeus Kiwi SA

7. Examples of product category rules (PCR) related to food

In the EPD framework, the comparability among rules and results can be guaranteed by following a methodological pattern consistent with the requirements of the environmental labels (Schau and Fet, 2007; Del Borghi, 2012).

The structure of the PCRs related to the food sector is reported in Fig. 3.2.

In a product-oriented perspective, industrial products can be classified through Central Product Classification (CPC) classes. The ultimate version of the CPC (UN, 2015) defines a comprehensive scheme for the classification of all goods and services available on the market, providing an international standard for the harmonization of product detail collection and comparison. The CPC defines categories for all goods and services that can be the object of domestic or international transactions or that can be entered into stocks, generally following the definition of products within the System of National Accounts (SNA).

The overall set of products is subdivided into several categories and the CPC coding system—an equivalent set of five-digit numbers—defines a five-level structure of mutually exclusive categories, facilitating data collection, presentation and analysis at detailed levels of the economy in an internationally comparable, standardized way (Fig. 3.3).

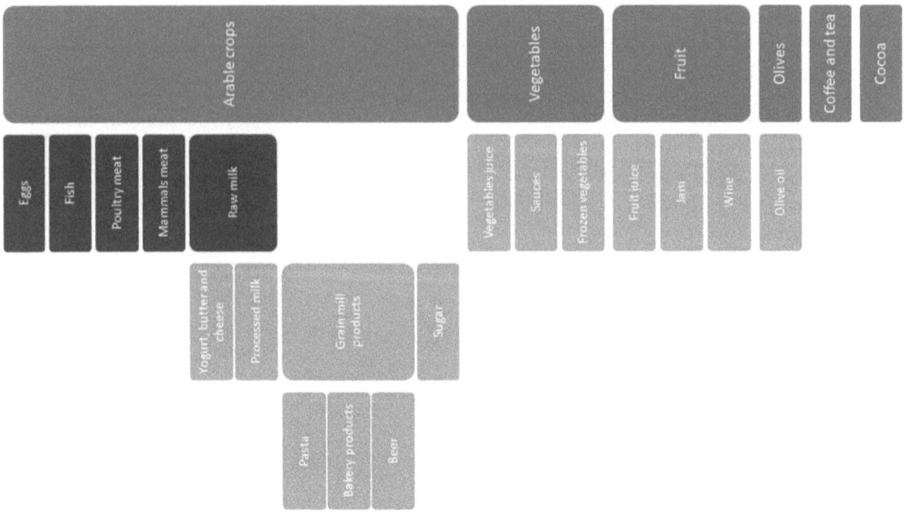

Figure 3.2 PCR structure for the food sector.

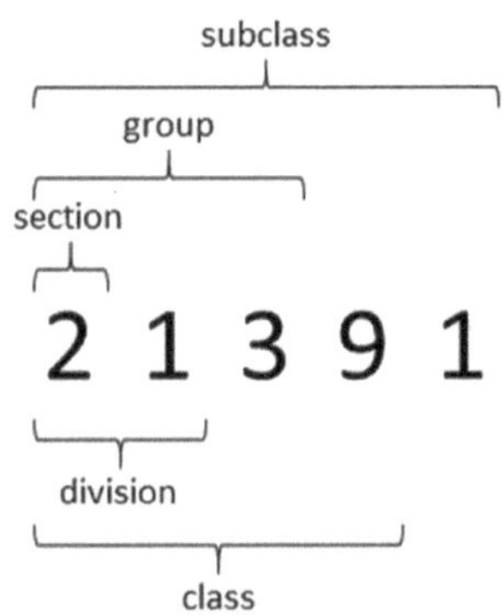

Figure 3.3 CPC code example.

The categories at the highest level are defined by the first digit of the CPC code and are called "sections", which are numerically coded categories. The existing sections are, in order:

- Section 0: Agriculture, forestry and fishery products
- Section 1: Ores and minerals; electricity, gas and water
- Section 2: Food products, beverages and tobacco; textiles, apparel and leather products
- Section 3: Other transportable goods, except metal products, machinery and equipment
- Section 4: Metal products, machinery and equipment
- Section 5: Constructions and construction services
- Section 6: Distributive trade services; accommodation, food and beverage serving services; transport services; and electricity, gas and water distribution services
- Section 7: Financial and related services; real estate services; and rental and leasing services

- Section 8: Business and production services
- Section 9: Community, social and personal services

The classification is then organized into successively more detailed categories, which are numerically coded:

- two-digit "divisions";
- three-digit "groups";
- four-digit "classes";
- and, at the greatest level of detail, five-digit "subclasses".

The categories of interest for the food sector belong to Section 0 and Section 2. The existing PCRs for the food sector are listed in Table 3.9.

Table 3.9 Existing PCRs in the "Food and beverage" category.

CPC code	CPC description	PCR name
01	Products of agriculture, horticulture and market gardening	Products of agriculture, horticulture and market gardening
011	Cereals	Arable crops
012	Vegetables	Vegetables
013	Fruits and nuts	Under development - Fruit and nuts
01342	Pears and quinces	Kiwifruit (being replaced)
02	Live animals and animal products (excluding meat)	Live animals and animal products (excluding meat)
022	Raw milk	Raw milk
02310	Hen eggs in shell, fresh	Hen eggs in shell, fresh
0293	Edible products of animal origin n.e.c.	Edible products of animal origin
04	Fish and other fishing products	Fish and other fishing products
21	Meat, fish, fruits, vegetables, oils and fats	Meat, fish, fruit, vegetables, oils and fats
2111	Meat of mammals, fresh or chilled	Meat of mammals
2112	Meat of poultry, fresh or chilled	Meat of poultry
2117	Other meat and edible offal, fresh, chilled or frozen	Preserves and preparations of meat
2132	Vegetable juices	Vegetable juice, Plant milk, plant milk based products, and Other prepared and preserved vegetables, pulses and potatoes
2143	Fruit juices	Fruit juices

Continued

Table 3.9 Existing PCRs in the "Food and beverage" category.—cont'd

CPC code	CPC description	PCR name
21491	Pineapples, otherwise prepared or preserved	Other prepared and preserved fruit
21494	Jams, fruit jellies, marmalades, fruit or nut purree and fruit or nut pastes	Jams, fruit jellies, marmalades, etc.
21537	Olive oil, crude	Being updated - Virgin olive oils and its fractions (expired 2018-04-01)
22	Dairy products and egg products	Dairy products and egg products
221	Processed liquid milk, cream and whey	Processed liquid milk and cream
2223	Yogurt and other fermented or acidified milk and cream	Yogurt, butter and cheese
23	Grain mill products, starches and starch	Grain mill products, starches and starch products; other food products
231	Grain mill products	Grain mill products
233	Preparations used in animal feeding; lucerne (alfalfa) meal and pellets	Preparations used in animal feeding for food-producing animals
234	Bakery products	Being updated - Bakery products (valid until 2019-05-29)
2351	Raw cane or beet sugar	Raw sugar, refined sugar, and molasses
2371	Uncooked pasta, not stuffed or otherwise prepared	Uncooked pasta, not stuffed or otherwise prepared
2372	Pasta, cooked, stuffed or otherwise prepared; couscous	Being updated - Pasta, cooked, stuffed or otherwise prepared; couscous (expired 2018-11-13)
23912	Coffee substitutes containing coffee; extracts, essences and concentrates of coffee, and preparations with a basis thereof or with a basis of coffee; roasted chicory and other roasted coffee substitutes, and extracts, essences and concentrates thereof	Espresso coffee
23912	Coffee substitutes containing coffee; extracts, essences and concentrates of coffee, and preparations with a basis thereof or with a basis of coffee; roasted chicory and other roasted coffee substitutes, and extracts, essences and concentrates thereof	Under development - Moka coffee

Table 3.9 Existing PCRs in the "Food and beverage" category.—cont'd

CPC code	CPC description	PCR name
23913	Green tea (not fermented), black tea (fermented) and partly fermented tea, in immediate packings of a content not exceeding 3 kg	Under development - Tea
23995	Sauces; mixed condiments; mustard flour and meal; prepared mustard	Sauces, mixed condiments and mustard
24	Beverages	Beverages
24211	Sparkling wine of fresh grapes	Sparkling wine of fresh grapes (expired)
24212	Wine of fresh grapes, except sparkling wine; grape must	Wine
2431	Beer made from malt	Beer made from malt
24410	Bottled waters, not sweetened or flavored	Bottled waters, not sweetened or flavored

8. Conclusions

In order to develop and implement a sustainable food supply chain, a system perspective is essential. For these reasons, the food sector needs to involve all the different stakeholders, who must jointly cooperate with the aim to develop and implement sustainable food production processes and environmentally-friendly products. Indeed, the synergic utilization of a life cycle perspective and LCA-based tools and labeling schemes may be part of the winning path to pursue in order to reduce the environmental impacts of the food industry, provided that recognized and reliable guidelines (e.g., PCRs, etc.) need to be used in order to obtain reliable and comparable results. In particular, all the processes, data and impacts related to the upstream processes (i.e., agriculture and livestock farming) have to be clearly defined in order to avoid erroneous and misleading calculations and to allow the comparison among different products. The opportunity to use several consolidated databases containing numerous processes related to the food industry provides a good basis for the start of an environmental improvement in the sector that can only be encouraged by the collection of new data and by comparison with the new companies of the sector that are gradually beginning to verify the environmental impact of their products within the existing labeling schemes. Finally, the debate regarding the definition of functional units more suited to the food sector should stimulate the scientific research for the identification of the most sustainable foods and dietary patterns from the point of view of an healthy nutrition.

References

Abeliotis, K., Detsis, V., Pappia, C., 2013. Life cycle assessment of bean production in the Prespa National Park, Greece. J. Clean. Prod. 41, 89–96. https://doi.org/10.1016/j.jclepro.2012.09.032.

Accorsi, R., Versari, L., Manzini, R., 2015. Glass vs. Plastic: life cycle assessment of extra-virgin olive oil bottles across global supply chains. Sustainability 7, 2818–2840. https://doi.org/10.3390/su7032818.

Al-Ansari, T., Korre, A., Nie, Z., Shah, N., 2015. Development of a life cycle assessment tool for the assessment of food production systems within the energy, water and food nexus. Sustain. Prod. Consum. 2, 52–66. https://doi.org/10.1016/j.spc.2015.07.005.

Amienyo, D., Azapagic, A., 2016. Life cycle environmental impacts and costs of beer production and consumption in the UK. Int. J. Life Cycle Assess. 21, 492–509. https://doi.org/10.1007/s11367-016-1028-6.

Amienyo, D., Camilleri, C., Azapagic, A., 2014. Environmental impacts of consumption of Australian red wine in the UK. J. Clean. Prod. 72, 110–119. https://doi.org/10.1016/j.jclepro.2014.02.044.

Amienyo, D., Gujba, H., Stichnothe, H., Azapagic, A., 2012. Life cycle environmental impacts of carbonated soft drinks. Int. J. Life Cycle Assess. 18, 77–92. https://doi.org/10.1007/s11367-012-0459-y.

Andersson, K., Ohlsson, T., Olsson, P., 1998. Screening life cycle assessment (LCA) of tomato ketchup: a case study. J. Clean. Prod. 6, 277–288. https://doi.org/10.1016/S0959-6526(98)00027-4.

Atallah, S.S., Gómez, M.I., Björkman, T., 2014. Localization effects for a fresh vegeTable 4.product supply chain: broccoli in the eastern United States. Food Policy 49, 151–159. https://doi.org/10.1016/j.foodpol.2014.07.005.

Avraamides, M., Fatta, D., 2008. Resource consumption and emissions from olive oil production: a life cycle inventory case study in Cyprus. J. Clean. Prod. 16, 809–821. https://doi.org/10.1016/j.jclepro.2007.04.002.

Benedetto, G., 2013. The environmental impact of a Sardinian wine by partial life cycle assessment. Wine Econ. Policy 2, 33–41. https://doi.org/10.1016/j.wep.2013.05.003.

Bertoluci, G., Leroy, Y., Olsson, A., 2014. Exploring the environmental impacts of olive packaging solutions for the European food market. J. Clean. Prod. 64, 234–243. https://doi.org/10.1016/j.jclepro.2013.09.029.

Biermann, G., Geist, J., 2019. Life cycle assessment of common carp (*Cyprinus carpio L.*) – a comparison of the environmental impacts of conventional and organic carp aquaculture in Germany. Aquaculture 501, 404–415. https://doi.org/10.1016/j.aquaculture.2018.10.019.

Bjelkemyr, M., Shahbazi, S., Jönsson, C., Wiktorsson, M., 2015. Individuals' perception of which materials are most important to recycle. In: Umeda, S., Nakano, M., Mizuyama, H., et al. (Eds.), Advances in Production Management Systems: Innovative Production Management Towards Sustainable Growth. Springer International Publishing, pp. 723–729.

Boulay, A.M., Jane, B., Benini, L., Berger, M., Michael, J.L., Alessandro, M., Manuele, M., et al., February 2018. The WULCA Consensus Characterization Model for Water Scarcity Footprints: Assessing Impacts of Water Consumption Based on Available Water Remaining (AWARE). The International Journal of Life Cycle Assessment 23 (2), 368–378. https://doi.org/10.1007/s11367-017-1333-8.

Bouwman, A.F., Boumans, L.J.M., Batjes, N.H., 2002. Modeling global annual N_2O and NO emissions from fertilized fields. Glob. Biogeochem. Cycles 16, 1080. https://doi.org/10.1029/2001GB001812.

Cesari, V., Zucali, M., Bava, L., et al., 2018. Environmental impact of rabbit meat: the effect of production efficiency. Meat Sci. 145, 447–454. https://doi.org/10.1016/j.meatsci.2018.07.011.

Cleary, J., 2013. Life cycle assessments of wine and spirit packaging at the product and the municipal scale: a Toronto, Canada case study. J. Clean. Prod. 44, 143–151. https://doi.org/10.1016/j.jclepro.2013.01.009.

Coltro, L., Mourad, A.L., Kletecke, R.M., et al., 2009. Assessing the environmental profile of orange production in Brazil. Int. J. Life Cycle Assess. 14, 656–664. https://doi.org/10.1007/s11367-009-0097-1.

COM 21, 2011. A Resource-Efficient Europe – Flagship Initiative Under the Europe 2020 Strategy. Communication from the Commission to the European Parliament, the Council, the European Economic and Social Committee and the Committee of the Regions.

COM 196, European Commission, 2013. Communication From the Commission to the European Parliament and the Council. Building the Single Market for Green Products. Facilitating Better Information on the Environmental Performance of Products and Organizations.

COM 571, 2011. Roadmap to a Resource Efficient Europe. Communication from the Commission to the European Parliament, the Council, the European Economic and Social Committee and the Committee of the Regions.

Cordella, M., Tugnoli, A., Spadoni, G., et al., 2007. LCA of an Italian lager beer. Int. J. Life Cycle Assess. 13, 133–139. https://doi.org/10.1065/lca2007.02.306.

de Figueirêdo, M.C.B., de Boer, I.J.M., Kroeze, C., et al., 2014. Reducing the impact of irrigated crops on freshwater availability: the case of Brazilian yellow melons. Int. J. Life Cycle Assess. 19, 437–448. https://doi.org/10.1007/s11367-013-0630-0.

Del Borghi, A., 2012. LCA and communication: environmental product declaration. Int. J. Life Cycle Assess. 18, 293–295. https://doi.org/10.1007/s11367-012-0513-9.

Del Borghi, A., Gaggero, P.L., Gallo, M., Strazza, C., 2008. Development of PCR for WWTP based on a case study. Int. J. Life Cycle Assess. 13 (6), 512–521.

Del Borghi, A., Gallo, M., Magrassi, F., 2016. Glass packaging design and life cycle assessment: deep review and guideline for future developments. Reference Module in Food Sciences. Elsevier, pp. 1–9. https://doi.org/10.1016/B978-0-08-100596-5.21008-1.

Del Borghi, A., Gallo, M., Strazza, C., Del Borghi, M., 2014. An evaluation of environmental sustainability in the food industry through life cycle assessment: the case study of tomato products supply chain. J. Clean. Prod. 78, 121–130. https://doi.org/10.1016/j.jclepro.2014.04.083.

Del Borghi, A., Strazza, C., Magrassi, F., et al., 2018. Life cycle assessment for eco-design of product–package systems in the food industry—the case of legumes. Sustain. Prod. Consum. 13, 24–36. https://doi.org/10.1016/j.spc.2017.11.001.

Detzel, A., Mönckert, J., 2009. Environmental evaluation of aluminium cans for beverages in the German context. Int. J. Life Cycle Assess. 14, 70–79. https://doi.org/10.1007/s11367-008-0057-1.

EC, 2002. Regulation (EC) No 1774/2002 of the European Parliament and of the Council of 3 October 2002 Laying Down Health Rules Concerning Animal By-Products Not Intended for Human Consumption.

EC, 2007. Council Regulation (EC) No 834/2007 of 28 June 2007 on Organic Production and Labelling of Organic Products and Repealing Regulation (EEC) No 2092/91.

EC, 2008. Commission Regulation (EC) No 889/2008 of 5 September 2008 Laying Down Detailed Rules for the Implementation of Council Regulation (EC) No 834/2007 on Organic Production and Labelling of Organic Products with Regard to Organic Production, Labelling and Control.

EC, 2013. European Commission JRC – Cattle Model Working Group, 2013.

Ecoinvent Centre, 2018. Ecoinvent Data v3.5. Swiss Centre for Life Cycle Inventories, Dübendorf.

Eide, M.H., 2002. Life cycle assessment (LCA) of industrial milk production. Int. J. Life Cycle Assess. 7, 115–126. https://doi.org/10.1007/BF02978855.

EMEP/EEA, 2013. Air Pollutant Emission Inventory Guidebook.

Eriksson, O., Jonsson, D., Hillman, K., 2016. Life cycle assessment of Swedish single malt whisky. J. Clean. Prod. 112 (1), 229–237. https://doi.org/10.1016/j.jclepro.2015.07.050.

Espinoza-Orias, N., Stichnothe, H., Azapagic, A., 2011. The carbon footprint of bread. Int. J. Life Cycle Assess. 16, 351–365. https://doi.org/10.1007/s11367-011-0271-0.

European Commission, 2012. The Common Agricultural Policy A partnership between Europe and Farmers. http://ec.europa.eu/agriculture/cap-overview/2012_en.pdf.

European Parliament and Council, 2013. Regulation (EU) No 1310/2013 of the European Parliament and of the Council. http://eur-lex.europa.eu/LexUriServ/LexUriServ.do?uri=OJ:L:2013:347:0865:0883:en:PDF.

EXPO, 2015. The Milan Chart - EXPO 2015. http://carta.milano.it/en/.

FAO, 2014. Greenhouse Gas Emissions and Fossil Energy Demand from Poultry Supply Chains. http://www.fao.org/3/a-mj752e.pdf.

FIL-IDF, 2015. Bulletin of the IDF N° 479/2015: A Common Carbon Footprint Approach for the Dairy Sector – the IDF Guide to Standard Life Cycle Assessment Methodology. https://www.fil-idf.org/wp-content/uploads/2016/09/Bulletin479-2015_A-common-carbon-footprint-approach-for-the-dairy-sector.CAT.pdf.

Flagship Initiative Under the Europe 2020 Strategy, 2011. Communication from the Commission to the European Parliament, the Council, the European Economic and Social Committee and the Committee of the Regions.

Flysjö, A., 2011. Potential for improving the carbon footprint of butter and blend products. J. Dairy Sci. 94, 5833–5841. https://doi.org/10.3168/jds.2011-4545.

Foteinis, S., Chatzisymeon, E., 2016. Life cycle assessment of organic versus conventional agriculture. A case study of lettuce cultivation in Greece. J. Clean. Prod. 112, 2462–2471. https://doi.org/10.1016/j.jclepro.2015.09.075.

Fusi, A., Guidetti, R., Benedetto, G., 2014. Delving into the environmental aspect of a Sardinian white wine: from partial to total life cycle assessment. Sci. Total Environ. 472, 989–1000. https://doi.org/10.1016/j.scitotenv.2013.11.148.

Galatola, M., Pant, R., 2014. Reply to the editorial "Product environmental footprint—breakthrough or breakdown for policy implementation of life cycle assessment?" written by Prof. Finkbeiner (Int J Life Cycle Assess 19(2):266–271). Int. J. Life Cycle Assess. 19, 1356–1360. https://doi.org/10.1007/s11367-014-0740-3.

Gallo, M., Borghi, A.D., Strazza, C., 2015a. Analysis of potential GHG emissions reductions from methane recovery in livestock farming. Int. J. Glob. Warming 8, 516. https://doi.org/10.1504/IJGW.2015.073053.

Gallo, M., Del Borghi, A., Strazza, C., 2015b. Analysis of potential GHG emissions reductions from methane recovery in livestock farming. Int. J. Glob. Warm. 8, 519–533.

Grahl, B., Schmincke, E., 2007. The part of LCA in ISO type III environmental declarations. Int. J. LCA 12 (1), 38–45.

Hollins, O., 2011. The Further Benefits of Business Resource Efficiency. Research report completed for the Department for Environment, Food and Rural Affairs.

Hospido, A., Vazquez, M.E., Cuevas, A., et al., 2006. Environmental assessment of canned tuna manufacture with a life-cycle perspective. Resour. Conserv. Recycl. 47, 56–72. https://doi.org/10.1016/j.resconrec.2005.10.003.

Huang, C.-C., Ma, H.-W., 2004. A multidimensional environmental evaluation of packaging materials. Sci. Total Environ. 324, 161–172. https://doi.org/10.1016/j.scitotenv.2003.10.039.

Humbert, S., Rossi, V., Margni, M., et al., 2009. Life cycle assessment of two baby food packaging alternatives: glass jars vs. plastic pots. Int. J. Life Cycle Assess. 14, 95–106. https://doi.org/10.1007/s11367-008-0052-6.

Ibanez-Fores, V., Pacheco-Blanco, B., Capuz-Rizo, S.F., Bovea, M.D., 2016. Environmental product declarations: exploring their evolution and the factors affecting their demand in Europe. J. Clean. Prod. 116, 157–169.

IES, 2017. General Programme Instructions for the International EPD System. https://www.environdec.com/The-International-EPD-System/General-Programme-Instructions/.

Ingrao, C., Matarazzo, A., Tricase, C., et al., 2015. Life cycle assessment for highlighting environmental hotspots in Sicilian peach production systems. J. Clean. Prod. 92, 109–120. https://doi.org/10.1016/j.jclepro.2014.12.053.

Ingwersen, W.W., Stevenson, M.J., 2012. Can we compare the environmental performance of this product to that one? An update on the development of product category rules and future challenges toward alignment. J. Clean. Prod. 24, 102–108.

IPCC, 2006a. Indirect N2O Emissions From Agriculture. https://www.ipcc-nggip.iges.or.jp/public/gp/bgp/4_6_Indirect_N2O_Agriculture.pdf.

IPCC, 2006b. Guidelines for National Greenhouse Gas Inventories.

ISO, 2000. Environmental Labels and Declarations – General Principles. European Standard EN ISO 14020. International Organization for Standardization.

ISO, 2006a. Environmental Labels and Declarations - Type III Environmental Declarations - Principles and Procedures. European Standard EN ISO 14025. International Organization for Standardization.

ISO, 2006b. Environmental Management – Life Cycle Assessment – Principles and Framework. European Standard EN ISO 14040. International Organization for Standardization.

ISO, 2006c. Environmental Management – Life Cycle Assessment – Requirements and Guidelines. European Standard EN ISO 14044. International Organization for Standardization.

ISO, 2016. Environmental Labels and Declarations – Self-Declared Environmental Claims (Type II Environmental Labelling). European Standard EN ISO 14021. International Organization for Standardization.

ISO, 2017. Environmental Management – Water Footprint – A Practical Guide for SMEs. European Standard EN ISO 14046. International Organization for Standardization.

ISO, 2018. Environmental Labels and Declarations – Type I Environmental Labelling – Principles and Procedures. European Standard EN ISO 14024. International Organization for Standardization.

Jensen, J.D., Andersen, M., 2003. Marginale producenter af udvalgte landbrugsprodukter. FØI Working paper no. 08/2003 (in Danish). www.foi.dk. COM 21, A Resource-Efficient Europe, A Resource-Efficient Europe.

Jeswani, H.K., Burkinshaw, R., Azapagic, A., 2015. Environmental sustainability issues in the food–energy–water nexus: breakfast cereals and snacks. Sustain. Prod. Consum. 2, 17–28. https://doi.org/10.1016/j.spc.2015.08.001.

Klugman, J., 2011. Human Development Report 2011, Sustainability and Equity: A Better Future for All. Sustainability and Equity: A Better Future for All. UNDP-HDRO.

Manfredi, M., Vignali, G., 2014. Life cycle assessment of a packaged tomato puree: a comparison of environmental impacts produced by different life cycle phases. J. Clean. Prod. 73, 275–284. https://doi.org/10.1016/j.jclepro.2013.10.010.

Marsh, K., Bugusu, B., 2007. Food packaging—roles, materials, and environmental issues. J. Food Sci. 72, R39–R55. https://doi.org/10.1111/j.1750-3841.2007.00301.x.

Martins, A.A., Araújo, A.R., Graça, A., et al., 2018. Towards sustainable wine: comparison of two Portuguese wines. J. Clean. Prod. 183, 662–676. https://doi.org/10.1016/j.jclepro.2018.02.057.

Masset, G., Vieux, F., Darmon, N., 2015. Which functional unit to identify sustainable foods? Publ. Health Nutr. 18, 2488–2497. https://doi.org/10.1017/S1368980015000579.

Mata, T.M., Costa, C.A.V., 2001. Life cycle assessment of different reuse percentages for glass beer bottles. Int. J. Life Cycle Assess. 6, 307–319. https://doi.org/10.1007/BF02978793.

Meylan, G., Ami, H., Spoerri, A., 2014. Transitions of municipal solid waste management. Part II: hybrid life cycle assessment of Swiss glass-packaging disposal. Resour. Conserv. Recycl. 86, 16–27. https://doi.org/10.1016/j.resconrec.2014.01.005.

Milà i Canals, L., Burnip, G.M., Cowell, S.J., 2006. Evaluation of the environmental impacts of apple production using Life Cycle Assessment (LCA): case study in New Zealand. Agric. Ecosyst. Environ. 114, 226–238. https://doi.org/10.1016/j.agee.2005.10.023.

Milà i Canals, L., Chapagain, A., Orr, S., et al., 2010. Assessing freshwater use impacts in LCA, part 2: case study of broccoli production in the UK and Spain. Int. J. Life Cycle Assess. 15, 598–607. https://doi.org/10.1007/s11367-010-0187-0.

Miranda-Ackerman, M.A., Azzaro-Pantel, C., 2017. Extending the scope of eco-labelling in the food industry to drive change beyond sustainable agriculture practices. J. Environ. Manag. 204, 814–824. https://doi.org/10.1016/j.jenvman.2017.05.027.

Morris, J., 2004. Comparative LCAs for curbside recycling versus either landfilling or incineration with energy recovery (12 pp). Int. J. Life Cycle Assess. 10, 273–284. https://doi.org/10.1065/lca2004.09.180.10.

Mouron, P., Heijne, B., Naef, A., et al., 2012. Sustainability assessment of crop protection systems: SustainOS methodology and its application for apple orchards. Agric. Syst. 113, 1–15. https://doi.org/10.1016/j.agsy.2012.07.004.

Naudin, C., van der Werf, H.M.G., Jeuffroy, M.-H., Corre-Hellou, G., 2014. Life cycle assessment applied to pea-wheat intercrops: a new method for handling the impacts of co-products. J. Clean. Prod. 73, 80–87. https://doi.org/10.1016/j.jclepro.2013.12.029.

Nemecek, T., Bengoa, X., Lansche, J., Mouron, P., Rossi, V., Humbert, S., 2014. Methodological Guidelines for the Life Cycle Inventory of Agricultural Products. Version 2.0, July 2014. World Food LCA Database (WFLDB). Quantis and Agroscope, Lausanne and Zurich, Switzerland.

Nemecek, T., Bengoa, X., Lansche, J., Mouron, P., Rossi, V., Humbert, S., 2014. Methodological Guidelines for the Life Cycle Inventory of Agricultural Products. Version 2.0, July 2014. World Food LCA Database (WFLDB). Quantis and Agroscope. Lausanne and Zurich, Switzerland.

Nemecek, T., Weiler, K., Plassmann, K., et al., 2011. Modular extrapolation approach for crop LCA MEXALCA: global warming potential of different crops and its relationship to the yield. In: Finkbeiner, M. (Ed.), Towards Life Cycle Sustainability Management. Springer Netherlands, Dordrecht, pp. 309–317.

Neto, B., Dias, A.C., Machado, M., 2012. Life cycle assessment of the supply chain of a Portuguese wine: from viticulture to distribution. Int. J. Life Cycle Assess. 18, 590−602. https://doi.org/10.1007/s11367-012-0518-4.

Nielsen, P.H., Nielsen, A.M., Weidema, B.P., Dalgaard, R., Halberg, N., 2003. LCA Food Data Base. www.lcafood.dk.

Nikkhah, A., Royan, M., Khojastehpour, M., Bacenetti, J., 2017. Environmental impacts modeling of Iranian peach production. Renew. Sustain. Energy Rev. 75, 677−682. https://doi.org/10.1016/j.rser.2016.11.041.

Nunez, Y., Fermoso, J., Garcia, N., Irusta, R., 2005. Comparative life cycle assessment of beef, pork and ostrich meat: a critical point of view. Int. J. Agric. Resour. Gov. Ecol. 4, 140−151. https://doi.org/10.1504/IJARGE.2005.007196.

OMAFRA, 2009. "Introduction to organic farming". Ontario ministry of agriculture, food and rural affairs. http://www.omafra.gov.on.ca/english/crops/facts/09-077.htm.

Pairotti, M.B., Cerutti, A.K., Martini, F., et al., 2015. Energy consumption and GHG emission of the Mediterranean diet: a systemic assessment using a hybrid LCA-IO method. J. Clean. Prod. 103, 507−516. https://doi.org/10.1016/j.jclepro.2013.12.082.

Pattara, C., Raggi, A., Cichelli, A., 2012. Life cycle assessment and carbon footprint in the wine supply-chain. Environ. Manag. 49, 1247−1258. https://doi.org/10.1007/s00267-012-9844-3.

Point, E., Tyedmers, P., Naugler, C., 2012. Life cycle environmental impacts of wine production and consumption in Nova Scotia, Canada. J. Clean. Prod. 27, 11−20. https://doi.org/10.1016/j.jclepro.2011.12.035.

Ponstein, H.J., Meyer-Aurich, A., Prochnow, A., 2019. Greenhouse gas emissions and mitigation options for German wine production. J. Clean. Prod. 212, 800−809. https://doi.org/10.1016/j.jclepro.2018.11.206.

Prasuhn, V., 2006. Erfassung der PO_4 − Austräge für die Ökobilanzierung − SALCA − Phosphor. Agroscope FAL Reckenholz, Zürich, p. 22. Available at: www.agroscope.admin.ch.

Proietti, S., Sdringola, P., Desideri, U., et al., 2014. Carbon footprint of an olive tree grove. Appl. Energy 127, 115−124. https://doi.org/10.1016/j.apenergy.2014.04.019.

Rinaldi, S., Barbanera, M., Lascaro, E., 2014. Assessment of carbon footprint and energy performance of the extra virgin olive oil chain in Umbria. Italy. Sci. Total Environ. 482−483, 71−79. https://doi.org/10.1016/j.scitotenv.2014.02.104.

Roches, A., Nemecek, T., Gaillard, G., Plassmann, K., Sim, S., King, H., LMi, C., 2010. MEXALCA: a modular method for the extrapolation of crop LCA. Int. J. Life Cycle Assess. 15, 842−854.

Romero-Gámez, M., Suárez-Rey, E.M., Antón, A., et al., 2012. Environmental impact of screenhouse and open-field cultivation using a life cycle analysis: the case study of green bean production. J. Clean. Prod. 28, 63−69. https://doi.org/10.1016/j.jclepro.2011.07.006.

Roy, P., Nei, D., Orikasa, T., et al., 2009. A review of life cycle assessment (LCA) on some food products. J. Food Eng. 90, 1−10. https://doi.org/10.1016/j.jfoodeng.2008.06.016.

Schau, E.M., Fet, A.M., 2007. LCA studies of food products as background for environmental product declarations. Int. J. Life Cycle Assess. 13, 255−264. https://doi.org/10.1065/lca2007.12.372.

SETAC, 2013. Scoping Workshop − Glasgow, 2013. 23rd SETAC Europe Annual Meeting.

SETAC, 2014. Basel, 2014. 24th SETAC Europe Annual Meeting.

Simon, B., Amor, M.B., Földényi, R., 2016. Life cycle impact assessment of beverage packaging systems: focus on the collection of post-consumer bottles. J. Clean. Prod. 112 (1), 238−248. https://doi.org/10.1016/j.jclepro.2015.06.008.

Sonesson, U., Davis, J., Flysjö, A., et al., 2017. Protein quality as functional unit – a methodological framework for inclusion in life cycle assessment of food. J. Clean. Prod. 140, 470–478. https://doi.org/10.1016/j.jclepro.2016.06.115.

Sørensen, J.T., Østergaard, S., Houe, H., Hindhede, J., 10 September, 2002. Expert opinions of strategies for milk fever control. Preventive Veterinary Medicine 55 (1), 69–78.

Strazza, C., Del Borghi, A., Blengini, G.A., Gallo, M., 2010. Definition of the methodology for a Sector EPD (Environmental Product Declaration): case study of the average Italian cement. Int. J. Life Cycle Assess. 15, 540–548.

Strazza, C., Del Borghi, A., Gallo, M., 2013. Development of specific rules for the application of life cycle assessment to carbon capture and storage. Energies 6, 1250–1265.

Strazza, C., Magrassi, F., Gallo, M., Del Borghi, A., 2015. Life cycle assessment from food to food: a case study of circular economy from cruise ships to aquaculture. Sustain. Prod. Consum. 2, 40–51. https://doi.org/10.1016/j.spc.2015.06.004.

Tecco, N., Baudino, C., Girgenti, V., Peano, C., 2016. Innovation strategies in a fruit growers association impacts assessment by using combined LCA and s-LCA methodologies. Sci. Total Environ. 568, 253–262. https://doi.org/10.1016/j.scitotenv.2016.05.203.

United Nations – Department of Economics and social affairs, 2015. Central Product Classification (CPC), Version 2.1 – CPCv2.1. http://unstats.un.org/unsd/cr/downloads/CPCv2.1_complete%28PDF%29_English.pdf.

UNOPS, 2009. A Guide to Environmental Labels for Procurement Practitioners of the United Nations System. http://www.ungm.org.

van der Werf, H.M.G., Salou, T., 2015. Economic value as a functional unit for environmental labelling of food and other consumer products. J. Clean. Prod. 94, 394–397. https://doi.org/10.1016/j.jclepro.2015.01.077.

van Dooren, C., 2016. Proposing the nutrient density unit as the functional unit in LCAs of foods. In: 10th International Conference on Life Cycle Assessment of Food, Dublin.

Vellini, M., Savioli, M., 2009. Energy and environmental analysis of glass container production and recycling. Energy 34, 2137–2143. https://doi.org/10.1016/j.energy.2008.09.017.

Vitali, A., Grossi, G., Martino, G., et al., 2018. Carbon footprint of organic beef meat from farm to fork: a case study of short supply chain: GHG emissions from short organic beef supply chain. J. Sci. Food Agric. 98, 5518–5524. https://doi.org/10.1002/jsfa.9098.

World Bank, 2009. The World Bank Annual Report 2009: Year in Review, vol. 1. © World Bank, Washington, DC. https://openknowledge.worldbank.org/handle/10986/4354 License: CC BY 3.0 IGO.

Xu, Z., Xu, W., Peng, Z., et al., 2018. Effects of different functional units on carbon footprint values of different carbohydrate-rich foods in China. J. Clean. Prod. 198, 907–916. https://doi.org/10.1016/j.jclepro.2018.07.091.

Yuttitham, M., Gheewala, S.H., Chidthaisong, A., 2011. Carbon footprint of sugar produced from sugarcane in eastern Thailand. J. Clean. Prod. 19, 2119–2127. https://doi.org/10.1016/j.jclepro.2011.07.017.

Microbes and the environment: fermented products

4

Maria Aspri, Dimitrios Tsaltas
Department of Agricultural Sciences, Biotechnology and Food Science, Cyprus University of Technology, Lemesos, Cyprus

Chapter outline

1. Introduction 119
2. Interaction of cheese with the dairy environment 120
3. House microbiota of dairy industry 121
4. Examples of cheeses with characteristic in house microbiota 122
4.1 Pit cheese 122
4.2 Cheeses ripening in wooden vats 122
4.3 Cheese ripening in goat's skin bag 130
4.4 Cheese ripening in caves 130
5. Interaction of fermented sausages with meat industry 131
6. House microbiota of meat industry 131
7. Examples of traditional sausages with characteristic house microbiota 134
7.1 Serbia 134
7.2 Portugal 134
7.3 Spain 134
7.4 Greece 138
7.5 Italy 138
7.6 Methods to study the microbial ecology of fermented foods 139
8. Conclusion 141
Funding 141
References 141

1. Introduction

Food industry environments can play a major role in the development of microbial consortia of the final food product. Food processing environments contain many distinct microbial communities which can originate from raw materials, processing equipment, air and staff (Montel et al., 2014). Once microbes are introduced into these facilities, they may occupy specific niches and persist and may affect not only the microbiota of food but also the physicochemical characteristics of the food product.

The Interaction of Food Industry and Environment. https://doi.org/10.1016/B978-0-12-816449-5.00004-7

For instance, many studies have proved that food processing environments can harbor microbiota which affects beneficially the final food products such as dairy, alcoholic beverages, sausages and sourdough (Bokulich et al., 2012; 2016, Bokulich and Mills, 2013; Calasso et al., 2016; Minervini et al., 2015; Stellato et al., 2015). (See Figs. 4.1 and 4.2).

2. Interaction of cheese with the dairy environment

Cheese is one of the oldest fermented food products. The history of cheese production is a "natural history" in which dairy animals, microorganisms, and the processing environment interact to yield food for human consumption (Marcellino and Benson, 2013). Cheese is a complex biological ecosystem, harboring diverse microbial communities. The microbiota of any type of cheese can be originated from starter or adjunct cultures, the source and the treatment of the milk, the indigenous microorganisms that originate from the equipment and the cheese processing environments, the manufacturing process and ripening. The contribution of the cheese microbiota to development of flavor and texture characteristics of the cheese is very critical, as many of the final characteristics of the cheese are due to the complex dynamics and interactions between the cheese microorganisms, growth substrates and proteins in milk and the cheese environment (Afshari et al., 2018). In addition, microbial communities of cheese can act as bio-preservatives against foodborne pathogens and spoilage bacteria (Franck et al., 2008; Marcellino and Benson, 2013).

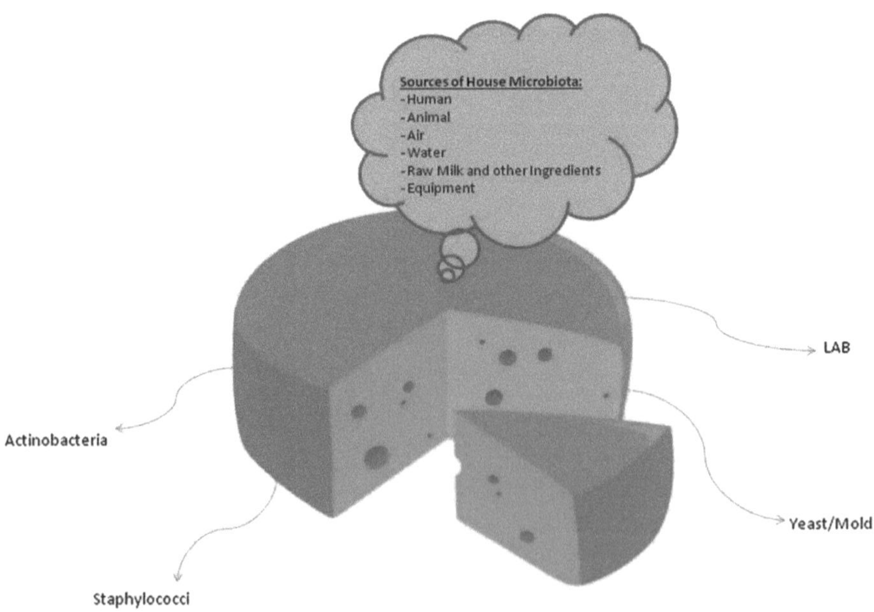

Figure 4.1 Schematic depiction of the sources of house microbiota and dominant microbial groups in cheese and cheese making environment.

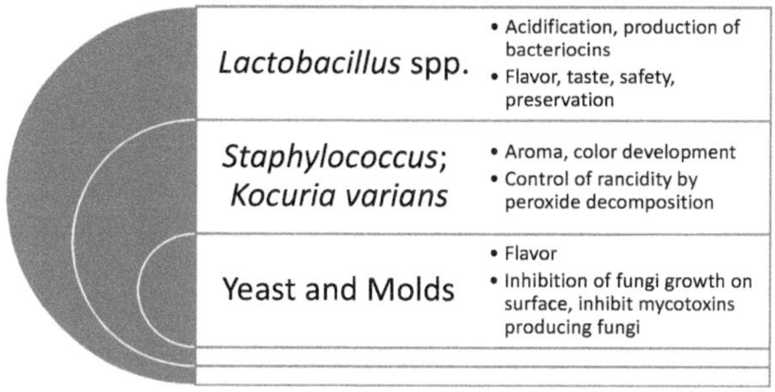

Figure 4.2 Main microbial consortia and their role in sausage characteristics.

This is more obvious at traditional cheeses known as artisanal cheeses. Artisan cheeses are defined as cheeses produced by hand on a small scale, normally using milk from heritage breed animals in a closed herd and using traditional practices or traditional utensils such as wood (Donnelly, 2014). There are over 100 cheeses that are registered under the Protection Designation of Origin (PDO). The main characteristic of PDO cheeses is that are harbor a complex microbiota due to the fact that are produced, processed and prepared from raw or heat-treated milk (Dugat-bony et al., 2016). Several studies demonstrated that each cheese has a distinct microbiota which can be composed from single to several species. According to Goerges et al. (2008) more than half of the microorganisms detected in cheeses were not inoculated as starter cultures but originated from the milk and the processing environment.

Cheese terroir: In France, *terroir* is defined as "an area or terrain, usually rather small, whose soil and microclimate import distinctive qualities to food products" (Barham, 2003). Although the term *terroir* is used to describe the organoleptic characteristics of the wine, now it can be used to describe other foods such as cheese (Bokulich et al., 2016). In the case of cheeses, Grappin and Coulon (1996) define the terroir as "a geographic area characterized by environmental conditions and types of animals, that when exploited by humans, lead to specific products."

3. House microbiota of dairy industry

Cheeses and dairy processing environments are colonized by many microbial communities. During cheese production, the product is coming into contact with many equipment, brine tanks and ripening rooms on its journey from milk to curd to cheese which can all act as potential vectors for microorganisms deriving from raw milk, starter cultures, water, air, salt and human skin (Montel et al., 2014).

The microbial communities colonized the processing environment and equipment is defined as house microbiota, which interacts with the microorganisms present in raw milk, primary and with the secondary and/or adjunct starter cultures inoculated during cheese production (Bokulich and Mills, 2013). Raw milk can act either as a direct source of microbes such as NSLAB and as an indirect source, by enriching the microbiota of the cheese processing environment and equipment (Montel et al., 2014). The successful colonization of these autochthonous microorganisms is due to their ability to cope with environmental conditions such as low temperature, high salt content, humidity, pH, moisture control (brining, pressing), cleaning procedures and competition against other microbes (Bokulich and Mills, 2013; Mounier et al., 2006). The dairy environment mainly consists of bacteria belonging to Firmicutes, Actinobacteria and Proteobacteria phyla, yeasts and molds (Bokulich and Mills, 2013; Montel et al., 2014; Stellato et al., 2015). According to Stellato et al. (2015), there is an equilibrium between dairy products and processing environment where microbial transfer occurs from both parts. Therefore, the dairy-environment relationship can play an important role for the quality and the development of specific characteristics of the final product.

4. Examples of cheeses with characteristic in house microbiota (Table 4.1)

4.1 Pit cheese

An example of Pit cheese is Fossa, which is produced by placing the cheese in a pit of tuffaceous rocks in the Emilia- Romagna and Marche regions of Central Italy (Barbieri et al., 2012). It is a crumbly cheese with a sweet and sometimes piquant taste. The microbial diversity of Fossa cheese is characterized by *Lactococcus lactis* subsp. *lactis*, *Lc. lactis* subsp. *cremoris*, *Streptococcus thermophiles*, *Streptococcus bovis*, *Lb. plantarum*, *Lb. casei*, *Lb. paracasei*, *Lb. rhamnosus* and *Lb. fermentum* (Barbieri et al., 2012; Fontana et al., 2010). Moreover, the pit environment is responsible for the fungal microflora on cheese surface such as *Penicillium* and *Aspergillus* spp. (Santi et al., 2010).

4.2 Cheeses ripening in wooden vats

Wooden vats and shelves are still used for the ripening of some PDO cheeses such as Sicilian Ragusano, French Cantal, Comté, and Reblochon. Many studies have been carried out regarding the biofilms of wooden vats (Didienne et al., 2012; Licitra et al., 2007; Lortal et al., 2009; Scatassa et al., 2015; Settanni et al., 2012). Some examples of cheeses ripening in wooden vats are discussed below.

Ragusano cheese is a semihard brine salted pasta-filata cheese from the Hyblean region of Sicily (Licitra et al., 2007). Raw milk is placed directly in the traditional wooden vat called tina, without the addition of any starter. The fermentation is carried out by indigenous lactic acid bacteria (LAB) present in raw milk, in the inner surface of wooden vats and the environment (Licitra et al., 2007; Lortal et al., 2009). Molecular characterization of tinas showed the predominance of lactic acid bacteria, in particular *Streptococcus thermophilus*, and the presence of thermophilic lactobacilli, lactococci,

Table 4.1 Cheeses microbial ecology studies.

Cheese type	Microbial ecology	References
Austrian Vorarlberger Bergkäse	*Coryneforms, Staphylococcus equorum* and *Halomonas*	Quijada et al. (2018)
Blue-veined Cabrales cheese	*Lactocococcus lactis* subsp. *Lactis*, *Lactococcus garvieae* and *Lactococcus raffinolactis*, *Penicillium roqueforti* and *Geotrichum candidum.*	Flórez and Mayo (2006a)
Bola de Ocosingo	*Streptococcus thermophilus*, *Lactococcus lactis*, *Lactobacillus helveticus*, *L. delbrueckii* and *L. plantarum*	Aldrete-Tapia et al. (2018)
Bukuljiac cheese	*Lactobacillus paracasei* subsp. *paracasei*, *Lactococcus lactis* subsp. *lactis, Enterococcus faecalis, Leuconostoc mesenteroides*	Nikolic et al. (2008)
Caciocavallo of Castelfranco	*S. thermophilus, L. casei*	Giello et al. (2016)
Caciocavallo Pugliese Cheese	*St. thermophilus, Lactobacillus parabuchneri, L. casei,* and *L. paracasei* subsp. *Paracasei*	(Aquilanti et al., 2006; Pasquale et al., 2014)
Caciocavallo Silano PDO cheese	*Lactobacillus paracasei*, *L. fermentum*, and *L. plantarum, Lactobacillus delbrueckii*, *L. helveticus* and *Streptococcus thermophilus*	(Ercolini et al., 2008; Piraino et al., 2005)
Calenzana	*L. plantarum, L. acidophilus*	Casalta et al. (2009)
Camembert cheese	*L. paracasei, Lactobacillus plantarum, Lactobacillus parabuchneri, Lactobacillus fermentum, Lactobacillus acidophilus, Lactobacillus helveticus, Lactobacillus psittaci/ delbrueckii* subsp. *bulgaricus/gallinarum/crispatus* group, *Lactobacillus rhamnosus, L. delbrueckii* subsp. *lactis, Lactobacillus brevis, Lactobacillus kefiri,* and *Lactobacillus perolens, Geotrichum candidum*	(Henri-Dubernet et al., 2008, 2004; Marcellino et al., 2001)

Continued

Table 4.1 Cheeses microbial ecology studies.—cont'd

Cheese type	Microbial ecology	References
Canestrato Pugliese	*Lactococcus* (*Lactococcus cremoris* subsp. *cremoris*), *Lactobacillus* (*Lactobacillus plantarum* and *Lactobacillus sakei*), and *Leuconostoc*	(Aquilanti et al., 2006; Pasquale et al., 2014)
Casín	*Lactococcus lactis* subsp. *lactis*, *Lactobacillus plantarum*, *Corynebacterium variabilis*	Alegría et al. (2009)
Castelmagno PDO cheese	*Streptococcus thermophiles*, *S. agalactiae*, *Lactococcus lactis*, *Lactobacillus plantarum* and *Lb. casei*, *Enterococcus faecium*, *Macrococcus caseolyticus*, *Enterococcus faecalis*, *Streptococcus agalactiae*	(Dolci et al., 2010, 2008)
Cheddar cheese	*Lactobacillus paracasei*, *Lact. plantarum*, *Lact. Rhamnosus*, *Lb. casei*, and *Lb. curvatus*	Fitzsimons et al. (2001)
Comté cheese	*Lactobacillus paracasei* and *Lactobacillus rhamnosus*	Berthier et al. (2001)
Danish Danbo	*Lactococcus* spp., *Corynebacterium Staphylococcus* and *Brevibacterium*, *Debaryomyces*	Ryssel et al. (2015)
Divle Cave	*Psychrobacter glacincola*, *Brevibacterium antiquum*, *Brachybacterium tyrofermentas*, *Micrococcus luteus*, *Kocuria salcicia*, *Microbacterium oryzae*, *Arthrobacter arilaitensis*, *Microbacterium gubbenense*, *Microbacterium halotolerans*, *Lactobacillus paraplantarum*, *Lactococcus lactis* subsp. *lactis*, *Enterococcus faecium*, *Staphylococcus warneri* and *Lactobacillus brevis*, *Penicillium polonicum*, *Penicillium biforme*, *Penicillium roqueforti*, *Penicillium chrysogenum* and *Debaryomyces hansenii.*	Budak et al. (2015)
Feta cheese	*Lactococcus* spp. *Streptococcus macedonicus*, *Lactobacillus plantarum*	Bozoudi et al. (2016)

Gouda cheese	*Lactobacillus paracasei,* Lactobacillus plantarum, *Lactobacillus brevis, Lactobacillus rhamnosus, Pediococcus pentosaceus, Streptococcus salivarius Enterococcus faecalis, Weissella paramesenteroides*	Hoorde et al. (2008)
Grana Padano cheese	*Lactobacillus helveticus, Lactobacillus delbrueckii* subsp. *lactis, Streptococcus thermophilus*, and *Lactobacillus fermentum.*	(Rossetti et al., 2008; Santarelli et al., 2013)
Graviera Kritis	*Lb. brevis* and *Lb. paracasei*	Tsafrakidou et al. (2016)
Herve cheese	*Corynebacterium casei, Psychrobacter* spp., *Lactococcus* lactis ssp. *cremoris, Staphylococcus equorum, Vagococcus salmoninarum, Psychroflexus casei, Lactococcus lactis* ssp. *cremoris, Psychrobacter* spp.,	Delcenserie et al. (2014)
Istrian cheese	*Lactococcus lactis* subsp. *lactis* and *Enterococcus* spp.	Fuka et al. (2010)
Kaşar	*Lactobacillus casei Lactobacillus plantarum* and *Pediococcus* acidilactici	Aydemir et al. (2014)
Liqvan cheese	*Lactococcus lactis, Leuconostoc mesenteroides, L. raffinolactis, Lb. plantarum, L. garvieae, Lb. sakei, Lb. casei and paracaei, Streptococcus* sp. and *Enterobacter* sp.	Ramezani et al. (2017)
Livarot	*Geotrichum candidum, Arthrobacter, Brevibacterium, Corynebacterium,* and *Staphylococcus.*	Larpin-laborde et al. (2011)

Continued

Table 4.1 Cheeses microbial ecology studies.—cont'd

Cheese type	Microbial ecology	References
May bryndza cheese	*Acinetobacter calcoaceticus, Acinetobacter guillouiae, Acinetobacter* sp., *Acinetobacter johnsonii, Citrobacter braakii, Clostridium bartlettii, Corynebacterium callunae, Corynebacterium maris, Enterobacter aerogenes, Enterobacter asburiae, Enterobacter hormaechei, Enterococcus faecium, Enterococcus pallens, Escherichia coli, Haemophilus haemolyticus, Hafnia alvei, Kluyvera cryocrescens, Lactobacillus helveticus, Lactococcus garvieae, Lc. lactis subsp. cremoris, Lc. lactis subsp. lactis, Leuconostoc garlicum, Mannheimia glucosida, Mannheimia haemolytica, Pseudomonas* sp., *Ps. fluorescens, Ps. reactans, Raoultella ornithinolytica, R. terrigena, Rothia arfidiae, Staphylococcus aureus, Staph. epidermidis, Staph. felis, Staph. pasteuri, Staph. sciuri, Staph. xylosus, Streptococcus parauberis, Str. thermophilus Variovorax paradoxus, Alternaria alternata, Ascomycete* sp., *Aspergillus fumigatus, Beauveria brongniartii, Candida xylopsoci, C. inconspicua, Cladosporium cladosporioides, Debaromyces hansenii, Fomes fomentarius, Galactomyces candidus, Gymnoascus reesii, Chaetomium globosum, Kluyveromyces marxianus, Metarhizium anisopliae, Penicillium aurantiogriseum, P. camemberti, P. freii, P. polonicum, P. viridicatum, Pichia kudriavzevii, Sordaria alcina, Trichosporon lactis and Yarrowia lipolytica.*	Pangallo et al. (2014)
Moroccan soft white cheese (Jben)	*Lactobacillus plantarum, Lactobacillus rhamnosus, Lactobacillus paracasei, Lactobacillus brevis, Lactobacillus buchneri, Lactococcus lactis, Lactococcus garvieae, Lactococcus raffinolactis, Leuconostoc pseudomesenteroides, Leuconostoc mesenteroides, Leuconostoc citreum, Enterococcus durans, Enterococcus faecalis, Enterococcus faecium, Enterococcus saccharominimus* and *Streptococcus* spp.	Ouadghiri et al. (2005)

Nicastrese goat's cheese	*Lactococcus lactis, Lactobacillus plantarum, Lactobacillus casei, Lactobacillus brevis, Leuconostoc mesenteroides,* and *Enterococcus faecalis.*	Pino et al. (2017)
Parmigiano Reggiano	*L. paracasei, L. rhamnosus,* and *P. acidilactici. L. helveticus, L. delbrueckii* subsp. *lactis,* and *L. delbrueckii* subsp. *bulgaricus*	Gala et al. (2008)
Pecorino Abruzzese cheese	*Enterococcus faecium* and *Ent. faecalis, Lactobacillus plantarum, Lactobacillus brevis, Lactococcus lactis* subsp. *cremoris.*	Centi et al. (2017)
Pecorino Crotonese	*Lactococcus* lactis and *Streptococcus thermophilus, Lactobacillus rhamnosus*	Randazzo et al. (2010)
Pecorino di Filiano PDO	*Brevibacterium linens, Lactobacillus delbrueckii* subsp. *Bulgaricus, Lactobacillus paracasei* subsp. *paracasei, Leuconostoc lactis* and *Leuconostoc mesenteroides* subsp. *mesenteroides*	Bonomo and Salzano (2012)
Pecorino di Tramonti	*Enterococcus faecium, Lactococcus lactis* subsp. *lactis, L. lactis* subsp. *cremoris, Enterococcus faecalis* and *Enterococcus durans.*	Mormile et al. (2016)
Pecorino Siciliano cheese	*Lactococcus lactis, Streptococcus thermophilus, Enterococcus faecalis* and *Leuconostoc mesenteroides*	Randazzo et al. (2006)
Pico cheese	*Lactococcus, Streptococcus, Acinetobacter, Enterococcus, Lactobacillus, Staphylococcus, Rothia, Pantoea, Lactobacillus paracasei, Leuconostoc pseudomesenteroides, and Enterococcus faecalis, L. lactis, L. casei,*	(Dapkevicius et al., 2015; Domingos-lopes et al., 2017; Riquelme et al., 2015)
Plaisentif cheese	*Lactococcus, Lactobacillus* and *Streptococcus*	Dalmasso et al. (2016)

Continued

Table 4.1 Cheeses microbial ecology studies.—cont'd

Cheese type	Microbial ecology	References
Provolone del Monaco	*Lactococcus lactis, Lactococcus lactis,* *Streptococcus thermophilus,* *Enterococcus faecalis/durans/faecium* *Macrococcus caseolyticus,* *Streptococcus parauberis, Morexella osloensis* *Streptococcus haemolyticus/croceolyticus/warneri/pasteuria* *Weisella species* *Shigella boydiia* *Lactobacillus helveticus/delbrueckii* subsp. *lactis* *Staphylococcus aureus, Rahnella* species *Aeromonas simiae* *Staphylococcus aureus/haemolyticus* *Lactobacillus paracasei/rhamnosus* *Lactococcus garvieae* *Leuconostoc mesenteroides* subsp. *lactis,* *Pediococcus acidilactici*	Aponte et al. (2008)
Ragusano cheese	*Leuconostoc mesenteroidesa Leuconostoc mesenteroides* *Lactobacillus fermentum/plantarum/casei, Lactobacillus fermentum/plantarum/casei/Pediococcus acidilacticia delbrueckii* subsp. *bulgaricus* *Enterococcus sulfurans/faecalis Pediococcus acidilactici* *Streptococcus thermophilus, Enterococcus faecalis/hirae/sulfurans* *Enterococcus hirae, Macrococcus caseolyticus*	(Licitra et al., 2007; Randazzo et al., 2002)

Salers cheese	*Leuconostoc pseudomesenteroides, Lactococcus lactis, Lactobacillus casei, Lactobacillus plantarum, Lactobacillus hilgardii Kluyveromyces Kluyveromyces Candida zeylanoides/parapsilosis/silvae/intermedia/ tropicalis/rugose Candida, zeylanoides/parapsilosis/silvae/intermedia Debaryomyces hansenii Debaryomyces hansenii Saccharomyces cerevisiae/unisporus Saccharomyces cerevisiae/unisporus Pichia guilliermondii*	(Callon et al., 2006; Didienne et al., 2012)
San Simón da Costa cheese	*Lactococcus lactis* subsp. *lactis, Lactobacillus casei* subsp. *casei, Lb. paracasei* subsp. *paracasei, Leuconostoc mesenteroides* and *Enterococcus faecalis.*	González et al. (2015)
Serpa Cheese	*Lactobacillus paracasei/Lactobacillus casei, Debaromyces hansenii* and *Kluveromyces* spp.	(Dos Santos et al., 2017; Gonçalves et al., 2018)
Silter cheese	*St. thermophilus*, *Lc. lactis*, *Ln. mesenteroides, Lb. paracasei.*	Silvetti et al. (2017)
Stilton Cheese	*Lactococcus lactis*, *Enterococcus faecalis*, *Lactobacillus curvatus*, *S. equorum*, and *Lactobacillus plantarum*, *Yarrowia lipolytica*, *Penicillium roqueforti*, *Debaryomyces hansenii*	(Ercolini et al., 2003; Gkatzionis et al., 2014)
Tomme d'Orchies	*Yarrowia lipolytica, Lactococcus* spp., *Streptococcus* spp., *Lactobacillales*, *Corynebacterium*, *BrevibacteriumPsychrobacter*	(Ceugniez et al., 2017a,b)
Vastedda della Valle del Belice" cheese	*Lactococcus lactis and Leuconostoc mesenteroides, Streptococcus thermophilus and Streptococcus gallolyticus* subsp. *macedonicus*	(Cruciata et al., 2019; Reale et al., 2016)

and a few high-GC-content bacteria, such as coryneform bacteria (Licitra et al., 2007; Lortal et al., 2009).

Cantal is also a PDO cheese produced in central France. The raw milk is directly placed into a traditional wooden vat called a gerle without the addition of any starter cultures (Licitra et al., 2017). The gerle biofilm is dominated by lactobacilli, leuconostocs, yeasts and molds. Both wooden vats were free from food-borne pathogens such as *Listeria, Salmonella, Escherichia coli* O157, and *Staphylococcus aureus* even after artificially contaminations of the vat with high levels of *Listeria* and *Staphylococcus* (Lortal et al., 2009). These results demonstrated that wooden vat surfaces are safe and able to inhibit and limit the foodborne pathogens.

Reblochon is a cow's milk smear-ripened cheese produced in France. This cheese is ripening on wooden shelves. Mariani et al. (2007), showed that the biofilm from the wooden shelve is mainly composed of micrococci, corynebacteria, yeasts, and molds but also contained leuconostocs, lactobacilli, enterococci, coagulase-negative staphylococci, and *Pseudomonas*. Wooden shelves have two roles, the first one is related to the microbial diversity of the cheese surface and the second one is related to hydric exchange with the cheese curd and the cellar air humidity (Lortal et al., 2014).

Caciocavallo Palermitano cheese, is a "pasta-filata" product, produced in Palermo region of Sicily, Italy. The traditional production of caciovallo cheese is carried out employing the wooden dairy equipment without the addition of LAB (Di Grigoli et al., 2015). Its LAB biodiversity is characterized by *P. acidilactici, Pediococcus pentosaceus, E. casseliflavus, E. gallinarum, E. faecalis, L. rhamnosus, L. casei, Lactobacillus delbrueckii, L. fermentum* and *Lactobacillus paracasei.*

4.3 Cheese ripening in goat's skin bag

Wood is not the only the container for the production of artisanal cheeses. Containers made from goat or sheep skin are also used for the production of traditional cheeses such as Tulum cheese produced in Turkey. Tulum cheese has a white or creamy color, a high fat content, and a crumbly, semihard texture (Hayaloglu et al., 2007). It takes its name "tulum" from goat's or sheep's skin bags in which the cheese is ripened which are responsible for the characteristic texture and flavor of the cheese. According to the study carried out by Hayaloglu et al. (2007) and Karabey et al. (2018) the microbial diversity of the cheese is characterized by *Lactobacillus, Streptococcus, Enterococcus, Lactococcus, Leuconostoc, Weissella* and *Pediococcus.*

4.4 Cheese ripening in caves

Cabrales is one of the most famous Spanish blue-veined cheeses. The cheese is ripened in natural caves with constant humidity (90%–95%) and temperature (9–12°C) (Flórez and Mayo, 2006a). This allows *Penicillium roqueforti* to enter and grow in the cheese matrix, providing the product its characteristic appearance and sensory properties. Microbiologically, this cheese offers a complex habitat in which prokaryotic and eukaryotic populations interact and develop throughout manufacturing and ripening. The microbial diversity of the cheese is characterized by *Lactocococcus lactis* subsp. *lactis, Lactococcus garvieae, Lactococcus raffinolactis, Penicillium roqueforti, Geotrichum candidum, D. hansenii, K. lactis* and *P. fermentans* (Álvarez-Martín et al., 2007; Flórez et al., 2006; Flórez and Mayo, 2006a, 2006b).

Divle Cave (DC) cheese is a semihard cheese made from raw ewes' milk. It took its name due to the fact that it is ripening in Divle cave in Karaman, where the average temperature is 5−10°C with a humidity of 85%−90%. The microbial diversity of Divle Cave cheese is characterized by *Staphylococcus equorum* subsp. *equorum, Lactobacillus paraplantarum, Lactococcus lactis* subsp. *lactis, Enterococcus faecium, Staphylococcus warneri, Lactobacillus brevis, Penicillium polonicum, Penicillium biforme, Penicillium roqueforti, Penicillium chrysogenum* and *Debaryomyces hansenii* (Budak et al., 2015; Budak and Aykas, 2018).

5. Interaction of fermented sausages with meat industry

Fermented sausages are part of traditional products which are characterized by great diversity between different countries in terms of manufacturing methods and organoleptic characteristics. Typical or traditional foods of any region have their own peculiar characteristics that are deeply rooted in tradition and linked to the territory and which arise from the use of local ingredients and specific production techniques (Aquilanti et al., 2007). Fermented sausages have a long history in Europe, originating from Mediterranean countries during Roman times. Production then spread to Germany, Hungary and others countries including the United States, Argentina and Australia (Talon et al., 2007b). They can be defined as products made by mixing minced meat and fat, salt, spices and other seasonings and herbs. The mixture is then stuffed into casings and then ripened and dried (cured) and, sometimes, smoked. The fermentation is driven by indigenous microflora, without the addition of starter cultures. The ripening time of sausages affects the type of microflora that develops. Sausages with a short ripening time have more lactobacilli from the early stages of fermentation, and at the end of ripening while sausages with longer ripening times contain higher numbers of coagulase-negative cocci (CNC) in the early stages of fermentation (Urso et al., 2006). These two groups of bacteria are determined the several biochemical and physical reactions taking place during the fermentation and ripening of sausages (Rantsiou and Cocolin, 2006).The organoleptic characteristics of fermented sausages are dependent on the quality of the raw ingredients materials, the specific conditions of the processing and ripening environment, and the composition of the microbial population (Aquilanti et al., 2007).

6. House microbiota of meat industry

The fermentation and ripening of traditional sausages only relies on indigenous microflora whose composition is variable. This indigenous microbial is affected by pH, redox potential, water activity, salt concentration, nutrient availability and by relationships among the different bacterial populations (Rantsiou et al., 2006). Many authors support the belief that the microorganisms present in traditional sausages relies on natural contamination of raw materials, the manufacturing environment and the equipment (Talon et al., 2007b). For instance, the contamination of raw ingredients and the environmental conditions during production increase the population of microbial flora (Cocolin et al., 2001a; Rebecchi et al., 1998). Greppi et al. (2015) showed that microorganisms isolated from environmental samples were also detected either in the raw materials or in the final product. Results of this study demonstrate that raw materials are a source of microorganisms that can enter in the production facility and also that the processing environment is a source of continuous "inoculation," during fermentation and ripening, with strains that may have important technological characteristics and influence the organoleptic characteristics of the final product. Each processing plant has a distinct house flora, consist of technological strains for the fermentation and flavor development of sausage, as well as of spoilage bacteria that can cause negative changes in appearance, odor, flavor and consistency of the final product due to their metabolic activity and pathogenic flora. Therefore, the characterization of the house microbiota is important because safety (food-borne pathogens), acceptability (spoilage bacteria) and sensorial quality/organoleptic characteristics (technological strains) of the final sausage product depend totally on it.

Most of the studies that have been carried out have dealt with the microbial ecology of traditional fermented sausages and only few studies have focused on the microbial diversity of the processing environment such as surfaces and equipment (Chevallier, 2006; Morot-bizot et al., 2006; Talon et al., 2007b). Results of different studies showed that the surfaces of processing environment as well as the equipment use for the production of fermented foods are contaminated by resident microflora of *Staphylococcus, Kocuria* and LAB (Chevallier, 2006; Morot-bizot et al., 2006; Talon et al., 2007b). Furthermore Lebert (2007) was also demonstrated that cold rooms and mixing machines have the lowest levels of contamination, while knives, table surfaces and stuffing machines have higher levels of contamination.

The first studies on the microbial ecology of fermented sausages date back to 1970 (Rantsiou et al., 2005). According to the results of several studies, two main groups of microorganisms as being the main organisms of technological interest responsible for the transformations involved during fermentation and ripening of sausages and contribute for the formation of the sensory profile of the final product. These are the lactic acid bacteria (LAB) and the gram-positive catalase-positive cocci (GCC), which includes both coagulase-negative staphylococci (CNS) and *Kocuria*. Molecular analysis of fermented sausages during ripening showed that after 3 days of ripening, LAB and CNS were the most abundant group of microorganisms (Cocolin et al., 2001a).

LAB are responsible for the decrease in pH due to the production of lactic acid, which contributes to the firmness of the dry fermented sausages and ensures their stability and safety of the final product by preventing the growth of pathogens and spoilage microorganisms (De Vuyst and Leroy, 2007). In addition, LAB are responsible for the production of volatile compounds through the fermentation of carbohydrates and for the "tangy" flavor of sausages (Kumar et al., 2015). In addition, results of different studies showed that *Lactobacillus* spp. isolated from fermented sausages possess proteolytic activity on muscle sarcoplasmatic proteins (Fadda et al., 1998; Pereira et al., 2001). Furthermore, some LAB isolated from fermented sausages are able to produce bacteriocins which are active against foodborne pathogens such as *Listeria monocytogenes,* thereby increasing the competitiveness of the producer cells and the safety of the final products (De Vuyst and Leroy, 2007). Within the group of LAB, *L. sakei, L. curvatus*, and *L. plantarum* are the main species most frequently isolated from fermented sausages, while other lactobacilli, such as *Lb. versmoldensis, Lb. brevis, Lb. farciminis, Lb. alimentarius, Weissella* species, pediococci, enterococci and leuconostocs, usually occur in significantly lower numbers.

In fermented sausages, CNS are mainly contribute to the development and stability of the red color through nitrate reductase activity that leads to the formation of nitrosomyoglobin and avoid the rancidity of sausages due to their antioxidant activities (Barrière et al., 2001, 2002). They enhance the flavor of fermented sausages mainly through the release of amino acids, peptides and free fatty acids by proteolysis and lipolysis (Talon et al., 2007b). S. *xylosus, S. carnosus, S. equorum* and *S. saprophyticus* are the predominant species of CNC involved in the fermentation of naturally fermented sausages, but many other species have been associated with fermented sausages such as *S. succinus, S. warneri, S. vitulinus, S. pasteuri, S. epidermidis, S. lentus* and *S. haemolyticus, Staphylococcus carnosus, Staphylococcus simulans, Staphylococcus epidermidis, Staphylococcus pasteuri* (Aquilanti et al., 2007; Cocolin et al., 2001a; 2011, Mainar et al., 2016; Morot-bizot et al., 2006; Talon and Leroy, 2011). Regarding the Korucia species in fermented sausages, *Kocuria varians* and *Kocuria kristinae* were mainly isolated (Raghupathi et al., 2016).

Yeast and molds also play an important role in sausage fermentation and contributes to the formation of characteristic flavor due to their lipolytic and proteolytic activities and surface appearance through the formation of a superficial film, which exerts a protective action against both excessive dehydration and the oxidation of the lipid fraction due to oxygen and light and also prevents the development of sticky or slimy characteristics on the surface (Cocolin et al., 2006; Franciosa et al., 2018; Gardini et al., 2001). Some of the yeast that have been isolated from fermented sausages belongs to the following genera *Debaryomyces hansenii, Candida, Pichia, Rhodotorula, Yarrowia lipolytica, Williopsis, Hansenula, Torulopsis, Trichosporum* and *Cryptococcus* (Coppola et al., 2000; Rantsiou et al., 2005). Molds such as xerotolerant and xerophilic fungi are also able to grow on the surface of fermented meat products due to the low water activity and high salt concentration (Franciosa et al., 2018). The most frequently isolated mold from the surface of fermented sausages belongs to the genus Penicillum such as *P. nalgiovense* and *P. chrysogenum* (Franciosa et al., 2018).

7. Examples of traditional sausages with characteristic house microbiota (Table 4.2)

7.1 Serbia

The Petrovac sausage (Petrovská klobása), is a seasonal artisanal and indigeneous dry fermented sausage, which is produced in a traditional way in rural households in the Municipality of Bački Petrovac in Serbia (Talon et al., 2007a). The microbial diversity of Petrovac sausage is characterized by *Lactobacillus sakei, Leuconostoc mesenteroides, Pediococcus pentosaceus, Enterococcus durans* and *Enterococcus caseliflavus* (Danilovi et al., 2011; Jankovic et al., 2015; Talon et al., 2007a).

Another traditional fermented dry sausage produced in central Serbian region, Šumadija is called levačka. The microbial ecology of this product is characterized by *Pediococcus pentosaceus, Leuconostoc mesenteroides, Lb. curvatus, Lb. sakei and Ln. carnosum* (Borovic et al., 2015).

7.2 Portugal

The most famous traditional fermented sausage of Portugal is Alheira. Alheira has a characteristic shape, horse-shoe and is produced mainly in the Northern area of Trásos-Montes (Albano et al., 2008; Esteves et al., 2007; Ferreira et al., 2007). The microbial diversity of Alheira is characterized by *Lb. plantarum, Ec. faecalis, Lb. paraplantarum, Lb. brevis, Lb. rhamnosus, Lb. sakei, Lb. zeae, Lb. paracasei, Leuconostoc (Leuc.) mesenteroides, Pediococcus (Pc.) pentosaceus, Pc. acidilactici, Weissella (Ws.) cibaria, Ws. viridescens* and *Ec. faecium*. Regarding the CNC group only one specie was identified, *Macr. caseolyticus.*

7.3 Spain

The microbiota of Spanish fermented sausages was also investigated by many researchers. In a study carried out by Aymerich et al. (2006), three different Spanish artisan fermented meat products were studied chorizo, fuet and salsicho. Results of the study showed that the microbial diversity of all three sausages is characterized by *Lb. sakei, Lb. curvatus Lc. lactis, Lb. plantarum, Leuc. mesenteroides, Pediococcus pentosaceus, Pediococcus acidilactici, Staph. Xylosus, Staph. Intermedius, Staph. Saprophyticus, Staph. Carnosus, Staph. Epidermidis* and *K. varians*

Botillo and Androlla are also traditional dry fermented sausages produced mainly in North- Western Spain. The microbiota of these two sausages was studied by Fontán et al. (2007b, 2007a) and showed that both products have similar microbial diversity. The microbial ecology using culture dependent methods showed that *Lb. sakei was* predominated among the isolates, followed by *Lb. alimentarius, Lb. curvatus, Lb. plantarum, Lactobacillus farciminis*, and *Leuc. mesenteroides* subsp. *mesenteroides.* Regarding the CNC population, *Staphylococcus saprophyticus* was the most frequently isolated, followed by *Staph. Xylosus, Staph. Lentus, Staph. Cohnii, Staph.*

Table 4.2 Fermented sausages microbial ecology studies.

Type of fermented sausage	Microbial ecology		References
	LAB	CNC	
Italy			
Salame Milano	*Lb. sakei, Lb. plantarum*	*S. xylosus, S. sciuri*	Rebecchi et al. (1998)
Soppressa	*Lb. sakei, Lb. curvatus*		Andrighetto et al. (2001)
Piedmontese sausage	*Lb. curvatus, Lb. sakei, Lb. piscium, Lb. algidus, Leuc. Carnosum, Leu. Gelidum, Leu. mesenteroides*	*S. xylosus, S. succinus, S. equorum*	Greppi et al. (2015)
Friuli Venezia Giulia	*Lc. Lactis* spp *lactis, Lb. sakei, Lb. curvatus, Lb. plantarum, Leuconostoc* spp., *Pediococcus* spp.	*S. xylosus, S. warneri, S. pasteuri, S. saprophyticus, S. epidermidis, S. cohnii, S. equorum, S. carnosus, M. caseolyticus*	(Cocolin et al., 2001a, 2001b; Comi et al., 2005; Iacumin et al., 2006; Rantsiou et al., 2005; Urso et al., 2006)
Salame mantovano	*Lb. sakei, Lb. curvatus, Lb. plantarum, Lb. paraplantarum, Leu. Citreum, W. hellenica, Lb. algidus, Leu. Mesenteroides, Lb. paracasei, Lb. salivarius, Lb. fermentum*	*S. xylosus, S. saprophyticus, S. epidermidis, K. varians*	(Cocolin et al., 2009; Pisacane et al., 2015)
Salame piacentino	*Lb. acidophilus, Lb. helveticus, Lb. sakei, Lb. antri, Lb. oris, Lb. vaginalis, Lb. brevis, Lb. panis, Lb. versmoldensis, Lb. paracasei, Lb. curvatus, Lb. paralimentarius, Lb. frumenti, Lb. plantarum, Lb. reuteri*	*S. auricularis, S. arlettae, S. caseolyticus, S. carnosus, S. gallinarum, S. wameri, S. condiment, S. nepalensis, S. lentus, S. sciuri, S. saprophyticus, S. equorum, S. simulans, S. succinus, S. epidermidis, S. cohnii, S. hominis, S. pasteuri, S. xylosus*	(Di Cagno et al., 2008; Połka et al., 2015)
Ciauscolo	*Lb. sakei, Lb. curvatus, Lb. plantarum, P. pentosaceus*	*S. xylosus, S. equorum, S. saprophyticus*	(Aquilanti et al., 2007; Federici et al., 2014; Silvestri et al., 2007)

Continued

Table 4.2 Fermented sausages microbial ecology studies.—cont'd

Type of fermented sausage	Microbial ecology		References
	LAB	**CNC**	
Salame Napoli	*Lb. sakei, Lb. curvatus, Lb. alimentarius, Lb casei* spp *tolerans, Lb. plantarum*	*S. saprophyticus, S. chromogenes, S. hominis, S.warneri, S. epidermidis, S. capitis, S. simulans*	(Coppola et al., 1995, 2000; Mauriello et al., 2000; Moschetti et al., 1997)
Soppressata del Vallo di Diano	*Lb. sakei, Lb. curvatus*	*S. xylosus, S. equorum, S. succinus*	Villani et al. (2007)
Salame di Senise	*Lb. sakei, Lb. curvatus, Lb. casei, Lc. Lactis* subsp *lactis*	*S. saprophyticus, S. equorum, S. succinus*	Baruzzi et al. (2006)
Ventricina	*Lactobacillus* sakei, *Lb. plantarum, Weissella* hellenica and *Leuconostoc* mesenteroides		Tremonte et al. (2017)
Spain			
Chorizo	*Lb. brevis, Lb. curvatus, Lb. sakei, Lc. Lactis, P. acidilactici, P. pentosaceus, Leu. mesenteroides*	*S. xylosus, S. intermedius, S. equorum, S. saprophyticus*	(Aymerich et al., 2006, 2003; Benito et al., 2007; Fonseca et al., 2013; Garcı;a-Varona et al., 2000; Martín et al., 2006)
Salsichon	*Lb. brevis, Lb. curvatus, Lb. plantarum, Lb. sakei, lc. Lactis, P. acidilactici, P. pentosaceus, Leu. mesenteroides*	*S. epidermidis, S. xylosus, S. warneri, K. varians*	(Aymerich et al., 2006; Benito et al., 2007; Fernández-López et al., 2008; Martín et al., 2006)

Fuet	*Lb. sakei, Lb. plantarum, Lb. curvatus*	*S. carnosus, S. epidermidis, S. xylosus, S. warneri, K. varians*	(Aymerich et al., 2006, 2003; Martín et al., 2006)
Botillo	*Lb. alimentrarius, Lb. curvatus, Lb. plantarum, Lb. farcinimis, Leuc.*		Fontán et al. (2007a)
Androlla	*Lb. sake, Lb. curvatus, Lb. plantarum*		Fontán et al. (2007b)
Portugal			
Alheira	*Lb. sakei, Lb. curvatus, Lb. brevis, Lc. Lactis, Lb. plantarum, Lb. paraplantarum Lb. rhamnosus, Lb. zeae, Lb. paracasei, Ent. faecalis, Ent. faecium, Leuc. mesenteroides, Ped. pentosaceus, Ped. acidilactici, W. cibaria, W. viridescens*		Albano et al. (2008)

Equorum, Staph. epidermidis, Staph. sciuri and *Staph. Capitis, Staph. saprophyticus, M. luteus, Micrococcus lylae, K. varians* and *K. krist.*

7.4 Greece

Salami aeros is a naturally fermented Greek dry sausage. Results of different studies carried out regarding the diversity of Greek fermented meat products showed that most of the isolates belonged to *Lb. sakei/curvatus, Weissella genus, Lb. plantarum, Enterococcus faecium, Staph. saprophyticus Staph. xylosus* and *Staph. simulans* (Drosinos, 2005; Drosinos et al., 2007; Rantsiou et al., 2006; Samelis et al., 1994).

7.5 Italy

There are many fermented meat produces produced in Italy. Some examples of traditional sausages following the ancient traditions of the different regions are discussed below.

Salame Napoli, a traditional product from the Campania region. The microbiota of salame Napoli was studied by Coppola et al. (2000) and is characterized by *Lb. sakei Lb. curvatus, Lactobacillus alimentarius, Lactobacillus casei* ssp. *Tolerans, Lb. plantarum. Staph. xylosus* was the most frequently isolate of CNC family, followed by *Staph. saprophyticus, Staphylococcus chromogenes, Staphylococcus hominis, Staph. warneri, Staphylococcus lugdunensis*, and *Staph. epidermidis*.

Soppressata molisana is another traditional fermented meat product from the Molise region which has also been studied (Coppola et al., 1997, 1998). The LAB group was dominated by *Lb. sakei*, followed by *Lb. plantarum, Lactobacillus paracasei* subsp. *paracasei, Weissella viridescens, Lactobacillus coryniformis* ssp. *torquens, Lactobacillus paralimentarius, Lactobacillus brevis, Lactobacillus graminis* and *Lb. curvatus. Staph. Xylosus* was the most dominate specie of CNC group followed by *Staph. simulans, Staph. equorum* and *Staphylococcus kloosii.*

Soppressata del Vallo di Diano, a traditional fermented meat product manufactures in the Campania region was studied by Villani et al. (2007). The main LAB and CNS species identified were *Lb. sakei, Lb. curvatus, Staph. xylosus, Staph. succinus*, and *Staph. Equoru.*

Salame di Senise, salsiccia and soppressata produced in Basilicata, a region in Southern Italy were also investigated (Baruzzi et al., 2006; Bonomo et al., 2008; Parente et al., 2001).The results obtained revealed that lactobacilli were the dominant microbiota and is characterized by *Lb. sakei, Lb. curvatus* and *Lb. casei, Lb. plantarum Leu. carnosum, Leuconostoc gelidum* and *Leu. Pseudomesenteroides*. Regarding the CNS community, Blaiotta et al. (2004) showed that is characterized by *Staph. Xylosus, Staphylococcus pulvereri, Staphylococcus vitulus, Staph. saprophyticus, Staph. Equorum.*

Salsiccia sarda is a traditional fermented meat product produced in Sardinia. Mangia et al. (2007, 2008) studied the microbial diversity of this product. Results of both studies showed that the microbial ecology is consist of *Lb. sakei, Lb. curvatus, Lb.*

plantarum, *Staph. xylosus*, *Staphylococcus* spp., *Kocuria varians*, and *Micrococcus* spp.

Di Cagno et al. (2008) studied the microbial ecology of three fermented sausages produced in Northen Italy exclusively from pig breeds located in the same or adjacent Italian regions, Varzi, Brianza and Piacentino. The following species were identified for each sausage: Varzi, *Lactobacillus curvatus* and *Lactobacillus sakei*; Brianza, *L. sakei, L. curvatus* and *Pediococcus pentosaceus*; and Piacentino, *L. sakei, L. curvatus* and *Lactobacillus coryneformis*. Moreover S. xylosus was present in all three sausages.

Ciauscolo is a fermented sausage manufactured in Central Italy. The microbial diversity of ciauscolo is characterized by to *Lb. sakei* and *Lb. curvatus, Lb. plantarum, Pediococcus pentosaceus, Staph. xylosus, Staph. equorum, Staph. Saprohyticus* (Aquilanti et al., 2007; Silvestri et al., 2007).

7.6 Methods to study the microbial ecology of fermented foods

There is an increase interest to study and preserve the diversity of microorganisms involved in the fermentation of food products and try to correlate the occurrence of certain microorganisms with specific flavor or organoleptic characteristics of the final product. Therefore, due to the technological and economic importance of these strains there are different methods to the study to the microbial diversity of fermented food products by using conventional culturing and culture-independent molecular techniques.

The main approaches used today to study the microbial ecology of fermented food products and processing environments can be divided into three subcategories (1) cultivation and phenotypic characterization, (2) cultivation followed by molecular characterization, and (3) culture-independent methods (Cocolin et al., 2013; Quigley et al., 2011). Fig. 11.3 shows a schematic outline of most frequently approaches used to study the microbial diversity of fermented food products and processing environment. Traditional microbiological methods such as plate counts, biochemical identification cannot give the correct view of microbial diversity and therefore new molecular approaches have been introduced, allowing scientists to overcome the limitations of classical methods (Cocolin et al., 2017; Filippis et al., 2018; Rantsiou and Cocolin, 2006). Most of the studies were based on RAPD (Random Amplified Polymorphic DNA), 16S rDNA sequencing, PCR-DGGE (Denaturing Gradient Gel Electrophoresis) analysis and REA-PFGE (Restriction Endonucleases Analysis - Pulsed Field Gel Electrophoresis) analysis. Nowadays, high-throughput sequencing (HTS) techniques are mainly used to study the microbial ecology of food products (Filippis et al., 2017; Fontana et al., 2016). Compared to the tradition sequencing approach, Sanger, HTS could be performed in mixed nucleic acid molecules from a complex ecosystem and therefore can lead to detailed profile of the present microbial diversity (Cocolin et al., 2017). The molecular approaches based on the use of metagenomes combined with HTS technologies now offer an unique opportunity to profile dominant as well as subdominant microbial populations of fermented food products or any other environmental niche on a large scale (Ercolini, 2013; Parente et al., 2016).

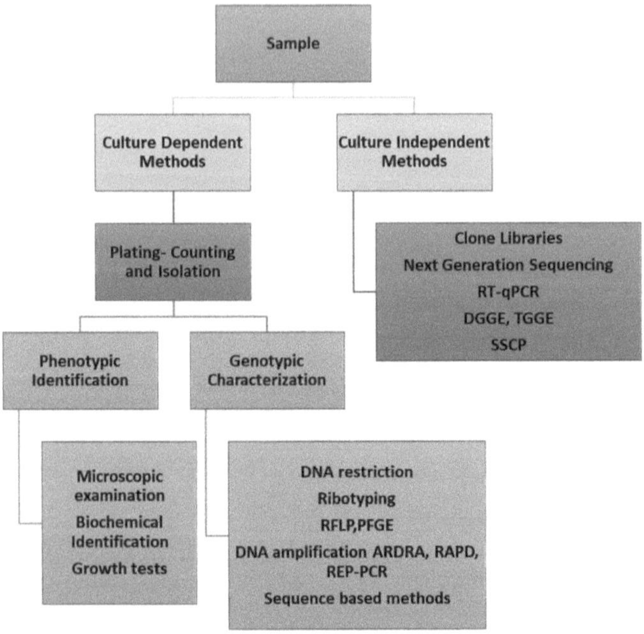

Figure 4.3 Schematic outline of most frequently approaches used to study the microbial diversity of fermented food products and processing environment.

For instance, Wolfe et al. (2014) and Quigley et al. (2012) used these technologies to study the microbial ecology of cheese rind communities in different countries and from Irish artisanal cheeses, respectively. HTS technology can also be employed to study the resident microbial diversity in food processing environments and production facilities. For instance it can be used to investigate the beneficial microbes that may be involved in the fermentative process of cheese (Bokulich and Mills, 2013; Calasso et al., 2016; Stellato et al., 2015), sourdough (Minervini et al., 2015), and fermented beverages (Bokulich et al., 2013, 2012). On the other hand, the processing environment may also be a primary source of foodborne pathogens and spoilage microorganisms (Bokulich et al., 2015; Stellato et al., 2015). The knowledge on structure and dynamics of the microbial populations during fermentation stages, including maturation processes can give ample information for product optimization or new product development.

It has to be noted though that food fermentations represent live, usually highly active and continuously changing environments where dominant viable cells are most likely to play a key role in the development of the physicochemical characteristics of the matrix (milk, meat, must, etc.). Therefore, it is of equal importance for food scientists beside the scouting for new metagenomes, to isolate, identify and characterize the dominant microbiota. In addition, the metagenome analysis should be looked in view of new genera or even new species and to be followed by comprehensive studies/efforts to grow new, previously uncultured microbes of potential biotechnological usefulness. While metagenomics is a powerful technique, using it solely to study microbiomes is limited in value. Recently, studies seek further information by combinatorial analysis of metagenomic, metabolomic proteomic and metatranscriptomic data to network potentially linked properties and functions (for review see (Aguiar-Pulido et al., 2016)). Also Chong and Xia (2017) in their recent review conclude that bioinformatics integrating statistical analysis, visualization and biological information in user-friendly interfaces can greatly facilitate the integration of metabolomics within current microbiome studies.

All these studies receive further attention if applied to other microbial communities such as the human microbiome where compositional similarities as well as functional roles and interplays can be of paramount importance in health, wellbeing and longevity (Wolfe and Dutton, 2015).

8. Conclusion

This chapter described the microbial diversity of naturally fermented dairy and meat products, and the impact of processing environment and equipment on its growth. The production of traditional fermented meat or dairy products is a complex biological phenomenon affected by the action of indigenous microbes. These traditional manufacturing practices lead to a great variability in the quality and flavor of the products. Therefore, understanding the metabolic activities, diversity and behavior of microorganisms during food fermentations is a very interesting and challenging task. This will help the producers to reach high quality of their products in terms of safety, shelf life and organoleptic characteristics, as the most promising strains for starter cultures are those which are isolated from the indigenous microbiota of traditional products. These strains are well adapted to the processing environment and to the specific manufacturing process and thus are capable of dominating the microbiota of products.

Funding

Dr D. Tsaltas has been funded by INTERREG Greece—Cyprus 2014—20 program ("AGROID – ΑΓΡΟΤΑΥΤΟΤΗΤΑ" Project). The project is funded by the European Union and National Funds of the participating countries.

References

Afshari, R., Pillidge, C.J., Dias, D.A., Osborn, A.M., Afshari, R., Pillidge, C.J., Dias, D.A., Osborn, A.M., 2018. Cheesomics : the future pathway to understanding cheese flavour and quality. Crit. Rev. Food Sci. Nutr. 0, 1—15. https://doi.org/10.1080/10408398.2018.1512471.

Aguiar-Pulido, V., Huang, W., Suarez-ulloa, V., Cickovski, T., Mathee, K., Narasimhan, G., 2016. Metagenomics, metatranscriptomics, and metabolomics approaches for microbiome analysis. Evol. Bioinf. 12, 5–16. https://doi.org/10.4137/EBO.S36436.TYPE.

Albano, H., Henriques, I., Correia, A., Hogg, T., Teixeira, P., 2008. Characterization of microbial population of 'Alheira' (a traditional Portuguese fermented sausage) by PCR-DGGE and traditional cultural microbiological methods. J. Appl. Microbiol. 105, 2187–2194. https://doi.org/10.1111/j.1365-2672.2008.03947.x.

Aldrete-tapia, A., Escobar-ramírez, C.M., Tamplin, M.L., 2018. Characterization of bacterial communities in Mexican artisanal raw milk "Bola de Ocosingo" cheese by high-throughput sequencing bioinformatic analysis of sequences. Front. Microbiol. 9, 1–7. https://doi.org/10.3389/fmicb.2018.02598.

Alegría, Á., Álvarez-martín, P., Sacristán, N., Fernández, E., Delgado, S., Mayo, B., 2009. International Journal of Food Microbiology Diversity and evolution of the microbial populations during manufacture and ripening of Casín, a traditional Spanish, starter-free cheese made from cow's milk. Int. J. Food Microbiol. 136, 44–51. https://doi.org/10.1016/j.ijfoodmicro.2009.09.023.

Álvarez-Martín, P., Flórez, A.B., López-Díaz, T.M., Mayo, B., 2007. Phenotypic and molecular identification of yeast species associated with Spanish blue-veined Cabrales cheese. Int. Dairy J. 17, 961–967. https://doi.org/10.1016/j.idairyj.2006.11.005.

Andrighetto, C., Zampese, L., Lombardi, A., 2001. RAPD-PCR characterization of lactobacilli isolated from artisanal meat plants and traditional fermented sausages of Veneto region (Italy). Lett. Appl. Microbiol. 33, 26–30.

Aponte, M., Fusco, V., Andolfi, R., Coppola, S., 2008. Lactic acid bacteria occurring during manufacture and ripening of Provolone del Monaco cheese: detection by different analytical approaches. Int. Dairy J. 18, 403–413. https://doi.org/10.1016/j.idairyj.2007.10.011.

Aquilanti, L., Aquila, L.D., Zannini, E., Zocchetti, A., Clementi, F., 2006. Resident lactic acid bacteria in raw milk Canestrato Pugliese cheese. Lett. Appl. Microbiol. 43, 161–167. https://doi.org/10.1111/j.1472-765X.2006.01935.x.

Aquilanti, L., Santarelli, S., Silvestri, G., Osimani, A., Petruzzelli, A., Clementi, F., 2007. The microbial ecology of a typical Italian salami during its natural fermentation. Int. J. Food Microbiol. 120, 136–145. https://doi.org/10.1016/j.ijfoodmicro.2007.06.010.

Aydemir, O., Harth, H., Weckx, S., Dervişoğlu, M., Vuyst, L. De, 2014. Microbial communities involved in Kaşar cheese ripening. Food Microbiol. 46, 587–595. https://doi.org/10.1016/j.fm.2014.10.002.

Aymerich, T., Martín, B., Garriga, M., Hugas, M., 2003. Microbial quality and direct PCR identification of lactic acid bacteria and nonpathogenic staphylococci from artisanal low-acid sausages. Appl. Environ. Microbiol. 69, 4583–4594. https://doi.org/10.1128/AEM.69.8.4583.

Aymerich, T., Martin, B., Garriga, M., Vidal-Carou, M.C., Bover-Cid, S., Hugas, M., 2006. Safety properties and molecular strain typing of lactic acid bacteria from slightly fermented sausages. J. Appl. Microbiol. 100, 40–49. https://doi.org/10.1111/j.1365-2672.2005.02772.x.

Barbieri, E., Schiavano, G.F., Santi, M. De, Vallorani, L., Casadei, L., Guescini, M., Gioacchini, A.M., Rinaldi, L., Stocchi, V., Brandi, G., 2012. Bacterial diversity of traditional Fossa (pit) cheese and its ripening environment. Int. Dairy J. 23, 62–67. https://doi.org/10.1016/j.idairyj.2011.09.009.

Barham, E., 2003. Translating terroir: the global challenge of French AOC labeling. J. Rural Stud 19, 127–138.

Barrière, C., Centeno, D., Lebert, A., Leroy-Setrin, S., Berdague, J.L., Talon, R., 2001. Roles of superoxide dismutase and catalase of *Staphylococcus xylosus* in the inhibition of linoleic acid oxidation. FEMS Microbiol. Lett. 201, 181–185.

Barrière, C., Brückner, R., Centeno, D., Talon, R., 2002. Characterisation of the katA gene encoding a catalase and evidence for at least a second catalase activity in *Staphylococcus xylosus*, bacteria used in food fermentation. FEMS Microbiol. Lett. 216, 277–283.

Baruzzi, F., Matarante, A., Caputo, L., Morea, M., 2006. Molecular and physiological characterization of natural microbial communities isolated from a traditional Southern Italian processed sausage. Meat Sci. 72, 261–269. https://doi.org/10.1016/j.meatsci.2005.07.013.

Benito, M.J., Martín, A., Aranda, E., Pérez-Nevado, F., Ruiz-Moyano, S., Córdoba, M.G., 2007. Characterization and selection of autochthonous lactic acid bacteria isolated from traditional Iberian dry-fermented salchich on and chorizo sausages. J. Food Sci. 72, 193–201. https://doi.org/10.1111/j.1750-3841.2007.00419.x.

Berthier, F., Beuvier, E., Dasen, A., Grappin, R., 2001. Origin and diversity of mesophilic lactobacilli in Comt e as revealed by PCR with repetitive and species-specific primers. Int. Dairy J. 11, 293–305.

Blaiotta, G., Pennacchia, C., Villani, F., Ricciardi, A., Tofalo, R., Parente, E., 2004. Diversity and Dynamics of Communities of Coagulase-Negative Staphylococci in Traditional Fermented Sausages, pp. 271–284. https://doi.org/10.1111/j.1365-2672.2004.02298.x.

Bokulich, N.A., Mills, D.A., 2013. Facility-specific "house" microbiome drives microbial landscapes of artisan cheesemaking plants. Appl. Environ. Microbiol. 79, 5214–5223. https://doi.org/10.1128/AEM.00934-13.

Bokulich, N.A., Bamforth, C.W., Mills, D.A., 2012. Brewhouse-resident microbiota are responsible for multi-stage fermentation of American Coolship Ale. PLoS One 7. https://doi.org/10.1371/journal.pone.0035507.

Bokulich, N.A., Ohta, M., Richardson, P.M., Mills, D.A., 2013. Monitoring seasonal changes in Winery-resident microbiota. PLoS One 8. https://doi.org/10.1371/journal.pone.0066437.

Bokulich, N.A., Bergsveinson, J., Ziola, B., Mills, D.A., 2015. Mapping Microbial Ecosystems and Spoilage-Gene Flow in Breweries Highlights Patterns of Contamination and Resistance, pp. 1–21. https://doi.org/10.7554/eLife.04634.

Bokulich, N.A., Lewis, Z.T., Boundy-mills, K., Mills, D.A., 2016. A new perspective on microbial landscapes within food production. Curr. Opin. Biotechnol. 37, 182–189. https://doi.org/10.1016/j.copbio.2015.12.008.

Bonomo, M.G., Salzano, G., 2012. Microbial diversity and dynamics of Pecorino di Filiano PDO, a traditional cheese of Basilicata region (Southern Italy). Int. J. Dairy Technol. 65, 1–11. https://doi.org/10.1111/j.1471-0307.2012.00860.x.

Bonomo, M.G., Ricciardi, A., Zotta, T., Parente, E., Salzano, G., 2008. Molecular and technological characterization of lactic acid bacteria from traditional fermented sausages of Basilicata region (Southern Italy). Meat Sci. 80, 1238–1248. https://doi.org/10.1016/j.meatsci.2008.05.032.

Borovic, B., Velebit, B., Veskovic, S., Lakicevic, B., 2015. Molecular characterization of lactic acid bacteria in leva č ka sausage. Ital. Oral Surg. 5, 14–17. https://doi.org/10.1016/j.profoo.2015.09.004.

Bozoudi, D., Torriani, S., Zdragas, A., Litopoulou-Tzanetaki, E., 2016. LWT – food science and technology assessment of microbial diversity of the dominant microbiota in fresh and mature PDO Feta cheese made at three mountainous areas of Greece. LWT – Food Sci. Technol. 72, 525–533. https://doi.org/10.1016/j.lwt.2016.04.039.

Budak, S.O., Figge, M.J., Houbraken, J., Vries, R.P. De, 2015. The diversity and evolution of microbiota in traditional Turkish Divle Cave cheese during ripening. Int. Dairy J. https://doi.org/10.1016/j.idairyj.2015.09.011.

Calasso, M., Ercolini, D., Mancini, L., Stellato, G., Minervini, F., Di, R., Angelis, M. De, Gobbetti, M., 2016. Relationships among house, rind and core microbiotas during manufacture of traditional Italian cheeses at the same dairy plant. Food Microbiol. 54, 115−126. https://doi.org/10.1016/j.fm.2015.10.008.

Callon, C., Delbès, C., Duthoit, F., Montel, M.C., 2006. Application of SSCP − PCR fingerprinting to profile the yeast community in raw milk Salers cheeses. Syst. Appl. Microbiol. 29, 172−180. https://doi.org/10.1016/j.syapm.2005.07.008.

Casalta, E., Sorba, J., Aigle, M., Ogier, J., 2009. International Journal of Food Microbiology Diversity and dynamics of the microbial community during the manufacture of Calenzana, an artisanal Corsican cheese. Int. J. Food Microbiol. 133, 243−251. https://doi.org/10.1016/j.ijfoodmicro.2009.05.022.

Centi, V., Matteucci, F., Lepidi, A., Gallo, M. Del, Ercole, C., 2017. Microbiological and biochemical aspects of inland Pecorino Abruzzese cheese. Heliyon e00258. https://doi.org/10.1016/j.heliyon.2017.e00258.

Ceugniez, A., Taminiau, B., Coucheney, F., Jacques, P., Delcenserie, V., Daube, G., 2017a. Fungal diversity of "Tomme d'Orchies" cheese during the ripening process as revealed by a metagenomic study. Int. J. Food Microbiol. 258, 89−93. https://doi.org/10.1016/j.ijfoodmicro.2017.07.015.

Ceugniez, A., Taminiau, B., Daube, G., Drider, D., 2017b. Use of a metagenetic approach to monitor the bacterial microbiota of "Tomme d'Orchies" cheese during the ripening process. Int. J. Food Microbiol. 247, 65−69. https://doi.org/10.1016/j.ijfoodmicro.2016.10.034.

Chevallier, I., 2006. Microbial ecology of a small-scale facility producing traditional dry sausage. Food Control 17, 446−453. https://doi.org/10.1016/j.foodcont.2005.02.005.

Chong, J., Xia, J., 2017. Computational approaches for integrative analysis of the metabolome and microbiome. Metabolites 7, 62. https://doi.org/10.3390/metabo7040062.

Cocolin, L., Manzano, M., Aggio, D., Cantoni, C., Comi, G., 2001a. A novel polymerase chain reaction (PCR) Ð denaturing gradient gel electrophoresis (DGGE) for the identication of Micrococcaceae strains involved in meat fermentations. Its application to naturally fermented Italian sausages. Meat Sci. 57, 59−64.

Cocolin, L., Manzano, M., Cantoni, C., Comi, G., Veterinaria, M., 2001b. Denaturing gradient gel electrophoresis analysis of the 16S rRNA gene V1 region to monitor dynamic changes in the bacterial population during fermentation of Italian sausages. Appl. Environ. Microbiol. 67, 5113−5121. https://doi.org/10.1128/AEM.67.11.5113.

Cocolin, L., Urso, R., Rantsiou, K., Cantoni, C., Comi, G., 2006. Dynamics and characterization of yeasts during natural fermentation of Italian sausages. FEMS Yeast Res. 6, 692−701. https://doi.org/10.1111/j.1567-1364.2006.00050.x.

Cocolin, L., Dolci, P., Rantsiou, K., Urso, R., Cantoni, C., Comi, G., 2009. Lactic acid bacteria ecology of three traditional fermented sausages produced in the North of Italy as determined by molecular methods. Meat Sci. 82, 125−132. https://doi.org/10.1016/j.meatsci.2009.01.004.

Cocolin, L., Dolci, P., Rantsiou, K., 2011. Biodiversity and dynamics of meat fermentations : the contribution of molecular methods for a better comprehension of a complex ecosystem. MESC 89, 296−302. https://doi.org/10.1016/j.meatsci.2011.04.011.

Cocolin, L., Alessandria, V., Dolci, P., Gorra, R., 2013. Culture independent methods to assess the diversity and dynamics of microbiota during food fermentatio. Int. J. Food Microbiol. 167, 29−43. https://doi.org/10.1016/j.ijfoodmicro.2013.05.008.

Cocolin, L., Mataragas, M., Bourdichon, F., Doulgeraki, A., Pilet, M., Jagadeesan, B., Rantsiou, K., Phister, T., 2017. International Journal of Food Microbiology Next generation microbiological risk assessment meta-omics : the next need for integration. Int. J. Food Microbiol. 0−1. https://doi.org/10.1016/j.ijfoodmicro.2017.11.008.

Comi, G., Urso, R., Iacumin, L., Rantsiou, K., Cattaneo, P., Cantoni, C., Cocolin, L., 2005. Characterisation of naturally fermented sausages produced in the North East of Italy. Meat Sci. 69, 381−392. https://doi.org/10.1016/j.meatsci.2004.08.007.

Coppola, R., Marconi, E., Rossi, F., Dellaglio, F., 1995. Artisanal production of Naples-type salami: chemical and microbiological aspects. Ital. J. Food Sci. 7, 57−62.

Coppola, R., Iorizzo, M., Saotta, R., Sorrentino, E., Grazia, L., 1997. Characterization of micrococci and staphylococci isolated from soppressata molisana, a Southern Italy fermented sausage. Food Microbiol. 14, 47−53.

Coppola, R., Giagnacovo, B., Iorizzo, M., Grazia, L., 1998. Characterization of lactobacilli involved in the ripening of soppressata molisana, a typical southern Italy fermented sausage. Food Microbiol. 15, 347−353.

Coppola, S., Mauriello, G., Aponte, M., Moschetti, G., Villani, F., 2000. Microbial succession during ripening of Naples-type salami, a southern Italian fermented sausage. Meat Sci. 56, 321−329.

Cruciata, M., Gaglio, R., Todaro, M., Settanni, L., 2019. Ecology of Vastedda della valle del Belìce cheeses : a review and recent findings to stabilize the traditional production Ecology of Vastedda della valle del Belìce cheeses : a review. Food Rev. Int. 35, 90−103. https://doi.org/10.1080/87559129.2018.1469142.

Dalmasso, A., Dolores, M.D.L., Civera, T., Pattono, D., Cardazzo, B., Teresa, M., 2016. Characterization of microbiota in Plaisentif cheese by high-throughput sequencing. LWT − Food Sci. Technol. 69, 490−496. https://doi.org/10.1016/j.lwt.2016.02.004.

Danilovi, B., Petrovi, L., Veljovi, K., Tolina, M., Savi, D., 2011. The characterisation of lactic acid bacteria during the fermentation of an artisan Serbian sausage (Petrovská Klobása). Meat Sci. 88, 668−674. https://doi.org/10.1016/j.meatsci.2011.02.026.

Dapkevicius, A., Henrique, J.D., Dapkevicius, E., 2015. Physicochemical, biochemical, microbiological and safety aspects of Pico cheese : assessment throughout maturation and on the final product. Int. J. Dairy Technol. 70, 1−14. https://doi.org/10.1111/1471-0307.12424.

De Vuyst, L., Leroy, F., 2007. Bacteriocins from lactic acid bacteria : production, purification, and food. J. Mol. Microbiol. Biotechnol. 13, 194−199. https://doi.org/10.1159/000104752.

Delcenserie, V., Taminiau, B., Delhalle, L., Nezer, C., Doyen, P., Crevecoeur, S., Roussey, D., Korsak, N., 2014. Microbiota characterization of a Belgian protected designation of origin cheese, Herve cheese, using metagenomic analysis. J. Dairy Sci. 97, 6046−6056. https://doi.org/10.3168/jds.2014-8225.

Di Cagno, R., Lòpez, C.C., Tofalo, R., Gallo, G., De Angelis, M., Paparella, A., Hammes, W.P., Gobbetti, M., 2008. Comparison of the compositional, microbiological, biochemical and volatile profile characteristics of three Italian PDO fermented sausages. Meat Sci. 79, 224−235. https://doi.org/10.1016/j.meatsci.2007.09.006.

Di Grigoli, A., Francesca, N., Gaglio, R., Guarrasi, V., Moschetti, M., Scatassa, M.L., Settanni, L., Bonanno, A., 2015. The in fluence of the wooden equipment employed for cheese manufacture on the characteristics of a traditional stretched cheese during ripening. Food Microbiol. 46, 81−91. https://doi.org/10.1016/j.fm.2014.07.008.

Didienne, R., Defargues, C., Callon, C., Meylheuc, T., Hulin, S., Montel, M., 2012. International Journal of Food Microbiology Characteristics of microbial bio fi lm on wooden vats ('gerles') in PDO Salers cheese. Int. J. Food Microbiol. 156, 91–101. https://doi.org/10.1016/j.ijfoodmicro.2012.03.007.

Dolci, P., Alessandria, V., Rantsiou, K., Rolle, L., Zeppa, G., Cocolin, L., 2008. Microbial dynamics of Castelmagno PDO, a traditional Italian cheese, with a focus on lactic acid bacteria ecology. Int. J. Food Microbiol. 122, 302–311. https://doi.org/10.1016/j.ijfoodmicro.2007.12.018.

Dolci, P., Alessandria, V., Rantsiou, K., Bertolino, M., Cocolin, L., 2010. Microbial diversity, dynamics and activity throughout manufacturing and ripening of Castelmagno PDO cheese. Int. J. Food Microbiol. 143, 71–75. https://doi.org/10.1016/j.ijfoodmicro.2010.07.007.

Domingos-lopes, M.F.P., Stanton, C., Ross, P.R., Dapkevicius, M.L.E., Silva, C.C.G., 2017. Genetic diversity, safety and technological characterization of lactic acid bacteria isolated from artisanal Pico cheese. Food Microbiol. 63, e9–e10. https://doi.org/10.1016/j.fm.2016.11.014.

Donnelly, C.W., 2014. From pasteur to probiotics: a historical overview of cheese and microbes. In: Press, A. (Ed.), Cheese and Microbes, pp. 1–17.

Dos Santos, M.T.P.G., Benito, M.J., de Guía Córdoba, M., Alvarenga, N., de Herrera, S.R.M.S., 2017. Yeast community in traditional Portuguese Serpa cheese by culture- dependent and -independent DNA approaches. Int. J. Food Microbiol. 262, 63–70. https://doi.org/10.1016/j.ijfoodmicro.2017.09.013.

Drosinos, E.H., 2005. Characterization of the microbial flora from a traditional Greek fermented sausage. Meat Sci. 69, 307–317. https://doi.org/10.1016/j.meatsci.2004.07.012.

Drosinos, E.H., Paramithiotis, S., Kolovos, G., Tsikouras, I., Metaxopoulos, I., 2007. Phenotypic and technological diversity of lactic acid bacteria and staphylococci isolated from traditionally fermented sausages in Southern Greece. Food Microbiol. 24, 260–270. https://doi.org/10.1016/j.fm.2006.05.001.

Dugat-bony, E., Garnier, L., Denonfoux, J., Ferreira, S., Sarthou, A., Bonnarme, P., Irlinger, F., 2016. International journal of food microbiology highlighting the microbial diversity of 12 French cheese varieties. Int. J. Food Microbiol. 238, 265–273. https://doi.org/10.1016/j.ijfoodmicro.2016.09.026.

Ercolini, D., 2013. High-throughput sequencing and metagenomics : moving forward in the culture-independent analysis of food microbial ecology. Appl. Environ. Microbiol. 79, 3148–3155. https://doi.org/10.1128/AEM.00256-13.

Ercolini, D., Hill, P.J., Dodd, C.E.R., Ii, F., 2003. Bacterial community structure and location in stilton cheese. Appl. Environ. Microbiol. 69, 3540–3548. https://doi.org/10.1128/AEM.69.6.3540.

Ercolini, D., Frisso, G., Mauriello, G., Salvatore, F., Coppola, S., 2008. Microbial diversity in natural whey cultures used for the production of Caciocavallo silano PDO cheese. Int. J. Food Microbiol. 124, 164–170. https://doi.org/10.1016/j.ijfoodmicro.2008.03.007.

Esteves, A., Patarata, L., Saraiva, C., 2007. Assessment of the microbiological characteristics of industrially produced Alheira, with particular reference to foodborne pathogens. J. Food Saf. 28, 88–102.

Fadda, S., Vignolo, G., Holgado, A.P.R., Oliver, G., 1998. Proteolytic activity of LactobaciZlus strains isolated from dry- fermented sausages on muscle sarcoplasmic proteins. Meat Sci. 49, 11–18.

Federici, S., Ciarrocchi, F., Campana, R., Ciandrini, E., Blasi, G., Baffone, W., 2014. Identification and functional traits of lactic acid bacteria isolated from Ciauscolo salami produced in Central Italy. MESC 98, 575–584. https://doi.org/10.1016/j.meatsci.2014.05.019.

Fernández-López, J., Sendra, E., Sayas-Barberá, E., Navarro, C., Pérez-Alvarez, J.A., 2008. MEAT Physico-chemical and microbiological profiles of ''salchicho (Spanish dry-fermented sausage) enriched with orange fiber. Meat Sci. 80, 410–417. https://doi.org/10.1016/j.meatsci.2008.01.010.

Ferreira, V., Barbosa, J., Silva, J., Felício, M.T., Hogg, T., Gibbs, P., Teixeira, P., Mena, C., 2007. Characterisation of alheiras, traditional sausages produced in the North of Portugal, with respect to their microbiological safety. Food Control 18, 436–440. https://doi.org/10.1016/j.foodcont.2005.11.011.

Filippis, F. De, Parente, E., Ercolini, D., 2017. Metagenomics insights into food fermentations. Microb. Biotechnol. 10, 91–102. https://doi.org/10.1111/1751-7915.12421.

Filippis, F. De, Parente, E., Ercolini, D., 2018. Recent past, present, and future of the food microbiome. Annu. Rev. Food Sci. Technol. 9, 589–608.

Fitzsimons, N.A., Cogan, T.M., Condon, S., Beresford, T., 2001. Spatial and temporal distribution of non-starter lactic acid bacteria in Cheddar cheese. J. Appl. Microbiol. 90, 600–608.

Flórez, A.B., Mayo, B., 2006a. Microbial diversity and succession during the manufacture and ripening of traditional, Spanish, blue-veined Cabrales cheese, as determined by PCR-DGGE. Int. J. Food Microbiol. 110, 165–171. https://doi.org/10.1016/j.ijfoodmicro.2006.04.016.

Flórez, A.B., Mayo, B., 2006b. PCR – DGGE as a tool for characterizing dominant microbial populations in the Spanish blue-veined Cabrales cheese. Int. Dairy J. 16, 1205–1210. https://doi.org/10.1016/j.idairyj.2005.11.008.

Flórez, A.B., López-Díaz, T.M., Álvarez-Martín, P., Mayo, B., 2006. Microbial characterisation of the traditional Spanish blue-veined Cabrales cheese : identification of dominant lactic acid bacteria. Eur. Food Res. Technol. 223, 503–508. https://doi.org/10.1007/s00217-005-0230-8.

Fonseca, S., Cachaldora, A., Gómez, M., Franco, I., Carballo, J., 2013. Monitoring the bacterial population dynamics during the ripening of Galician chorizo, a traditional dry fermented Spanish sausage. Food Microbiol. 33, 77–84. https://doi.org/10.1016/j.fm.2012.08.015.

Fontán, M.C.G., Lorenzo, J.M., Martínez, S., Franco, I., Carballo, J., 2007a. Microbiological characteristics of Botillo, a Spanish traditional pork sausage. LWT-Food Sci. Technol. 40, 1610–1622. https://doi.org/10.1016/j.lwt.2006.10.007.

Fontán, M.C.G., Lorenzo, J.M., Parada, A., Franco, I., Carballo, J., 2007b. Microbiological characteristics of '"androlla"', a Spanish traditional pork sausage. Food Microbiol. 24, 52–58. https://doi.org/10.1016/j.fm.2006.03.007.

Fontana, C., Cappa, F., Rebecchi, A., Cocconcelli, P.S., 2010. Surface microbiota analysis of Taleggio, Gorgonzola, Casera, scimudin and formaggio di Fossa Italian cheeses. Int. J. Food Microbiol. 138, 205–211. https://doi.org/10.1016/j.ijfoodmicro.2010.01.017.

Fontana, C., Bassi, D., López, C., Pisacane, V., Claudia, M., Puglisi, E., Rebecchi, A., Sandro, P., Vignolo, G., 2016. Microbial ecology involved in the ripening of naturally fermented llama meat sausages. A focus on lactobacilli diversity. Int. J. Food Microbiol. 236, 17–25. https://doi.org/10.1016/j.ijfoodmicro.2016.07.002.

Franciosa, I., Alessandria, V., Dolci, P., Rantsiou, K., Cocolin, L., 2018. International Journal of Food Microbiology Sausage fermentation and starter cultures in the era of molecular biology methods. Int. J. Food Microbiol. 279, 26–32. https://doi.org/10.1016/j.ijfoodmicro.2018.04.038.

Franck, G., Susanne, M., Leo, M., Christophe, L., 2008. Recent developments in cheese cultures with protective and probiotic functionalities. Dairy Sci. Technol. 88, 421–444.

Fuka, M.M., Engel, M., Skelin, A., Redžepović, S., Schloter, M., 2010. Microbiology bacterial communities associated with the production of artisanal Istrian cheese. Int. J. Food Microbiol. 142, 19–24. https://doi.org/10.1016/j.ijfoodmicro.2010.05.008.

Gala, E., Landi, S., Solieri, L., Nocetti, M., Pulvirenti, A., Giudici, P., 2008. Diversity of lactic acid bacteria population in ripened Parmigiano Reggiano cheese. Int. J. Food Microbiol. 125, 347–351. https://doi.org/10.1016/j.ijfoodmicro.2008.04.008.

Garcıa-Varona, M., Santos, E.M., Jaime, I., Rovira, J., 2000. Characterisation of Micrococcaceae isolated from different varieties of chorizo. Int. J. Food Microbiol. 54, 189–195.

Gardini, F., Suzzi, G., Lombardi, A., Galgano, F., Antonietta, M., Andrighetto, C., Schirone, M., Tofalo, R., 2001. A survey of yeasts in traditional sausages of southern Italy. FEMS Yeast Res. 1, 161–167.

Giello, M., La Storia, A., Masucci, F., Francia, A.Di, Villani, F., 2016. Dynamics of bacterial communities during manufacture and ripening of traditional Caciocavallo of Castelfranco cheese in relation to cows' feeding. Food Microbiol. https://doi.org/10.1016/j.fm.2016.11.016.

Gkatzionis, K., Yunita, D., Linforth, R.S.T., Dickinson, M., Dodd, C.E.R., 2014. Diversity and activities of yeasts from different parts of a Stilton cheese. Int. J. Food Microbiol. 177, 109–116. https://doi.org/10.1016/j.ijfoodmicro.2014.02.016.

Goerges, S., Rea, M.C., Gelsomino, R., Heise, V., Cogan, T.M., Vancanneyt, M., Scherer, S., Universita, T., 2008. Commercial ripening starter microorganisms inoculated into cheese milk do not successfully establish themselves in the resident microbial ripening consortia of a south German red smear cheese. Appl. Environ. Microbiol. 74, 2210–2217. https://doi.org/10.1128/AEM.01663-07.

González, L., Cuadrillero, A.F., Castro, J.M., Bernardo, A., Tornadijo, M.E., 2015. Selection of lactic acid bacteria isolated from San Simón da Costa Cheese (PDO) in order to develop an autochthonous starter culture. Adv. Microbiol. 5, 748–759.

Gonçalves, M.T.P., Benito, M.J., Córdoba, M.D.G., Egas, C., Merchán, A.V., Galván, A.I., Ruiz-Moyano, S., 2018. Bacterial communities in serpa cheese by culture dependent techniques, 16S rRNA gene sequencing and high-throughput sequencing analysis. J. Food Sci. 83, 1333–1341. https://doi.org/10.1111/1750-3841.14141.

Grappin, R., Coulon, J.B., 1996. No title, in: "terroir" milk and cheese production: some comments. In: 3. Rencontres Autour Des Recherches Sur Les Ruminants. Paris.

Greppi, A., Ferrocino, I., La, A., Rantsiou, K., Ercolini, D., Cocolin, L., 2015. Monitoring of the microbiota of fermented sausages by culture independent rRNA-based approaches. Int. J. Food Microbiol. 1–9. https://doi.org/10.1016/j.ijfoodmicro.2015.01.016.

Hayaloglu, A.A., Fox, P.F., Guven, M., Cakmakci, S., Hayaloglu, A.A., Fox, P.F., Guven, M., Cakmakci, S., 2007. Cheeses of Turkey : 1. Varieties ripened in goat-skin bags. Lait 87, 79–95. https://doi.org/10.1051/lait.

Henri-Dubernet, S., Desmasures, N., Guéguen, M., 2004. Culture-dependent and culture-independent methods for molecular analysis of the diversity of lactobacilli in "Camembert de Normandie" cheese. Lait 84, 179–189. https://doi.org/10.1051/lait.

Henri-Dubernet, S., Desmasures, N., Guéguen, M., 2008. Diversity and dynamics of lactobacilli populations during ripening of RDO Camembert cheese. Can. J. Microbiol. 54, 218–228. https://doi.org/10.1139/W07-137.

Hoorde, K. Van, Verstraete, T., Vandamme, P., Huys, G., 2008. Diversity of lactic acid bacteria in two Flemish artisan raw milk Gouda-type cheeses. Food Microbiol. 25, 929–935. https://doi.org/10.1016/j.fm.2008.06.006.

Iacumin, L., Comi, G., Cantoni, C., Cocolin, L., 2006. Ecology and dynamics of coagulase-negative cocci isolated from naturally fermented Italian sausages. Syst. Appl. Microbiol. 29, 480–486. https://doi.org/10.1016/j.syapm.2005.11.006.

Jankovic, V., Lakicevic, B., Saric, L., 2015. Microbial ecosystem of processing units during production process of Petrovská klobása. Tehnol. Mesa 55, 123–129. https://doi.org/10.5937/tehmesa1402123J.

Karabey, B., Eroglu, D., Vural, C., Ozdemir, G., 2018. Determination of the microbial flora in traditional İzmir tulum cheeses by denaturing gradient gel electrophoresis. J. Food Sci. Technol. 55, 956–963. https://doi.org/10.1007/s13197-017-3003-z.

Kumar, P., Chatli, M.K., Verma, A.K., Mehta, N., Malav, O.P., 2015. Quality, functionality and shelf life of fermented meat and meat Products : a review. Crit. Rev. Food Sci. Nutr. 57, 2844–2856. https://doi.org/10.1080/10408398.2015.1074533.

Larpin-laborde, S., Imran, M., Bonaïti, C., Bora, N., Gelsomino, R., Goerges, S., Irlinger, F., Goodfellow, M., Ward, A.C., Vancanneyt, M., Swings, J., Scherer, S., Guéguen, M., Desmasures, N., 2011. Surface microbial consortia from Livarot, a French smear-ripened cheese. Can. J. Microbiol. 57, 651–660. https://doi.org/10.1139/W11-050.

Lebert, I., 2007. Diversity of microorganisms in the environment and dry fermented sausages of small traditional French processing units. Meat Sci. 76, 112–122. https://doi.org/10.1016/j.meatsci.2006.10.019.

Licitra, G., Ogier, J.C., Parayre, S., Pediliggieri, C., Carnemolla, T.M., Falentin, H., Madec, M.N., Carpino, S., Lortal, S., 2007. Variability of bacterial biofilms of the "Tina" wood vats used in the Ragusano cheese-making process. Appl. Environ. Microbiol. 73, 6980–6987. https://doi.org/10.1128/AEM.00835-07.

Licitra, G., Caccamo, M., Valence, F., Lortal, S., 2017. Traditional wooden equipment used for cheesemaking and their effect on quality. In: Global Cheesemaking Technology. Wiley Ltd., pp. 157–172

Lortal, S., Di, A., Madec, M., Pediliggieri, C., Tuminello, L., Tanguy, G., Fauquant, J., Lecuona, Y., Campo, P., Carpino, S., Licitra, G., 2009. Tina wooden vat biofilm : a safe and highly ef fi cient lactic acid bacteria delivering system in PDO Ragusano cheese making. Int. J. Food Microbiol. 132, 1–8. https://doi.org/10.1016/j.ijfoodmicro.2009.02.026.

Lortal, S., Licitra, G., Valence, F., 2014. Wooden tools: reservoirs of microbial diversity in traditional cheesemaking. In: ASM (Ed.), Cheese and Microbes, pp. 167–176.

Mainar, M.S., Stavropoulou, D.A., Leroy, F., 2016. Exploring the metabolic heterogeneity of coagulase-negative staphylococci to improve the quality and safety of fermented meats : a review. Int. J. Food Microbiol. 247, 24–37. https://doi.org/10.1016/j.ijfoodmicro.2016.05.021.

Mangia, N., Murgia, M.A., Garau, G., Deiana, P., 2007. Microbiologia e valutazione igienico-sanitaria della salsiccia sarda Documento digitalizzato dallo Staff di UnissResearch.

Mangia, N.P., Murgia, M.A., Garau, G., Sanna, M.G., Ã, P.D., 2008. Influence of selected lab cultures on the evolution of free amino acids, free fatty acids and Fiore Sardo cheese microflora during the ripening. Food Microbiol. 25, 366–377. https://doi.org/10.1016/j.fm.2007.09.009.

Marcellino, S.N., Benson, D.R., 2013. The good, the bad, and the ugly : tales of mold-ripened cheese. Am. Soc. Microbiol. 1−27. https://doi.org/10.1128/microbiolspec.CM-0005-2012. Correspondence.

Marcellino, N., Beuvier, E., Grappin, R., Benson, D.R., 2001. Diversity of *Geotrichum candidum* strains isolated from traditional cheesemaking fabrications in France. Appl. Environ. Microbiol. 67, 4752−4759. https://doi.org/10.1128/AEM.67.10.4752.

Mariani, C., Briandet, R., Chamba, J., Notz, E., Eyoug, R.N., Oulahal, N., 2007. Biofilm ecology of wooden shelves used in ripening the French raw milk smear cheese reblochon de Savoie. J. Dairy Sci. 90, 1653−1661. https://doi.org/10.3168/jds.2006-190.

Martín, B., Garriga, M., Hugas, M., Bover-Cid, S., Veciana-Nogués, M.T., Aymerich, T., 2006. Molecular, technological and safety characterization of Gram-positive catalase-positive cocci from slightly fermented sausages. Int. J. Food Microbiol. 107, 148−158. https://doi.org/10.1016/j.ijfoodmicro.2005.08.024.

Mauriello, G., Moschetti, G., Villani, F., Blaiotta, G., Coppola, S., 2000. Antibiotic resistance of coagulase-negative staphylococci isolated from artisanal Naples-type salami. Int. J. Food Sci. Nutr. 51, 19−24. https://doi.org/10.1080/096374800100868.

Minervini, F., Lattanzi, A., Angelis, M. De, Celano, G., Gobbetti, M., 2015. House microbiotas as sources of lactic acid bacteria and yeasts in traditional Italian sourdoughs. Food Microbiol. 52, 66−76. https://doi.org/10.1016/j.fm.2015.06.009.

Montel, M., Buchin, S., Mallet, A., Delbes-paus, C., Vuitton, D.A., Desmasures, N., Berthier, F., 2014. Traditional cheeses : rich and diverse microbiota with associated benefits. Int. J. Food Microbiol. 177, 136−154. https://doi.org/10.1016/j.ijfoodmicro.2014.02.019.

Mormile, A., Barile, M., Mercogliano, R., Johansson, P., Björkroth, K.J., Aponte, M., Murru, N., 2016. Dynamics of lactic acid bacteria in B Pecorino di Tramonti Ewe's milk cheese — with particular emphasis on enterococci : a preliminary study. Ann. Microbiol. 66, 179−185. https://doi.org/10.1007/s13213-015-1094-1.

Morot-bizot, S.C., Leroy, S., Talon, R., 2006. Staphylococcal community of a small unit manufacturing traditional dry fermented sausages. Int. J. Food Microbiol. 108, 210−217. https://doi.org/10.1016/j.ijfoodmicro.2005.12.006.

Moschetti, G., Mauriello, G., Villani, F., 1997. Differentiation of *Staphylococcus xylosus* strains from Italian sausages by antibiotyping and low frequency restriction fragment analysis of genomic DNA. Syst. Appl. Microbiol. 20, 432−438. https://doi.org/10.1016/S0723-2020(97)80012-4.

Mounier, J., Goerges, S., Gelsomino, R., Vancanneyt, M., Vandemeulebroecke, K., Hoste, B., 2006. Sources of the adventitious microflora of a smear-ripened cheese. J. Appl. Microbiol. 101, 668−681. https://doi.org/10.1111/j.1365-2672.2006.02922.x.

Nikolic, M., Terzic-vidojevic, A., Jovcic, B., Begovic, J., Golic, N., Topisirovic, L., 2008. Characterization of lactic acid bacteria isolated from Bukuljac, a homemade goat's milk cheese. Int. J. Food Microbiol. 122, 162−170. https://doi.org/10.1016/j.ijfoodmicro.2007.11.075.

Ouadghiri, M., Amar, M., Vancanneyt, M., Swings, J., 2005. Biodiversity of lactic acid bacteria in Moroccan soft white cheese (Jben). FEMS Microbiol. Lett. 251, 267−271. https://doi.org/10.1016/j.femsle.2005.08.012.

Ozturkoglu-budak, S., Aykas, D.P., 2018. Temporal microbiota and biochemical pro fi les during production and ripening of Divle Cave cheese. J. Dairy Sci. 71, 99−106. https://doi.org/10.1111/1471-0307.12493.

Pangallo, D., Nikoleta, Š., Kore, J., Pu, A., Kraková, L., Valík, L., 2014. Microbial diversity and dynamics during the production of May bryndza cheese. Int. J. Food Microbiol. 170, 38−43. https://doi.org/10.1016/j.ijfoodmicro.2013.10.015.

Parente, E., Grieco, S., Crudele, M.A., 2001. Phenotypic diversity of lactic acid bacteria isolated from fermented sausages produced in Basilicata (Southern Italy). J. Appl. Microbiol. 90, 943–952.

Parente, E., Cocolin, L., Filippis, F. De, Zotta, T., Ferrocino, I., Sullivan, O.O., Neviani, E., Angelis, M. De, Cotter, P.D., Ercolini, D., 2016. International Journal of Food Microbiology FoodMicrobionet : a database for the visualisation and exploration of food bacterial communities based on network analysis. Int. J. Food Microbiol. 219, 28–37. https://doi.org/10.1016/j.ijfoodmicro.2015.12.001.

Pasquale, I. De, Calasso, M., Mancini, L., Ercolini, D., Storia, L., De Angelis, M., Cagno, R. Di, 2014. Causal relationship between microbial ecology dynamics and proteolysis during manufacture and ripening of protected designation of origin (PDO) cheese Canestrato Pugliese. Appl. Environ. Microbiol. 80, 4085–4094. https://doi.org/10.1128/AEM.00757-14.

Pereira, C.I., Crespo, M.T.B., Romao, M.V.S., 2001. Evidence for proteolytic activity and biogenic amines production in *Lactobacillus curvatus* and *L. homohiochi*. Int. J. Dairy Technol. 68, 211–216.

Pino, A., Liotta, L., Randazzo, C.L., Todaro, A., Mazzaglia, A., 2017. Polyphasic approach to study physico-chemical, microbiological and sensorial characteristics of artisanal Nicastrese goat's cheese. Food Microbiol. 70, 143–154. https://doi.org/10.1016/j.fm.2017.09.005.

Piraino, P., Zotta, T., Ricciardi, A., Parente, E., 2005. Discrimination of commercial Caciocavallo cheeses on the basis of the diversity of lactic microflora and primary proteolysis. Int. Dairy J. 15, 1138–1149. https://doi.org/10.1016/j.idairyj.2004.12.006.

Pisacane, V., Luisa, M., Puglisi, E., Dallolio, G., Rebecchi, A., 2015. International Journal of Food Microbiology Microbial analyses of traditional Italian salami reveal microorganisms transfer from the natural casing to the meat matrix. Int. J. Food Microbiol. 207, 57–65. https://doi.org/10.1016/j.ijfoodmicro.2015.04.029.

Połka, J., Rebecchi, A., Pisacane, V., Morelli, L., Puglisi, E., 2015. Ecology of antibiotic resistant coagulase-negative staphylococci isolated from the production chain of a typical Italian salami. Food Control 53, 14–22. https://doi.org/10.1016/j.foodcont.2015.01.001.

Quigley, L., O'Sullivan, O., Beresford, T.P., Ross, R.P., Fitzgerald, G.F., Cotter, P.D., 2011. Molecular approaches to analysing the microbial composition of raw milk and raw milk cheese. Int. J. Food Microbiol. 150, 81–94. https://doi.org/10.1016/j.ijfoodmicro.2011.08.001.

Quigley, L., Sullivan, O.O., Beresford, T.P., Ross, R.P., Fitzgerald, G.F., Cotter, P.D., 2012. A comparison of methods used to extract bacterial DNA from raw milk and raw milk cheese. J. Appl. Microbiol. 113, 96–105. https://doi.org/10.1111/j.1365-2672.2012.05294.x.

Quijada, N.M., Mann, E., Wagner, M., Rodríguez-lázaro, D., Hernández, M., Schmitz-esser, S., 2018. International Journal of Food Microbiology Autochthonous facility-specific microbiota dominates washed-rind Austrian hard cheese surfaces and its production environment. Int. J. Food Microbiol. 267, 54–61. https://doi.org/10.1016/j.ijfoodmicro.2017.12.025.

Raghupathi, P.K., Herschend, J., Røder, H.L., Sørensen, S.J., Burmølle, M., 2016. Genome sequence of Kocuria varians G6 isolated from a slaughterhouse in Denmark, 2006–2007 Genome Announc. 4. Copyright. https://doi.org/10.1128/genomeA.00076-16.

Ramezani, M., Hosseini, S.M., Ferrocino, I., Amoozegar, M.A., Cocolin, L., 2017. Molecular investigation of bacterial communities during the manufacturing and ripening of semi-hard Iranian Liqvan cheese. Food Microbiol. 66, 64–71. https://doi.org/10.1016/j.fm.2017.03.019.

Randazzo, C.L., Torriani, S., Akkermans, A.D.L., Vos, W.M. De, Vaughan, E.E., 2002. Diversity, dynamics, and activity of bacterial communities during production of an artisanal Sicilian cheese as evaluated by 16S rRNA analysis. J. Appl. Microbiol. 68, 1882−1892. https://doi.org/10.1128/AEM.68.4.1882.

Randazzo, C.L., Vaughan, E.E., Caggia, C., 2006. Artisanal and experimental Pecorino Siciliano cheese : microbial dynamics during manufacture assessed by culturing and PCR − DGGE analyses. Int. J. Food Microbiol. 109, 1−8. https://doi.org/10.1016/j.ijfoodmicro.2005.11.002.

Randazzo, C.L., Pitino, I., Ribbera, A., Caggia, C., 2010. Pecorino Crotonese cheese : study of bacterial population and fl avour compounds. Food Microbiol. 27, 363−374. https://doi.org/10.1016/j.fm.2009.11.010.

Rantsiou, K., Cocolin, L., 2006. New developments in the study of the microbiota of naturally fermented sausages as determined by molecular methods : a review. Int. J. Food Microbiol. 108, 255−267. https://doi.org/10.1016/j.ijfoodmicro.2005.11.013.

Rantsiou, K., Urso, R., Iacumin, L., Cantoni, C., Cattaneo, P., Comi, G., Cocolin, L., 2005. Culture-dependent and -independent methods to investigate the microbial ecology of Italian fermented sausages. Appl. Environ. Microbiol. 71, 1977−1986. https://doi.org/10.1128/AEM.71.4.1977.

Rantsiou, K., Drosinos, E.H., Gialitaki, M., Metaxopoulos, I., Comi, G., Cocolin, L., 2006. Use of molecular tools to characterize *Lactobacillus* spp. isolated from Greek traditional fermented sausages. Int. J. Food Microbiol. 112, 215−222. https://doi.org/10.1016/j.ijfoodmicro.2006.09.001.

Reale, S., Vitale, F., Scatassa, M.L., Caracappa, S., Todaro, V.C.M., Vitale, F., Scatassa, M.L., Caracappa, S., Todaro, V.C.M., 2016. Molecular characterization of dominant bacterial population in "Vastedda della Valle del Belice" cheese : preliminary investigation Molecular characterization of dominant bacterial population in "Vastedda della Valle del Belice" cheese : Ital. J. Anim. Sci. 6, 20−23. https://doi.org/10.4081/ijas.2007.1s.595.

Rebecchi, A., Crivori, S., Sarra, P.G., Cocconcelli, P.S., 1998. Physiological and molecular techniques for the study of bacterial community development in sausage fermentation. J. Appl. Microbiol. 84, 1043−1049.

Riquelme, C., Câmara, S., Lurdes, M. De, Dapkevicius, N.E., Vinuesa, P., Costa, C., Malcata, F.X., Rego, O.A., 2015. Characterization of the bacterial biodiversity in Pico cheese (an artisanal Azorean food). Int. J. Food Microbiol. 192, 86−94. https://doi.org/10.1016/j.ijfoodmicro.2014.09.031.

Rossetti, L., Emanuela, M., Gatti, M., Lazzi, C., Neviani, E., Giraffa, G., 2008. Grana Padano cheese whey starters : microbial composition and strain distribution. Int. J. Food Microbiol. 127, 168−171. https://doi.org/10.1016/j.ijfoodmicro.2008.06.005.

Ryssel, M., Johansen, P., Al-soud, W.A., Sørensen, S., Arneborg, N., Jespersen, L., 2015. Microbial diversity and dynamics throughout manufacturing and ripening of surface ripened semi-hard Danish Danbo cheeses investigated by culture-independent techniques. Int. J. Food Microbiol. 215, 124−130. https://doi.org/10.1016/j.ijfoodmicro.2015.09.012.

Samelis, J., Maurogenakis, F., Metaxopoulos, J., 1994. Characterisation of lactic acid bacteria isolated from naturally fermented Greek dry salami. Int. J. Food Microbiol. 23, 179−196.

Santarelli, M., Bottari, B., Lazzi, C., Neviani, E., Gatti, M., 2013. Survey on the community and dynamics of lactic acid bacteria in Grana Padano cheese. Syst. Appl. Microbiol. 36, 593−600. https://doi.org/10.1016/j.syapm.2013.04.007.

Santi, M. De, Sisti, M., Barbieri, E., Piccoli, G., Brandi, G., Stocchi, V., 2010. A combined morphologic and molecular approach for characterizing fungal microflora from a traditional Italian cheese (Fossa cheese). Int. Dairy J. 20, 465–471. https://doi.org/10.1016/j.idairyj.2010.02.004.

Scatassa, M.L., Cardamone, C., Miraglia, V., Lazzara, F., Fiorenza, G., Macaluso, G., Arcuri, L., Settanni, L., Mancuso, I., 2015. Characterisation of the microflora contaminating the wooden vats used for traditional Sicilian cheese production. Ital. J. Food Saf. 4. https://doi.org/10.4081/ijfs.2015.4509.

Settanni, L., Di Grigoli, A., Tornambé, G., Bellina, V., Francesca, N., Moschetti, G., Bonanno, A., 2012. Persistence of wild *Streptococcus thermophilus* strains on wooden vat and during the manufacture of a traditional Caciocavallo type cheese. Int. J. Food Microbiol. 155, 73–81. https://doi.org/10.1016/j.ijfoodmicro.2012.01.022.

Silvestri, G., Santarelli, S., Aquilanti, L., Beccaceci, A., Osimani, A., Tonucci, F., Clementi, F., 2007. Investigation of the microbial ecology of Ciauscolo, a traditional Italian salami, by culture-dependent techniques and PCR-DGGE. Meat Sci. 77, 413–423. https://doi.org/10.1016/j.meatsci.2007.04.015.

Silvetti, T., Capra, E., Morandi, S., Cremonesi, P., Decimo, M., Gavazzi, F., Giannico, R., Noni, I. De, Brasca, M., 2017. Microbial population profile during ripening of Protected Designation of Origin (PDO) Silter cheese, produced with and without autochthonous starter culture. LWT – Food Sci. Technol. https://doi.org/10.1016/j.lwt.2017.06.022.

Stellato, G., Filippis, F. De, Storia, A. La, Ercolini, D., 2015. Coexistence of lactic acid bacteria and potential spoilage microbiota. Appl. Environ. Microbiol. 81, 7893–7904. https://doi.org/10.1128/AEM.02294-15.

Talon, R., Leroy, S., 2011. Diversity and safety hazards of bacteria involved in meat fermentations. MESC 89, 303–309. https://doi.org/10.1016/j.meatsci.2011.04.029.

Talon, R., Lebert, I., Lebert, A., Leroy, S., Garriga, M., Aymerich, T., 2007a. Traditional dry fermented sausages produced in small-scale processing units in Mediterranean countries and Slovakia. 1 : microbial ecosystems of processing environments. Meat Sci. 77, 570–579. https://doi.org/10.1016/j.meatsci.2007.05.006.

Talon, R., Leroy, S., Lebert, I., 2007b. Microbial ecosystems of traditional fermented meat products : the importance of indigenous starters. Meat Sci. 77, 55–62. https://doi.org/10.1016/j.meatsci.2007.04.023.

Tremonte, P., Sorrentino, E., Pannella, G., Tipaldi, L., Sturchio, M., Masucci, A., Maiuro, L., Coppola, R., Succi, M., 2017. Detection of different microenvironments and *Lactobacillus sakei* biotypes in Ventricina, a traditional fermented sausage from central Italy. Int. J. Food Microbiol. https://doi.org/10.1016/j.ijfoodmicro.2016.11.009.

Tsafrakidou, P., Bozoudi, D., Pavlidou, S., Kotzamanidis, C., Hatzikamari, M., Zdragas, A., Litopoulou-Tzanetaki, E., 2016. Technological, phenotypic and genotypic characterization of lactobacilli from Graviera Kritis PDO Greek cheese, manufactured at two traditional dairies. LWT – Food Sci. Technol. 68, 681–689. https://doi.org/10.1016/j.lwt.2016.01.002.

Urso, R., Comi, G., Cocolin, L., 2006. Ecology of lactic acid bacteria in Italian fermented sausages : isolation, identification and molecular characterization. Syst. Appl. Microbiol. 29, 671–680. https://doi.org/10.1016/j.syapm.2006.01.012.

Villani, F., Casaburi, A., Pennacchia, C., Filosa, L., Russo, F., Ercolini, D., Ii, N.F., Universita, V., 2007. Microbial ecology of the soppressata of Vallo di Diano, a traditional dry fermented sausage from southern Italy, and in vitro and in situ selection of autochthonous starter cultures. Appl. Environ. Microbiol. 73, 5453−5463. https://doi.org/10.1128/AEM.01072-07.

Wolfe, B.E., Dutton, R.J., 2015. Fermented foods as experimentally tractable microbial ecosystems. Cell 161, 49−55. https://doi.org/10.1016/j.cell.2015.02.034.

Wolfe, B.E., Button, J.E., Santarelli, M., Dutton, R.J., 2014. Cheese rind communities provide tractable systems for in situ and in vitro studies of microbial diversity. Cell 158, 422−433. https://doi.org/10.1016/j.cell.2014.05.041.

Biorefineries for the valorization of food processing waste

5

Antonio D. Moreno, Mercedes Ballesteros, María José Negro
CIEMAT, Department of Energy, Biofuels Unit, Madrid, Spain

Chapter outline

1. Introduction 156
2. The food industry: wastes and valuable by-products 157
2.1 Animal-derived waste 157
2.2 Plant-derived waste 158
2.3 Market potentials from food-derived by-products 159
3. Biorefineries based on food processing waste 160
3.1 Biorefinery based on wine grape waste 160
3.1.1 Vine shoots 160
3.1.2 Grape marc or pomace 163
3.1.3 Grape seeds 164
3.1.4 Wine lees 164
3.2 Biorefinery based on brewer's spent grain 165
3.3 Biorefinery based on olive waste 169
3.3.1 Olive pomace 169
3.3.2 Olive stones 172
3.3.3 Olive leaves 172
3.3.4 Olive mill wastewater 173
3.3.5 Olive tree pruning 173
3.4 Biorefinery based on potato peels 174
3.5 Other biowaste-based biorefineries 176
3.5.1 Tomato pomace 176
3.5.2 Apple pomace 177
3.5.3 Citrus waste 177
3.5.4 Dairy by-products 178
4. Legislation on waste management 178
5. Concluding remarks 179
Acknowledgments 180
References 180

The Interaction of Food Industry and Environment. https://doi.org/10.1016/B978-0-12-816449-5.00005-9

1. Introduction

The increasing global demand for food production to meet the needs of an exponentially growing population is claiming for sustain and effective strategies for food waste management, especially in terms of treatment and waste disposal. The Food and Agriculture Organization has estimated the production of food waste in about 1.3 billion tons/year worldwide (FAO, 2011). Only in Europe, 173 kg/person of food (corresponding to ca. 90 million tons in total) are discarded every year as waste, even though much of this food is still suitable for human consumption (Stenmarck et al., 2016). These figures come from households (53%), the food manufacturing industry (19%), the food service sector (12%; including ready-to-eat food, catering and restaurants), primary production (11%), and losses derived from the distribution chain (5%). Today, these waste fractions are usually recovered or disposed for composting, crops plowed in/not harvested, anaerobic digestion, bioenergy production, cogeneration, incineration, disposal to sewer, landfill, or discarded to sea, having tremendous economic and environmental impacts. For instance, Europe loses about €143 billion annually due to food waste (Stenmarck et al., 2016). Furthermore, these residues promote severe environmental damage since they contribute to the greenhouse gas emissions, may cause groundwater contamination due to landfilling, and represent an excellent nutrient-rich niche for the growth of disease-causing organisms (Ravindran and Jaiswal, 2016).

As an alternative to the traditional methods for food waste management, food waste is nowadays considered a very attractive raw material for the so-called biorefineries. This is due to its high organic content and the presence of several valuable compounds such as oils, sugars, proteins and phenolic compounds. Biorefineries are industries capable of converting biomass feedstocks (including industrial wastes, the organic fraction of municipal solid wastes, lignocellulose, and algal biomass) into a wide range of renewable bioproducts such as biofuels, bioplastics, fertilizers, nutraceuticals, and/or phytochemicals (Ravindran and Jaiswal, 2016). In this context, the use of food waste as biorefinery feedstock will contribute to cut down the amounts of residues discarded, offering huge potentials and opportunities for a better economic and environmental performance of the food sector. This strategy will also aid at developing and implementing a "zero waste" bioeconomy, where the industrial symbiosis—the use of wastes from one process as raw materials in another one—works as the central core.

As a concept, circular economy started back in the 1970s, including contributors such as U.S. professor John Lyle, his student William McDonough, the German chemist, Michael Braungart, and architect and economist Walter Stahel (Winans et al., 2017). Now, sustainability criteria, material and energy efficiency, and the needs for reducing waste production and greenhouse gas emissions are the main driving forces for the circular economy to go beyond the research perspective (Kaur et al., 2018). Circular economy aims at preserving product values and to extent the materials and resources lives as long as possible. In this way, efficient production and consumption are prioritized, following continuous and regenerative cycles. This leads to a reduced consumption of raw materials and energy, which in turn, results in less amount of waste and greenhouse gas emissions during the production process. Within the framework of a sustainable economy, biorefineries will integrate different technologies (physical, thermochemical, chemical and biological) for the conversion of the residual biomass, obtaining several energy and chemical products of industrial interest for the development and implementation of a sustainable bioeconomy.

One of the major challenges for the use of food waste as biorefinery feedstock is sample heterogeneity. Food waste includes residues from cereals, root and tubers, oil crops and pulses, fruit and vegetables, meat, fish, and dairy products. Households and the food manufacturing sector are the largest contributors to the generation of food waste. However, while households provide complex heterogeneous and not segregated mixtures, the food manufacturing industries offers the possibility of collecting more homogenous materials. This is key for the optimal valorization of these residues, since feedstock variations lead to differences in the chemical composition of wastes. Thus, the implementation of proper collection and storage strategies at each stage of the production chain (food processing, packaging, transportation, and storage) would contribute to the suitable sorting of these materials for their subsequent use.

The present chapter reviews the use of food waste derived from the food manufacturing sector as feedstocks for the production of relevant bioproducts (excluding energy-related products), highlighting the most recent advances for waste conversion from the bioeconomy point of view.

2. The food industry: wastes and valuable by-products

Considering its biochemical nature, food waste can be classified into animal-derived waste and plant-derived waste. In addition to these main groups, food waste can also be divided into seven subcategories according to the waste origin (Galanakis, 2012).

2.1 *Animal-derived waste*

Meat, fish, and dairy products are the three main subgroups included in the animal-derived waste. These food products have largely increased their production and consumption worldwide, specifically in developed countries (Uçak, 2007). The meat industry mainly discards slaughterhouse by-products including skin, blood, bones, meat trimmings, fatty tissues, horns, hoofs, feet, skull, and internal organs. These by-products represent about 60%–70% of the slaughtered carcass, of which 40% forms edible and 20% inedible (Bhaskar et al., 2007). Meat by-products must be treated under strict hygiene conditions to avoid any risk for human health. Furthermore, some diseases such as Bovine spongiform encephalopathy and/or scrapie require special attention and particular treatment rules. Apart from these considerations, the main applications of meat by-products include purification of proteins as food/feed additives (e.g., plasma proteins isolated from blood show gelation and emulsification properties), production of bioactive peptides with antihypertensive, antioxidant or opioid activities, generation of antimicrobial peptides, purification of fat and fatty acids for cosmetic and plastic/rubber polymerization, and the use of hides for leather products (Toldrá et al., 2012).

The fish and seafood industry mainly produces fish leftovers (skin, head, and bones), and shrimp and crab shells. These by-products are mainly used to obtain protein hydrolysates as seafood flavors for soups or surimi, bioactive peptides, collagen, oils rich in omega-3 fatty acids (including eicosapentaenoic acid and docosahexaenoic acid), pepsin, trypsin, chymotrypsin, and collagenase enzymes, and chitin and chitosan polymers for biomedical applications (Bastidas-Oyanedel et al., 2016).

Finally, the dairy industry is a major and important contributor to food waste. The main by-product derived from the dairy industry is cheese whey, which is produced in large quantities (i.e., 85%–90% of the milk becomes whey after cheese processing) (Kosseva, 2011). Whey is mainly used to obtain proteins, salts, fatty substances, lactose, lactic acid, and cleaning chemicals. In addition to cheese whey, other important subproducts from the dairy industry are casein-derived biopeptides.

2.2 Plant-derived waste

The food manufacturing industries include cereals, root and tubers, oil crops and pulses, and fruit and vegetables as the main subgroups from plant-derived waste. Wheat and rice are among the most dominant crops worldwide. These cereals produce bran, husk and/or straw as major by-products from grinding and dehulling, which are rich in high nutritional proteins, dietary fibers, and glucuronoarabinoxylans (Galanakis, 2012). Brewer's spent grain (BSG), the insoluble residue generated from the production of wort in the brewing industry, is another important plant-derived by-product. It contains significant amounts of valuable components including several biologically active compounds such as fatty acids, polyphenols, flavonoids, and antioxidants (Fărçaş et al., 2015). Furthermore, the volatile fingerprint of BSG is comparable to the composition of barley, malt and wheat flour samples.

With regards to roots and tubers, potato is widely consumed over the world. After processing (mainly for the production of chips or French fires), potato generates peels or cull as by-products (Arapoglou et al., 2010). These residues can be subjected to extraction processes for antioxidants recovery (Pathak et al., 2018; Wu, 2016). Cassava is also very popular in South Asia and America, and cassava peels have been also investigated as substrate for microbial protein enrichment (Falade and Akingbala, 2011).

From oil crops and pulses subgroup, sunflower, soybean, and olive are among the most important crops in terms of waste production. Seeds and olive mill pomace are the main by-products and have been identified as a source for phytosterols extraction and bioactive phenols and pectin, respectively (Velasco et al., 2014; Roig et al., 2006).

Fruits and vegetables wastes such as pomace (apple, peach, grape, tomato) and peels (mainly from citrus) are considered the most attractive by-products from the food manufacturing industry and have been widely investigated. These residues are good substrates for the extraction of different types of antioxidants, dietary fibers, pectin, flavonoids, essential oils, and carotenoids (Galanakis, 2012).

2.3 Market potentials from food-derived by-products

There is no doubt about the great potential of food waste as biorefinery feedstock for placing a large number of value-added products in the market. Nonetheless, existing technologies are still at an early stage of development. Techno-economic and profitability analyses must be therefore performed to assess the economic feasibility of the proposed process schemes at large scale. With this aim, different elements can be highlighted for evaluating process viability within a biorefinery context. These elements are (Cristóbal et al., 2018):

(1) estimating the food waste available as feedstock,
(2) evaluating process performance of waste-based biorefineries in terms of resource utilization, and comparing it with the production process from virgin material,
(3) comparing alternative pathways for the valorization of food by-products from the economic and environmental point of view.

Among these parameters, knowing the amounts of food-derived by-products is of utmost important to assess the economic viability of these process. These numbers are usually estimated and more detailed and exact quantities should be therefore provided to reduce uncertainty about the volume of food waste available (Cristóbal et al., 2018; Pfaltzgraff et al., 2013).

Food waste biorefineries are expected to play a key role in the development and implementation of a circular economy. However, they must show to be independent in terms of profitability at the same time that reducing their environmental impact. There are some examples of food waste–based biorefinery that have evaluated the production of value-added chemicals from the economic and environmental point of view (Dávila et al., 2015, 2017; Shahzad et al., 2017; Hernández et al., 2014; Lohrasbi et al., 2010). In most cases, process viability is usually assessed by integrating fuel and energy production with products of high potential revenues (Dávila et al., 2017; Lohrasbi et al., 2010). Notwithstanding, few studies have also assessed the economic viability of proposed plants focusing on the production of value-added chemicals as standalone scenario (Dávila et al., 2015; Hernández et al., 2014). Overall, profitability of evaluated food waste biorefineries are inevitably linked to the market price of the obtained product(s). This element makes crucial an exhaustive analysis of the market prior to the implementation of the corresponding biorefinery. In this context, it is important to keep in mind that the chemical composition of food waste highly influence feedstock potential, since it ultimately determines the compound(s) of interest to be produced. Moreover, feedstock must show high levels of homogeneity and be easily accessible for collection.

Logistics is also an important variable to quantify the profitability of specific food waste biorefineries. Here, plant size (including both processing and storage volume), plant location, and transportation are the main parameters to be considered. The plant size and plant location are directly linked to feedstock availability. Plants will be located in places where larger amount of the corresponding food waste is produced, while the plant size will be directly correlated with the volume of waste produced. Plant size is also limited to feedstock transportation, since large plants demand larger distances for the collection of wastes (de Jong and Jungmeier, 2015). This is decisive for a biorefinery based on waste by-products since long distances increase the risk of spoilage and microbial contaminations of feedstocks with high water content and high organic loadings.

3. Biorefineries based on food processing waste

3.1 Biorefinery based on wine grape waste

Grapes (*Vitis* spp.) are one of the largest fruit crops in the world. In 2017, the International Organization of Vine and Wine (OIV, 2018) has reported a total dedicated area for the production of grapes in about 7.6 million ha (including wine grapes, table grapes and dried grapes). In Europe, grape crop represents 44% of the total cultivation in EU-28, leaded by Spain (13%), France (11%) and Italy (11%). In 2016, about 76 million tons of grapes (excluding intermediate and juices) were produced worldwide, being half of them used for wine production. In 2017, wine production was estimated in 250 million hL, with Italy (42.5 million hL), France (36.7 million hL) and Spain (36.7 million hL) as the main wine producers (OIV, 2018).

The wine sector involves the production of large amounts of residues and by-products including vine shoots, grape marc or pomace, lees and wastewater. Fig. 5.1 summarizes the simplified process for wine production, indicating the points at which these by-products are obtained.

3.1.1 Vine shoots

The main by-product from viticulture activity is the trimming vine shoots, which are generated during vine pruning. Vine pruning is an essential operation to improve the quality and quantity of vine production. Vineyards require a substantial pruning of all plants every year, producing significant amounts of this residue. In the Mediterranean area, the residual biomass available from normal pruning of vineyards ranges from 1.7 to 2.15 tons dry matter/ha/year, showing that the shape of the structure of plantation has a strong influence on the quantity of biomass produced, and therefore on the quantity of pruning material (Velázquez-Martí et al., 2011). In agreement with these values, a recent study has reported the biomass availability and quality produced by vineyard management within a 15-year period, showing yields of 1850–5360 kg fresh matter per ha (moisture content of about 50%) (Manzone et al., 2016). Similar to other lignocellulosic biomass, the chemical composition of vine shoots mainly include cellulose (33.2%–34.1%), hemicelluloses (19%–27%), and lignin (26.0%–24.1%) (Dávila et al., 2017a; Moldes et al., 2007). In addition to these major components, vine shoots contain significant amounts of soluble compounds (i.e., extractives) and ash. The high extractives content, which mainly include phenolic and volatile compounds, is an attractive feature of this biomass. Stilbenoids are an important group from vine shoots extractives. These components have great economic potential as source of high-value phytochemicals such as resveratrol and other secondary metabolites (Cebrián et al.,

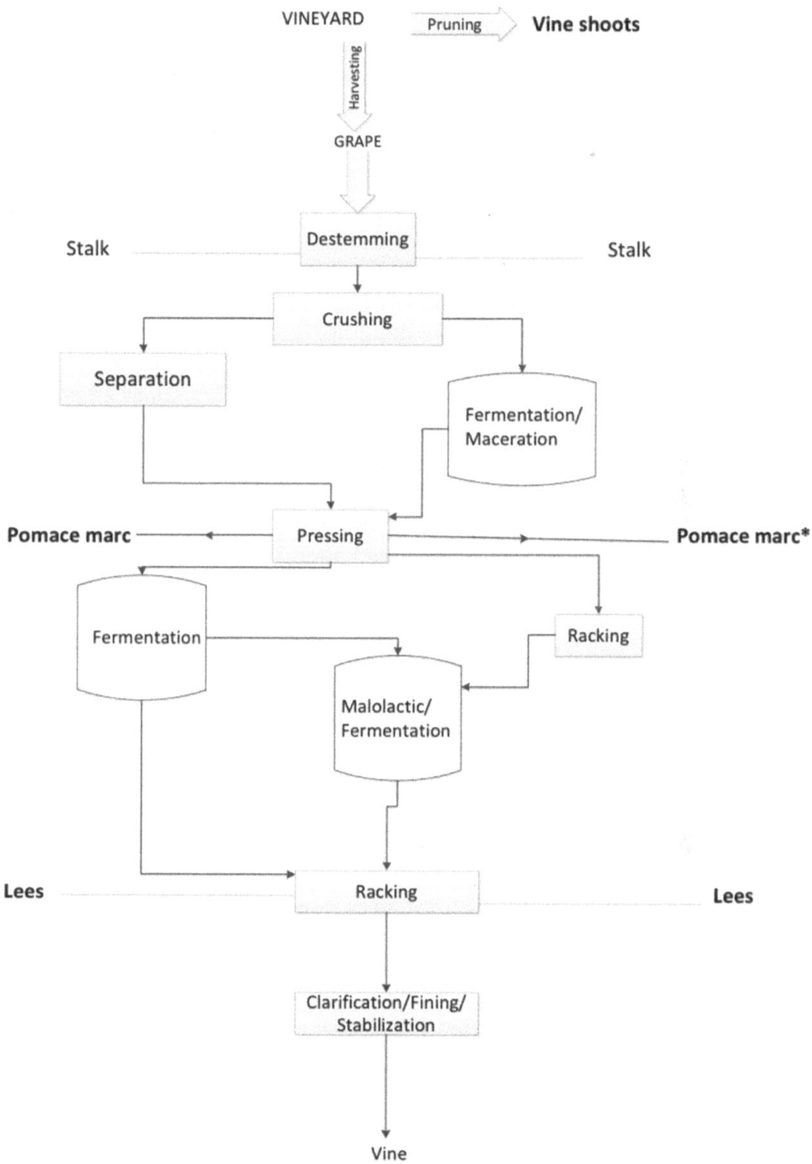

Figure 5.1 Simplified process for wine production.

2017; Gorena et al., 2014). Stilbenes compounds are intensively studied due their beneficial effect on the health. The biological bioactivities of stilbenoids has been reviewed by Akinwumi et al. (2018).

Vine shoots by-products are considered an important energy source due to their composition, offering possibilities for valorization in a multiproduct processing industry. Traditionally, vine shoots have been uncontrolled burned, or mulched and left in the field to increase the organic matter. Burning of these residues in the field has serious environmental concerns and is therefore prohibited in many countries. In this context, this by-product has been investigated as feedstock for the production of stilbenes and sugars within a biorefinery context (Gorena et al., 2014; Devesa-Rey et al., 2011).

The concentration of stilbenes in vine shoots is greatly variable, ranging from 0.8 to 7.9 g/kg of dry matter (Gorena et al., 2014). Such differences are attributed to both the plant variety and the time lag between pruning and processing of vine shoots. In addition, differences in the conditions of the extraction technique, the type and proportion of the solvent used, and the methodology used for the identification and quantification of these compounds, also contributes to variations in the concentration of stilbenes. For instance, the concentration of the monomeric form E-resveratrol showed a strong increase (approximately 40-fold) in eight studied grape varieties canes after six weeks of storage at 20°C, without any further change in the oligomer concentrations (Houillé et al., 2015). This accumulation of E-resveratrol was temperature-dependent, showing an optimal range of 15–20°C, and might a result from the de novo biosynthesis of this molecule after an induction in the expression of the stilbene synthase during storage. Induction of stilbene synthase gene is promoted after the pruning process as a stress-defense response (Billet et al., 2018). Thus, mechanical wounding of freshly pruned cane has been proposed to maximize stilbenoid accumulation while reducing storage time (from 6 to 2 weeks). This storage period therefore represents one of the major considerations for scaling up the process, since it is limited by the time and space required for the preparation of stilbenoid-enriched canes before the grinding and extraction steps.

In addition to stilbenes, carbohydrates are also an attractive fraction of vine shoots, which can be used as building block for their subsequent conversion into a wide of products (Dávila et al., 2017a; Devesa-Rey et al., 2011). Dávila et al. (2017b) studied the suitability of combining different processing technologies for the fractionation of vine shoots, allowing recovering each biomass component in single streams. This strategy strongly facilitates the subsequent conversion and valorization processes of each collected stream. First, vine shoots were hydrothermally pretreated (~200°C, S/L ratio 8 g/g) to collect oligosaccharides and antioxidants. Then, the resulting solid fraction was subjected to delignification with alkali (124°C, S/L ratio 10 kg/kg, 12% NaOH, 105 min). Finally, solubilized lignin was collected by precipitation with sulfuric acid, while the remaining solid fraction was enzymatically hydrolyzed for collecting glucose. Alkaline delignification showed to be a crucial step in this process, yielding 67% lignin removal and allowing collecting a glucan-rich solid easily hydrolysable (hydrolysis yields were close to 100%) (Dávila et al., 2017b). The environmental impact of vine shoots valorization into value-added products has been assessed by Gullón et al. (2018). This study considered different potential valorization scenarios with diverse extraction and delignification steps to identify the most sustainable biorefining route. From the scenarios under study, alkali delignification showed higher lignin removal (about 60%) when compared with other treatments such as organosolv (12%–26%). In contrast, Life Cycle Assessment highlighted worse environmental profiles when using routes involving higher energy and chemical requirements. Notwithstanding, a better process balance might be obtained by introducing "closing the loop" strategies for the optimal reuse and recycling of waste streams (Gullón et al., 2018). Furthermore, as an alternative to delignification processes, Jesús et al. (2017) proposed a sequential two-step autohydrolysis process, that allows lignin recovery after sugar fermentation processes. Briefly, vine shoots are subjected to autohydrolysis (130°C, 60 min) for the recovery of oligosaccharides and phenolic compounds. Then, the remaining solid fraction is subjected to a second autohydrolysis step (200°C, 40 min). The pretreated solid material is then subjected to simultaneous hydrolysis and fermentation for sugar conversion (ethanol was considered as fermentation product in this study). Finally, the spent solid residue can be recovered after the fermentation process for lignin recovery. This strategy allowed the recovery of 13.7 kg of xylooligosaccharides, 3.1 kg of phenolic compounds, 13.1 kg of ethanol and 27 kg of lignin per 100 kg of vine pruning residue (Jesús et al., 2017). Among the major phenolic compounds found in the liquid fraction, these authors identified different hydroxycinnamic acids such as caffeic acid, ferulic acid, chlorogenic acid, and rosmarinic acid, hydroxybenzoic acids such as gallic acid, vanillic acid and syringic acid, flavonoids (e.g., catequine), and stilbenes (e.g., resveratrol).

3.1.2 Grape marc or pomace

Grape pomace or grape marc is the major solid by-product generated during wine making process. This residue usually constitutes about 10%–30% of the processed wine grape weight, depending on the grape variety, viticulture practices, and winemaking process (Beres et al., 2017; Dwyer et al., 2014; Toscano et al., 2013). Pomace is comprised of skins, seeds and any other solid remaining after pressing. It is important to highlight that the type of process and the conditions under which winemaking is carried out markedly influence grape pomace composition. For instance, red grape pomace exhibits a different sugar profile when compared to withe grape pomace, since red grape pomace undergoes a period of fermentation whereas white grape pomace is removed before alcoholic fermentation.

Grape pomace is characterized by high water content (55%–75%), which varies depending on the pressing conditions. Its chemical composition includes dietary fiber as the main component (43%–75%), proteins (6%–15%), lipids, and a wide diversity of phenolic compounds (García-Lomillo and González-San José, 2017). Anthocyanins (present in red pomace) and flavonols are the most abundant polyphenols in grape pomace, which concentrate in the skin and seeds, respectively (García-Lomillo and González-San José, 2017). Great pomace can be sent to distillation, animal feeding and composting, or being used as a source of tartaric acid and coloring agents (e.g., enocyanine E163) (Spigno et al., 2017). As alternative to these conventional practices, this by-product has been extensively investigated due to its large number of healthy and technological compounds with applications in the food, cosmetic and pharmaceutical industries (Beres et al., 2017).

The potential of grape pomace as feedstock for biorefineries and its uses in the food industry and other sectors have been widely reviewed and discussed (Beres et al., 2017;

García-Lomillo and González-San José, 2017). Martínez et al. (2016) have proposed a multipurpose biorefinery for the valorization of red pomace with a four step-cascading biorefinery scheme. This multi-purpose strategy is dedicated to the recovery of phenols by supercritical CO_2 extraction, using 10% ethanol-water mixture as cosolvent, the production of volatile fatty acids by anaerobic acidogenic digestion, which are subsequently converted into polyhydroxyalkanoates, and the production of biogas by subjecting the solid collected from the acidogenic process to anaerobic digestion. The potential advantages offered by this multipurpose strategy are the better exploitation of the equipment, incomes diversification, minimization of waste, and the self-sufficiency from the energy point of view (Martínez et al., 2016). The latest advances on polyphenol extraction from grape pomace, focusing on the use of green solvents and technologies, has been reviewed by Makris (2018). That study highlights the high number of extraction methods developed for the effective extraction of polyphenols from grape pomace, including eco-friendly procedures such as green solvent extraction, ultrasound-assisted extraction, microwave-assisted extraction, pressurized liquid extraction, and supercritical fluid extraction. These processes are considered environmentally benign compared to conventional methods, thus representing an actual choice toward a sustainable industrial production. Nonetheless, most of these solvents and technologies have only been tested at the laboratory scale, and further studies are needed for scaling up these processes (Makris, 2018).

3.1.3 Grape seeds

Grape seeds have been also considered a potential biorefinery feedstock for obtaining value-added compounds before using residual biomass for energy purposes (Lucarini et al., 2018). Grape seeds represent the portion of fruit the highest antioxidant activity and can be easily recovered from pomace by separation and sifting technologies. This wine by-product has shown great potential as a source of nutraceuticals products such as the already mentioned phenolic compounds (e.g., gallic acid, hydroxybenzoic acid, cinnamic acid, quercetin, catechin), unsaturated fatty acids, vitamin E, carotenoids, and phytosterols (Garavaglia et al., 2016; Giannini et al., 2016; Xu et al., 2010). As an attempt to maximize product recovery, after a drying process, dried grape seeds can be first subjected to a solvent-free oil mechanical extraction, rendering a 15% yield (Lucarini et al., 2018). Subsequently, the oil-extracted residue can be then extracted with hydroalcoholic solvent (ethanol:water 70:30, pH 2.5 by HCOOH addition) to recover a phenol-rich solution, containing gallic acid among many other compounds, which has shown high antioxidant and antimicrobial properties (Sorrentino et al., 2018).

3.1.4 Wine lees

Wine lees are the residue formed at the bottom of the corresponding fermentation vessel. It is generated during fermentation and aging of wine. After racking, lees are collected at the winery and then disposed (Fig. 5.1). Wine lees represent 2%–6% of the total wine production (Spigno et al., 2017). The lees consist of a solid phase mainly containing yeast cell, insoluble carbohydrates from the cellulosic and hemicellulose fractions, lignin, proteins, and organic acids (e.g., tartrates), and a liquid phase rich in ethanol, and organic acids (e.g., lactic acid and acetic acid) (Pérez-Bibbins et al., 2015a).

The conventional uses of lees include pigment extraction, nutritional supplements, tartaric acid recovery, land-spreading, and landfill (Spigno et al., 2017). Different processes have been proposed for the valorization of wine lees, including the production of antioxidant extracts. In this context, Dimou et al. (2015) suggested a biorefinery approach for valorization of wine lees. First, wine lees are subjected to a separation process for collecting liquid and solid phases. The liquid fraction is distilled for the production of ethanol, and an alcohol-free nutrient-rich liquid (which is used in the subsequent enzymatic hydrolysis of yeast cells). On the other hand, the solid fraction is used for the recovery of antioxidants and tartaric acid, and to obtain a yeast extract hydrolysate, that is supplemented with glycerol for the production of polyhydroxybutyrate (Dimou et al., 2015).

Another emerging alternative is the use of wine lees as nutrients extract for supplementation of fermentation media. Here, Pérez-Bibbins et al. (2015a) reviewed the recent applications of wine lees with special attention to their use as a sustainable source of nutrients, and for the preparation of yeast extract. Wine less contains residual carbohydrates, nitrogen compounds and yeast-derived compounds such as essential vitamins that makes this by-product and excellent nutrient-rich supplement, similar to yeast extract or corn steep liquor, with potential uses in microbial growth (e.g., lactic acid bacteria) (Salgado et al., 2009). In addition, wine lees might be utilized for the preparation of yeast extract by breaking down yeast cells through autolysis and/or cellular disruption (Pérez-Bibbins et al., 2015b).

3.2 Biorefinery based on brewer's spent grain

Among other agroindustrial side streams suggested as new biorefinery feedstock for the production of potential high value-added compounds, those generated at large amount from the beer-brewing process are seen as interesting feedstock for biorefinery process (Mussatto, 2014). These by-products are readily available throughout the year and they show a chemical composition with different attractive compounds.

In 2017, the global beer production reached around 1909 million hL. From this production, Europe had a total share of 27%, with Germany (93 million hL), United Kingdom (44 million hL), Poland (40.4 million hl) and Spain (37.2 million hl) as the main four EU-28 beer producers (Kirin Beer University Report, 2018). The BSG is the most abundant by-product generated from the beer-brewing process. It is generated during the mashing process and it is removed from the mash filter before the boiling step occurs. About 20 kg wet BSG are usually obtained per 100 L of beer produced (Mussatto et al., 2013). Therefore, an estimated production of 38.2 million tons can be collected worldwide. BSG is considered as a lignocellulosic material. Its chemical composition includes proteins (15%–31%), cellulose (12%–33%), hemicellulose (19%–42%), lignin (12%–28%), starch (2%–12%), lipids (6%–10%), ash (1%–4%), and phenolic compounds (0.7%–2%) (Lynch et al., 2016). However, this composition greatly varies depending on intrinsic factors (e.g., cereal variety, harvesting period, etc.) and extrinsic factors (e.g., type of hops added, configuration for malting and mashing processes, etc.). Also, BSG has high moisture content (70%–80%), which combined to a high carbohydrate and protein content make it susceptible to microbial growth and spoilage, being one of the main limitations for its successful exploitation (Lynch et al., 2016).

The main conventional applications of BSG are animal feed (mainly for cattle) and cocomposting (Nigam, 2017). Research on new process and new products have been suggested for the valorization of BSG, including its use as food additive in human diet, being a substrate for the cultivation of microorganisms and/or enzyme production, as a source of value-added products (e.g., arabino-xylooligosacharides, hydroxycinnamic acids), or as feedstock for xylitol, ethanol, or lactic acid production (Lynch et al., 2016; Mussatto et al., 2013).

Proteins are major components in BSG (Lynch et al., 2016). Among them, hordeins (A, B, and C) constitute over 50% of the total amount of proteins, followed by glutenins (Vieira et al., 2014). This high protein content has high potential for the production of protein hydrolysates with functional features and biological properties of pharmaceutical interest. For instance, Connolly et al. (2015) identified 12 out of 34 peptides isolated from a BSG protein hydrolysate, which worked as antihypertensive agent by exhibiting inhibitory activity of the agiotensin converting enzyme. On the other hand, Vieira et al. (2017) produced a protein hydrolysate with antioxidant properties by subjecting BSG to hydrolysis with brewer's spent yeast proteases. Such protein hydrolysate exerted protective effect against the induced oxidative damage in the human intestinal epithelial cell line Caco-2 and the human hepatocarcinoma cell line HepG2.

Arabinoxylan are the main constituent of the hemicellulose fraction of BSG. The structure of arabinoxylans is built up of a backbone of β-(1,4)-linked xylose residues, which can either be unsubstituted, mono-substituted or di-substituted to a variable extent with L-α-arabinose residues. Also, ferulic acid can be ester-linked to arabinose residues. The presence of other substituents different than arabinose in arabinoxylans backbone of (β-1-4)-linked xylose residues has been described by Coelho et al. (2016). The arabinoxylan from BSG simultaneously contains (methylated) uronic acids, hexose, and acetyl groups as side chain. Such structural variety makes arabinoxylooligossaccharides derived from BSG an excellent source for a large number of prebiotic compounds. Furthermore, these characteristics offer the possibility to provide molecules with both fast and slow probiotic fermentation rates (Coelho et al., 2016). However, valorization of BSG for the production of prebiotic oligosaccharides is challenged by the arabinoxylans extraction methods and the arabinoxylan-to-oligosaccharides conversion processes. In this sense, the enzymatic processing of arabinoxylan with xylanases has been proposed to hydrolyze arabinoxylan and produce arabino-xylooligosaccharides of the desired chain length, with potential to be used as prebiotics (Sajib et al., 2018).

BSG is also rich in phenolic compounds. Both hydroxbenzoic acids and hydroxycinnamic acids have been reported to be present in BSG. Hydrobenzoic acids are usually found in lower concentrations than those observed for hydroxycinnamic acids.

Ferulic acid and *p*-coumaric are the most abundant phenolic compounds found in BSG (Moreira et al., 2013). These hydroxycinnamic acids have shown antioxidant properties with important industrial applications in the food, cosmetic, and pharmaceutical sectors (Mussatto, 2014). Several extraction techniques have been proposed to recover these phenolic compounds from BSG. Guido and Moreira (2017) have reviewed and discussed different extraction techniques for the recovery of the phenolic fraction from BSG, highlighting major strengths and weaknesses of each technique. Traditional methods include solid-liquid extraction procedures, with the use of organic solvents, alkaline or enzyme pretreatment among the most studied. Furthermore, microwave-assisted extraction has been also proposed. This technique increased by five-fold the resulting recovery yields when compared to conventional extraction methods (Moreira et al., 2012).

Based on its composition and the great diversity of compounds that can be obtained from this by-product, several integrative approaches have been proposed for the valorization of BSG. For instance, Vieira et al. (2014) proposed a simultaneous extraction of proteins and arabinoxylans from BSG by using alkali reagents. A sequential extraction process with increasing alkali (KOH or NaOH) concentrations was performed for the recovery of proteins and arabinoxylans from BSG. Subsequently, alkaline extracts were acidified with citric acid to obtain the protein-rich fraction with 82%–85% yields (Vieira et al., 2014). On the other hand, arabinoxylans were recovered by ethanol precipitation with 66%–73% yields. This system also considered citric acid and ethanol recycling, and allowed the recovery of a cellulose-rich fraction that could be further processed via glucose-based conversion strategies.

The implementation of a biorefinery based on BSG in Brazil has been assessed from the techno-economic and environmental point of view. This plant was intended to produce lactic acid, xylitol, activated carbon and phenolic acid compounds as value-added products (Mussatto et al., 2013). In brief, the process consisted in a diluted acid pretreatment stage for the recovery of hemicellulosic monomeric sugars (mainly xylose), which was further subjected to xylitol production by microbial fermentation. The pretreated solid fraction was subjected to a soda process, and then enzymatically hydrolyze to produce a glucose-rich hydrolysate that was used as substrate for the production of lactic acid. Finally, the black liquor resulting from soda pulping was sent to the phenolic acids and activated carbon production plant. The corresponding assessment highly recommended the combination of energy and mass integration for the implementation of the suggested scheme in Brazil (Mussatto et al., 2013). A simplified block diagram describing the proposed biorefinery is schematically depicted in Fig. 5.2.

Dávila et al. (2016) proposed an alternative process for converting BSG into xylitol, ethanol, and polyhydroxybutyrate, which is shown in Fig. 5.3. Xylitol is obtained by fermentation from the liquid fraction collected after pretreatment with diluted acid. Ethanol is obtained through fermentation of the corresponding enzymatic hydrolysate. The remaining solid residue (mainly lignin) is recovered together with cell biomass subjected to gasification for generation of electricity. Polyhydroxybutyrate is produced from the amount of glucose that remains after ethanol fermentation. When evaluating techno-economic and environmental analyses, two biorefinery scenarios (with and without heat integration) were presented. Heat integration strategy had both

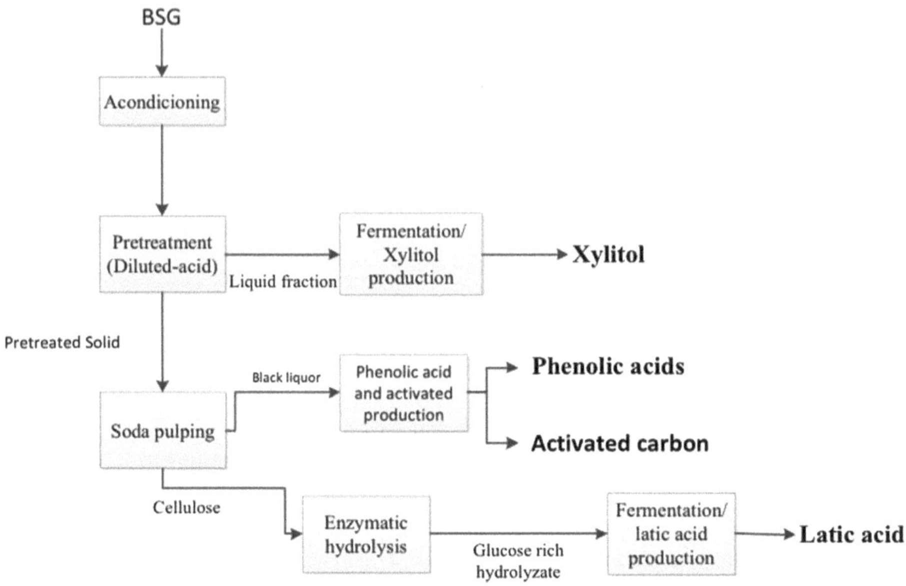

Figure 5.2 Simplified scheme for the production of xylitol, lactic acid, activated carbon and phenolic acids of a biorefinery based on brewer's spent grain (BSG).

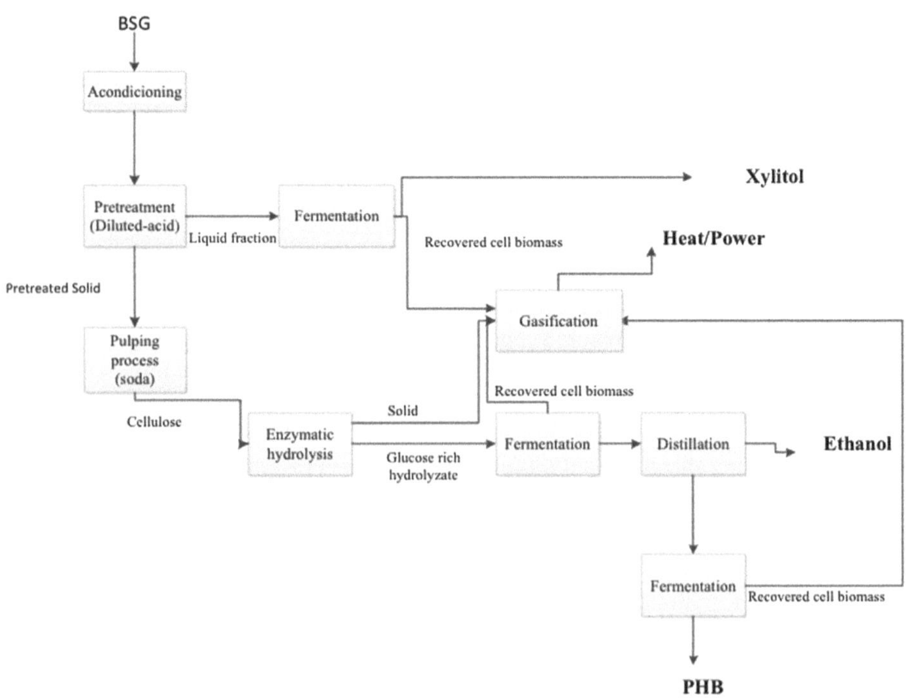

Figure 5.3 Simplified scheme describing a proposed biorefinery based on brewer's spent grain (BSG) for ethanol, xylitol, and polyhydroxybutyrate (PHB) production.

environmental and economic implications, contributing to lowering the potential environmental impact and reducing by 43% total production costs. Furthermore, this study clearly indicates the unfeasibility of ethanol and xylitol production in standalone manner, but attractive value-added products within the biorefinery context (Dávila et al., 2016).

3.3 Biorefinery based on olive waste

The health benefits attributed to olive oil consumption have spread the cultivation of olive trees from the Mediterranean region to all over the world. This crop has a total dedicated area of 10.5 Mha in more than 40 countries, producing about 200 million tons of olives annually (FAOSTAT, 2016). The global olive market is led by Spain, which produced 6.6 million tons of olives in 2016. In Spain, olives are mainly processed through a mill technology that include a two-phase centrifugation system. This system was introduced in 1992 to reduce the tremendous amounts of wastewater generated from the three-phase centrifugation method, since this residue was illegally dumped to the soil or into a nearby stream or river for many years (Roig et al., 2006).

From the olive industry, olive pomace, olive stones, olive leaves, wastewater and olive tree pruning can be obtained as major wastes. With the aim of reducing the environmental impact and their quantity, different value-added compounds may be obtained within a biorefinery context, which are summarized in Fig. 5.4.

3.3.1 Olive pomace

Independently of the method used, olive pomace is recovered as the main solid by-product after collecting olive oil, with yields of 0.80 kg/kg (two-phase system) and 0.55 kg/kg (three-phase system) (Stamatakis, 2010). The chemical composition of this residue is however different depending on the production process. One of the major differences is moisture content. Olive pomace from the two-phase process shows 55%–70% moisture content, compared to the 40%–45% observed for the analogous three-phase by-product (Borja et al., 2006). Furthermore, the two-phase pomace has higher carbohydrate content and possesses high chemical oxygen demand and biological oxygen demand due to its high organic proportion.

The first approach for valorization of olive pomace, which is in general performed by the mill industry, is a second oil extraction with organic solvents. Olive pomace is first dried and then subjected to a solid-liquid extraction with hexane. The higher water content of the two-phase processing olive pomace is a great limitation for this extraction approach, since in addition to require special storage tanks, mass pumps and tank trucks for the proper management (Roig et al., 2006), the drying process has a high energy demand, thus increasing operational costs. Furthermore, the high temperatures required during the drying process lead to the formation of benzopyrenes that may limit the subsequent commercialization of the extracted oil. It is also important to have in mind that olive oil is produced in a short period of time, what force the storage of olive pomace in large open-air ponds, being a source of different environmental problems (Romero-García et al., 2014).

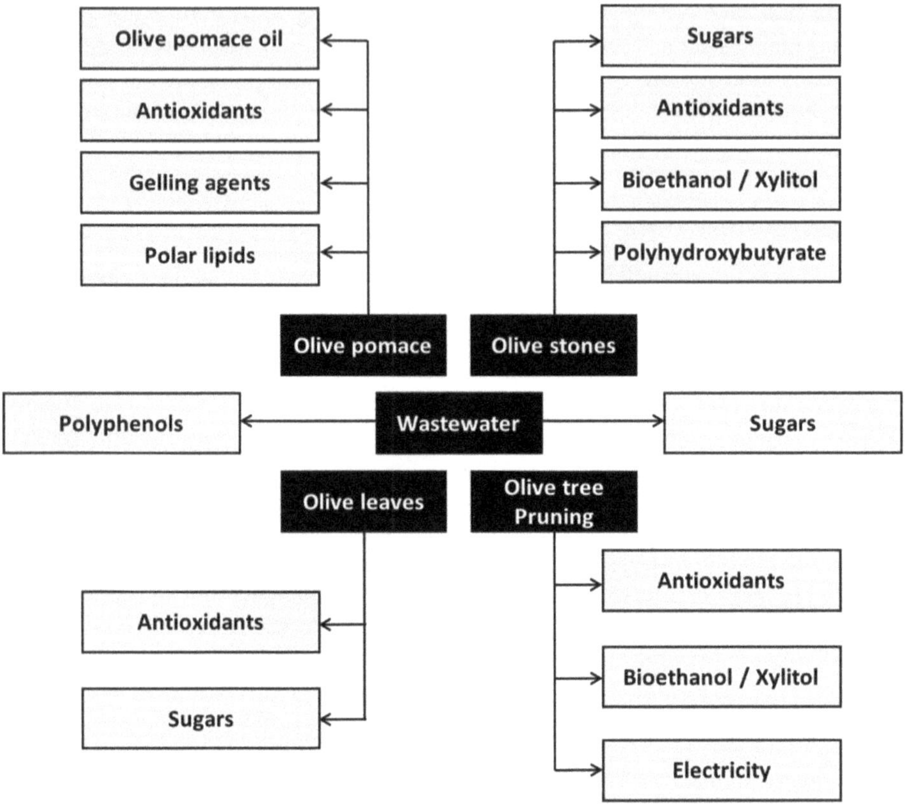

Figure 5.4 Value-added compounds obtained from olive pomace, olive stones, olive leaves, wastewater, and olive tree pruning by-products within a biorefinery context.

Several studies have been performed, mainly at the laboratory scale, as an attempt to reduce the environmental impact of olive pomace and use it as a source of value-added compounds within a biorefinery context (Negro et al., 2017). Olive pomace is characterized for having slightly acidic pH, high organic content, high sugar and lipid concentration, and high C/N ratio (Roig et al., 2006). It is also rich in potassium and has a low concentration of toxic heavy metals (e.g., Pb, Cd, Cr and Hg). Olive pomace has been investigated as livestock feed (after protein supplementation), and as feedstock for bioethanol, biogas and compost production through biotechnological processes (Tayeh et al., 2014; Rincón et al., 2013; Federici et al., 2011). It can also be used for soil as heavy metals sorbent and/or to extend and strengthen sorption of some herbicides and insecticides (e.g., simazine, imidacloprid), thus retarding biodegradation and leaching and reducing the risk of groundwater contamination (Albarrán et al., 2004; Cox et al., 2004; Pagnanelli et al., 2002). Furthermore, olive pomace can be mixed with thermoplastic polymers to obtain new materials for the manufacture of containers (Siracusa et al., 2001).

Besides all those applications, olive pomace has been considered a great source of value-added products with biological and pharmaceutical interest due to the presence of a wide range of phenolic compounds (Negro et al., 2017; Ruiz et al., 2017). Lozano-Sánchez et al. (2011) identified the presence of several polyphenols such as phenolic acids and alcohols, secoiridoids, lignans, and flavones. These compounds have great interest as food additives and nutraceutical ingredients due to their antioxidant and pharmaceutical activities. Hydroxytyrosol is one of the most abundant phenols in olive pomace (about 50% of the total phenolic content) (Nunes et al., 2018). This molecule is known not only for having a strong antioxidant activity, but also possesses antiinflammatory, antimicrobial, and antitumor properties as well as a positive role in cardio and neurovascular diseases. For instance, Granados-Principal et al. (2014) treated a rat in vivo model with hydroxytyrosol, improving the antitumor response and showing a lower number of adverse effects associated with the doxorubicin-based treatment. This effect was attributed to the improvements of the cardiac disorders promoted by the oxidative stress and mitochondria damage caused by the therapy. Nunes et al. (2016) highlighted the use of hydroxytyrosol as food additive due to its potential to improved chemical, nutritional and sensory properties of food. Tyrosol, squalene, verbascoside, oleuropein, caffeic acid, β-sitosterol and pinoresinol have also been identified in olive pomace (Seçmeler and Güçlü Üstündağ, 2017; López-Biedma et al., 2016; Chandramohan et al., 2015; Ramos et al., 2013; Loru et al., 2009; Stavroulias and Panayiotou, 2005). These compounds have shown potential in the pharmaceutical industry, especially for preventing neuro- and cardiovascular-related diseases, and exert antiosteoporosis, antiinflammatory, antidiabetic, and antitumoral activities. Chandramohan et al. (2015) used tyrosol to treat diabetic rats, reverting all the analyzed biochemical parameter to near normal in a dose dependent manner. López-Biedma et al. (2016) showed that pinoresinol may act as adjuvant to prooxidative chemotherapies by promoting intracellular oxidative stress in breast cancer cells after H_2O_2 treatment. In contrast, pinoresinol exerts a chemical antioxidant capacity in human mammary epithelial cells, which may have a therapeutic potential to prevent breast cancer development. Similarly, Ramos et al. (2013) used phenolic extracts from a dry olive mill residue, showing its antiproliferative activity against a specific breast cancer cell line.

Several methodologies can be applied to recover the phenolic fraction from olive pomace. Solid-liquid extraction procedures are the most widely applied methods. This methodology requires fully agitation, long surface contact and stable temperatures. However, its industrial applicability is limited due to the high energy demand and the high amounts of solvents needed (Da Porto et al., 2013). As alternatives, ultrasound- and microwave-assisted extractions, and the use of supercritical fluids have also been investigated for the extraction of phenols from olive pomace (Chanioti and Tzia., 2017; Schievano et al., 2015; Sánchez de Medina et al., 2012).

In addition to phenolic compounds, olive pomace can be considered as a source of gelling components. Here, olive pomace contains cellulose, glucuronoxylans, and arabinan-rich polysaccharides, which have shown to work as gelling agents, stabilizers and emulsifiers, and can therefore be applied in the food industry (Coimbra et al., 2010). On the other hand, lipids and fatty acids from olive pomace are also attractive components. In particular, polar lipids have shown in vivo antiatherogenic properties in hyperlipidemic rabbits by inhibiting the development, and even promoting their regression, of atherosclerotic plaques (Tsantila et al., 2010; Karantonis et al., 2006).

3.3.2 Olive stones

After oil separation in the olive oil extraction process, crushed olive stones can be recovered (Romero-García et al., 2014). This by-product can be also collected from the table olive industry, since they are usually removed before further processing. Olive stones are lignocellulose-based by-products with cellulose (28.1%–40.4%), hemicelluloses (18.5%–32.2%), and lignin (25.3%–27.2%) as main components (Matos et al., 2010; Rodríguez et al., 2008). Stones represent 10%–15% of olives in dry weight basis and they have been traditionally used in domestic boilers and in larger plants for power cogeneration (Lama-Muñoz et al., 2014; Pattara et al., 2010). It can also serve as a solid fuel in public heating systems or as feedstock for gasification processes (Vera et al., 2013; López et al., 2010). With regards to the production of bio-based compounds, Lama-Muñoz et al. (2014) highlighted the use of olive stone as a source for antioxidants and sugars recovery. Ben Mansour et al. (2015) identified oleoside, oleoside 11-methyl ester, nuezhenide, oleoside 11-methyloleoside, nuezhenide 11-methyloleoside, oleuropein, and glycosides of tryosol and hydroxytyrosol glycosides as the major phenolic compounds extracted with aqueous methanol from Chemlali olive stones, with nuezhenide as the predominant phenol. On the other hand, glucose and xylose are the main recovered sugars, which can be further utilized as feedstock for the production of xylitol, furfural, ethanol, and polyhydroxybutyrate (Hernández et al., 2014). During processing of olive stone, a lignin-rich solid might be also left at the end of the processes (Hernández et al., 2014; Lama-Muñoz et al., 2014). This final residue can be also transformed into phenols, biopolymers or fibers, or directly burned for energy production.

3.3.3 Olive leaves

In addition to olive pomace and olive stones, leaves collected during olive harvesting (about 6% of olives by weight) are separated at the early stages of the process (Romero-García et al., 2016a). These leaves are usually discarded or utilized as animal feed. Nonetheless, they can be subjected to valorization as another olive-based by-product. Similar to olive stones, olive leaves have been investigated as a source of antioxidants and sugar compounds. So far, oleanoic acid, mannitol and oleuropein have been found at higher proportions in olive leaves (Guinda et al., 2015). Among them, oleanoic acid is of great interest due to its activity against diabetes and metabolic syndromes. Direct extraction methods can be used for antioxidants recovery, including the solid-liquid extraction methods, Soxhlet extraction, pressurized liquid extraction, extraction with supercritical fluids, or microwave- and ultrasound-assisted extractions (Rahmanian et al., 2015). Both sugar and antioxidant recovery can be simultaneously collected by subjecting olive leaves to steam explosion pretreatment. Under optimal process conditions (180°C, 8.3 min), steam-exploded hydrolysate may yield up to 70% of the initial sugar content and an interesting concentration of antioxidants, including oleuropein, hydroxytyrosol, and flavonoids (Romero-García et al., 2016a).

3.3.4 Olive mill wastewater

Together with solids by-products, olive mill wastewater is obtained from olive processing in large quantities. This liquid by-product is collected at different stages of the olive oil production: (1) during olive washing, (2) from the horizontal centrifuge (decanter) during the three-phase separation step, and (3) during the washing process of virgin olive oil (Ruiz et al., 2017). Olive mill wastewater represents a serious environmental problem since it exhibits high values for most pollution parameters: biological oxygen demand (40−95 g/L), chemical oxygen demand (50−180 g/L), lethal dose toxicity for fish (8.7%), and contains large amounts of suspended solids and high concentrations of polyaromatic compounds (0.5−24 g/L) (Ouzounidou et al., 2010). With the aim of reducing the potential environmental impact of olive mill wastewater and gain value-added compounds, this residue has been subjected to several treatment processes for isolation of major relevant products. Phenolic compounds, including hydroxytyrosol, verbascoside, and isoverbascoside, are among the most attractive olive mill wastewater compounds (Cardinali et al., 2012). Phenols from olive mill wastewater can be recovered by membrane filtration, liquid-liquid extraction, and ultrasonic-assisted extraction (Cassano et al., 2013; Klen and Vodopivec, 2011; Scoma et al., 2011). La Scalia et al. (2017) have shown the economic feasibility of coupling membrane filtration and reverse osmosis for phenol extraction, which can subsequently be used to enrich Extra Virgin Olive Oil during the malaxation phase. After phenol extraction, olive mill wastewater offers the possibility for sugars recovery, including glucose, galactose, arabinose, rhamnose, and galacturonic acid (Nadour et al., 2015). Also, pectins, arabinans, arabinose-rich glycoproteins, xyloglucans, glucuronoxylans, and glucans were found as major polysaccharides. These polysaccharides exert a potent prebiotic and antioxidant activities, and therefore offers good possibilities for their valorization.

3.3.5 Olive tree pruning

Similar to vine pruning, pruning of olive trees is performed every 2 years to regenerate the fruiting surface and eliminate unproductive old branches. These practices result in an annual production of 1.5−3.0 tons of olive tree pruning biomass per ha cultivated. Olive tree pruning is a lignocellulosic material, with cellulose (26.1%−36.6%), hemicellulose (20.0%−25.0%), lignin (17.9%−27.7%), and extractives (14.1%−31.4%) as main chemical components (Romero-García et al., 2014). Traditionally, this residue has been burned *on site* or used as soil vegetal additive. As alternative, olive tree pruning has been considered as feedstock for combustion, pyrolysis or gasification. Biotechnological applications of olive tree pruning include its conversion into ethanol and/or xylitol via pretreatment, enzymatic hydrolysis, and fermentation. Extraction of antioxidants from olive tree pruning has also been proposed, being a natural source of hydroxytyrosol, homovanillyl alcohol, and oleuropein (Conde et al., 2009). In this sense, Romero-García et al. (2016b) investigated the economic feasibility a biorefinery based on olive tree pruning biomass for antioxidants recovery and ethanol production. The plant combined a thermochemical technology for recovering antioxidants and an energy self-sufficient biochemical platform for ethanol production. Technoeconomic assessment based on the total production costs of each coproduct demonstrated the feasibility of this integrative strategy at medium scale (Romero-García et al., 2016b). However, it should be noted that the use of olive tree pruning biomass as a biorefinery feedstock is limited by the discontinuous supply and the high costs associated to biomass collection.

3.4 Biorefinery based on potato peels

Potato is one of the most commonly consumed vegetables worldwide. Its chemical composition is characterized by a high carbohydrate content, mainly starch. Apart from starch, potato also contains proteins, vitamins, minerals and phytochemicals, including carotenoids and phenolic compounds. Potato processing industry (production of French fries, chips, and puree) produces peels as major by-product. The potato peeling process yields between 48 and 129 million tons annually of waste (Van Dyk et al., 2013). This value directly depends on the peeling procedure used (e.g., abrasion, drum-shaped mechanized knife peelers, lye peeling, etc.).

Potato peels by-product can be used as fertilizer, for animal feeding, as substrate for microbial growth and/or enzyme production, and as feedstock for the extraction of chemicals such as antioxidants, steroidal alkaloids, polysaccharides, and for the production of biopolymer films. The latest advances about the potential uses of potato peels as raw material for the production of value-added compounds has been reviewed by Pathak et al. (2018). Fig. 5.5. summarizes the main processes described for the production of value-added compounds from potato peels within a biorefinery context.

The antioxidant capacity of potato peels extracts is one of the most widely studied applications of this by-products. Sabeena Farvin et al. (2012) supplemented Mackerel mince with an ethanol extract of potato peel for preventing lipid and protein oxidation. In addition to demonstrate its antioxidant capacity, this extract exerted a protective activity against the loss of α-tocopherol and tryptophan and tyrosine residues from the minced Mackerel. Traditional methods for phenol recovery from potato peels mainly include the use of organic solvents (Proestos and Komaitis, 2008). However, these methods exhibit high solvent consumption and high toxicity, and require long extraction times which might promote degradation of thermo-labile compounds. The use of alternative green solvents such as ethanol or glycerol might represent an actual choice to produce extracts of potato peels enriched in polyphenolic antioxidants (Paleologou et al., 2016). Pressurized liquid extraction, microwave-assisted extraction and subcritical water extraction also represent attractive, greener methods for the extraction of polyphenols from potato peels (Paleologou et al., 2016; Singh et al., 2011; Singh and Saldaña, 2011). Major phenolic compounds extracted from potato peels

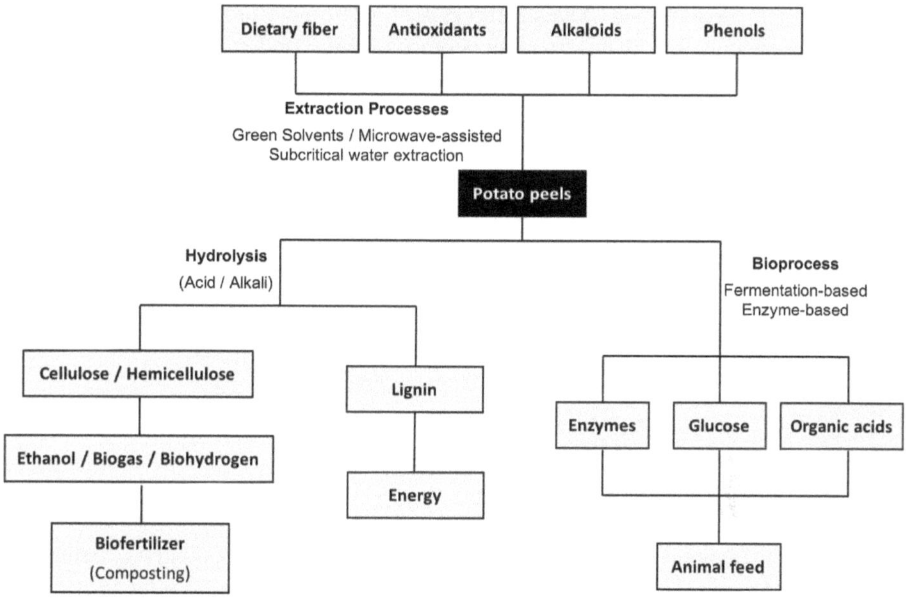

Figure 5.5 Proposed scheme for a biorefinery based on potato peels.

by-products include chlorogenic acid, caffeic acid, gallic acid, hydroxybenzoic acid, and syringic acid among others (Singh and Saldaña, 2011).

Potato peels also represent a good source for the extraction of dietary fibers (cellulose, hemicellulose, lignin, pectin, etc.). Dietary fibers from potato peels have shown superior properties than those isolated from wheat bran in terms of water-holding capacity, mineral content, and total dietary fiber, and have no phytate and the amounts of starchy components are lower (Toma et al., 1979). These dietary fibers can be extracted by combining enzymatic and chemical methods (Dhingra et al., 2012).

The high starch content of potato peels and the presence of steroidal alkaloids make this by-product attractive for medicinal use (Survase et al., 2016). For instance, potato peels have shown to be effective to prepare bandages to treat burn wound (Keswani and Patil, 1985), to ameliorate the oxidative stress in streptozotocin diabetic rats (Singh et al., 2005), or to significantly offsets CCl_4-induced liver injury in rats (Singh et al., 2008). Steroidal alkaloids are interesting potato-peel-derived compounds, since they serve as precursors to agents with anticancer and antiinflammatory properties. Maldonado et al. (2014) proposed an integrative extraction process for the simultaneous recovery of antioxidants (neochlorogenic acid, chlorogenic acid and caffeic acid) and glycoalkaloids (α-solanine, α-chaconine, and α-solanidine) with a solvent mixture containing water (46%), ethanol (51%), and acetic acid (3%). Hossain et al. (2014) reported and environmentally friendly, low-cost extraction method based on ultrasonic extraction to recover α-solanine and α-chaconine from potato peel. Steroidal alkaloids can be also potent cytotoxins (promote cell lysis), depending on the dose, and require chemical modifications for reducing this cytotoxicity by maintaining or even improving their bioactivities. Alves-Filho et al. (2018) have proposed an alternative ultrasound-assisted method for extraction of α-solanine and α-chaconine and their simultaneous conversion into β-solanine and β-chaconine in a single step without the use of organic solvents.

Within the biorefinery context, potato peels also serve as an excellent substrate for the production of value-added compounds by fermentation-based processes. This includes the production of biofertilizers (Pandit et al., 2012), industrially-relevant enzymes (e.g., amylases, polyphenol oxidase, etc.) (Mukherjee et al., 2012; Shukla and Kar, 2006), organic acids (e.g., lactic acid) (Liang et al., 2015), sugars (mainly glucose from starch), adsorbents for water treatment (Pathak et al., 2015), biofilms/biocomposites (Wei et al., 2015), corrosion inhibitors (Ibrahim et al., 2011), or cationized starch (Lappalainen et al., 2015).

3.5 Other biowaste-based biorefineries

3.5.1 Tomato pomace

Tomato has an annual production of 100 million tons in 144 countries, being the second most important vegetable crop after potato (Kalogeropoulos et al., 2012). The tomato processing industry yields between 4.3 and 10.2 million tons of waste in the form of tomato pomace, which mainly includes tomato peels and seeds, and small fraction of pulp. Several value-added compounds of economic interest can be obtained from tomato peels and seeds, including carotenoids (mainly lycopene, β-carotene) oil, proteins, or polyphenols. Kalogeropoulos et al. (2012) isolated significant amounts of bioactive phytochemicals (e.g., tocopherols, polyphenols, carotenoids, terpenes, sterol, etc.) from tomato by-products that are known to exert antioxidant activities. Tacking advantage of these antioxidant activities, Ćetković et al. (2012) used tomato waste extracts to scavenge hydroxyl radicals, showing antiproliferative effects in HeLa, MCF7, and MRC-5 cell lines. Tomato by-products are also a good source for dietary fiber and macronutrients, therefore having applications in the food industry as ingredient of new functional foods.

Tomato pomace has been significantly studied as a source of lycopene and β-carotene. These compounds can be extracted through several methods, including green organic solvents such as ethyl acetate (Strati and Oreopoulou, 2011), enzyme-assisted extraction (Zuorro et al., 2011), and supercritical fluids (Machmudah et al., 2012).

A potential biorefinery based on tomato pomace was suggested by Kehili et al. (2016). In a first step, carotenoids were extracted by supercritical CO_2 (80°C, 400 bar, 4 g CO_2/min for 2 h), yielding 60% of the initial lycopene and β-carotene in tomato pomace. This yield represents 58.8% of the initial carotenoid content of peels, and 80%−90% of the initial lycopene and β-carotene content of seeds. The resulting extracted residue was then subjected to protein extraction, and subsequently to liquid hot water treatment for hemicellulose and cellulose recovery (Kehili et al., 2016). These results demonstrate, at the laboratory scale, the capacity of tomato pomace as a biorefinery feedstock. Nonetheless, further research is needed to evaluate the full potential of the system at higher scale.

3.5.2 Apple pomace

The apple juice industry produces a large amount of apple pomace (peels, seeds, stems, calyces, and flesh), which represents 25%–35% of the apple after the juicing process. Van Dyk et al. (2013) reported a total production of about 17–24 million tons of apple pomace in 2010. Apple pomas has high water content and its main constituents are simple sugars (glucose, fructose, and sucrose) and carbohydrates (cellulose, hemicellulose), pectin, proteins, vitamins, and minerals.

Several routes for the recovery of valuable compounds from apple pomace have been already established, even at industrial scale (Schieber, 2017). Also, apple pomace can be used as substrate for microbial growth and/or the production of fermentation-based products. Pectin is considered the most important fraction in apple pomace, having applications as thickening agent in foods, cosmetics, and pharmaceuticals. Pectin is usually extracted using diluted mineral acids and is then concentrated and precipitated to obtain the alcohol insoluble substance. As alternative to extraction with mineral acids, microwave-assisted extraction (Wang et al., 2007), and subcritical water extraction (Wang et al., 2014) has also been propose for pectin recovery from apple pomace. Phenolic compounds from apple pomace are attractive product as well. Apple pomace extracts with high phenolic content might support the prevention of chronic diseases, such as cardiovascular diseases and cancer (Wolfe and Liu, 2003). Also, extracts of apple pomace have great potential as food additive due to the antioxidant activity against lipid oxidation (Huber and Rupasinghe, 2009). Apple seed oil, extracted from the apple seeds contained in apple pomace, is another important product with application in the cosmetic, food, and pharmaceutical sectors (Walia et al., 2014). As feedstock in bioconversion processes, apple pomace represents and good source for the production of organic acids or aroma compounds among others (Madrera et al., 2015; Alonso et al., 2009).

3.5.3 Citrus waste

The edible portion of citrus fruits is smaller than other types of fruits. Therefore, large amounts of wastes (peels and seeds) are discarded during juice processing. Considering that only 1/3 of the fruits citrus is used for juice production, the annual production of citrus waste can be estimated in 15 million tons (Ledesma-Escobar and Luque de Castro, 2014). Citrus waste is mainly composed of water, soluble sugars, fiber, organic acids, proteins, minerals, oils, and flavonoids and vitamins (Fernández-López et al., 2004). The potential uses of citrus waste have been reviewed by Mamma and Christakopoulos (2014).

Similar to apple pomace, pectin is an attractive and important product from citrus waste. This gelling agent is usually extracted with nitric acid, although environmentally friendly alternatives such as enzymatic extraction are also considered (Dominiak et al., 2014). Essential oils, mainly limonene, can be obtained from citrus peel, having direct applications in cleaning agents such as degreasers, release agents, part washers, and dip bath (Ledesma-Escobar and Luque de Castro, 2014). In addition to the solid residues, wastewater generated during juicing (e.g., the orange juice wastewater) may be considered as an interesting source of citrus fiber. This fiber can be used as food additives to increase the dietary fiber content in processes food (e.g., cooked and dry-cured sausages) (Fernández-López et al., 2004), or as fat replacer in ice cream (Crizel de Moraes et al., 2013).

3.5.4 Dairy by-products

The dairy industry is an important part of the food industry and greatly contributes to generates liquid wastes. One of the main wastes from the dairy industry is cheese whey. About 9 kg of cheese whey is obtained per kg of cheese produced (Prazeres et al., 2012). In Europe, it has an estimated annual production of about 50 million m^3 (Kosseva, 2011). Cheese whey contains 93%–94% of water, lactose, soluble proteins, minerals, lactic acid, and fats (Prazeres et al., 2012). This residue represents one of the biggest reservoirs of food proteins and sugars (mainly lactose) and has been extensively investigated for recovery purposes (Galanakis, 2012). In general, cheese whey can be used directly in the liquid form, or treated and transformed into various food products, including powdered cheese whey, lactose, lactose-free by-products, and protein concentrates.

Ultrafiltration (three to five bars, 1–1000 kDa), nanofiltration (three to five bars, 100–500 Da), or ultrasound-assisted ultrafiltration (~20 kHz, 3–5 bars, 1–1000 kDa) have been described as efficient methods for the recovery of valuable products from whey (Gajendragadkar and Gogate, 2016). By ultrafiltration, whey yields a retentate rich in proteins and some fats, and a permeate with lactose, mineral and water. The retentate can be used to produce whey protein concentrate powder, while the permeate is used as lactose source. Furthermore, by subjecting ultrafiltered-permeate to nanofiltration, a salt-depleted lactose concentrate, and a salt-enrich permeate with low organic content are obtained (Minhalma et al., 2007). Cheese whey concentrates (0.5%, 1.5% and 3%) have been utilized as natural sanitizing agent for washing lettuce or carrots, showing effective decontamination of fresh-cut vegetables (Martin-Diana et al., 2006). Regarding permeate, Paseephol et al. (2008) reported the potential use of milk permeate as feedstock for the production of lactulose, using $CaCO_3$-based catalysts. This compound has excellent commercial applicability in food and pharmaceutical sectors since it reduces the intestinal pH, improves the intestinal microbiota, and helps to smooth fecal excretion. Lactose-rich permeates, together with yogurt whey from expired products and cheese whey, can also be employed as feedstock for lactic acid production (Alonso et al., 2010), which is widely used in the food and pharmaceutical sectors, and as raw material for biopolymers production.

4. Legislation on waste management

Legislation is an important aspect for waste management since directs and regulates the production, handling, storage, transport, and disposal of waste with the aim of minimizing the health and environmental risks. Food waste must be treated as nonhazardous whenever it does not manifest any evidence to be considered as "hazardous" (Ravindran and Jaiswal, 2016). This does not apply to animal-derived waste, which is subjected to more strict hygiene considerations and important constraints to prevent dissemination of serious diseases such as the Bovine spongiform encephalopathy (Toldrá et al., 2012).

In Europe, Directive 2008/98/EC (2008) rules waste handling and management hierarchically according to the following priority order: prevention of waste generation, reuse, recycling, energy-recovery, and disposal (e.g., landfilling). In the particular case of food waste, this directive aims at collecting biowaste to ensure maximum environmental protection by promoting composting and anaerobic digestion of residues and/or by treating by-products in biorefineries for their valorization within the bioeconomy concept. In this context, the European Parliament has published a new set of rules for amending this Directive, targeting at the separate collection of biowaste from household, to be implemented by 2023, thus allowing a better sorting from the main food waste producer (Directive 2018/851, 2018). The European Parliament and Council also regulates the production and marketing of newly produced chemicals according to the Regulation (EC) N° 1907/2006 (2006). This Regulation, also known as REACH, may however limit the production of new chemicals, especially those from small-scale producers, as it establishes a threshold of 1 ton of product per year as the minimal amount above which the manufacturer shall register the product. Nonetheless, it must be highlighted that development and implementation of a sustainable circular bioeconomy is one of the central pillars for the European Commission to face important issues such as the increasing food demand, depletion of natural resources, and climate change (E4tech Report, 2015).

5. Concluding remarks

Strategies for waste management in the food processing industries have identified a bunch of value-added compounds with applications in the food, cosmetic, industrial, and pharmaceutical sectors, thus replacing "waste label" by "by-products." Despite the great research efforts for developing novel effective conversion processes, most of these systems are still far from their commercialization. Small and/or large scale biorefineries will definitely contribute to the implementation of this new economic model, taking advantages of the industrial symbiosis for boosting process integration and the cooperation between different facilities. These biorefineries will also minimize their environmental impact by using green solvent chemicals and sustainable emerging technologies through the whole production process.

Methodology for separation, purification, and identification of targeted product(s) is another important aspect to be considered. Most of current extraction protocols lead to the recovery of a wide range of products, usually of interest (e.g., phenolic compounds). However, nondesired products can be simultaneously collected in the product stream, which might therefore limit its applicability.

Legislation and logistics are also key for process development. Variables such as plant location, by-product stabilization, substrate storage, or substrate and product volume, must be considered for choosing an appropriate conversion technology and estimating actual operational costs.

Acknowledgments

Authors thank the Spanish "Agencia Estatal de Investigación" (AEI) and "Fondos Europeos de Desarrollo Regional" (FEDER) for funding this work via project ACMIBIO: Ref. ENE2017-86864-C2-1-R (AEI/FEDER, UE).

References

Akinwumi, B.C., Bordun, K.A.M., Anderson, H.D., 2018. Biological activities of stilbenoids. Int. J. Mol. Sci. 19, 792.

Albarrán, A., Celis, R., Hermosín, M.C., López- Piñeiro, A., Cornejo, J., 2004. Behaviour of simazine in soil amended with the final residue of the olive oil extraction process. Chemosphere 54, 717–724.

Alves-Filho, E.G., Sousa, V.M., Ribeiro, P.R.V., Rodrigues, S., de Brito, E.S., Tiwari, B.K., Fernandes, F.A.N., 2018. Single-stage ultrasound-assisted process to extract and convert α-solanine and α-chaconine from potato peels into β-solanine and β-chaconine. Biomass Convers. Biorefin. 8 (3), 689–697.

Alonso, J.L., Garrote, G., Domínguez, H., Santos, V., Parajó, J.C., 2009. Lactic acid from apple pomace: a laboratory experiment for teaching valorization of wastes. CyTA J. Food 7, 83–88.

Alonso, S., Herrero, M., Rendueles, M., Díaz, M., 2010. Residual yoghurt whey for lactic acid production. Biomass Bioenergy 34, 931e938.

Arapoglou, D., Varzakas, T., Vlyssides, A., Israilides, C., 2010. Ethanol production from potato peel waste (PPW). Waste Manag. 30 (10), 1898–1902.

Bastidas-Oyanedel, J.-R., Fang, C., Almardeai, S., Javid, U., Yousuf, A., Schmidt, J.E., 2016. Waste biorefinery in arid/semi-arid regions. Bioresour. Technol. 215, 21–28.

Ben Mansour, A., Porter, E.A., Kite, G.C., Simmonds, M.S.J., Abdelhedi, R., Bouaziz, M., 2015. Phenolic profile characterization of chemlali olive stones by liquid chromatography-ion trap mass spectrometry. J. Agric. Food Chem. 63, 1990–1995.

Beres, C., Costa, G.N.S., Cabezudo, I., da Silva-James, N.K., Teles, A.S.C., Crus, A.P.G., Mellinger-Silva, C., Tonon, R.V., Cabral, L.M.C., Freitas, S.P., 2017. Towards integral utilization of grape pomace from winemaking process: a review. Waste Manag. 68, 581–594.

Bhaskar, N., Modi, V.K., Govindaraju, K., Radha, C., Lalitha, R.G., 2007. Utilization of meat industry by products: protein hydrolysate from sheep visceral mass. Bioresour. Technol. 98 (2), 388–394.

Billet, K., Houillé, B., Besseau, S., Mélin, S., Oudin, A., Papon, N., Courdavault, V., Clastre, M., Giglioli-Guivarc'h, N., Lanoue, A., 2018. Mechanical stress rapidly induces E-resveratrol and E-piceatannol biosynthesis in grape canes stored as a freshly-pruned byproduct. Food Chem. 240, 1022–1027.

Borja, R., Raposo, F., Rincón, B., 2006. Treatment technologies of liquid and solid wastes from two-phase olive oil mills. Grasas Aceites 57, 32−46.

Cardinali, A., Pati, S., Minervini, F., D'Antuono, I., Linsalata, V., Lattanzio, V., 2012. Verbascoside, isoverbascoside, and their derivatives recovered from olive mill wastewater as possible food antioxidants. J. Agric. Food Chem. 60, 1822e1829.

Cassano, A., Conidi, C., Giorno, L., Drioli, E., 2013. Fractionation of olive mil wastewaters by membrane separation techniques. J. Hazard. Mater. 248−249, 185−193.

Cebrián, C., Sánchez-Gómez, R., Salinas, M.R., Alonso, G.,L., Zalacaín, A., 2017. Effect of post-pruning vine-shoots storage on the evolution of high-value compounds. Ind. Crops Prod. 109, 730−736.

Ćetković, G., Savatović, S., Čanadanović-Brunet, J., Djilas, S., Vulić, J., Mandić, A., Četojević-Simin, D., 2012. Valorisation of phenolic composition, antioxidant and cell growth activities of tomato waste. Food Chem. 133 (3), 938−945.

Chandramohan, R., Pari, L., Rathinam, A., Sheikh, B.A., 2015. Tyrosol, a phenolic compound, ameliorates hyperglycemia by regulating key enzymes of carbohydrate metabolism in streptozotocin induced diabetic rats. Chem. Biol. Interact. 229, 44−54.

Chanioti, S., Tzia, C., 2017. Optimization of ultrasound-assisted extraction of oil from olive pomace using response surface technology: oil recovery, unsaponifiable matter, total phenol content and antioxidant activity. LWT Food Sci. Technol. 79, 178−189.

Coelho, E., Rocha, M.A., Moreira, A.S.P., 2016. Revisiting the structural features of arabinoxylans from brewer's spent grain. Carbohydr. Polym. 139, 167−176.

Coimbra, M.A., Cardoso, S.M., Lopes-da-Silva, J.A., 2010. Olive pomace, a source for valuable arabinan-rich pectic polysaccharides. Top. Curr. Chem. 294, 129−141.

Conde, E., Cara, C., Moure, A., Ruiz, E., Castro, E., Domínguez, H., 2009. Antioxidant activity of the phenolic compounds released by hydrothermal treatments of olive tree pruning. Food Chem. 114, 806−812.

Connolly, A., O'Keeffe, M.B., Piggott, C.O., Nongonierma, A.B., FitzGerald, R.J., 2015. Generation and identification of angiotensin converting enzyme (ACE) inhibitory peptides from a brewers' spent grain protein isolate. Food Chem. 176, 64−71.

Cox, L., Hermosín, M.C., Cornejo, J., 2004. Influence of organic amendments on sorption and dissipation of imidacloprid in soil. Int. J. Environ. Anal. Chem. 84, 95−102.

Cristóbal, J., Caldeira, C., Corrado, S., Sala, S., 2018. Techno-economic and profitability analysis of food waste biorefineries at European level. Bioresour. Technol. 259, 244−252.

Crizel de Moraes, T., Jablonski, Rios de Oliveira, A., Rech, R., Flôres Hickmann, S., 2013. Dietary fiber from orange byproducts as a potential fat replacer. LWT Food Sci. Technol. 53 (1), 9e14.

Da Porto, C., Porretto, E., Decorti, D., 2013. Comparison of ultrasound-assisted extraction with conventional extraction methods of oil and polyphenols from grape (*Vitis vinifera* L.) seeds. Ultrason. Sonochem. 20, 1076−1080.

Dávila, J.A., Rosenber, M., Cardona, C.A., 2015. Techno-economic and environmental assessment of p-cymene and pectin production from orange peel. Waste Biomass Valorization 6, 253−261.

Dávila, J.A., Rosenberg, M., Cardona, C.A., 2016. A biorefinery approach for the production of xylitol, ethanol and polyhydroxybutyrate from brewer's spent grain. Agr. Food 1, 52−66.

Dávila, J.A., Rosenber, M., Cardona, C.A., 2017. A biorefinery for efficient processing and utilization of spent pulp of Colombian Andes Berry (*Rubus glaucus Benth.*): experimental, techno-economic and environmental assessment. Bioresour. Technol. 223, 227−236.

Dávila, I., Robles, E., Egües, I., Labidi, J., Gullón, P., 2017a. The biorefinery concept for the industrial valorization of grape processing by-products. In: Galanakis, C.M. (Ed.), Handbook of Grape Processing by-Products: Sustainable Solutions. Academic Press, pp. 29−53.

Dávila, I., Robles, E., Andrés, M.A., Gullón, P., 2017b. Coproduction of lignin and glucose from vine shoots by eco-friendly strategies: towards the development of an integrated biorefinery. Bioresour. Technol. 244, 328−337.

de Jong, E., Jungmeier, G., 2015. Biorefinery concepts in comparison to petrochemical refineries. In: Pandey, A., Höfer, R., Taherzadeh, M., Nampoothiri, K.M., Larroche, C. (Eds.), Industrial Biorefineries and White Biotechnology. Elsevier., pp. 3−33

Devesa-Rey, R., Vecino, X., Varela-Alende, J.L., Barral, M.T., Cruz, J.M., Moldes, A.B., 2011. Valorization of winery waste vs. the cost of not recycling. Waste Manag. 31, 2327−2335.

Dhingra, D., Michael, M., Rajput, H., 2012. Physico-chemical characteristics of dietary fibre on potato peel and its effect on organoleptic characteristics of biscuits. J. Agri. Eng. 49, 25−32.

Dimou, C., Kopsahelis, N., Papadaki, A., Papanikolaou, S., Koskos, I.K., Mandala, I., Koutinas, A.A., 2015. Wine lee valorization: biorefinery development including production of a generic fermentation feed-stock employed for poly(3-hydroxybutyrate) synthesis. Food Res. Int. 73, 81−87.

Directive 2008/98/EC of the European Parliament and of the Council of 19 November 2008 on waste and repealing certain Directives. Available at: http://data.europa.eu/eli/dir/2008/98/oj.

Directive (EU) 2018/851 of the European Parliament and of the Council of 30 May 2018 amending Directive 2008/98/EC on waste. Available at: http://data.europa.eu/eli/dir/2018/851/oj.

Dominiak, M., Søndergaard, K.M., Wichmann, J., Vidal-Melgosa, S., Willats, W.G.T., Meyer, A.S., Mikkelsen, J.D., 2014. Application of enzymes for efficient extraction, modification, and development of functional properties of lime pectin. Food Hydrocoll 40, 273−282.

Dwyer, K., Hosseinian, F., Rod, M., 2014. The market potential of grape waste alternatives. J. Food Res. 3, 91−106.

E4tech, RE-CORD, WUR, 2015. From the Sugar Platform to Biofuels and Biochemicals. Final report for the European Commission, Contract No. ENER/C2/423-2012/SI2.673791. Available at: http://www.e4tech.com.

Falade, K.O., Akingbala, J.O., 2011. Utilization of cassava for food. Food Rev. Int. 27, 51e83.

FAO, 2011. Global Food Losses and Food Waste − Extent, Causes and Prevention (Rome).

FAOSTAT, 2016. Crop production worldwide. Available at: http://www.fao.org/faostat/en.

Federici, E., Pepi, M., Esposito, A., Scargetta, S., Fidati, L., Gasperini, S., Cenci, G., Altieri, R., 2011. Two-phase olive mill waste composting: community dynamics and functional role of the resident microbiota. Bioresour. Technol. 102, 10965−10972.

Fernández-López, J., Fernández-Ginés, J.M., Aleson-Carbonell, L., Sendra, E., Sayas-Barberá, E., Pérez-Alvarez, J.A., 2004. Application of functional citrus byproducts to meat products. Trends Food Sci. Technol. 15, 176e185.

Fărcaş, A.C., Socaci, S.A., Dulf, F.V., Tofană, M., Mudura, E., Diaconeasa, Z., 2015. Volatile profile, fatty acids composition and total phenolics content of brewers' spent grain by-product with potential use in the development of new functional foods. J. Cereal Sci. 64, 34−42.

Gajendragadkar, C.N., Gogate, P.R., 2016. Intensified recovery of valuable products from whey by use of ultrasound in processing steps − a review. Ultrason. Sonochem. 32, 102−118.

Galanakis, C.M., 2012. Recovery of high added-value components from food wastes: conventional, emerging technologies and commercialized applications. Trends Food Sci. Technol. 26 (2), 68–87.

Garavaglia, J., Markoski, M.M., Oliveira, A., Marcadenti, A., 2016. Grape seed oil compounds: biological and chemical actions for health. Nutr. Metab. Insights 9, 59–64.

García-Lomillo, J., González-San José, M.J., 2017. Application of wine pomace in the food industry: approaches and functions. Compr. Rev. Food Sci. Food Saf. 16, 3–22.

Giannini, B., Mulinacci, N., Pasqua, G., Innocenti, M., Valletta, A., Cecchini, F., 2016. Phenolics and antioxidant activity in different cultivars/clones of *Vitis vinifera* L. seeds over two years. J. Plant Biosyst. 150, 1408–1416.

Gorena, T., Sáez, V., Mardones, C., Vergara, C., Winterhalter, P., Von Baer, D., 2014. Influence of post-pruning storage on stilbenoid levels in *Vitis vinifera* L. canes. Food Chem. 155, 256–263.

Granados-Principal, S., El-Azem, N., Pamplona, R., Ramirez-Tortosa, C., Pulido-Moran, M., Vera-Ramirez, L., Quiles, J.L., Sanchez-Rovira, P., Naudí, A., Portero-Otin, M., Perez-Lopez, P., Ramirez-Tortosa, M., 2014. Hydroxytyrosol ameliorates oxidative stress and mitochondrial dysfunction in doxorubicin-induced cardiotoxicity in rats with breast cancer. Biochem. Pharmacol. 90 (1), 25–33.

Guido, L.F., Moreira, M.M., 2017. Techniques for extraction of brewer's spent grain polyphenols: a review. Food Bioprocess Technol. 10, 1192–1209.

Guinda, A., Castellano, J.M., Santos-Lozano, J.M., Delgado-Hervás, T., Gutiérrez-Adánez, P., Rada, M., 2015. Determination of major bioactive compounds from olive leaf. LWT Food Sci. Technol. 64, 431–438.

Gullón, P., Gullón, B., Dávila, I., Labidi, J., González-García, S., 2018. Comparative environmental life cycle assessment of integral revalorization of vine shoots from a biorefinery perspective. Sci. Total Environ. 624, 225–240.

Hernández, V., Romero-García, J.M., Dávila, J.A., Castro, E., Cardona, C.A., 2014. Techno-economic and environmental assessment of an olive stone based biorefinery. Resour. Conserv. Recycl. 92, 145–150.

Hossain, M.B., Tiwari, B.K., Gangopadhyay, N., O'Donnell, C.P., Brunton, N.P., Rai, D.K., 2014. Ultrasonic extraction of steroidal alkaloids from potato peel waste. Ultrason. Sonochem. 21 (4), 1470–1476.

Houillé, B., Besseau, S., Courdavault, V., Oudin, A., Glevarec, G., Delanoue, G., Guerin, L., Simkin, A.J., Papon, N., Clastre, M., Giglioli-Guivarc, N., Lanoue, A., 2015. Biosynthetic origin of E-Resveratrol accumulation in grape canes during postharvest storage. J. Agric. Food Chem. 63, 1631–1638.

Huber, G.M., Rupasinghe, H.P.V., 2009. Phenolic profiles and antioxidant properties of apple skin extracts. J. Food Sci. 74, C693eC700.

Ibrahim, T.H., Chehade, Y., Zour, M.A., 2011. Corrosion inhibition of mild steel using potato peel extract in 2M HCl solution. Int. J. Electrochem. Sci 6 (12), 6542–6556.

Jesús, M.S., Romaní, A., Genisheva, Z., Teixeira, J., Domínguez, L., 2017. Integral valorization of vine pruning residue by sequential autohydrolysis stages. J. Clean. Prod. 168, 74–86.

Kalogeropoulos, N., Chiou, A., Pyriochou, V., Peristeraki, A., Karathanos, V.T., 2012. Bioactive phytochemicals in industrial tomatoes and their processing byproducts. LWT Food Sci. Technol. 49, 213e216.

Karantonis, H.C., Antonopoulou, S., Perrea, D.N., Sokolis, D.P., Theocharis, S.E., Kavantzas, N., Iliopoulos, D.G., Demopoulos, C.A., 2006. *In vivo* antiatherogenic properties of olive oil and its constituent lipid classes in hyperlipidemic rabbits. Nutr. Metabol. Cardiovasc. Dis. 16 (3), 174–185.

Kaur, G., Uisan, K., Ong, K.L., Lin, C.S.K., 2018. Recent trends in green and sustainable chemistry & waste valorization: rethinking plastics in a circular economy. Curr. Opin. Green Sustain. Chem. 9, 30–39.

Kehili, M., Schmidt, L.M., Reynolds, W., Zammel, A., Zelzl, C., Smirnova, I., Allouche, N., Sayadi, S., 2016. Biorefinery cascade processing for creating added value on tomato industrial by-product from Tunisia. Biotechnol. Biofuels 9, 261.

Keswani, M.H., Patil, A.R., 1985. The boiled potato peel as a burn wound dressing: a preliminary report. Burns Incl. Therm. Inj. 11, 220–224.

Kirin Beer University Report, 2018. Global Beer Production by Country in 2017. Available at: https://www.kirinholdings.co.jp/english/news/2018/0809_01.html.

Klen, T.J., Vodopivec, B.M., 2011. Ultrasonic extraction of phenols from olive mill wastewater. Comparison with conventional methods. J. Agric. Food Chem. 59, 12725–12731.

Kosseva, M.R., 2011. Management and processing of food wastes. In: Moo-Young, M. (Ed.), Comprehensive Biotechnology, Environmental Biotechnology and Safety, second ed., vol. 6. Elsevier, pp. 557–593.

La Scalia, G., Micale, R., Cannizzaro, L., Marra, F.P., 2017. A sustainable phenolic compound extraction system from olive oil mill wastewater. J. Clean. Prod. 142, 3782–3788.

Lama-Muñoz, A., Romero-García, J.M., Cara, C., Moya, M., Castro, E., 2014. Low energy-demanding recovery of antioxidants and sugars from olive stones as preliminary steps in the biorefinery context. Ind. Crops Prod. 60, 30–38.

Lappalainen, K., Kärkkäinen, J., Joensuu, P., Lajunen, M., 2015. Modification of potato peel waste with base hydrolysis and subsequent cationization. Carbohydr. Polym. 132, 97–103.

Ledesma-Escobar, C.A., Luque de Castro, M.D., 2014. Towards a comprehensive exploitation of citrus. Trends Food Sci. Technol. 39, 63–75.

Liang, S., Han, Y., Wei, L., McDonald, A.G., 2015. Production and characterization of bio-oil and bio-char from pyrolysis of potato peel wastes. Biomass Convers. Biorefin. 5 (3), 237–246.

Lohrasbi, M., Pourbafrani, M., Niklasson, C., Taherzadeh, M.J., 2010. Process design and economic analysis of a citrus waste biorefinery with biofuel and limonene as products. Bioresour. Technol. 101, 7382–7388.

López, F.J., Pinzi, S., Ruiz, J.J., López, A., Dorado, M.P., 2010. Economic viability of the use of olive tree pruning as fuel for heating systems in public institutions in south Spain. Fuel 89, 1386–1391.

López-Biedma, A., Sánchez-Quesada, C., Beltrán, G., Delgado-Rodríguez, M., Gaforio, J.J., 2016. Phytoestrogen (+)-pinoresinol exerts antitumor activity in breast cancer cells with different oestrogen receptor statuses. BMC Complement Altern. Med. 16, 350.

Loru, D., Incani, A., Deiana, M., Corona, G., Atzeri, A., Melis, M., Rosa, A., Dessì, M.A., 2009. Protective effect of hydroxytyrosol and tyrosol against oxidative stress in kidney cells. Toxicol. Ind. Health 25, 301–310.

Lozano-Sánchez, J., Giambanelli, E., Quirantes-Piné, R., Cerretani, L., Bendini, A., Segura-Carretero, A., Fernández-Gutiérrez, A., 2011. Wastes generated during the storage of extra virgin olive oil as a natural source of phenolic compounds. J. Agric. Food Chem. 59, 11491–11500.

Lucarini, L., Durazzo, A., Romani, A., Campo, M., Lombardi-Boccia, G., Cecchini, F., 2018. Bio-based compounds from grape seeds: a biorefinery approach. Molecules 23, 1888.

Lynch, K.M., Steffen, E.J., Arendt, K., 2016. Brewers's spent grain: a review with an emphasis on food and health. J. Inst. Brew. 122, 553–568.

Machmudah, S., Winardi, S., Sasaki, M., Goto, M., Kusumoto, N., Hayakawa, K., 2012. Lycopene extraction from tomato peel by-product containing tomato seed using supercritical carbon dioxide. J. Food Eng. 108, 290e296.

Madrera, R.R., Bedrinana, R.P., Valles, B.S., 2015. Production and characterization of aroma compounds from apple pomace by solid-state fermentation with selected yeasts. LWT Food Sci. Technol. 64, 1342–1353.

Makris, D.P., 2018. Green extraction processes for the efficient recovery of bioactive polyphenols from wine industry solid waste: recent progress. Curr. Opin. Green Sus. Chem. 13, 30–35.

Maldonado, A.F.S., Mudge, E., Gänzle, M.G., Schieber, A., 2014. Extraction and fractionation of phenolic acids and glycoalkaloids from potato peels using acidified water/ethanol-based solvents. Food Res. Int. 65, 27–34.

Mamma, D., Christakopoulos, P., 2014. Biotransformation of citrus by-products into value added products. Waste Biomass Valorization 5, 529–549.

Manzone, M., Paravidino, E., Bonifacino, G., Balsari, P., 2016. Biomass availability and quality produced by vineyard management during a period of 15 years. Renew. Energy 99, 465–471.

Martin-Diana, A.B., Rico, D., Frias, J., Mulcahy, J., Henehan, G.T.M., Barry-Ryan, C., 2006. Whey permeate as a bio-preservative for shelf life maintenance of fresh-cut vegetables. Innov. Food Sci. Emerg. 7, 112e123.

Martínez, G.A., Rebecchi, S., Decorti, D., Domingos, J.M.B., Natolino, A., Del, D., Bertin, L., Da Porto, C., Faca, F., 2016. Towards multi-purpose biorefinery platforms for the valorisation of red grape pomace: production of polyphenols, volatile fatty acids, polyhydroxyalkanoates and biogas. Green Chem. 16, 261–270.

Matos, M., Barreiro, M.F., Gandin, A., 2010. Olive stone as a renewable source of biopolyols. Ind. Crops Prod. 32, 7–12.

Minhalma, M., Magueijo, V., Queiroz, D.P., De Pinho, M.N., 2007. Optimization of "Serpa" cheese whey nanofiltration for effluent minimization and by-products recovery. J. Environ. Manag. 82, 200e206.

Moldes, A.B., Bustos, G., Torrado, A., Domínguez, J.M., 2007. Comparison between different hydrolysis processes of vine-trimming waste to obtain hemicellulosic sugars for further lactic acid conversion. Appl. Biochem. Biotechnol. 143, 244–256.

Moreira, M.M., Morais, S., Barros, A.A., Deleurue-Matos, C., Guido, L.F., 2012. A novel application of microwave-assisted extraction of polyphenols from brewer's spent grain with HPLC-DAD-MS analysis. Anal. Bioanal. Chem. 403, 1019–1029.

Moreira, M.M., Morais, S., Carvalho, D.O., Barros, A.A., Deleure-Matos, C., Guido, L.F., 2013. Brewer's spent grain from different types of malt: evaluation of the antioxidant activity and identification of the mayor phenolic compounds. Food Res. Int. 54, 382–388.

Mukherjee, S., Bandyopadhayay, B., Basak, B., Mandal, N., Dey, A., Mondal, B., 2012. An improved method of optimizing the extraction of polyphenol oxidase from potato (*Solanum tuberosum* L.) peel. Not. Sci. Biol. 4 (1), 98–107.

Mussatto, S.I., 2014. Brewer's pent grain: a valuable feedstock for industrial applications. J. Sci. Food Agric. 94, 1264–1275.

Mussatto, S.I., Moncada, J., Roberto, I.C., Cardona, C.A., 2013. Techno-economic analysis for brewer's spent grain use on a biorefinery concept. Bioresour. Technol. 148, 302–310.

Nadour, M., Laroche, C., Pierre, G., Delattre, C., Moulti-Mati, F., Michaud, P., 2015. Structural characterization and biological activities of polysaccharides from olive mill wastewater. Appl. Biochem. Biotechnol. 177 (2), 431–445.

Negro, M.J., Manzanares, P., Ruiz, E., Castro, E., Ballesteros, M., 2017. The biorefinery concept for the industrial valorization of residues from olive oil industry. In: Galanakis, C.M. (Ed.), Olive Mill Waste: Recent Advances for Sustainable Management, pp. 57–78.

Nigam, P.S., 2017. An overview: recycling of solid barley waste generated as a by-product in distillery and brewery. Waste Manag. 62, 255–261.

Nunes, M.A., Pimentel, F.B., Costa, A.S.G., Alves, R.C., Oliveira, M.B.P.P., 2016. Olive byproducts for functional and food applications: challenging opportunities to face environmental constraints. Innov. Food Sci. Emerg. Technol. 35, 139–148.

Nunes, M., Costa, A.S.G., Bessada, S., Santos, J., Puga, H., Alves, R.C., Freitas, V., Oliveira, M.B.P.P., 2018. Olive pomace as a valuable source of bioactive compounds: a study regarding its lipid- and water-soluble components. Sci. Total Environ. 644, 229–236.

OIV, 2018. Organization Internationale de la Vigne et du Vin, State of the Vitiviniculture Word Market. April 2018.

Ouzounidou, G., Zervakis, G.I., Gaitis, F., 2010. Raw and microbiologically detoxified olive mill waste and their impact on plant growth. Terr. Aquat. Environ. Toxicol. 4 (special Issue 1), 21–38.

Pagnanelli, F., Toro, L., Veglio, F., 2002. Olive mill solid residues as heavy metal sorbent material: a preliminary study. Waste Manag. 22, 901–907.

Paleologou, I., Vasiliou, A., Grigorakis, S., Makris, D.P., 2016. Optimisation of a green ultrasound-assisted extraction process for potato peel (*Solanum tuberosum*) polyphenols using bio-solvents and response surface methodology. Biomass Convers. Biorefin. 6 (3), 289–299.

Pandit, N.P., Ahmad, N., Maheshwari, S.K., 2012. Vermicomposting biotechnology: an eco-loving approach for recycling of solid organic wastes into valuable biofertilizers. J. Biofert. Biopestic. 3, 113.

Paseephol, T., Small, D.M., Sherkat, F., 2008. Lactulose production from milk concentration permeate using calcium carbonate-based catalysts. Food Chem. 111, 283e290.

Pathak, P.D., Mandavgane, S.A., Kulkarni, B.D., 2015. Fruit peel waste as a novel low-cost bio adsorbent. Rev. Chem. Eng. 31 (4), 361–381.

Pathak, P.D., Mandavgane, S.A., Puranik, N.M., Jambhulkar, S.J., Kulkarni, B.D., 2018. Valorization of potato peel: a biorefinery approach. Crit. Rev. Biotechnol. 38 (2), 218–230.

Pattara, C., Cappelletti, G.M., Cichelli, A., 2010. Recovery and use of olive stones. Commodity, environmental and economic assessment. Renew. Sustain. Energy Rev. 14, 1484–1489.

Pérez-Bibbins, B., Torrado-Agrasar, A., Salgado, J.M., Pinheiro de Souza Oliveira, R., Dominguez, J.M., 2015a. Potential of lees from wine, beer and cider manufacturing as a source of economic nutrient: an overview. Waste Manag. 40, 72–81.

Pérez-Bibbins, B., Torrado, A., Pérez-Rodríguez, N., Aguilar-Uscanga, M.G., Domínguez, J.M., 2015b. Evaluation of the liquid, solid and total fractions of beer, cider and wine lees as economic nutrient for xylitol production. J. Chem. Technol. Biotechnol. 90, 1027–1039.

Pfaltzgraff, L.A., De bruyn, M., Cooper, E.C., Budarin, V., Clark, J.H., 2013. Food waste biomass: a resource for high-value chemicals. Green Chem. 15, 307–314.

Prazeres, A.R., Carvalho, F., Rivas, J., 2012. Cheese whey management: a review. J. Environ. Manag. 110, 48e68.

Proestos, C., Komaitis, M., 2008. Application of microwave-assisted extraction to the fast extraction of plant phenolic compounds. LWT Food Sci. Technol. 41 (4), 652–659.

Rahmanian, N., Jafari, S.M., Wani, T.A., 2015. Bioactive profile, dehydration, extraction and application of the bioactive components of olive leaves. Trends Food Sci. Technol. 42, 150–172.

Ramos, P., Santos, S.A.O., Guerra, Â.R., Guerreiro, O., Felício, L., Jerónimo, E., Silvestre, A.J.D., Neto, C.P., Duarte, M., 2013. Valorization of olive mill residues: antioxidant and breast cancer antiproliferative activities of hydroxytyrosol-rich extracts derived from olive oil by-products. Ind. Crops Prod. 46, 359–368.
Ravindran, R., Jaiswal, A.K., 2016. Exploitation of food industry waste for high-value products. Trends Biotechnol. 34 (1), 58–69.
Regulation (EC) No 1907/2006 of the European Parliament and of the Council of 18 December 2006 concerning the Registration, Evaluation, Authorisation and Restriction of Chemicals (REACH), establishing a European Chemicals Agency, amending Directive 1999/45/EC and repealing Council Regulation (EEC) No 793/93 and Commission Regulation (EC) No 1488/94 as well as Council Directive 76/769/EEC and Commission Directives 91/155/EEC, 93/67/EEC, 93/105/EC and 2000/21/EC, 2006. Available at: http://data.europa.eu/eli/reg/2006/1907/2014-04-10.
Rincón, B., Bujalance, L., Fermoso, F.G., Martín, A., Borja, R., 2013. Biochemical methane potential of two-phase olive mill solid waste: influence of thermal pretreatment on the process kinetics. Bioresour. Technol. 140, 249–255.
Rodríguez, G., Lama, A., Rodríguez, R., Jiménez, A., Guillén, R., Fernández-Bolaños, J., 2008. Olive stones an attractive source of bioactive and valuable compounds. Bioresour. Technol. 99, 5261–5269.
Roig, A., Cayuela, M.L., Sánchez-Monedero, M.A., 2006. An overview on olive mill wastes and their valorisation methods. Waste Manag. 26 (9), 960–969.
Romero-García, J.M., Niño, L., Martínez-Patiño, C., Álvarez, C., Castro, E., Negro, M.J., 2014. Biorefinery based on olive biomass. State of the art and future trends. Bioresour. Technol. 159, 421–432.
Romero-García, J.M., Lama-Muñoz, A., Rodríguez-Gutiérrez, G., Moya, M., Ruiz, E., Fernández-Bolaños, J., Castro, E., 2016a. Obtaining sugars and natural antioxidants from olive leaves by steam-explosion. Food Chem. 210, 457–465.
Romero-García, J.M., Sanchez, A., Rendón-Acosta, G., Martínez-Patiño, J.C., Ruiz, E., Magaña, G., Castro, E., 2016b. An olive tree pruning biorefinery for co-producing high value-added bioproducts and biofuels: economic and energy efficiency analysis. Bioenerg. Res. 9, 1070–1086.
Ruiz, E., Romero-García, J.M., Romero, I., Manzanares, P., Negro, M.J., Castro, E., 2017. Olive-derived biomass as a source of energy and chemicals. Biofuel Bioprod. Bior. 11 (6), 1077–1094.
Sabeena Farvin, K.H., Grejsen, H.D., Jacobsen, C., 2012. Potato peel extract as a natural antioxidant in chilled storage of minced horse mackerel (*Trachurus trachurus*): effect on lipid and protein oxidation. Food Chem. 131 (3), 843–851.
Sajib, M., Falck, P., Sardari, R.R.R., Mathew, S., Grey, C., Karlsson, E.N., Adlercreutz, P., 2018. Valorization of Brewer's spent grain to prebiotic oligosaccharide: production, xylanase catalyzed hydrolysis, *in-vitro* evaluation with probiotic strains and in a batch human fecal fermentation model. J. Biotechnol. 268, 61–70.
Salgado, J.M., Rodríguez, N., Cortés, S., Domínguez, J.M., 2009. Development of cost-effective media to increase the economic potential for larger-scale bioproduction of natural food additives by *Lactobacillus rhamnosus, Debaryomyces hansenii* and *Aspergillus niger*. J. Agric. Food Chem. 57, 10414–10428.
Sánchez de Medina, V., Priego-Capote, F., Luque De Castro, M.D., 2012. Characterization of refined edible oils enriched with phenolic extracts from olive leaves and pomace. J. Agric. Food Chem. 60 (23), 5866–5873.

Schieber, A., 2017. Side streams of plant food processing as a source of valuable compounds: selected examples. Annu. Rev. Food Sci. Technol. 8, 97−112.

Schievano, A., Adani, F., Buessing, L., Botto, A., Casoliba, E.N., Rossoni, M.R., Goldfarb, J.L., 2015. An integrated biorefinery concept for olive mill waste management: supercritical CO_2 extraction and energy recovery. Green Chem. 17, 2874−2887.

Scoma, A., Bertin, L., Zanaroli, G., Fraraccio, S., Fava, F., 2011. A physicochemical-biotechnological approach for an integrated valorization of olive mill wastewater. Bioresour. Technol. 102, 10273−10279.

Seçmeler, Ö., Güçlü Üstündağ, Ö., 2017. Behavior of lipophilic bioactives during olive oil processing. Eur. J. Lipid Sci. Technol. 119 (9), 1600404.

Shahzad, K., Narodoslawsky, M., Sagir, M., Ali, N., Ali, S., Rashid, M.I., Ismail, I.M.I., Koller, M., 2017. Techno-economic feasibility of waste biorefinery: using slaughtering waste streams as starting material for biopolyester production. Waste Manag. 67, 73−85.

Shukla, J., Kar, R., 2006. Potato peel as a solid state substrate for thermostable α-amylase production by thermophilic *Bacillus* isolates. World J. Microbiol. Biotechnol. 22, 417−422.

Singh, P.P., Saldaña, M.D.A., 2011. Subcritical water extraction of phenolic compounds from potato peel. Food Res. Int. 44, 2452−2458.

Singh, N., Kamath, V., Rajini, P.S., 2005. Attenuation of hyperglycemia and associated biochemical parameters in STZ-induced diabetic rats by dietary supplementation of potato peel powder. Clin. Chim. Acta 353, 165−175.

Singh, N., Kamath, V., Narasimhamurthy, K., Rajini, P.S., 2008. Protective effect of potato peel extract against carbon tetrachloride-induced liver injury in rats. Environ. Toxicol. Pharmacol. 26, 241−246.

Singh, A., Sabally, K., Kubow, S., Donnelly, D.J., Gariepy, Y., Orsat, V., Raghavan, G.S., 2011. Microwave-assisted extraction of phenolic antioxidants from potato peels. Molecules 16, 2218−2232.

Siracusa, G., La Rosa, A.D., Siracusa, V., Trovato, M., 2001. Eco-compatible use of olive husk as filler in thermoplastic composites. J. Polym. Environ. 9 (4), 157−161.

Sorrentino, E., Succi, M., Tipaldi, L., Pannella, G., Maiuro, L., Sturchio, M., Coppola, R., Tremonte, P., 2018. Antimicrobial activity of gallic acid against food-related *Pseudomonas* strains and its use as biocontrol tool to improve the shelf life of fresh black truffles. Int. J. Food Microbiol. 266, 183−189.

Spigno, G., Marinoni, L., Garrido, G.D., 2017. State of the art in grape processing by-products. In: Galanakis, C.M. (Ed.), Handbook of Grape Processing By-Products: Sustainable Solutions. Academic Press, pp. 1−27.

Stamatakis, G., 2010. Energy and geo-environmental applications for olive mill wastes. A review. Hellenic J. Geosci. 45, 269−282.

Stavroulias, S., Panayiotou, C., 2005. Determination of optimum conditions for the extraction of squalene from olive pomace with supercritical CO_2. Chem. Biochem. Eng. Q. 19, 373−381.

Stenmarck, Å., Jensen, C., Quested, T., Moates, G., 2016. Estimates of European food waste levels. Fusions: Stockholm, Sweden. Available at: https://www.eu-fusions.org/index.php/publications/266-establishing-reliable-data-on-food-waste-and-harmonising-quantification-methods.

Strati, I.F., Oreopoulou, V., 2011. Effect of extraction parameters on the carotenoid recovery from tomato waste. Int. J. Food Sci. Technol. 46, 23e29.

Survase, S.A., Singh, J., Singhal, R.S., 2016. The role of potatoes in biomedical/pharmaceutical and fermentation applications. In: Singh, J., Kaur, L. (Eds.), Advances in Potato Chemistry and Technology, second ed. Elsevier, pp. 603–625.

Tayeh, H.A., Najami, N., Dosoretz, C., Tafesh, A., Azaizeh, H., 2014. Potential of bioethanol production from olive mill solid wastes. Bioresour. Technol. 152, 24–30.

Toldrá, F., Aristoy, M.-C., Mora, L., Reig, M., 2012. Innovations in value-addition of edible meat by-products. Meat Sci. 92 (3), 290–296.

Toma, R.B., Orr, P.H., D'appolonia, B., Dlntzis, F.R., Tabekhia, M.M., 1979. Physical and chemical properties of potato peel as a source of dietary fiber in bread. J. Food Sci. 44, 1403–1407.

Toscano, G., Riva, G., Duca, D., Foppa Pedretti, E., Corinaldesi, F., Rossini, G., 2013. Analysis of the characteristics of the residues of the wine production chain finalized to their industrial and energy recovery. Biomass Bioenergy 55, 260–267.

Tsantila, N., Karantonis, H.C., Perrea, D.N., Theocharis, S.E., Iliopoulos, D.G., Iatrou, C., Antonopoulou, S., Demopoulos, C.A., 2010. Atherosclerosis regression study in rabbits upon olive pomace polar lipid extract administration. Nutr. Metabol. Cardiovasc. Dis. 20 (10), 740–747.

Uçak, H., 2007. Trends in meat industry – production, consumption and trade. Acta Sci. Pol. Oeconomia 6 (4), 125–131.

Van Dyk, J.S., Gama, R., Morrison, D., Swart, S., Pletschke, B.I., 2013. Food processing waste: problems, current management and prospects for utilization of the lignocellulose component through enzyme synergistic degradation. Renew. Sustain. Energy Rev. 26, 521–531.

Velasco, L., Fernández-Cuesta, A., Fernández-Martínez, J.M., 2014. New sunflower seeds with high contents of phytosterols. OCL 21 (6), D604.

Velázquez-Martí, B., Fernández-González, E., López-Cortés, I., Salazar-Hernández, D.M., 2011. Quantification of the residual biomass obtained from pruning of vineyards in Mediterranean area. Biomass Bioenegy 35, 3453–3464.

Vera, D., Mena, B., Jurado, F., Schories, G., 2013. Study of a downdraft gasifier and gas engine fueled with olive oil industry wastes. Appl. Therm. Eng. 51, 119–129.

Vieira, E., Rocha, M.A.M., Coelho, E., Pinho, O., Saraiva, J.A., Ferreira, I.M.P.L.V.O., Coimbra, M.A., 2014. Valuation of brewer's spent grain using a fully recyclable integrated process for extraction of proteins and arabinoxylans. Ind. Crops Prod. 52, 136–143.

Vieira, E.F., da Silva, D.D., Carmo, H., Ferreira, I.M.P.L.V.O., 2017. Protective ability against oxidative stress of brewers'spent protein hydrolysates. Food Chem. 228, 602–609.

Walia, M., Rawat, K., Bhushan, S., Padwad, Y.S., Singh, B., 2014. Fatty acid composition, physicochemical properties, antioxidant and cytotoxic activity of apple seed oil obtained from apple pomace. J. Sci. Food Agric. 94, 929–934.

Wang, S.J., Chen, F., Wu, J.H., Wang, Z.F., Liao, X.J., Hu, X.S., 2007. Optimization of pectin extraction assisted by microwave from apple pomace using response surface methodology. J. Food Eng. 78, 693–700.

Wang, X., Chen, Q.R., Lu, X., 2014. Pectin extracted from apple pomace and citrus peel by subcritical water. Food Hydrocoll 38, 129–137.

Wei, L., Liang, S., McDonald, A.G., 2015. Thermophysical properties and biodegradation behavior of green composites made from polyhydroxybutyrate and potato peel waste fermentation residue. Ind. Crops Prod. 69, 91–103.

Winans, K., Kendall, A., Deng, H., 2017. The history and current applications of the circular economy concept. Renew. Sustain. Energy Rev. 68, 825–833.

Wolfe, K.L., Liu, R.H., 2003. Apple peels as a value-added food ingredient. J. Agric. Food Chem. 51, 1676e1683.

Wu, D., 2016. Recycle technology for potato peel waste processing: a review. Procedia Environ. Sci. 31, 103−107.

Xu, C., Zhang, Y., Zhang, Y., Jun, W., Lu, J., 2010. Extraction, distribution and characterization of phenolic compounds and oil in grape seeds. Food Chem. 122, 688−694.

Zuorro, A., Fidaleo, M., Lavecchia, R., 2011. Enzyme-assisted extraction of lycopene from tomato processing waste. Enzym. Microb. Technol. 49, 567e573.

Packaging and food sustainability

6

Fabio Licciardello[1,3], *Luciano Piergiovanni*[2,3]
[1]DSV, University of Modena and Reggio Emilia, Reggio Emilia, Italy; [2]DeFENS, University of Milan, Milan, Italy; [3]GSICA, Italian Scientific Group of Food Packaging, Italy

Chapter outline

1. Introduction 191
2. Environmental aspects of packaging materials for food use 192
3. Impact of food wastes on the overall sustainability 194
4. The contribution of packaging to the sustainability of food chains 196
5. Sustainability-oriented strategies through the improvement of packaging efficiency 200
5.1 Packaging lightweighting 200
5.2 The choice of packaging materials with a focus on recyclability 201
5.3 The shift to biobased plastic materials 202
6. Sustainability-oriented strategies through the improvement of packaging effectiveness 208
6.1 Redesign product-packaging configurations 210
6.2 Maximize packaging protective role 210
6.3 Nanotechnologies for packaging improvement 211
6.4 Correct shelf life assignment 214
7. Conclusion 215
References 216

1. Introduction

Food packaging is a crucial unit of operation within the food production processes, as it provides the food products with suitable conditions able to prevent mechanical damage and slow-down biochemical deterioration, as well as changes caused by microorganisms. Food-packaging's primary protective role is mainly exerted by controlling gas and vapor exchanges with the external environment and by preventing microbial and chemical contamination, thus extending shelf life and preventing food safety issues. In other words, packaging can maintain the beneficial effects of processing after the process is complete (Marsh and Bugusu, 2007) and increase the storage life of fresh foods, such as meats, fruits, or vegetables.

The Interaction of Food Industry and Environment. https://doi.org/10.1016/B978-0-12-816449-5.00006-0

Much research has been dedicated to the improvement of packaging materials performances which has made available a wide range of materials for any purpose, able to respond to the specific requirements of different food products. Indeed, packaging systems should be designed specifically for each food product, taking into consideration: the food characteristics (specific quality parameters), the events which lead to quality loss (i.e., gas exchange, light transmittance, etc.), distribution specifications, and marketing needs; food-packaging systems should comply with food-contact materials regulations. Modern food packaging must face up to another requirement: it must be environmentally sustainable. Indeed, packaging contributes to the sustainability of a food product and should be designed holistically together with that product to optimize the overall environmental performance and minimize the risk of product spoilage, damage and wastage. A great deal of effort must be addressed to make the packaging system "greener", but still able to protect food and to retard degradation phenomena. To date, the concept of sustainability of packaging has been markedly influenced by the evil reputation of plastics, being one of the major concerns in environmental pollution. Actually, there is a strong need for a more objective consideration of packaging, which balances the undoubtful environmental drawbacks linked to packaging production and end-of-life, with the enormous benefits in terms of food products safeguard and waste reduction potential. Indeed, consumers have the common belief that food packaging represents an additional economic and environmental cost, an unnecessary solid waste rather than a valuable means for waste reduction. According to Cox and Downing (2007), 75%–90% consumers agree that discarded packaging is a greater environmental issue than the wasted food. Sustainability is one of the factors which guide consumers in food products choice. Recent studies (Grunert et al., 2014), however, have pointed out that packaging-related sustainability factors (i.e., the amount of packaging used, food waste and the recyclability of packaging) represent elements of choice for specific products categories, such as ready meals and soft drinks. In general, there is growing awareness on packaging role and what is clear to everyone is that modern food cannot exist without its packaging. Nevertheless, significant improvements are possible, and the food industry should target packaging systems which are more efficient and more effective, i.e., more performing and more sustainable at the same time.

2. Environmental aspects of packaging materials for food use

Plastic packaging, which represents the largest application of plastics, accounts for about 40% of the total volume (PlasticsEurope, 2017). The continuous increase in the share of plastic packaging within the global packaging volumes is to be attributed to the several advantages of plastics as packaging materials: they are inexpensive, lightweight, versatile and high performing.

The assessment of the environmental impact of packaging has recently become a hot topic. This is because packaging materials, especially plastic, have been blamed as primary responsible for environmental pollution. A case which has recently shocked the public opinion is that of plastic recovered in the oceans and in the tissues of fish. Plastic packaging which leak out of collecting and sorting systems, can end up in the soil and in the sea, slowly degrading to micro- and nanoplastics which can reach the food chain through the accumulation into marine organisms, with dramatic long-term consequences (United Nations Environment Program (UNEP), 2011) (European commission, 2011). It is estimated that about 32% of plastic, which is the fraction not suitably recovered by the collection system, pollutes the environment and that 8 million tons end up in the oceans every year: this figure is equivalent to one garbage truck dumping in the ocean every minute, and is estimated to grow to four trucks per minute by 2050 (Ellen MacArthur Foundation, 2016). In Europe, the total packaging waste collected has increased by 12% in the last 10 years. Of the collected plastic packaging waste, 41% is recycled, 39% ends up to energy recovery, while around 20% is still landfilled, however this last fraction has been halved (−53%) between 2006 and 2016 (PlasticsEurope, 2017).

However, the environmental responsibility of packaging materials is not only related with its end-of-life destiny, but also with the fossil resources depletion: indeed, conventional plastic packages (i.e., polyehtylene, PE; polypropylene, PP; polyethylene therephthalate, PET, polyvinyl chloride, PVC, etc.) are made from fossil oil-derived raw materials, implying the consumption of resources which are nonrenewable (actually, they are renewable but in a very long timeframe), thus creating an overall resource imbalance. However, we must be aware that the overall plastic production consumes about 4%–6% of the fossil resources and, therefore, plastic packaging accounts for much less (about 1%–3%); on the other hand, electricity, heating and transport are responsible for 87% of fossil resources depletion (PlasticsEurope, 2017). The new concepts of "green packaging," "eco-friendly packaging," or "sustainable packaging" have, thus, permeated the society, even if they have not been precisely defined and may cause some confusion in the consumers (Herbes et al., 2018).

Life Cycle Assessment (LCA) is, worldwide, the reference tool for the identification and quantification of the environmental impacts associated with a product life cycle (Ingrao et al., 2015). This methodology is ruled by the International Standards 14040 and 14044 (ISO, 2006a,b) and considers representative environmental indicators, such as global warming, nonrenewable energy, respiratory inorganic matter, acidification and eutrophication, eco-toxicities, and land occupation. The adoption of Life Cycle Thinking, i.e., the approach to becoming really aware of how and to what extent everyday life affects the environment, in the food chain particularly has strongly impacted the availability of information on the actual burdens associated with the various phases and such data represent a reference for environmentally-responsible choices. LCA studies have successfully been applied to packaging, highlighting the impacts associated with its production and disposal (Büsser and Jungbluth, 2009; Ingrao et al., 2017; Siracusa et al., 2014; Zabaniotou and Kassidi, 2003).

A comparison with results from LCA studies of packaging revealed a number of discrepancies between consumer perception and facts on the environmental impact of different packaging options: indeed, apparently less environmentally friendly packaging options may be, in fact, more sustainable because they prevent food losses better than other solutions (Herbes et al., 2018). Moreover, several organizations, active in the packaging field, have put on the market, or made available to their associates, simplified and specifically packaging-oriented software, useful for an impact assessment of different packaging options. It is worth mentioning, for instance, the Compass (comparative packaging assessment) offered by the Sustainable Packaging Coalition [https://greenblue.org/work/compass/] and Piqet (packaging impact quick evaluation tool) provided by the Sustainable Packaging Alliance [http://piqet.com/].

3. Impact of food wastes on the overall sustainability

The environmental load of the food industry is function of the resources and emissions associated with both the food and the packaging, including the amount of food that is lost. Indeed, resources and emissions associated with nonconsumed food account for unnecessary impact, whose prevention might significantly improve the overall sustainability. Thus, food loss and waste is not just an important ethical issue, but also a matter of sustainability. Food losses occurring at the end of the food chain (retail and final consumption) are rather called "food waste," which relates to retailers' and consumers' behavior (Parfitt et al., 2010).

A study by the FAO (Gustavsson et al., 2011) highlighted that in medium- and high-income countries a high fraction of food is thrown away, even if it is still suitable for human consumption. Much more food is wasted in the industrialized countries than in developing ones, in particular, food waste at the consumer level in industrialized countries (222 Mton) is almost as high as the total net food production in sub-Saharian countries (230 Mton).

Food wastage is always bad, but some types of waste are even worse than others: this depends on two factors. First, the stage of the value chain at which they occur: indeed, a possible alternative use is possible for losses occurring at the processing stage, while food wasted after distribution is often lost without any alternative. Moreover, as the environmental impacts build up with each step in the production cycle, wastage of finished products implies the maximum level of unnecessary impact and, as a consequence, the reduction of losses occurring in the consumer phase is of paramount importance (Williams and Wikström, 2011). Secondly, the food category determines the importance of the waste: it is well known that certain food types (i.e., animal and dairy) are characterized by higher production impacts, as a consequence, their wastage results in higher resources consumption and higher emissions compared with the wastage of lower-impact foods (vegetables, beverages). For instance, Venkat (2012) estimated that beef accounted for the largest impact contribution of wasted food (16% of the total emissions), even though the amount of waste for this product was less than 2% (by weight) of total waste. Recently, a study conducted on the US food system, with the aim of estimating the actual environmental weight of food wastes, highlighted that the emissions associated with just producing wasted food equal the emissions from 33 million vehicles driven for 1 year (Heller and Keoleian, 2015).

It has been estimated that 35%–60% of food waste is avoidable and that 45% of the avoidable losses (which, in turn, represent 26% of the wasted food in Switzerland), occurs at households (Beretta et al., 2013). In 2012, the European Commission (EC) set a target to reduce by 50% the food waste in Europe by 2025. Indeed, it is widely accepted that food waste reduction must be attempted both for reducing the environmental burden of food chains and as a food security strategy, which would allow to increase the availability of food worldwide.

Wikström and Williams (2010) established the need to utilize a functional unit based on the food eaten in order to account for consumer-level food losses in the assessment of the environmental performances of foods, this concept was also adopted by (Heller et al., 2018). The same authors proposed that it could be environmentally reasonable to increase the impact of packaging with the adoption of new design and innovation that contribute to reduce food losses. Following studies (Silvenius et al., 2014; Verghese et al., 2015; Williams and Wikström, 2011) strengthened this achievement and it is now consolidated that an increase in packaging impact is acceptable when it is counterbalanced by an impact reduction due to shelf life extension and/or improved protection along the product chain.

Fig. 6.1 reproduces some of the results from (Williams and Wikström, 2011) showing the maximum increase in environmental impact from the packaging as a function of the reduction of food losses. T is the environmental impact to produce the package for one unit of purchased food to the consumer, inclusive the waste handling of the packaging, and T_2 and T_1 are the impacts referred to two different packaging systems, the former with increased impact. As it can be inferred, a reduction

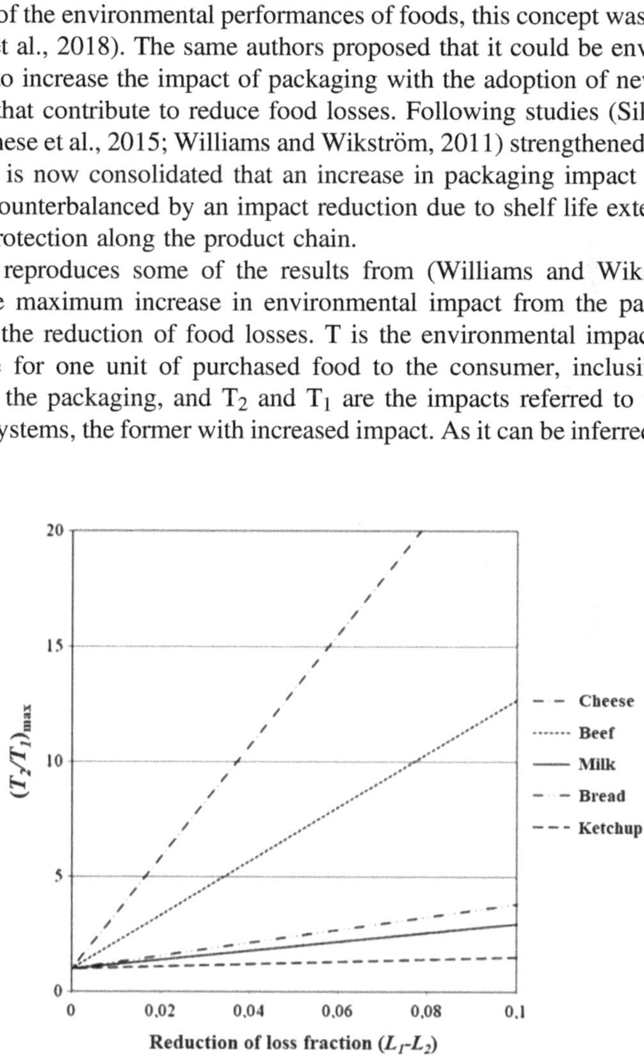

Figure 6.1 Maximum increase in environmental impact from the packaging as a function of the reduction of food losses.
Reproduced from Williams, H., Wikström., F., 2011. Environmental impact of packaging and food losses in a life cycle perspective: a comparative analysis of five food items. J. Clean. Prod. 19 (1), 43–48 https://doi.org/10.1016/j.jclepro.2010.08.008.

of food losses justifies the impact increase of packaging, especially for cheese and beef which are characterized by much higher production impacts and higher loss rates.

Wikström et al. (2018) recently contributed to consolidate the growing awareness on that the impact of packaging is not as high as that of food waste for different products; in their paper, the authors showed the greenhouse gas distribution between the consumed food, wasted food and packaging materials of: (1) meat, fish and eggs; (2) dairy; (3) fruits, vegetables, and nuts. The greenhouse gas of consumed food was 80%, 75%, and 60% for the three categories, respectively; the impact of wasted food was as high as 18%, 13%, and 22%, much higher than the impact arising from packaging materials (2%, 10%, and 12%, respectively). This study confirmed that, for the considered food categories, packaging accounts for the least environmental impact compared to the food itself.

Overall, it should be noted that it is possible to measure the environmental load of food and packaging, while the correct estimation of the waste level for specific products remains the major challenge within this area. Those few studies which consider food wastes usually refer to general data, while the actual waste level is very product-specific and is strongly dependent on the specific food-packaging configuration and socioeconomic factors.

Consumer behavior has the largest potential to decrease food waste, but packaging may influence behavior through new information and/or technical solutions. Some of the reasons for food to turn into waste at the consumer level are packaging-related: packaging role to prevent food from turning into a waste is linked to the main packaging functions (i.e., protection, convenience, communication), which should be further implemented in the direction of waste reduction. In particular, packaging should protect the food from physical damage and biochemical degradation; it should be easy to reseal, thus extending secondary shelf life; it should be easy to empty completely, it should be provided in suitable sizes that avoid leftovers and it should also provide the correct information (content, best-before-date, etc.) to avoid that food that is still suitable for consumption is thrown away (Williams and Wikström, 2011).

The wastage of packaged food, occurring during the distribution and at the household level, brings about another issue: not only the food is wasted, with unnecessary impact arising from resource consumption and emissions, but also this waste is enclosed in its package and a correct disposal would imply the package opening and separation of the food from its package, which should follow two separate recovery paths (i.e., the organic fraction and the plastic fraction, respectively). When the two items are not correctly separated, which is the current scenario due to the lack of suitable sorting plants, this waste is collected as unsorted waste and is likely to end up in landfill or incinerated (Garcia-Garcia et al., 2015; Vitale et al., 2018).

4. The contribution of packaging to the sustainability of food chains

Packaging reputation by consumers can be inferred from a study by Tanner and Kast (2003) who report that an environment-friendly food product is, ideally, "domestically produced rather than imported from abroad; furthermore, it is organically grown, seasonal, fresh (rather than frozen), and unwrapped". The role of packaging in the sustainability of food chains has been controversial and it appears that the different considerations of packaging under the sustainability point of view raise from the attempt to generalize its role. As an example, in one paper on the theme (Silvenius et al., 2014) it was stated that "the significance of packaging production and post-consumer life on environmental impact is low and represents 1%–10% of the total environmental impact generated by the food chains", while, on the other hand, another work (Manfredi and Vignali, 2015) reported that "the impacts of the packaging materials life cycle are very often among the most relevant environmental burdens". The understanding of the contribution of packaging to overall food sustainability, however, should consider each specific food case and should be based on considerations on the environmental impact connected to the production of both packaging and food.

Recently the Packaging Relative Environmental Impact (PREI) was proposed as an indicator able to assess the role of packaging in the sustainability of a specific product (Licciardello, 2017). PREI is the ratio of the environmental impact related to the packaging, $\mathbf{EI}_{(pack)}$, to the overall environmental impact of the food, $\mathbf{EI}_{(food)}$. Impacts are conveniently expressed as Global Warming Potential (GWP), since this impact category is the most widely used and the less affected by other factors.

$$\mathbf{PREI} = \mathbf{EI}_{(pack)}/\mathbf{EI}_{(food)}$$

This simple index represents a balance between the actual environmental load of packaging and that of a specific food product and could be used as a tool to support improvement strategies. In particular, high-PREI values suggest that actions aimed at reducing the environmental impact of packaging materials may result in the sustainability improvement of the system, while low-PREI values suggest that food wastes, for the specific product, have a significant impact on the environmental performances: therefore, investments aimed at reducing food wastage may be positive, even when they imply some increase of the packaging impact.

PREI can assist in packaging design and optimization and does not require to perform full LCA studies. However, the source of data (i.e., LCA studies on food products) should display the impact of packaging production, while studies which group the impact of packaging together with other phases (usually, transportation, storage, distribution) are not suitable for this aim.

In some cases, which are hereafter exemplified, may be useful to understand the potential of this indicator. Based on data collected by the author (Licciardello, 2017), beverages such as beer, wine and carbonated soft drinks are characterized by a high-PREI, ranging from 45 to 80, as to testify the higher environmental impact generated by the packaging production, usually glass bottle, PET bottle or aluminum can, compared with the impact generated by the production process of the beverage itself. For such products, actions aimed at minimizing the impact of packaging, with special regards for the material reduction (lighter bottles and caps, new shapes) and at improving the recyclability of materials represent the right strategies for sustainability improvement. A similar consideration applies to canned vegetables and fish or tomatoes in glass jars, whose impact is highly affected by the packaging. On the other hand, dairy product (milk, cheese, butter) and beef meat are characterized by low and very low-PREI values, ranging from about 1 to less than 15, as a result of high impacts related to the food production and raw materials primary production, compared to which the environmental load of packaging becomes, in some cases, negligible. For such products, packaging is usually polymeric (HDPE or PS bottles and jars, PS or PET trays with lid film) or composite (multilayer laminated carton) and environmental improvement strategies should be based on actions aimed at extending the product shelf life and reducing product losses. Indeed, food losses, however generated, represent unnecessary impact. A previous study (Williams and Wikström, 2011) compared the environmental impact from food (F) and from packaging (T) for five different food categories. The environmental impact related to cheese and beef was much higher than the impact of packaging, thus resulting in the highest F/T ratio (or lowest PREI). Similar considerations arose from this study: large increases from the cheese packaging could be justified for a new packaging reducing cheese losses, while the low F/T ratio for ketchup indicates that it is just as important to find packaging systems with less environmental impact as it is to develop packaging that lead to lower losses (Williams and Wikström, 2011).

Similarly to PREI, the Food-to-Packaging (FTP) environmental impact indicator proposed by (Heller et al., 2018) gives useful information for assessing the actual weight of packaging and food on overall food sustainability, but its interpretation is, of course, inverse compared with PREI. FTP ratio is calculated as [("agricultural (farm-gate) production/kg food" + "food processing/kg food")/"packaging materials/kg food".

Based on the established awareness of the impacts of food waste, packaging environmental assessment needs to be reconsidered to include the indirect effects due to packaging-related food waste. Both PREI and FTP can be also used to estimate the influence of food waste on the overall system environmental performance. When FTP ratios are high, food waste is likely to have a strong influence on system environmental performance and changes in packaging design that lead to food waste reduction are likely to determine net decreases in environmental impact; on the other hand, low FTP ratios indicate that investments should be better addressed at reducing the impact of the packaging itself. Based on a survey of LCA studies performed by (Heller et al., 2018), the FTP ratio (for GHG emissions) ranges from 0.06 in wine to 780 in beef. High ratios for cereals, dairy, seafood, and meats, which correspond to the low-PREI values estimated by Licciardello (2017), suggest that net impact reductions can be achieved through packaging-based food waste reduction innovations.

Often, the agricultural production stage is responsible for the highest impacts in food life cycles (Ingrao et al., 2018), however processing or packaging can also play a significant role, depending on the product. A survey of LCA studies on food products has brought to light large differences in the environmental load caused by the packaging and by the production (primary production + processing) of the food. Fig. 6.2 reports different food products with their calculated PREI, as an estimate of the weight of packaging on the overall environmental impact generated in the product chain.

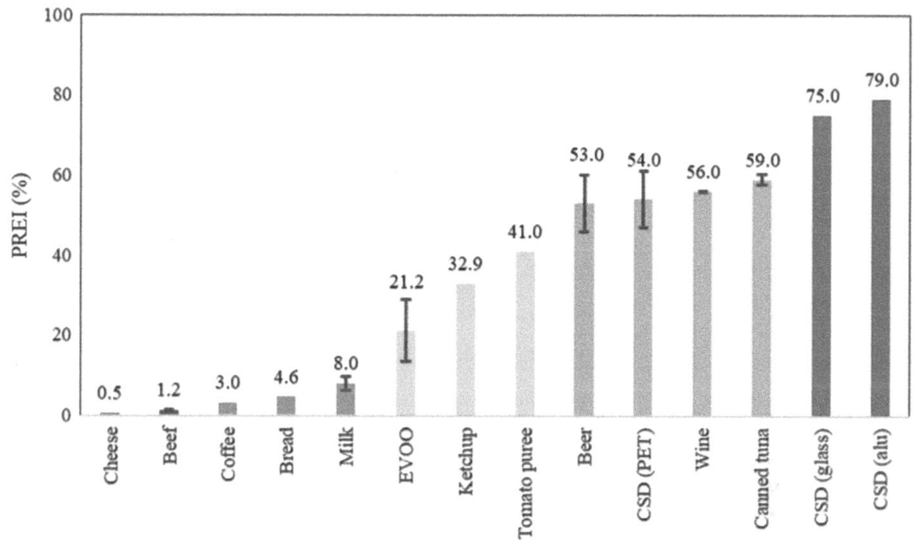

Figure 6.2 Packaging relative environmental impact (PREI) calculated for different food products. *CSD*, Carbonated soft drinks; *EVOO*, Extra virgin olive oil. Where multiple data are available, the mean calculated value is reported with standard deviation. Sources of data for PREI calculation are reported in Table 5.1.

As it can be inferred, based on the literature data considered (Table 6.1), food products can be classified into very low-PREI (cheese, meat, coffee, bread), low-PREI (milk), moderate-PREI (extra virgin olive oil, ketchup, tomato puree), high-PREI (beer, wine, carbonated soft drinks bottled in PET and canned tuna), and very high-PREI products (carbonated soft drinks packed in aluminum cans and in glass bottles).

Based on the above consideration that for systems characterized by a low-PREI (or a high FTP ratio) the adoption of packaging systems able to reduce food waste would improve the overall sustainability, in some cases the best and simplest solutions could be to pack the products in smaller package sizes, despite the increase in packaging materials.

There remains the challenge to estimate food wastes related to specific scenarios. Only a few studies have attempted to consider food waste in the sustainability assessment of packaging systems and have calculated the waste reduction that can be achieved with a new packaging, while in most cases waste reduction scenarios were only hypothesized. The study by (Heller et al., 2018) analyzes a group of foods with the aim of demonstrating the influence of food waste on product system environmental performance. This paper considered 13 foods in their typical packaging configurations applied waste rates provided at the food commodity level as an estimate for product-specific waste rates. The environmental balance between food waste and food packaging can be delicate, and a careful measurement of waste levels, which is the actual challenge in the theme, is the fundamental step for demonstrating the environmental benefits which can be achieved.

Table 6.1 Data sources (LCA studies) for the PREI calculation (Fig. 5.1).

Food product	Data source
Cheese	Williams & Wikström (2011)
Beef	Williams and Wikström (2011); Zhang et al. (2015)
Coffee	Büsser & Jungbluth (2009)
Bread	Williams & Wikström (2011)
Milk	Williams and Wikström (2011); Hospido et al. (2003); Manfredi et al. (2015)
EVOO	Pattara et al. (2016)
Ketchup	Williams and Wikström (2011)
Tomato puree	Manfredi and Vignali, 2014
Beer	Cimini and Moresi, 2016
CSD	Amienyo et al., 2013
Wine	Fusi et al., 2014; Bonamente et al., 2016
Canned tuna	Avadì et al., 2015; Hospido et al. (2003)

5. Sustainability-oriented strategies through the improvement of packaging efficiency

5.1 *Packaging lightweighting*

Packaging materials, as well as other resources used in the production of a food product, should be minimized. This does not mean that packaging should be avoided, as it is undoubtful that packaging represents an essential element for the quality preservation of food products, and modern foods could not be conceived without their packaging. Packaging should be optimized in order to use as minimum materials as necessary to fulfill its protective functions. Packaging lightweighting, i.e., the adoption of thinner and/or lighter packaging materials, reduces the use of resources and, obviously, represents the first simple strategy toward waste prevention. In the last decades, glass, aluminum and steel containers have been significantly lightweighted, as well as PET bottles and paperboard (Marsh and Bugusu, 2007). Since packaging lightweighting also implies a reduction of costs for the producer, this concept might sound obvious and packaging minimization should be taken for granted; however, many commercial cases demonstrate that this theme still carries significant improvement potential (Licciardello, 2017). Overpackaging, that is the unnecessary use of packaging materials, is, in fact, an underestimated issue. Packaging systems are often oversized, and, in some cases, it has been demonstrated that a significant reduction of thickness is possible, with noticeable economic and environmental benefits, which do not compromise the package performance. The optimization of packaging systems implies the verification of such performances by comparative shelf life testing, whose aim is to prove that the alternative system guarantees the same shelf life standards as the conventional system. As an example, a recent study (Licciardello et al., 2017) evaluated two alternative packaging systems for industrial durum wheat bread, commercially available in a two-piece form-fill-seal packaging system made of a thermoformed bottom and a lid. The packaging system identical to the one in use, but with lower thickness (about 20% lightweighting) was able to guarantee the same shelf life standard, as evaluated by instrumental and sensory analyses. In particular, the two packaging systems had the same barrier to gases, thus allowing the preservation of the modified atmosphere conditions. The other candidate was a flow-pack system with a 62 micron-thick co-extruded film, allowing an even higher lightweighting: in this case the performances of the system were slightly lower, causing a shelf life reduction which could still meet the requirements for a local distribution of the product. Soft drinks companies keep on trying to reduce the preform weight, which might be possible with new PET grades and bottle design (Coriolani et al., 2006; Licciardello et al., 2011): in this case, the comparative shelf life tests are made easier by the mere consideration of carbon dioxide retention inside the bottle, that is the key quality parameter determining the shelf life of such products. Packaging lightweighting remains an underestimated strategy for many food products, but it represents the easiest and more affordable measure for sustainability improvement of food packaging. Lightweighting could be achieved by reducing thicknesses for the same material and/or adopting composite structures with higher performances: however, sustainability-driven packaging reduction should be addressed under an end-of-life perspective, preferring mono-materials instead of composite multilayer materials (wherever possible), which bring higher performances but are less recyclable. Indeed, recyclability could be the downside of lightweighting when alternative multilayer/composite materials are considered.

5.2 The choice of packaging materials with a focus on recyclability

According to Peelman et al. (2013), sustainability of food packaging can be achieved at three levels: (1) at the raw materials level; (2) at the production level, through more energy-efficient processes; (3) at the waste management level, considering reuse, recycling and biodegradation. Recycling diverts packaging from the waste stream (incineration or landfill) to recovery, through collection, sorting, processing, and manufacturing into recycled materials and products. Recyclability of plastic packaging depends, in part, on the efficiency of the collection and sorting systems. To these regards, recent data on the end-of-life of plastic postconsumer waste (PlasticsEurope, 2017) demonstrate that in 2016, for the first time, the level of recycling (31.1%) in Europe overcame that of landfilling (27.3%), while the remaining major fraction (41.6%) was addressed to energy recovery. According to these data (very uneven across the EU countries), from 2006 to 2016 the total amount of plastic waste collected increased by 11%, while recycling grew by about 80%. Similarly, the levels of plastic packaging recycling increased by about 75% in the same decade, and the average plastic packaging recycling rate across the EU countries was about 41%, well above the requested 22.5% of the EU Packaging Waste Directive.

The ambitious target to further increase plastic packaging recyclability through improved selection and recovery implies redesigning packaging through an end-of-life thinking approach. The adoption of high-performance composite materials, in the last decades, has contributed to reduce the packaging amounts but, on the other hand, has worsened the end-of-life scenario for plastic packages, whose recyclability is limited to PET and HDPE. Composite multilayer films, which are made up of two or more layers, and are designed to combine the performances of different materials into one single structure: they include materials made of different plastic layers, or combining plastic (especially, polyethylene) with aluminum and paperboard (multilayer cartons). In particular, multilayer structures are designed to improve the gas and moisture barrier of films for application to modified atmosphere- or vacuum-packaged products and to moisture-sensitive foods, allowing the extension of the shelf life by keeping the modified atmosphere (or vacuum) for a longer time and by minimizing the moisture transfer, respectively. Sorting and recycling plants are equipped with optical devices (usually, Fourier Transform Near-Infrared, FT-NIR) able to identify different polymers, however, the coupling of different materials into one structure makes it impossible, at the actual state of technology, to correctly process each component (Ragaert et al., 2017), thus addressing the material to the unsorted fraction and to energy recovery (pyrolysis), in the best scenario.

Lightweighting design of packaging has widespread at the expense of recyclability, in the wrong understanding that reducing the weight of a package is so resource-efficient that its recyclability can be neglected (Scriba and Germany, 2016). However, the most recent innovations in the field (i.e., nanotechnologies) have paved the way to the improvement of packaging performances which might not affect recyclability. The plastic packaging sector, where possible, has the responsibility to revert to materials which guarantee good recyclability, primarily mono-PET. Similarly, the choice to adopt black-dyed plastic trays for some applications (such as meat), driven by market needs, has not taken into consideration the end-of-life of such packages, which cannot be sorted by the current optical sorting equipment due to their color. Based on the above considerations, the choice of mono-materials, with improved recycling potential, as well as lightweighting design, could significantly contribute to the improvement of resource efficiency for packaging materials.

5.3 The shift to biobased plastic materials

The search for packaging materials sources alternative to oil-based materials is another target toward sustainability, and this objective was included among the three levels to tackle sustainability of food packaging proposed by Peelman et al. (2013). The use of synthetic plastic has posed serious ecological problems due to their nondegradability and to the depletion of fossil resources. Bioplastics have recently been proposed as alternatives, at least for certain applications, to fossil-based, nonbiodegradable, plastics. According to the European Bioplastics Organization

(EUBP, www.european-bioplastics.org/), "bioplastic" is the word to be used to indicate plastics (1) which are "biodegradable", (2) which come from renewable resources ("bio-based"), or (3) having both these features, i.e., are biobased and biodegradable. Indeed, bioplastics are not a single kind of polymer but rather a very heterogeneous family of materials.

(1) There actually exist some fossil-based polymers which are biodegradable: it is the case of some PVOH formulas (Polyvinyl alcohol, also named as PVA), an array of water-soluble polymers showing high gas barrier properties, PCL (Polycaprolactone), a polyester with a low melting point widely used in polyurethane adhesive, PBS (Polybutylene succinate), a thermoplastic polymer with properties comparable to polypropylene and few other minor resins. All these polymers, according to the EUBP's nomenclature, deserve to be named as "bioplastics", though they are oil-based.

(2) "Biobased" bioplastics are, instead, materials based on renewable resources and they can be biodegradable or not. The second EUBP class of bioplastics corresponds to biobased, nonbiodegradable, plastics. Nowadays it has become possible to synthetize conventional plastic from precursors obtained from renewable sources. For instance, PP (polypropylene), PE (polyethylene) and other useful resins, can be produced starting from ethanol and other small molecules obtained by biomasses and sugar fermentation. Ethylene, for instance, is obtained by the catalytic dehydratation of bioethanol, followed by normal polymerization to produce low and high-density polyethylene (LDPE, HDPE). The ethylene obtained from bioethanol may be also dimerized to produce n-butene, which reacting with ethylene can lead to PP. These Bio-PP and Bio-PE are exactly alike the conventional polymers obtained through oil cracking and chemical synthesis; hence they are not biodegradable but, coming from renewable resources, they are very interesting anyhow. It is worth reminding that these polyolefinic polymers, together, currently account for about 55% of all plastic and 70% of plastics for packaging. Bio-PP and Bio-PE, due to their origin, show carbon neutrality and do not contribute to fossil depletion, making energy recovery even more attractive, and contribute to a circular economy. Among durable plastics obtained from renewable resources, biopolyethylene terephthalate (bio-PET) represents a case of success, which has achieved wide commercial exploitation. PET is a polymer (polyester) obtained mainly by direct esterification of 30% mono-ethylene glycol (EG, $C_2H_6O_2$) and 70% terephthalic acid (TA, $C_8H_6O_4$) or by transesterification of dimethyl terephthalate with EG (Xiao et al., 2015). PET is one of the most widely produced polymers, whose main application is in the food and beverage sector (bottles, trays, films). EG, which represents 30% of the PET composition, can be obtained through direct conversion of lignocellulose in the one-pot process with tungstic catalysts under hydrothermal conditions (Wang and Zhang, 2013): in this process, the carbohydrates of cellulose, hemicellulose, and sugars are selectively transformed into EG with high yields of 60–76%. Unlike the conventional oil-derived EG, the biomass-derived EG (bio-EG) can contain small amount of other diols, as the byproducts in the catalytic conversion of biomass. However, it has been demostrated that bio-PET prepared with bio-EG at purity higher than 95% could be used as widely as the conventional oil-PET resin without significant changes in their performance (Xiao et al., 2015). TA, which accounts for the remaining 70%, is normally obtained from paraxylene, which typically is oil-based, but new technologies (not yet on the market) converts biomasses into paraxylene or muconic acids to obtain bio-TA (Brandvold, 2012) that might make possible a 100% bio-based PET. In 2009 The Coca-Cola Company launched the PlantBottle program, aimed at distributing beverages in fully recyclable PET bottles obtained in part from plant material. By 2015 (Ren et al., 2015) the use of PlantBottle packaging since launch had helped save the equivalent annual emissions of more than 315,000 metric tons of carbon dioxide. Together with the challenge to commercialize 100% plant-based PET, the Coca-Cola Company set the target of having all the new PET packaging containing PlantBottle packaging material by 2020. In every case, being the bioplastic chemically identical to the oil-based polymer, performances and uses remain the same as for the conventional polymer. Currently, the bio-PET market is already quite interesting worldwide, accounting for more than one fourth of the overall bio-based bioplastics (see Fig. 6.3).

Other biobased, nonbiodegradable materials can be obtained, as well as bio-PET, from monomers produced by bioprocesses employing metabolically engineered microorganisms (so-called "biorefineries") (Chung et al., 2015). These include polyamide precursors (diamines, dicarboxylic acids, amino acids), and polyester precursors (diols and hydroxy acids).

It is worth reminding that all the biobased plastics (biodegradable or not) nominally belong to the category of "bioproducts", which have been already properly defined by various institutions. According to the European Committee for Standardizations (EN 16575) they are "products that are wholly or partly derived from biomass of biological origin, such as from trees, plants or animals", which "may have undergone some kind of physical, chemical or biological treatment before being turned into the intermediate, material, semifinished or final product, called bio-product". In other words, the extent of "biorefinery concept" is huge, and its potential is continuously increasing (Cherubini, 2010). When the production of potential biobased chemicals, and particularly bioplastics, is concerned, elevated costs and long implementation times must be taken into account. In such cases, computational tools can be very useful to explore the biological and technical spectrum of feasibility (Campodonico et al., 2018).

The wide definition of bioproducts, as well as the fast growth of this technological area, poses also a problem of authenticity, i.e., how can we be sure about the biobased nature of the packaging material selected? Indeed, a significant contribution has been offered by the assessment of biogenic carbon derived from biomass, implemented in various standard. A well-established method for measuring the biobased carbon content in a product is the ^{14}C-method, present in official standards in EU and in USA (EN 15440, ASTM 6866). A bioplastic may also be specified by the percentage of the overall biobased mass. This method, complementary to the ^{14}C-method, takes into account other chemical elements such as oxygen, nitrogen, and/or hydrogen. Some certification schemes and derived product labels based on European and US standards are already available (Kale et al., 2007; Tellnes et al., 2017).

(3) The third category, according to European Bioplastics Organization's classification, refers to biobased plastics that are also biodegradable or compostable. It should be emphasized that biodegradability is different from compostability. The term "biodegradability" refers to the possible biodegradation phenomenon, which is a chemical process during which microorganisms in the environment, or other biological means, convert materials into natural and simple substances such as water, carbon dioxide, methane and others. This phenomenon can occur both in aerobic and anaerobic conditions and a number of different factors determine the rate of the process and the final products. Almost all materials are subject to biodegradation processes, but the discriminants are the rate of the process (it can be too low for humans' expectations) and the consequences of such processes. Depending on the material, it is also possible that the biodegradation leads to harmful substances absorbed by the soil. The term "compostability," on the contrary, refers to aerobic process only and its definition, including the timeframe, the level of biodegradation, the surrounding conditions and the safety of final products, is much more precise and useful.

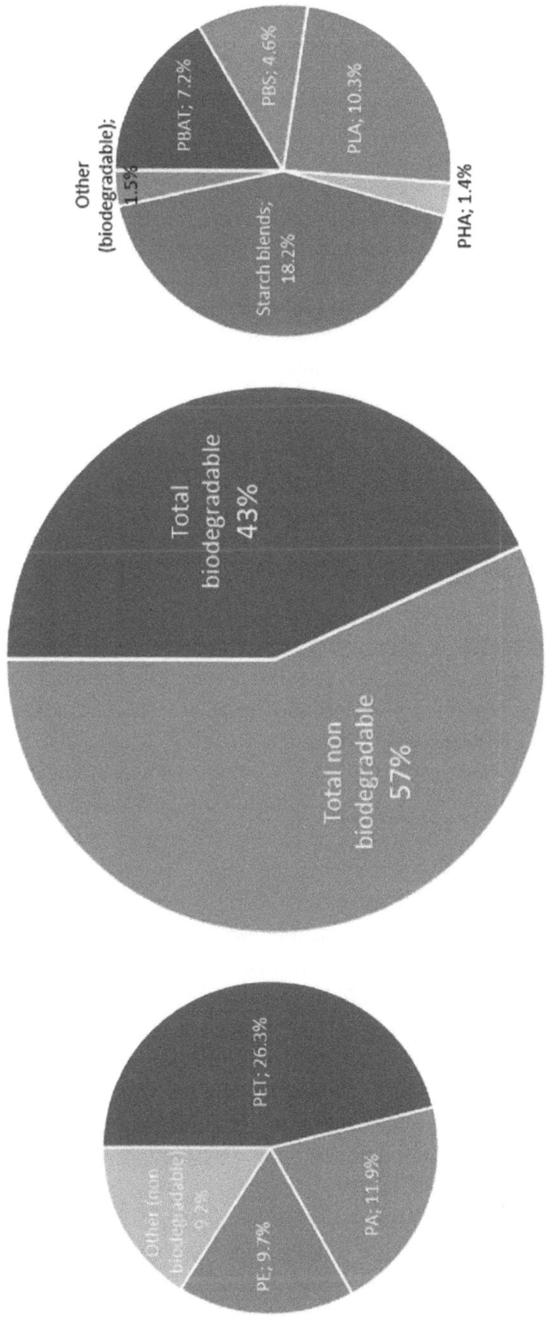

Figure 6.3 Global production capacities of bioplastics.
European Bioplastics—nova institute. http://www.european-bioplastics.org/market.

Moreover, compostability can rely on specific standard references, applicable both to packaging and plastics (ISO 17088, EN 13432, EN14995 or ASTM D6868). In the latter class of bioplastics (biobased and biodegradable) we find PLA (polylactic acid or polylactide) which is a very promising thermoplastic polyester derived from corn starch, cassava roots, or sugarcane. PLA is versatile, compostable and recyclable and it is characterized by very good optical properties, good processability, high resistance to water solubility (Siracusa et al., 2008). PLA also offers the possibility to be thermoformed, and the PLA granule can be expanded, offering a valuable alternative to expanded polystyrene (EPS) to produce trays addressed to the packaging of fresh products (meat, cheese, vegetables). The most common route to PLA is the ring-opening polymerization of lactide with various metal catalysts in solution but a huge number of studies have been carried out on this bioplastic (Farah et al., 2016; Bajpai et al., 2014), trying, in particular, to overcome its poor gas barrier properties by means of nanocomposite technology and green materials (Ortenzi et al., 2015; Gazzotti et al., 2017). A second class of biobased, biodegradable bioplastics, quite interesting for packaging applications, are the PHA polyesters (Polyhydroxyalkanoates), produced through bacterial fermentation of sugars or lipids. More than 150 different monomers are known with extremely different properties, but always biodegradable. In particular, the development of polyhydroxybutyrate-valerate (PHB−V) copolymers, is a promising route both under the technological point of view, since the materials have properties suitable for some food application, and under the sustainability aspect, as these biopolymers can be obtained through microbial fermentation of various substrates, including organic wastes, and are fully biodegradable (Guillard et al., 2018).

No less important is the class of thermoplastic starches (TPS). The characteristics of TPS bioplastics, in fact, can be easily tailored according to specific requirements. Starch-based bioplastics are often blended with PLA or PHA and to this class belongs the well-known brand of Mater-Bi products.

Whether simply biobased or also biodegradable, the use of these last bioplastics may result in environmental advantages in terms of safeguard of fossil resources, end-of-life, or both. Interestingly, it has been demonstrated recently that, also for compostable biopolymers, recycling represents the best end-of-life option showing the highest environmental benefits (Hottle et al., 2017).

Many applications of biodegradable/compostable, biobased bioplastics are already available, even if the commercial diffusion of such materials has been limited by costs and by their lower performances. Indeed, biopolymeric packages are usually more sensitive to high relative humidity, which acts as plasticizer thus compromising the structure and functionality. Anyway, technological limitations can be improved by tailoring the biomaterials to specific purposes. Other issues that limit a wider scaling up and market diffusion of biobased bioplastics are the raw material variability and the narrow processing window compared to conventional oil-based polymers (Guillard et al., 2018).

While the word "bioplastics" has been clearly defined by EUBP's classification (already discussed), there remains some ambiguity between the terms "biopolymers" and "bioplastics". "Biopolymer" relies on chemical and biological general nomenclature. Biopolymers are polymers, i.e., macromolecules containing arrays of monomeric units covalently bonded to form larger structures, produced by living organisms. There are various families of biopolymers, commonly classified according to the monomeric units; the most important and diffuse are: polynucleotides (RNA and DNA, composed of the 13 fundamental nucleotide monomers); proteins and polypeptides, which are polymers of amino acids, linked by peptide bonds; polysaccharides which are linear and branched carbohydrates polymers, such as starch, cellulose, chitin, glycogen. More complex structures (i.e., combined with molecules different from the main monomeric units) are also quite common in the wide category of biopolymers, such as rubber, suberin, melanin, lignin.

Based on the production process, biopolymers are generally classified into polyesters, starch-based polymers, and other polymers. Another classification depends on the source and includes polymers extracted directly from biomass (i.e., proteins, polysaccharides, lipids); materials chemically synthesized by polymerization from renewable monomers (e.g., polylactide, PLA); materials synthesized by microorganisms (polyhydroxyalkanoates, PHAs).

Unfortunately, the currently marketed bioplastics based on renewable resources compete with the food sources since they use corn starch, cassava roots or sugar cane, while a significant improvement, within a circular economy perspective, would be the development of processes for obtaining bioplastics from organic wastes (crop residues, sludges, agro-industrial by-products), fully biodegradable and which do not compete with food sources (European Parliament and Council, 2018a,b). Even though, the current estimates of the land use for bioplastics production (European Bioplastics, 2018) indicate a very small percentage (about 0.02%), less than 1/50 of what is used nowadays for biofuels which, in turn, is 1% of the global agricultural area, the debate about the use of limited resources such as the arable land is absolutely right and proper.

Beside starch, which is really a food biopolymer, our planet is a huge resource for two more carbohydrate biopolymers, which are chitin and cellulose, that mainly come from byproducts and biomass wastes. Both are extensively investigated for possible packaging application but not as such, rather as somehow modified forms of them, according to the bioproducts definition already proposed.

The great interest around the biopolymer chitin is related to its transformation into chitosan. Chitin, a long-chain polymer of N-acetylglucosamine, is a primary component of cell walls in fungi and of the exoskeletons of crustaceans and insects, it is also widely present in parts of molluscs, fish (scales) (Tang et al., 2015) and other living organisms. Chitosan is obtained by the deacetylation of chitin through an alkaline process; the degree of deacetylation ranges from 60% to 100%. Thus, chitosan is an etherogeneous linear polysaccharide, constituted by a randomly distributed β(1−4) linked D-glucosamine (deacetylated unit) and N-acetyl-D-glucosamine (acetylated unit). It has a number of commercial, even biomedical, uses but in the field of packaging it is worthy reminding its film forming ability and the antimicrobial activity which makes chitosan very interesting in active packaging materials. The antimicrobial function is related, very likely, to its positive charge: indeed, chitosan is one of the few natural polycationic polymers available.

The structures of chitin/chitosan are comparable to another polysaccharide, much more diffused and much more easily available than chitin, that is cellulose. Cellulose, unquestionably, is a true protagonist in the world of packaging materials. The packaging industry has always been using cellulose-based materials in huge amount. Currently, cellulosic packagings, include both wrapping materials and containers, primary and secondary packages, flexible and rigid packaging. The biggest part of the whole of packaging materials, is accounted, all around the world, for cellulosic materials with figures around 40% of the total (Li et al., 2015).

The biopolymer cellulose is generally recognized as safe, largely available, cheap and completely biodegradable. However, its most interesting packaging applications deal with modified cellulose. Regenerated cellulose (cellophane) had been the fundamental element in the development of flexible packaging. The clear, resistant and gas barrier thin films of cellophane dominated the flexible packaging scenario for many years, up to the development of plastic films and the affirmation of OPP and OPET films, in particular. The regenerated cellulose, however, in spite of its deep modification introduced by the manufacturing process (Kolpak and Blackwell, 1975), maintains its full biodegradability, at such an extent that cellophane has been used as benchmark in order to estimate the degradability of different packaging materials (Coma et al., 1994). Recently, very likely due to the environmental concerns, a revitalization of the cellophane sector has been noted and new products, certified as compostable, are available. Beside the market growing, also new processes with lower impacts on the environment have been proposed and extensively investigated for cellophane (Hudson and Cuculo, 1980).

However, the most significant innovations in the area of cellulosic materials come from nanotechnology and the production of cellulose nanocrystal (CNC), cellulose microfibrillated (MFC) and cellulose nanofiber (CNF). A discussion and some comments about these cellulose nanoparticles are proposed furtherly.

Both the improved recyclability of plastic packaging and the use of renewable resources for producing biobased and/or biodegradable packaging are complementary strategies targeting the sustainability improvement of packaging through the CO_2 emission reduction and the safeguard of fossil resources (Table 6.2).

6. Sustainability-oriented strategies through the improvement of packaging effectiveness

In the case of packaging, effectiveness can be declined into various meanings, under the common denominator of preventing food from turning into a waste. In this light, effective packaging should protect the food from physical damage and biochemical degradation, thus guaranteeing an adequate shelf life; it should allow an extended secondary shelf life and supply the suitable information to the consumer for domestic storage after opening; it should avoid leftovers and it should be easy to empty completely.

Table 6.2 Synopsis of commercial bioplastics and classification based on source and biodegradability.

Acronym	Full name	Source		Biodegradability
		Renewable	Fossil	
PVOH, PVA	Polyvinyl alcohol		CS	+
PCL, PLC	Polycaprolactone		CS	+
PBS	Polybutylene succinate		CS	+
PBAT	Polybutylene adipate terephthalate		CS	+
Bio-PP	Bio-polypropylene	CS		-
Bio-PE	Bio-polyethylene	CS		-
Bio-PET	Bio-polyethylene terephthalate	CS		-
PLA	Polylactic acid, Polylactide	CS		+
PHAs	Polyhydroxyalkanoates	SMO		+
TPS	Thermo-plastic starches	CS		+
Cellophane	Regenerated cellulose	CS		+
Chitosan	Chitosan	DFB		+

DFB: directly from biomass; CS: chemically synthesized from renewable material; SMO: synthesized by microorganisms

6.1 Redesign product-packaging configurations

In the last decades, family-sized packages have been proposed as the most convenient solutions, allowing the consumer to access food products at a lower cost per weight, and to use lower amounts of packaging. Actually, this trend has the hidden aim of increasing products purchase and consumption, since consumers are encouraged to buy larger pack sizes, probably exceeding their needs, because of quantity discounts. This commercial strategy had some merits, but it underestimated the waste increase potential. Indeed, the actual tendencies aiming at reducing food waste are proposing strategies able to minimize food waste potential: among these, the shift to smaller packages or single-serve formats and the adoption of reclosable packages may result in more packaging per serve, but the potential for food waste is reduced, meaning that, for some food categories, the overall environmental impact from the system will decrease (Wikström et al., 2014). Modern food industry is requested to redesign product-packaging configurations to help consumers reduce waste through convenience features (among which, package size); however, pricing policies need to support consumers to ensure they see value in the smaller pack sizes (and that they will not waste product) (Verghese et al., 2015).

6.2 Maximize packaging protective role

The design of packaging systems implies tailoring the packaging functions for the specific requirements of the food. Packaging protective function can be maximized exploiting the new concept of active packaging. This concept includes materials and devices able to offer additional protection to the packaged food by a tailored interaction with the food itself and/or with the package headspace. Active packaging has broadened the packaging potential to maintain food quality and is increasingly drawing the interest of the whole sector as a promising strategy for extending the shelf life of fresh and perishable foods. The European Regulation 450/2009 defines active materials as "*materials and articles that are intended to extend the shelf-life or to maintain or improve the condition of packaged food; they are designed to deliberately incorporate components that would release or absorb substances into or from the packaged food or the environment surrounding the food.*" A very high number of publications have focused on active packaging in the last decades, but the commercial exploitation of the findings has not been satisfactory, to date. An overview on this recent packaging innovation goes beyond the scope of this chapter but detailed information can be retrieved from valuable reviews on the theme (Fang et al., 2017; Francisco et al., 2015; Gómez-Estaca et al., 2014; Han et al., 2018; Ribeiro-Santos et al., 2017). The consolidated assumption that some increase in the impact of packaging could, in some cases, be justified by the achievement of significant waste reduction (Silvenius et al., 2014; Verghese et al., 2015; Wikström et al., 2018; Williams and Wikström, 2011) paves the way toward the commercial adoption of active packaging, especially for food products for which wastes account for a significant part of environmental impact.

Along with active packaging, "responsive packaging" represents an emerging field within packaging technologies, targeting the design of stimuli-response systems enabling real-time food quality and food safety monitoring or remediation (Brockgreitens and Abbas, 2016; Heller et al., 2018). The difference between this type of system and a typical active system based on diffusion and migration of an active compound into the food is subtle, but the distinction is important due to the engineered goal of the packaging. Active packaging systems based on the principles of diffusion are effective as soon as they come in contact with the food and do not respond to any change in food quality or food safety. On the other hand, responsive packaging operates "in response" to specific trigger mechanisms. An example of the former type could be a water-soluble film designed to release antimicrobials into the food only after the build-up of moisture inside the package, while an example of active packaging is a silver particle-embedded film, whose antimicrobial activity only depends on the diffusion of silver ions into the food (Brockgreitens and Abbas, 2016).

6.3 Nanotechnologies for packaging improvement

Nanotechnology deals with the manipulation of materials in the nanoscale (<100 nm). According to the EU definition (EU Recommendation, 2011), "*nanomaterial* means a natural, incidental or manufactured material containing particles, in an unbound state or as an aggregate or as an agglomerate and where, for 50% or more of the particles in the number size distribution, one or more external dimensions is in the size range 1−100 nm". The application of nanotechnology in polymers involve the design, manufacturing, processing and application of polymer materials including nanoparticles. Nanocomposites are multiphase materials resulting from the amalgamation of matrix (continuous phase) and a nano-dimensional material (discontinuous phase) (Arora and Padua, 2010), the latter generally consists of nanospheres or nanoparticles, nanowhiskers or nanorods, nanotubes and nanosheets, or nanoplatelets (Bratovcic et al., 2015). Nanomaterials are mainly inorganic, such as clay and silicate, organoclays, graphene, and metal nanoparticles (Sharma et al., 2017). They display specific and improved physicochemical properties, owing to the small size of nanoparticles which are characterized by a very high surface-to-volume ratio and surface activity. These nanomaterials applied to packaging result in improved electrical conductivity, optical properties, thermal stability, flame resistance, durability and processability and, especially, they improve the mechanical and barrier properties of food packages, along with offering active packaging functions (Han et al., 2018; Sharma et al., 2017).

Also, organic and biobased nanoparticles, however, are gaining more and more interest in packaging applications. Both chitin and cellulose, the two most abundant biopolymers on the earth, are greatly investigated for their properties at nanoscale; a dimension which is generally reached by a fragmentation-top-down process.

Chitosan is easily obtained following treatment of chitin with an alkaline substance. Chitosan is a linear polysaccharide composed of randomly distributed β-(1 → 4)-linked D-glucosamine (deacetylated unit) and N-acetyl-D-glucosamine (acetylated unit). Chitosan nanoparticles are of particular interest in packaging because they are biodegradable and bactericidal; thus, in the perspective of more sustainable functional packaging materials. Chitosan is already approved as food additives in many countries and this pushes for potential applications since it is expected that its features can be enhanced by the nano dimensions (De Lima et al., 2010; Woranuch and Yoksan, 2013).

Concerning cellulose nanoparticles (CNs), their potential in packaging applications is, very likely, much wider; however, it must be related to their specific properties, which depend on the CNs morphology, their origin and the process used for their obtainment. Two main approaches are used for obtaining CN from cellulosic raw materials which are acid hydrolysis and mechanical treatments. Both the two processes, particularly the mechanical (high pressure homogenization treatments), can be aided by different pretreatments applied on the raw materials, that can be chemical, enzymatic or mechanical refining. The methods based on acid hydrolysis have been extensively investigated for nanocellulose extraction from various natural sources (Dufresne, 2017). The fundamental method is to split the cellulose chains, cleaving preferentially the amorphous parts of cellulose. Since the acid attack occurs much faster on the less dense amorphous parts, the result is the isolation of denser and crystalline particles with shorter chains. The nanoparticles obtained may have different morphologies and dimensions, according to the source of native cellulose broken down, but they are always indicated as cellulose nanocrystals (CNCs). Different acids have been tested and used in CNCs production and together with the acid, quite often, also oxidant species are used to increase the charge density and enhance useful properties for packaging applications (Mascheroni et al., 2016). The prevalently crystalline nature of the nanoparticles obtained by acid hydrolysis must be underlined with reference to potential packaging applications. The crystalline portion, indeed, makes a fraction of polymeric volume practically unavailable for transport phenomena. High crystallinity in polymers means tortuous path and high density to the diffusion of possible permeating gases and migrant substances, resulting in a sharp decrease in permeability and migration. Even a thin layer of crystalline nanocellulose, applied as coating, can provide a significant improvement in gas barrier and in other properties, very useful in packaging applications (Li et al., 2013). As highly hydrophilic biopolymers, however, CNCs have a strong sensitivity to water that can be detrimental for packaging applications with fresh foods or in moist conditions, due to water absorption and to the consequent loss of barrier properties (Fotie et al., 2017; Fotie et al., 2018). However, identical issues concern synthetic polymers like polyvinyl alcohol (PVOH) or ethylene vinyl alcohol copolymer (EVOH), which even show a lower gas barrier than CNCs. Therefore, the same solutions used for these common barrier materials can be used to overcome the limitations of CNCs; i.e., the lamination or the coating with a more hydrophobic layer in order to avoid moisture absorption.

The most common mechanical process used for CN obtainment is high pressure homogenization. The CNs produced in this case are long and very thin fibers with a major presence of amorphous parts. For the higher aspect ratio and the different morphology their properties are quite different in comparison with nanocrystals; these nanoparticles are referred to, depending on the dimensions, as cellulose nanofibers (CNFs), or microfibrillated cellulose (MFC) and mainly addressed to possible applications, different from packaging (Dufresne, 2017).

Nanomaterials carry the potential to develop innovative materials as well as to enhance conventional materials properties. However, an analysis of the state of the art of research on nanomaterials falls beyond the scope of this chapter, which aims at highlighting the potential of this advancing technology in a sustainability perspective. Nanotechnology-based packaging has the potential to tackle both efficiency and effectiveness of packaging. One of the most important issues in food-packaging material science is the improvement of gas or water barrier properties of materials used to package food products: this objective has mainly been pursued with the development of laminated multilayer packaging and composite packaging combining the properties of different materials, but the downside of the creation of such materials consists in the higher production costs and on the difficulties related to their recycling. The use of nanocomposite materials has been proposed for improving film mechanical performances and barrier properties, both in conventional and biopolymeric materials (Angiolillo et al., 2017).

Nanomaterials that play a role in barrier improvement are also referred to as nanofillers and nanocoatings. Nanofillers are a heterogenous class of materials characterized by high ratio of largest to smallest dimension. Due to their specific surface area, nanofillers provide a tortuous path for gas and water molecules, which are forced to travel a longer path to diffuse through the material.

The possible issues of excessive dilution of nanofiller and incompatibility between the matrix material and the nanoparticles may be overcome by coating technology, using very thin coatings of nanoparticles on various conventional and biobased materials. Li et al. (2015) proposed the use of cellulose nanocrystals applied in the form of thin coating on various substrates, dramatically reducing the gas permeability and enhancing other useful properties, such as coefficient of friction, transparency, and antifog properties. These functional coatings may improve, as well, the sustainability of packaging, by reducing the thickness of oil-based conventional plastic films.

Thus, nanoengineered materials would allow packaging lightweighting and performance improvement at the same time, which are two strategies corresponding to the concepts of efficiency and effectiveness, respectively.

Nanotechnologies offer the possibility to compensate for some of the lower performances of biopolymers, which represent one of the factors limiting their widespread adoption. Nanotechnology would also allow to improve conventional packaging performances without worsening recyclability, unlike conventional composite and multilayer materials: this is another aspect which focuses nanobased packaging in the direction of efficiency (recyclability), and effectiveness (packaging performances in food protection), together.

In the European Regulation No 10/2011 on plastic materials and articles intended to come into contact with food, it is established that "substances in nano form shall only be used if explicitly authorized and mentioned in the specifications in Annex I." Indeed, since nanoparticles exhibit chemical and physical properties that significantly differ from those at a larger scale, they "may lead to different toxicological properties and therefore these substances should be assessed on a case-by-case basis by the authority as regards their risk until more information is known about such new technology" (EU 1282/2011). Safety is the major drawback of nanotechnology applied to food and to food packaging: a plethora of studies have highlighted some harmful effects of nanoparticles on human health. In particular, nanoparticles can penetrate the human body by inhalation, ingestion or cutaneous exposure and cause various negative effects. For instance, silver nanoparticles are known to be genotoxic, cytotoxic and carcinogenic, they cause oxidative stress responses and decrease of viability in model cell systems (Pradhan et al., 2015); silica nanoparticles may be toxic through oxidative stress generation, leading to DNA damage and induction of apoptosis (Tarantini et al., 2015), and they have been found to induce allergic immune responses in mice (Yoshida et al., 2011). While, apparently, the cellulose nanoparticles (CNs) maintain the biodegradability of the original biopolymer and show very low ecotoxicity, cytotoxicity and genotoxicity (Li et al., 2013). However, the possible release of whatever type of nanoparticles is still considered with suspect and strongly feared. Nevertheless, it has been recently demonstrated that measurable migration may occur only for nanoparticles up to approximately 3.5 nm in diameter and higher dimensions are practically excluded from any diffusional migration phenomena. For 10 nm diameter particles, an apparent diffusion coefficient (D) of 1.1E-35 $cm^2\ s^{-1}$ was calculated theoretically in an LDPE host matrix and such extremely low diffusivity results in almost null mobility of the migrants and undeterminable risk of migration (Bott et al., 2014).

Despite the numerous positive effects of nanotechnology on food-packaging sustainability, the controversial safety aspects and the unresolved concern lets this technology as a promising perspective.

6.4 Correct shelf life assignment

Shelf life can be defined as a finite length of time after production (in some cases after maturation or aging) and packaging during which the food product retains a required level of quality under well-defined storage conditions (Nicoli and Calligaris, 2018). Shelf life, however achieved, should be objectively determined and effectively communicated.

The correct shelf life assignment has commercial repercussions, since the estimated length of time during which the product retains the required level of quality results in the "use by" or minimum durability date, according to EU regulation (EU 1169/2011). Unfortunately, this timeframe is often estimated without proper objective studies, and the shelf life indicated on the package label does not usually correspond to the actual quality level of the product. Often, small and medium food industries do not perform this evaluation based on reliable and objective shelf life studies but estimate durability following a prudential approach, i.e., assigning a minimum durability date that is shorter than the actual shelf life based on the variation of quality indicators. While this approach is simple, cost-saving and guarantees that consumers purchase products with a high-quality standard, it results in a high amount of food turning into a waste, even when their quality is still within the limits of acceptability. Shelf life underestimation generates food waste and represents an important issue for the whole food system in a sustainability perspective. In this light, the conduction of shelf life studies based on systematic approaches in highly recommended (Nicoli, 2012; Robertson, 2009).

Moreover, consumers are often confused on the terms used on packages labels, and they throw away food right after the minimum durability date, in the belief that they will not be further suitable for consumption. A recent review of the history and current practices of date labeling concludes with a call to action to move toward uniformity in date labeling (Newsome et al., 2014). Food shelf life communication effectiveness and consumer education are as important for food waste reduction as any other strategy discussed above.

Another issue related to shelf life assessment and communication is the durability of products after opening, the so-called secondary shelf life. Again, this represents an underestimated issue (Nicoli and Calligaris, 2018) as far as food waste reduction is concerned. In a variety of situations, both at consumer level and at industrial level, the consumption of food (or the use of ingredients and semimanufactured products) is spread over a certain length of time after pack opening, hence a time period during which the product maintains an acceptable quality level should be defined. Moreover, suitable strategies for the extension of secondary shelf life, such as resealable packages and active packaging systems based on specific trigger mechanisms (moisture or gas changes after opening) would help save food from turning into waste and represent interesting actual research topics.

7. Conclusion

Sustainability of packaging is usually claimed in terms of resources (renewable, biobased) or with respect to the end-of-life (biodegradable, compostable), even if the environmental benefits are not always proved. Other than for the supposed or proved environmental benefit, packaging could be referred to as "sustainable" as a consequence of its improved effectiveness and for its potential to reduce the amount of food that turns into waste. The production of food, whether consumed or not, carries environmental impact, thus, reducing the amount of food that is wasted determines a net reduction of unnecessary impact. Since a significant fraction of food wastage is avoidable and occurs during the commercial phase and in households, there appears with evidence the potential of food-packaging design at reducing the amount of food that turns into waste. Sustainability of food packaging can be pursued through actions focusing on the improvement of efficiency and others targeting effectiveness. Ideally, a fully sustainable packaging should comply with all measures proposed within each category, however, priority strategies should be selected considering the balance between the environmental impact linked to the packaging and that arising from the food production. The estimation of this balance, whether the packaging relative environmental impact or the FTP ratio, gives an immediate indication on whether it is more "environmentally sound" to invest in the reduction of packaging impact, through lightweighting, shift to mono-PET and other fully recyclable materials and adopt biobased materials, or to mainly target food waste reduction, with packaging innovations guaranteeing higher protective performances and/or redesigning the package configuration, even if such measures imply some impact increase. The popular belief that packaging is responsible for high environmental impacts collides with scientific evidence of packaging benefits in terms of food waste reduction potential.

New technologies and materials are available today and offer enormous potential for sustainability improvement. Biodegradable and/or compostable materials could finally penetrate the markets thanks to the forthcoming legislation banning the use of single-use plastics in Europe, however, no less care should be addressed to the management of biobased materials waste, as recycling seems to be the best end-of-life option. Nanotechnologies which have emerged as a panacea in the food and packaging industry actually carry serious safety concern and, to date, cannot be considered as viable tools for sustainability improvement in the food system, however, nanotechnology-based packaging has the potential to tackle both efficiency and effectiveness of packaging.

Each single actor of the food system holds responsibility in terms of environmental sustainability: consumers are expected to reconsider their attitude toward food waste and packaging, and to adopt more responsible purchasing and waste sorting behaviors; food companies and packaging manufacturers are responsible for the packaging choice and design, especially preferring mono-material and biobased packaging, and for a correct shelf life assignment; the institutions have already undertaken measures to reduce plastic waste and to favor the commercial diffusion of bioplastics, but still have responsibility in consumers education and awareness-rising on the theme.

References

American Society for Testing and Materials, 2017. Standard Specification for Labeling of End Items that Incorporate Plastics and Polymers as Coatings or Additives with Paper and Other Substrates Designed to Be Aerobically Composted in Municipal or Industrial Facilities. ASTM D6868 − 17. Retrieved from: www.astm.org.

American Society for Testing and Materials, 2018. Standard Test Methods for Determining the Biobased Content of Solid, Liquid, and Gaseous Samples Using Radiocarbon Analysis. ASTM D6866-18. Retrieved from: www.astm.org.

Amienyo, D., Gujba, H., Stichnothe, H., Azapagic, A., 2013. Life cycle environmental impacts of carbonated soft drinks. Int. J. Life Cycle Assess. 18, 77−92.

Angiolillo, L., Spinelli, S., Conte, A., Del Nobile, M.A., 2017. Recent advances in food packaging with a focus on nanotechnology. Recent Pat. Eng. 11 (3). https://doi.org/10.2174/1872212111666170303110527.

Arora, A., Padua, G.W., 2010. Review: nanocomposites in food packaging. J. Food Sci. 75 (1), R43−R49. https://doi.org/10.1111/j.1750-3841.2009.01456.x.

Avadí, A., Bolaños, C., Sandoval, I., Ycaza, C., 2015. Life cycle assessment of Ecuadorian processed tuna. Int. J. Life Cycle Assess. 20 (10), 1415−1428.

Bajpai, P.K., Singh, I., Madaan, J., 2014. Development and characterization of PLA-based green composites: a review. J. Thermoplast. Compos. Mater. 27 (1), 52−81.

Beretta, C., Stoessel, F., Baier, U., Hellweg, S., 2013. Quantifying food losses and the potential for reduction in Switzerland. Waste Manag. 33 (3), 764−773. https://doi.org/10.1016/j.wasman.2012.11.007.

Bonamente, E., Scrucca, F., Rinaldi, R., Merico, M.C., Asdrubali, F., Lamastra, L., 2016. Environmental impact of an Italian wine bottle: carbon and water footprint assessment. Sci. Total Environ. 560–561, 274–283.

Bott, J., Störmer, A., Franz, R., 2014. A model study into the migration potential of nanoparticles from plastics nanocomposites for food contact. Food Packag. Shelf Life 2 (2), 73–80.

Brandvold, T.A., 2012. U.S. Patent No. 8, vol. 314. U.S. Patent and Trademark Office, Washington, DC, p. 267.

Bratovcic, A., Odobasic, A., Catic, S., Sestan, I., 2015. Application of polymer nanocomposite materials in food packaging. Croat. J. Food Sci. Technol. 7 (2), 86–94. https://doi.org/10.17508/CJFST.2015.7.2.06.

Brockgreitens, J., Abbas, A., 2016. Responsive food packaging: recent progress and technological prospects. Compr. Rev. Food Sci. Food Saf. 15 (1), 3–15. https://doi.org/10.1111/1541-4337.12174.

Büsser, S., Jungbluth, N., 2009. The role of flexible packaging in the life cycle of coffee and butter. Int. J. Life Cycle Assess. 14 (Suppl. 1), 80–91. https://doi.org/10.1007/s11367-008-0056-2.

Campodonico, M.A., Sukumara, S., Feist, A.M., Herrgard, M.J., 2018. Computational methods to assess the production potential of bio-based chemicals. In: Synthetic Metabolic Pathways. Methods in Molecular Biology, vol. 1671. Humana Press, New York, NY, pp. 97–116.

Cherubini, F., 2010. The biorefinery concept: using biomass instead of oil for producing energy and chemicals. Energy Convers. Manag. 51 (7), 1412–1421.

Chung, H., Yang, J.E., Ha, J.Y., Chae, T.U., Shin, J.H., Gustavsson, M., Lee, S.Y., 2015. Bio-based production of monomers and polymers by metabolically engineered microorganisms. Curr. Opin. Biotechnol. 36, 73–84. https://doi.org/10.1016/j.copbio.2015.07.003.

Cimini, A., Moresi, M., 2016. Carbon footprint of a pale lager packed in different formats: assessment and sensitivity analysis based on transparent data. J. Clean. Prod. 112, 4196–4213. https://doi.org/10.1016/j.jclepro.2015.06.063.

Coma, V., Couturier, Y., Pascat, B., Bureau, G., Guilbert, S., Cuq, J.L., 1994. Estimation of the biodegradability of packaging materials by a screening test and a weight-loss method. Packag. Technol. Sci. 7 (1), 27–37.

Coriolani, C., Rizzo, V., Licciardello, F., Muratore, G., 2006. Dependance of the shelf-life of carbonated soft drinks from PET performances as required by "The Coca Cola Company". Proceeding of SLIM 2006 (shelf life international meeting), Catania (Italy), June 21–23, 2006. Ital. J. Food Sci. 453–455. Special Issue.

Cox, J., Downing, P., 2007. Food Behavior Consumer Research: Quantitative Phase. WRAP.

De Lima, R., Feitosa, L., Pereira, A.D.E.S., De Moura, M.R., Aouada, F.A., Mattoso, L.H.C., Fraceto, L.F., 2010. Evaluation of the genotoxicity of chitosan nanoparticles for use in food packaging films. J. Food Sci. 75 (6), N89–N96.

Dufresne, A., 2017. Nanocellulose: From Nature to High Performance Tailored Materials. Walter de Gruyter GmbH & Co KG.

Ellen MacArthur Foundation, 2016. The New Plastics Economy RETHINKING The Future of Plastics. Retrieved from: https://www.ellenmacarthurfoundation.org/assets/downloads/EllenMacArthurFoundation_TheNewPlasticsEconomy_Pages.pdf.

EU Commission, 2011. Commission recommendation of 18 October 2011 on the definition of nanomaterial (2011/696/EU). Off. J. Eur. Communities – Legislation 275, 38.

European Bioplastics, 2018. Bioplastics, Facts and Figures. Retrieved from: https://www.european-bioplastics.org/.

European Commission, 2011. Plastic waste: ecological and human health impacts. Sci. Environ. Policy (November), 1–37. https://doi.org/KH-31-13-768-EN-N.

European Committee for Standardisation, 2000. Requirements for Packaging Recoverable through Composting and Biodegradation - Test Scheme and Evaluation Criteria for the Final Acceptance of Packaging. EN 13432:2000. Retrieved from: https://www.en-standard.eu/search/?q=EN13432.

European Committee for Standardisation, 2006. Plastics. Evaluation of Compostability. Test Scheme and Specifications. EN14995:2006. Retrieved from: https://www.en-standard.eu/search/?q=EN14995.

European Committee for Standardisation, 2011. Solid Recovered Fuels - Method for the Determination of Biomass Content. EN 15440:2011. Retrived from: https://www.en-standard.eu/search/?q=EN15440.

European Committee for Standardisation, 2014. Bio-based Products – Vocabulary. EN 16575: 2014. Retrieved from: https://www.en-standard.eu/search/?q=EN16575.

European Parliament and Council, 2018a. Report on a European Strategy for Plastics in a Circular Economy. Retrieved from: http://www.europarl.europa.eu/doceo/document/A-8-2018-0262_EN.html?redirect.

European Parliament and Council, 2018b. Directive (EU) 2018/851 of the European Parliament and of the Council of 30 May 2018 amending directive 2008/98/EC on waste. Off. J. Eur. Union (1907), 109–140. https://doi.org/10.1023/A:1009932427938.

Fang, Z., Zhao, Y., Warner, R.D., Johnson, S.K., 2017. Active and intelligent packaging in meat industry. Trends Food Sci. Technol. 61, 60–71. https://doi.org/10.1016/J.TIFS.2017.01.002.

Farah, S., Anderson, D.G., Langer, R., 2016. Physical and mechanical properties of PLA, and their functions in widespread applications—a comprehensive review. Adv. Drug Deliv. Rev. 107, 367–392.

Fotie, G., Amoroso, L., Muratore, G., Piergiovanni, L., 2018. Carbon dioxide diffusion at different relative humidity through coating of cellulose nanocrystals for food packaging applications. Food Packag. Shelf Life 18, 62–70.

Fotie, G., Rampazzo, R., Ortenzi, M., Checchia, S., Fessas, D., Piergiovanni, L., 2017. The effect of moisture on cellulose nanocrystals intended as a high gas barrier coating on flexible packaging materials. Polymers 9 (9), 415.

Francisco, B.B.A., Brum, D.M., Cassella, R.J., 2015. Determination of metals in soft drinks packed in different materials by ETAAS. Food Chem. https://doi.org/10.1016/j.foodchem.2015.04.020.

Fusi, A., Guidetti, R., Benedetto, G., 2014. Delving into the environmental aspect of a Sardinian white wine: from partial to total life cycle assessment. Sci. Total Environ. 472, 989–1000.

Garcia-Garcia, G., Woolley, E., Rahimifard, S., 2015. A framework for a more efficient approach to food waste management. ETP Int. J. Food Eng. 1 (1), 65–72. https://doi.org/10.18178/ijfe.1.1.65-72.

Gazzotti, S., Farina, H., Lesma, G., Rampazzo, R., Piergiovanni, L., Ortenzi, M.A., Silvani, A., 2017. Polylactide/cellulose nanocrystals: the in situ polymerization approach to improved nanocomposites. Eur. Polym. J. 94, 173–184.

Gómez-Estaca, J., López-de-Dicastillo, C., Hernández-Muñoz, P., Catalá, R., Gavara, R., 2014. Advances in antioxidant active food packaging. Trends Food Sci. Technol. 35 (1), 42–51. https://doi.org/10.1016/J.TIFS.2013.10.008.

Grunert, K.G., Hieke, S., Wills, J., 2014. Sustainability labels on food products: consumer motivation, understanding and use. Food Policy 44, 177–189.

Guillard, V., Gaucel, S., Fornaciari, C., Angellier-Coussy, H., Buche, P., Gontard, N., 2018. The next generation of sustainable food packaging to preserve our environment in a circular economy context. Front. Nutr. 5 (December), 1–13. https://doi.org/10.3389/fnut.2018.00121.

Gustavsson, J., Cederberg, C., Sonesson, U., van Otterdijk, R., 2011. Global Food Losses and Food Waste; Extent, Causes and Prevention. Swedish Institute for Food and Biotechnology (SIK), Gothenburg (Sweden), and FAO, Rome (Italy).

Han, J.W., Ruiz-Garcia, L., Qian, J.P., Yang, X.T., 2018. Food packaging: a comprehensive review and future trends. Compr. Rev. Food Sci. Food Saf. 17 (4), 860–877. https://doi.org/10.1111/1541-4337.12343.

Heller, M.C., Keoleian, G.A., 2015. Greenhouse gas emission estimates of U.S. Dietary choices and food loss. J. Ind. Ecol. 19 (3), 391–401. https://doi.org/10.1111/jiec.12174.

Heller, M.C., Selke, S.E.M., Keoleian, G.A., 2018. Mapping the influence of food waste in food packaging environmental performance assessments. J. Ind. Ecol. 00 (0), 1–16. https://doi.org/10.1111/jiec.12743.

Herbes, C., Beuthner, C., Ramme, I., 2018. Consumer attitudes towards biobased packaging – a cross-cultural comparative study. J. Clean. Prod. 194, 203–218. https://doi.org/10.1016/j.jclepro.2018.05.106.

Hospido, A., Moreira, M.T., Feijoo, G., 2003. Simplified life cycle assessment of Galician milk production. Int. Dairy J. 13 (10), 783–796. https://doi.org/10.1016/S0958-6946(03)00100-6.

Hottle, T.A., Bilec, M.M., Landis, A.E., 2017. Biopolymer production and end of life comparisons using life cycle assessment. Resour. Conserv. Recycl. 122 (July), 295–306. https://doi.org/10.1016/j.resconrec.2017.03.002.

Hudson, S.M., Cuculo, J.A., 1980. The solubility of cellulose in liquid ammonia/salt solutions. J. Polym. Sci. Polym. Chem. Ed. 18 (12), 3469–3481.

Ingrao, C., Gigli, M., Siracusa, V., 2017. An attributional Life Cycle Assessment application experience to highlight environmental hotspots in the production of foamy polylactic acid trays for fresh-food packaging usage. J. Clean. Prod. 150, 93–103. https://doi.org/10.1016/j.jclepro.2017.03.007.

Ingrao, C., Licciardello, F., Pecorino, B., Muratore, G., Zerbo, A., Messineo, A., 2018. Energy and environmental assessment of a traditional durum-wheat bread. J. Clean. Prod. 171, 1494–1509. https://doi.org/10.1016/j.jclepro.2017.09.283.

Ingrao, C., Tricase, C., Cholewa-Wójcik, A., Kawecka, A., Rana, R., Siracusa, V., 2015. Polylactic acid trays for fresh-food packaging: a carbon footprint assessment. Sci. Total Environ. 537 (August), 385–398. https://doi.org/10.1016/j.scitotenv.2015.08.023.

ISO (International Organization for Standardization), 2006a. 14040 - Environmental Management – Life Cycle Assessment – Principles and Framework.

ISO (International Organization for Standardization), 2006b. 14044 - Environmental Management - Life Cycle Assessment - Requirements and Guidelines.

ISO (International Organization for Standardization), 2012. Specifications for Compostable Plastics. ISO 17088:2012. Retrieved from: https://www.iso.org/standard/43373.html.

Kale, G., Kijchavengkul, T., Auras, R., Rubino, M., Selke, S.E., Singh, S.P., 2007. Compostability of bioplastic packaging materials: an overview. Macromol. Biosci. 7, 255–277.

Kolpak, F.J., Blackwell, J., 1975. The structure of regenerated cellulose. Macromolecules 8 (4), 563–564.

Li, F., Biagioni, P., Bollani, M., Maccagnan, A., Piergiovanni, L., 2013. Multi-functional coating of cellulose nanocrystals for flexible packaging applications. Cellulose 20 (5), 2491–2504.

Li, F., Mascheroni, E., Piergiovanni, L., 2015. The potential of NanoCellulose in the packaging field: a review. Packag. Technol. Sci. 28 (6), 475–508. https://doi.org/10.1002/pts.2121.

Licciardello, F., 2017. Packaging, blessing in disguise. Review on its diverse contribution to food sustainability. Trends Food Sci. Technol. https://doi.org/10.1016/j.tifs.2017.05.003.

Licciardello, F., Coriolani, C., Muratore, G., 2011. Improvement of CO_2 retention of PET bottles for carbonated soft drinks. Proceedings of SLIM 2010 (shelf life international meeting), Zaragoza (Spain), June 23–25, 2010. Ital. J. Food Sci. 115–117. Special Issue of Vol. 23.

Licciardello, F., Giannone, V., Del Nobile, M.A., Muratore, G., Summo, C., Giarnetti, M., et al., 2017. Shelf life assessment of industrial durum wheat bread as a function of packaging system. Food Chem. 224, 181–190. https://doi.org/10.1016/j.foodchem.2016.12.080.

Manfredi, M.m, Vignali, G., 2014. Life cycle assessment of a packaged tomato puree: a comparison of environmental impacts produced by different life cycle phases. J. Clean. Prod. 73, 275–284. https://doi.org/10.1016/j.jclepro.2013.10.010.

Manfredi, M., Vignali, G., 2015. Comparative Life Cycle Assessment of hot filling and aseptic packaging systems used for beverages. J. Food Eng. 147, 39–48. https://doi.org/10.1016/j.jfoodeng.2014.09.018.

Marsh, K., Bugusu, B., 2007. Food packaging – roles, materials, and environmental issues: scientific status summary. J. Food Sci. 72 (3). https://doi.org/10.1111/j.1750-3841.2007.00301.x.

Mascheroni, E., Rampazzo, R., Ortenzi, M.A., Piva, G., Bonetti, S., Piergiovanni, L., 2016. Comparison of cellulose nanocrystals obtained by sulfuric acid hydrolysis and ammonium persulfate, to be used as coating on flexible food-packaging materials. Cellulose 23 (1), 779–793.

Newsome, R., Balestrini, C.G., Baum, M.D., Corby, J., Fisher, W., Goodburn, K., et al., 2014. Applications and perceptions of date labeling of food. Compr. Rev. Food Sci. Food Saf. 13 (4), 745–769. https://doi.org/10.1111/1541-4337.12086.

Nicoli, M.C., 2012. Shelf Life Assessment of Food, vol. 20122242. CRC Press. https://doi.org/10.1201/b11871.

Nicoli, M.C., Calligaris, S., 2018. Secondary shelf life: an underestimated issue. Food Eng. Rev. 10 (2), 57–65. https://doi.org/10.1007/s12393-018-9173-2.

Ortenzi, M.A., Basilissi, L., Farina, H., Di Silvestro, G., Piergiovanni, L., Mascheroni, E., 2015. Evaluation of crystallinity and gas barrier properties of films obtained from PLA nanocomposites synthesized via "in situ" polymerization of l-lactide with silane-modified nanosilica and montmorillonite. Eur. Polym. J. 66, 478–491.

Parfitt, J., Barthel, M., MacNaughton, S., 2010. Food waste within food supply chains: quantification and potential for change to 2050. Philos. Trans. R. Soc. Biol. Sci. 365 (1554), 3065–3081. https://doi.org/10.1098/rstb.2010.0126.

Pattara, C., Salomone, R., Cichelli, A., 2016. Carbon footprint of extra virgin olive oil: a comparative and driver analysis of different production processes in Centre Italy. J. Clean. Prod. 127, 533–547. https://doi.org/10.1016/j.jclepro.2016.03.152.

Peelman, N., Ragaert, P., De Meulenaer, B., Adons, D., Peeters, R., Cardon, L., et al., 2013. Application of bioplastics for food packaging. Trends Food Sci. Technol. 32 (2), 128–141. https://doi.org/10.1016/J.TIFS.2013.06.003.

PlasticsEurope, 2017. Tworzywa Sztuczne – Fakty 2017. https://doi.org/10.1016/j.marpolbul.2013.01.015.

Pradhan, N., Singh, S., Ojha, N., Shrivastava, A., Barla, A., Rai, V., Bose, S., 2015. Facets of nanotechnology as seen in food processing, packaging, and preservation industry. BioMed. Res. Int. 2015 17. https://doi.org/10.1155/2015/365672.

Ragaert, K., Delva, L., Van Geem, K., 2017. Mechanical and chemical recycling of solid plastic waste. Waste Manag. 69, 24–58. https://doi.org/10.1016/j.wasman.2017.07.044.

Ren, H., Qiao, F., Shi, Y., Knutzen, M.W., Wang, Z., Du, H., Zhang, H., 2015. PlantBottle™ Packaging program is continuing its journey to pursue bio-mono-ethylene glycol using agricultural waste. J. Renew. Sustain. Energy 7 (4). https://doi.org/10.1063/1.4929336.

Ribeiro-Santos, R., Andrade, M., Melo, N. R. de, Sanches-Silva, A., 2017. Use of essential oils in active food packaging: recent advances and future trends. Trends Food Sci. Technol. 61, 132–140. https://doi.org/10.1016/j.tifs.2016.11.021.

Robertson, G., 2009. In: Robertson, G. (Ed.), Food Packaging and Shelf Life. CRC Press. https://doi.org/10.1201/9781420078459.

Scriba, M., Germany, N., 2016. Lightweighting Plastic Packaging and Recyclability.

Sharma, C., Dhiman, R., Rokana, N., Panwar, H., 2017. Nanotechnology: an untapped resource for food packaging. Front. Microbiol. 8 (SEP). https://doi.org/10.3389/fmicb.2017.01735.

Silvenius, F., Grönman, K., Katajajuuri, J., Soukka, R., Koivupuro, H., Virtanen, Y., 2014. The role of household food waste in comparing environmental impacts of packaging alternatives. Packag. Technol. Sci. 27, 277–292.

Siracusa, V., Ingrao, C., Lo Giudice, A., Mbohwa, C., Dalla Rosa, M., 2014. Environmental assessment of a multilayer polymer bag for food packaging and preservation: an LCA approach. Food Res. Int. 62 (September 2015), 151–161. https://doi.org/10.1016/j.foodres.2014.02.010.

Siracusa, V., Rocculi, P., Romani, S., Rosa, M.D., 2008. Biodegradable polymers for food packaging: a review. Trends Food Sci. Technol. 19 (12), 634–643. https://doi.org/10.1016/j.tifs.2008.07.003.

Tang, W.J., Fernandez, J.G., Sohn, J.J., Amemiya, C.T., 2015. Chitin is endogenously produced in vertebrates. Curr. Biol. 25 (7), 897–900.

Tanner, C., Kast, S., 2003. Promoting sustainable consumption: determinants of green purchases by Swiss consumers. Psychol. Mark. 20 (10), 883–902.

Tarantini, A., Lanceleur, R., Mourot, A., Lavault, M.T., Casterou, G., Jarry, G., Hogeveen, K., Fessard, V., 2015. Toxicity, genotoxicity and proinflammatory effects of amorphous nanosilica in the human intestinal Caco-2 cell line. Toxicol. In Vitro 29, 398–407. https://doi.org/10.1016/j.tiv.2014.10.023.

Tellnes, L.G., Ganne-Chedeville, C., Dias, A., Dolezal, F., Hill, C., Zea Escamilla, E., 2017. Comparative assessment for biogenic carbon accounting methods in carbon footprint of products: a review study for construction materials based on forest products. iFor. Biogeosci. For. 10 (5), 815.

United Nations Environment Programme (UNEP), 2011. Plastic Debris in the ocean. In: UNEP Year Book 2011: Emerging Issues in Our Global Environment, pp. 20–33. https://doi.org/10.1073/pnas.1314705111.

Venkat, K., 2012. ClimateChangeImpactofUSFoodWaste.pdf 2 (4), 431–446. https://doi.org/10.18461/ijfsd.v2i4.

Verghese, K., Lewis, H., Lockrey, S., Williams, H., 2015. Packaging's role in minimizing food loss and waste across the supply chain. Packag. Technol. Sci. 28 (7), 603–620.

Vitale, G., Mosna, D., Bottani, E., Montanari, R., Vignali, G., 2018. Environmental impact of a new industrial process for the recovery and valorisation of packaging materials derived from packaged food waste. Sustain. Prod. Consum. 14, 105–121. https://doi.org/10.1016/j.spc.2018.02.001.

Wang, A., Zhang, T., 2013. One-pot conversion of cellulose to ethylene glycol with multifunctional tungsten-based catalysts. Accounts Chem. Res. 46 (7), 1377–1386. https://doi.org/10.1021/ar3002156.

Wikström, F., Verghese, K., Auras, R., Olsson, A., Williams, H., Wever, R., Grönman, K., Pettersen, M.K., Møller, H., Soukka, R., 2018. Packaging strategies that save food: a research agenda for 2030. J. Ind. Ecol. 00 (0). https://doi.org/10.1111/jiec.12769.

Wikstrom, F., Williams, H., 2010. Potential environmental gains from reducing food losses through development of new packaging - a life cycle model. Packag. Technol. Sci. 23, 403–411. https://doi.org/10.1002/pts.906.

Wikström, F., Williams, H., Verghese, K., Clune, S., 2014. The influence of packaging attributes on consumer behaviour in food-packaging life cycle assessment studies - a neglected topic. J. Clean. Prod. 73, 100–108. https://doi.org/10.1016/J.JCLEPRO.2013.10.042.

Williams, H., Wikström, F., 2011. Environmental impact of packaging and food losses in a life cycle perspective: a comparative analysis of five food items. J. Clean. Prod. 19 (1), 43–48. https://doi.org/10.1016/j.jclepro.2010.08.008.

Woranuch, S., Yoksan, R., 2013. Eugenol-loaded chitosan nanoparticles: II. Application in bio-based plastics for active packaging. Carbohydr. Polym. 96 (2), 586–592.

Xiao, B., Zheng, M., Pang, J., Jiang, Y., Wang, H., Sun, R., Wang, A., Wang, X., Zhang, T., 2015. Synthesis and characterization of poly(ethylene terephthalate) from biomass-based ethylene glycol: effects of Miscellaneous diols. Ind. Eng. Chem. Res. 54 (22), 5862–5869. https://doi.org/10.1021/acs.iecr.5b00487.

Yoshida, T., Yoshioka, Y., Fujimura, M., Yamashita, K., Higashisaka, K., Morishita, Y., Kayamuro, H., Nabeshi, H., Nagano, K., Abe, Y., Kamada, H., Tsunoda, S., Itoh, N., Yoshikawa, T., Tsutsumi, Y., 2011. Promotion of allergic immune responses by intranasally- administrated nanosilica particles in mice. Nanoscale Res. Lett. 6, 1–6. https://doi.org/10.1186/1556-276X-6-195.

Zabaniotou, A., Kassidi, E., 2003. Life cycle assessment applied to egg packaging made from polystyrene and recycled paper. J. Clean. Prod. 11 (5), 549–559. https://doi.org/10.1016/S0959-6526(02)00076-8.

Zhang, H., Hortal, M., Dobon, A., Bermudez, J.M., Lara-Lledo, M., 2015. The effect of active packaging on minimizing food losses: life cycle assessment (LCA) of essential oil component-enabled packaging for fresh beef. Packag. Technol. Sci. 28, 761–774.

The concept of (virtual) water in the food industry

7

Camelia Adriana Bucatariu
Independent Researcher, Rome, Italy

Chapter outline

1. Introduction 223
1.1 Water (blue, green, gray): all virtual for the global food and agricultural system 224
1.2 Water on the global policy agenda 225
1.3 Data needs: blue, green, and gray (virtual) water 227
1.4 Water: a global vital flow impacted by climate change 231
1.5 Governance: from virtual water to a global food and agricultural system water requirement 233
2. Conclusions 236
References 239
Further reading 241

1. Introduction

Water security is directly linked with the global population dynamics and has four dimensions: availability, accessibility, stability, utilization. Population growth, urbanization (FAO, 2019), and dietary shifts are leading to food and agricultural systemic changes. These global forces generate an increasing and sophisticated demand of food and agricultural products. Water security is a vital requirement in order to reach food and nutrition security for all and this was acknowledged by the 2010 United Nations General Assembly Resolution no. 64/292 that recognized explicitly the human right to safe water for drinking and sanitation as essential to the realization of all human rights, including the Right to Adequate Food (UNESCO World Water Assessment Programme 2019).

Concurrently, literature provides reflection with regards the differentiation between human rights:

> *The human right to water is held by every individual, regardless of who he or she is, or where he or she lives, and safeguards his or her access to water for personal and domestic use. Water rights, on the other hand, are generally conferred to an individual or company through property rights or land rights, and are rights to access*

The Interaction of Food Industry and Environment. https://doi.org/10.1016/B978-0-12-816449-5.00007-2

or use a water resource. These are generally gained through land ownership or through an agreement with the State or landowner, and are granted for a variety of water uses, including for industry or agriculture.

De Albuquerque (2014)

A water right is a temporary right that can be provided to an individual and, importantly, that can be withdrawn from that individual. The human right to water is not temporary, it is not subject to state approval, and it cannot be withdrawn.

ap-Net/WaterLex/UNDP-SIWI WGF/Redica (2017)

Since the 1990s, Tony Allan enriched the public discourse with the concept of virtual water. From his perspective, virtual water is the water, both freshwater and soil water, needed to produce food and agricultural commodities that will support food security in water-stressed contexts. The same author argued that virtual water is a valuable tool when it comes to trading in foodstuffs between regions and nations in the context of water self-sufficiency. Others have argued that the term virtual water is challenging for real-life contextualization (Merrett, 2003) and it is not feasible to utilize it as a tool for policy makers.

Water has multiple uses at each scale: individual, local (rural, peri-urban, urban), national, regional, or global. Knowledge of the total amount of water utilized (blue, gree, gray) can have direct implications on strategies and policies, monitoring and evaluation systems, and practices.

According to the United Nations (UN), the world's population reached nearly 7.6 billion in 2017. Sixty percent live in Asia (4.5 billion), 17% in Africa (1.3 billion), 10% in Europe (742 million), 9% in Latin America and the Caribbean (646 million), and the remaining 6% in Northern America (361 million) and Oceania (41 million). China (1.4 billion) and India (1.3 billion) remain the two most populous countries of the world, comprising 19 and 18% of the global total, respectively. The world population is growing by 1.10% per year, yielding an additional 83 million people annually. Global population is projected to reach 8.6 billion in 2030 and increase further to 9.8 billion in 2050 and 11.2 billion by 2100 (UN, 2017).

The population growth drivers will impact the demand of food and agricultural products. In 2017, the Food and Agriculture Organization of the United Nations (FAO) estimated that more than 821 million people suffer from malnutrition. The same publication also reports that 50 million children under five suffer from wasting (i.e., 1 in 10 in Asia and Oceania). Overweight affected 38.3 million children and Africa and Asia account for 25% and 46% respectively of the world total. These two regions have the lowest percentage of children in excess of weight (5.0% in Africa and 4.8% in Asia). Oceania has 8.7% and Latin America and the Caribbean has 7.3% (FAO, 2018). Concurrently, it is striking to acknowledge that there is a global significant gap in knowledge and data on water availability and utilization for the food and agricultural system.

1.1 Water (blue, green, gray): all virtual for the global food and agricultural system

According to the UN (Bigas, 2013), water security has the following definition: "The capacity of a population to safeguard sustainable access to adequate quantities of and acceptable quality water for sustaining livelihoods, human well-being, and socio-economic development, for ensuring protection against water-borne pollution and water-related disasters, and for preserving ecosystems in a climate of peace and political stability." Therefore, it is relevant to clarify the concept of virtual water and consider how it could be updated and applied to decisions at policy level that refer not only to the water sector directly but also to food and agriculture.

Global supply and demand for foodstuffs and other agricultural products is influenced by climate change and it is expanding along with population growth. Water (blue, green, gray) becomes virtual when traded as embedded in foodstuffs and it is invisible to all actors of the food system, consumers, and relevant waste management stakeholders. This consumption-to-production systemic determinant brings significant challenges as any shift in the conditions of trade will impact food security and nutrition. Agriculture is accounting for 70% of global freshwater withdrawals and FAO estimated that irrigated food production will increase by more than 50% by 2050. Simultaneously, the amount of water withdrawn for agriculture could increase by only 10%, provided that irrigation practices are improved and yields increase (FAO, 2017a,b).

Moreover, it seems that when it comes to food and agricultural losses or wastes—the water impact may go unnoticed. Global estimates (FAO, 2011) indicate that a third of the total food production for human consumption is lost or wasted along the food supply chains. All this waste and loss of safe and nutritious food has a direct impact on the global food and agricultural system water requirement. The water wasted due to global food losses and waste is estimated by FAO to be around three times the volume of lake Geneva (3×89 km^3).

1.2 Water on the global policy agenda

In the context of the 2030 Agenda, where consumption is linked with production, the concept of virtual water should be reconsidered. The UN Sustainable Development Goals (SDGs) were launched in 2015. All 17 SDGs offer a guiding map toward a sustainable future. Eliminating hunger and malnutrition (Goal 2) will necessarily depend on water security (Goal 6 and Goal 14) and sustainable consumption and production (Goal 12) relies on eliminating poverty (SDG 1) and holistically addressing the global urbanization trends (Goal 11), educational developments (Goal 4), peace (Goal 16) and multi-stakeholders partnerships (Goal 17). A list of the SDGs working directly on water is provided below:

Goal 6: Ensure access to water and sanitation for all

6.5 By 2030, implement integrated water resources management at all levels, including through transboundary cooperation as appropriate

6.A By 2030, expand international cooperation and capacity-building support to developing countries in water- and sanitation-related activities and programmes, including water harvesting, desalination, water efficiency, wastewater treatment, recycling and reuse technologies

6.B Support and strengthen the participation of local communities in improving water and sanitation management

Goal 14: Conserve and sustainably use the oceans, seas and marine resources

14.1 By 2025, prevent and significantly reduce marine pollution of all kinds, in particular from land-based activities, including marine debris and nutrient pollution

14.C Enhance the conservation and sustainable use of oceans and their resources by implementing international law as reflected in UNCLOS, which provides the legal framework for the conservation and sustainable use of oceans and their resources, as recalled in paragraph 158 of The Future We Want

SDG 6 is included in the 2018 UN report produced for monitoring the 2030 Agenda. FAO, United Nations Environment Program (UNEP), United Nations Educational, Scientific and Cultural Organization (UNESCO), UN-Habitat, United Nations Children's Fund (UNICEF), World Health Organization (WHO) and World Meteorological Organization (WMO), operating under the UN-Water umbrella, established the Integrated Monitoring initiative (GEMI). GEMI has several work tracks, including: wastewater treatment and water quality (target 6.3), water stress and water-use efficiency (target 6.4), integrated water resources management and transboundary basins (target 6.5), and water-related ecosystems (target 6.6). AQUASTAT (FAO's global water information system, developed by the Land and Water Division) is the custodian for monitoring of SDG Target 6.4: "by 2030, substantially increase water-use efficiency across all sectors and ensure sustainable withdrawals and supply of freshwater to address water scarcity and substantially reduce the number of people suffering from water scarcity." For this exercise, FAO deploys two indicators: 6.4.1: Change in water-use efficiency over time and 6.4.2: Level of water stress: freshwater withdrawal in percentage of available freshwater resources.

The 2030 Agenda opens also the possibility to consider how traditional knowledge from global agro-ecological zones contributes toward global water security. Data collection is still in the primary phase as indicated in the FAO (2018) *Compendium of community and indigenous strategies for climate change ada*ptation with focus on addressing water scarcity in agriculture. However, for instance, there have been instances when indigenous populations are excluded from decisions concerning water that impact them directly (Jackson and Barber, 2015).

Water scarcity can be economical or physical. The main characteristics of economic scarcity are the lack of investments and human capacity, which lead to infrastructure challenges and inequitable access, for instance, in geographical areas such as Sub-Saharan Africa. Additionally, there is physical scarcity that can cause environmental degradation and a decline in groundwater, that can occur in both absence and abundance of water (if overcommitted due to the intensity of the hydraulic structure). Therefore, when addressing water infrastructure, a key component of any decision must be how to address premature obsolescence, ensure adaptation modules for shifting demographic and climatic changes, and how to incorporate flexibility for an expanding urbanization.

Regions worldwide have also been working on addressing water security issues, for instance, Northern Africa, Western Asia and Central and Southern Asia that are faced with severe water stress. In 2017, for the 22 surveyed countries for SDG 6, water stress level was above 70%, indicating a very high probability of future water scarcity (UN, 2018). The 2018 UN Sustainable Development Goals Report highlights also that in the 62 countries that provided data (out of 153 that share transboundary waters), the average percentage of national transboundary basins covered by an operational arrangement was only 59% in 2017. FAO (2014a,b) indicated that, globally, there are 276 transboundary rivers and approximately 300 transboundary aquifers.

1.3 Data needs: blue, green, and gray (virtual) water

Data on water must cover blue, green, and gray water. These three categories have the following definitions, according to FAO (2014a,b): (1) green water use, i.e., water from rainfall; (2) blue water use, i.e., water from groundwater or surface water; and (3) gray water use, i.e., dilution water required to reduce pollutant concentrations to acceptable values. The distinction among green, blue, and gray water footprints recognizes that these have different economic costs and environmental considerations. For instance, the cost of pumping blue water from boreholes needs to be factored in and it was estimated that, for instance, in Botswana, these costs are going to increase by 23% by 2050 (Turral et al., 2011). Access to available technologies and capacity development for water-use efficiency can make a significant contribution toward the right to water and the sustainable use of this vital resource. For instance, technologies for desalinated water that also adequately address their waste management. Some literature focused on projects that can generate evidence in support of measures tailored for small food producers. The AgWater Solutions Project was implemented from 2009 to 2012 in several Asian and African countries and worked on identifying investment opportunities for water management (Keller et al., 2014).

International Organizations address water from various viewpoints. In 2019, results were published from projects technically supported by the FAO and the International Fund for Agricultural Development (IFAD) in Côte d'Ivoire, the Gambia, Mali and Niger focused on adaptation of small-scale irrigation to climate change in West and Central Africa. The projects conducted used the SHARP (Self-evaluation and Holistic Assessment of climate Resilience of farmers and Pastoralists) tool and applied a climate resilience assessment for small holder farmers across 21 irrigation sites and collected from 691 households that indicated a reduction in the availability of water for irrigation (17%); an increase in crop failure (45%), a decrease in farm income (38%), and forced migration of family members (13%) (Mejias Moreno and Lagana, 2019).

Available estimates from FAO (2015a,b,c) indicate that, globally, groundwater withdrawal provide 36% of potable water, 42% of water for irrigated agriculture and 24% of direct industrial water supply. In 2010, it reached 26% of all freshwater withdrawn, and the highest levels are represented by China, India, Pakistan, Bangladesh, Iran, the US, Mexico and Europe.

Water quality and availability are leading priorities for the global food and agricultural system. According to the 2018 Environmental Outlook of the Organization for Economic Cooperation and Development (OECD), by 2050 over 40% of the world's population are likely to be living in river basins under severe water stress while overall water demand is projected to increase by 55%. In the majority of OECD countries water is under severe stress from pollution caused by agriculture (runoff). Water quality deterioration—as caused by nutrient flows from agriculture and a lack of adequate management—leads to biodiversity loss and eutrophication (OECD, 2018) and this, in turn, will increase pressure on water for food consumption and production.

The WHO estimated that in 2015, 71% of the global population (5.2 billion people) used a safely managed drinking-water service while 844 million people lack even a basic drinking-water service, including 159 million people who are dependent on surface water.

Water is vital for human health—as it is utilized as drinking water and for personal hygiene—and for food production, processing, distribution, preparation, intake, and management at home. WHO (2018) estimates that by 2025, half of the world's population will be living in water-stressed areas. This means available and accessible safe water for drinking, all hygiene related purposes, and food consumption and production will be affected.

The concept of virtual water was developed for utilization in the debate on food and nutrition. Nevertheless, if a region is under water stress, not only the food and agriculture sector is under pressure but also all the other sectors that rely on water. Water utilization in water-stressed areas is often in the public eye and becomes a policy and political debate for actors such as food producers. However, water efficiency and management for food and agriculture are critical in balancing the increasing pressure not only for the sector itself—from production to households' consumption and food losses and waste management—but also with other sectors that rely on this natural resource.

FAO (2017) stated that in order to balance demand and supply of water for the food and agricultural system (in the case of the Near East and North Africa, but this could also apply to all the other regions) several actions are needed. Firstly, all water sources need to be mapped and classified. Secondly, physical control needs to be ensured by the government or assigned agencies. Thirdly, interventions to effectively manage allocations should be identified, implemented, and monitored.

Water sustainability (social, economic, environmental) is based on primary aquifers that recharge naturally and feed surface flows. Primary aquifers can feed irrigation systems that, if not adequately managed, are a source of stress (Bredehoeft, 2002; Gleeson et al., 2012). Moreover, climate change causes shifts in water quality, groundwater recharge, and precipitation. Reduced precipitation and increased evapotranspiration will also need to be taken into account.

Today, approximately 80% of the global cropland is rain-fed (UNESCO World Water Assessment Program, 2019). With climate change, rain-fed crop production will then transfer stress to livestock production while the irrigated systems will heavily rely on basin availability and both will be impacted by drought. One of the sectors directly related to agriculture is employment. Water security will support agricultural developments that could also support employment—literature has indicated that, in developing countries, the moderate poor are less likely than the extreme poor to work in the agriculture sector (64.7% vs. 52.0%). (Castañeda et al., 2016).

Data on water should reflect the food and agricultural system needs and utilizations as well as the competing current demands and future potential trends.

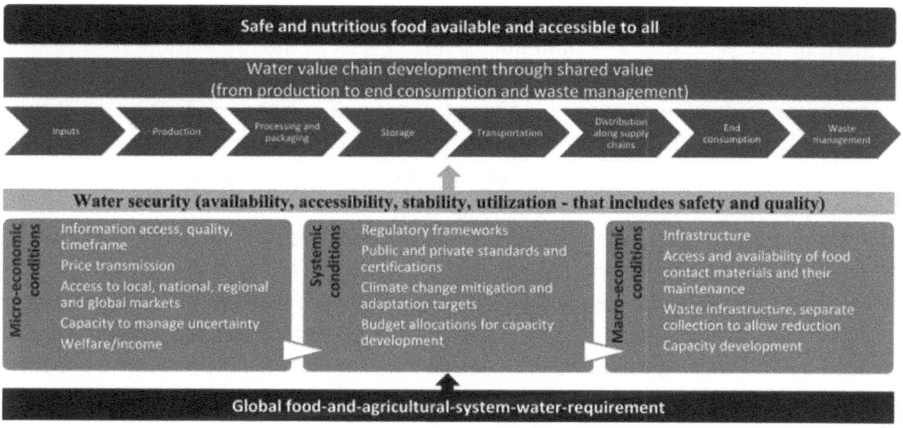

Figure 7.1 Water security for the global food and agricultural system requirement.

Fig. 7.1 indicates how a holistic view on water security (availability, accessibility, stability, utilization—that includes quality) could be mapped out for the global food and agricultural system.

There is a significant need to improve data quality and availability as well as the tools to compute and develop scenarios for future impacts of water utilization. Transparency is key when considering the concept of virtual water and data produced locally are essential. Nevertheless, accountability determines blue, green, and gray water data availability and quality. To date, it seems that policy priorities are oriented toward the demand and supply of the foodstuffs and agricultural commodities. The water discourse is contributing to the sustainability dimension of the global food and agricultural system through water security that sustains all operations.

The poultry and beef industry could be taken as examples for data availability and challenges. For instance, a broiler chicken takes around 15 weeks to reach 1 Kilogram. Approximately 0.5.l/bird/day of blue water for drinking is utilized for broilers in Confined Animal Feeding Operations (CAFOs) where up to 30,000 chickens can be housed at any given time. The blue water consumption is, however, very different for the production of pasture cattle that registers a global average 15.5 m^3 of water per 1 Kilogram of beef. Going further into the water utilization requirements, there are differences caused by the feed used and the timeframe necessary for each species to be brought to the desired size as well as by processing the meats. Additionally, the global increase in animal welfare standards and environmental regulations that introduced separation distances and quality standards for manure-laden water (OECD, 2018a,b) influence water demand.

In the case of rice, practices such as alternate wetting and drying of rice fields were identified as a valid solution to reduce methane emissions from paddies by 45%, while saving water and producing yields similar to those of fully flooded areas (FAO, 2016).

A kilogram of rice produced in Italy would have a different water impact than that of a kilogram of rice produced in China due to differences in the production systems and climate.

The sectors of fisheries and forestry need to be mentioned as well. In the case of fisheries, there is significant stress due to ocean acidification and overexploitation of stocks. Nonetheless, in the case of fisheries, water security is critical for aquaculture as well as marine exploitation. According to the UN (2018), as of January 2018, 16% (or over 22 million square kilometers) of marine waters are under national jurisdiction—that is, 0 to 200 nautical miles from shore—are covered by protected areas. However, literature is still not fully addressing the complexity of linkages between fisheries and food and nutrition security. Published literature (Dugan et al., 2006) stressed this gap and underlined the need to integrate specific water requirements for fisheries into water allocation decisions, including options that do not cause reduction in water availability for other purposes. Additionally, the use of ice for fish preservation from catch to end consumers also impacts the industry's' safe water requirement.

An aspect of paramount importance that often goes unmentioned is water availability and quality for postharvest operations (e.g., postharvest handling, packing house sanitation). The 2014 *On-Farm Decision Tree Project for Postharvest Water* from Cornell University provides useful guidance for postharvest operations key points such as: (1) adequate access to water of a quality equivalent to that for drinking—to which a sanitizer is added; (2) when contaminated, the bulk/batch tank water needs to be changed; (3) the containers need to be sanitized daily and pooled water must be eliminated.

These examples are among the many that can be considered when discussing methodological needs and limitations for water footprinting from local to global level. Whether it is protein production, or another food commodity, or postharvest operations—blue, green, and gray water percentages will vary from one context to another. Data transparency from local to global level would enable better decisions for investments for all private sector actors and better facilitation and coordination with policy processes, even more so, in the context of climate change.

Each agricultural commodity has a different water footprint. However, if considered at global level, this would be expressed through an average. Global averages for the water footprint in the food and agricultural sector should include fisheries and nonwood forestry products. Without reliable data—from global to local level, policy decisions concerning investments, for instance, for blue and green water, are especially challenging.

Methodological disagreement is not rare when performing calculations of the water footprint. Addressing the concept of virtual water can also start by addressing the water footprint (Hoekstra and Chapagain, 2008; Hoekstra and Mekonnen, 2012) that is measuring virtual water along food supply chains, including the level of pollution.

One of the areas that need immediate intervention is metering all irrigation wells for data on pumpage and improve the calculations on yields—directly impacted by climate change also in their demand for water—globally, not only in water-stressed areas. It was estimated in 2015 that the global irrigated area represents around 20% of the total agricultural land and contributes 40% of the total food production. The 2007 Comprehensive Assessment of Water Management in Agriculture estimated that 75% of the additional food we need over the next decades could be met by bringing the production levels of the world's low-yield farmers up to 80% of what high-yield farmers get from comparable land (FAO, 2015a,b,c).

A key requirement is to have better data on, for instance, extracted groundwater. The utilization of flow meters that provide information on pumping from aquifers could help. Groundwater storage depletion (overdraft) happes when intensive groundwater abstraction does not allow the aquifer systems to recharge and maintain the equilibrium with outflows (seepage or leakage). FAO (2015a,b,c) estimates that 200 km^3 per annum (nearly one-fifth of all groundwater) is causing permanent depletion.

The launch in 2011 of the "Groundwater Governance: A Global Framework for Action" initiative underscored a global lack of effective governance and scarce collective action as main causes of groundwater depletion and degradation of aquifers. As per FAO (2015a,b,c) definition, groundwater governance has four components: (1) a multiactor institutional framework; (2) an effective and articulate legal and regulatory framework; (3) available, accessible, and applicable knowledge; (4) policies, plans, finances, and incentives aligned with society's goals.

Another key gap to address is the lack or inadequate level of awareness for all the actors of the food system concerning their current impact on aquifers and how this impact translates into future scenarios. Data blindness and behavior affects not only the final consumer, receiving food for preparation, intake, and storage at household level, but also all the other actors before and after the step of final consumption.

1.4 Water: a global vital flow impacted by climate change

Being the most extracted natural resource globally with approximately 4000 billion tones, including 700 billion tones groundwater (Rockström et al., 2009; Zekster and Everett, 2004), water is considered for its availability and management (Cazcarro et al., 2015; Hoekstra and Chapagain, 2007; Chapagain et al., 2006). Ecosystems around the globe are based on land and water and they generate socio-economic and environmental development through human settlements. Countries such as Japan, USA, the European Union (28 Member States) import also water intensive foodstuffs from a wide number of geographical areas. Some of these trading partners include countries that face water stress, for instance, India.

However, the majority are both importers and exporters (Kuishuang and Hubacek, 2015). For the EU, exports of agricultural primary products to third countries represented EUR 129.1 billion in 2015 (11.9% of the value of EU production), i.e., +124% from 2005 (Agricultural and food trade, European Commission, 2017). In 2016, the EC hosted a Workshop on the Sustainability of EU's Livestock

Production Systems where it was stated (Ward et al., 2016) that creating and transporting meat and milk products from areas high on the water scarcity index to locations low on the water scarcity index, without addressing the "circularization" of water will only exacerbate scarcity issues at the point of production.

The Integrated Water Resources Management (IWRM) framework and the Life Cycle Analysis framework are two tools available to better understand the virtual water concept. However, both need to pay additional attention to the transboundary dimension. Data from the 2018 UN monitoring system for the 2030 Agenda indicated that in 2017 there were 157 countries reporting an average of less than 50% IWRM implementation. Additionally, data from 2015 estimated that about 2.1 billion people (29% of the world's population) lacked access to safely managed drinking-water supplies (UN, 2018).

AQUASTAT (FAO, 2014a,b) indicated that there are 118 countries that have rivers flowing into their territories from upstream countries: for instance, Amazon: 7 countries, Mekong: 8 countries, Nile: 11 countries, and Danube: 19 countries. The same dataset reports that there are 28 countries relying on freshwater resources that are generated from precipitation falling on the country itself (islands not included) and one country (Kuwait) with no internally generated freshwater resources.

Climate change is adding to the systemic pressures that can be addressed through strengthening management, agricultural, and manufacturing practices as well as wasteful consumption patterns. For instance, tropical forests are affected more by changes in the water availability and CO_2 fertilization than by temperature changes (FAO, 2016).

Policies, laws and private-set standards play a role in determining opportunities and recognizing the challenges that water security poses in the context of a changing climate that needs to support the food system to adapt also to shifting dietary preferences. Water governance needs reach out to communities whose behavior affects water demand and availability (for instance, farmers, city dwellers, energy suppliers) and to engage stakeholders in complex policy decisions (OECD, 2018a,b). In the same time, resource circularization is vital. For instance, sludge valorization for agriculture could be further explored (UNESCO World Water Assessment Programme, 2019).

Climate change impacts both land and water habitats; thus, the global food and agricultural system (including crops, livestock, fisheries and nonwood forestry products). On one hand, water availability through food trade redistribution was picked up in the literature as a means to address climate change challenges. Nevertheless, negative impacts were also identified (Konar et al., 2013). On the other hand, a very specific sector, that of fisheries, seems to get acknowledged more in the field of the food—water nexus. As data on freshwater availability in the context of climate change discussion becomes more and more transparent, requirements of freshwater for seafood production, including aquaculture need to be better understood (Gephart et al., 2017) and integrated into the climate decision making processes for the Paris Agreement. Water is a key factor in adapting to climate change and 93% of countries have identified it in their Intended Nationally Determined Contributions (INDCs) as stated in the 2017 Water and Climate Blue Book published by the World Water Council.

1.5 Governance: from virtual water to a global food and agricultural system water requirement

The UN (2018) estimates that more than two billion people are currently affected by water stress. Therefore, it is necessary to consider virtual water not only in the regions that are faced with severe shortages. Virtual water is relevant globally and a re-consideration of its terminology as well as boundaries must take place.

With an increase in human population and estimated urbanization trends, the demand for water in the food industry as well as at consumer level will continue to become more and more sophisticated and significant. In the same time, competition between the food and agriculture system and other sectors that utilize water is bound to add to the pressure placed on this renewable but limited resource. From smallholders to large food producers, water is needed to ensure that food is available, safe, and nutritious for all to access for intake.

Nevertheless, a large array of actors are co-sharing responsibility for water management, utilization, and consumption—from production, to processing, storage, distribution, consumption at household level and (food) waste management. Therefore, a first critical point that the concept of virtual water does not seem to fully capture is that of co-shared responsibility for water availability, accessibility, stability, utilization and quality—due to the fact that, according to its definition, it refers only to the water utilized for food production as traded. Another critical aspect that makes the virtual water concept incompatible with a food systems approach is the water impact of waste management. According to literature (Kibler et al., 2018), large-scale decisions require a more complete understanding of food waste and its management, particularly regarding postdisposal impacts related to water.

Water-related losses in agriculture, health, income and property could result in a decline by as much as 6% of GDP by 2050 and lead to sustained negative growth in some regions of the world (World Bank, 2016).

The major financing source for water is the public sector. However, it is the private sector that is engaged in food production. The global food and agricultural system water requirement that links consumption and production, reflects water as a shared global resource and opens the potential to explore shared investments, for instance, between crop production and fisheries (e.g., aquaponics and hydroponics).

Water infrastructure is capital intensive and highly vulnerable to climate change. Good water governance could mitigate and help balance competing sectors: energy, food, health and education. Therefore, water security should be comprehensive of all related demand and supply. Climate change will put additional pressure on the socio-economic and environmental costs that the global hydric resources will pay in order to meet the increasing food demand that comes with population growth. Complementary actions could be required in storage, dams, and water allocation systems in order to have a predictable and secured access in time. Currently there is no lack of technologies—the challenge is: (1) ensuring that all actors in the food systems have access to the technologies needed; (2) have the training and resources to ensure maintenance; (3) address environmental, economic, and social impact concurrently. For instance, it has been argued that dams have contributed to manage water resources (Chen et al., 2016) and have also generated displacement and severe impacts on biophysical systems. Nevertheless, literature also argues that local-scale dams managed at the community level can have significant positive impacts (Ryan and Elsner, 2016).

The 2030 Agenda and the Paris Agreement engage all countries in efforts to ensure that water is efficiently and effectively utilized and that this vital resource will continue to support our growing population. FAO (2016) has stressed that addressing the competing demands for water should not focus only on farmers for their management of water resources and on virtual water from trade to resource depleted areas. The food supply chain starting from production to processing, storage, and distribution is having a cumulative impact and the impact of food and agriculture related waste management must not go unnoticed.

In regions with intensive groundwater use, the OECD recommends employing a "tripod" combination of regulatory, economic, and collective action policy approaches, customized to local circumstances (OECD, 2018a,b).

Communities may externalize their water footprint and the concept of virtual water has helped to capture this dimension of consumption. Research (Zhuo et al., 2016a,b) has started to work on longitudinal studies that help see how yield improvements can decrease water footprint.

In the same time, literature argues that, when it comes to virtual water footprint for export production, trade should be considered at all levels (intranational or international). Policies can influence trade flows and may not have a birds-eye oversight on the global food and agricultural water requirement—thus, may also determine interregional water dependencies and may worsen rather than relieve the water scarcity in a country (Zhuo et al., 2016a,b).

The 2030 Agenda, the UN Decade of Action on Nutrition, and the Paris Agreement may have a contribution toward changing this current status by: (1) providing global data on consumption and production—through monitored indicators; (2) making the linkages between resources (i.e., water), all food supply chain operations, consumer behavior, and waste management. Wastewater is a resource. The food and agricultural system must consider it and enhance its safe utilization for nutrient cycling (e.g., for nitrogen, phosphorous, potassium).

Another gap in policy coherence is the lack of addressing the real costs of water depletion. If the price paid reflects only utilization, without adequately addressing the risks of bad management, then there is a lack of incentive to improve, for instance, extraction methods and rates, and there will be a lack of trust and ownership if other measures will be necessary in the future. Agricultural water productivity (the ratio of agricultural output to the amount of water used) in rain-fed and irrigated systems need better data and evidence-based policies and legislation that engages the public sector, the private sector, and civil society in a co-shared system of: assessing water availability, determining the criteria for its accessibility and adequate utilization, monitoring and evaluating its quality while ensuring its stability in time and waste management.

Variability in time has been highlighted (Zhuo et al., 2016a,b) for China for the period from 1978 to 2008. The paper analyzed crop yield improvements that helped reduce the national water footprint average. By 2008, China registered 28% of the total water consumption for export to other regions of the country. The country is also importing water due to soybean, which, in one of the scenarios developed in the paper, accounted for 5% of total water footprint of Chinese food consumption by 2050. For the authors, the premise to reach water security is water and cropland management, coherent policies, and infrastructure, and a shift to a diet containing less meat.

Driving forces for net virtual water export for China's provinces between 1995 and 2005 (Zhao et al., 2019) were analyzed by using a quantitative approach to the comparative advantage theory. The paper indicates that regional differences in land productivity between agricultural and nonagricultural sectors are the main forces determining the pattern of virtual water flows across major regions. Resources such as labor and water played only a limited role. The conclusion of the paper is that current market forces do not reflect water but land resources scarcity.

In 2011 Latin America and the Caribbean total freshwater withdrawal for agriculture accounted for 68% and the industrial (11%) and domestic sectors (21%) added to this natural resource cost bill. With substantial differences in these regions' countries, overall agricultural production increased by more than 50% from 2000 to 2012 and the majority of food produced is rain-fed (Mekkonen et al., 2015). Food and agricultural system water governance necessarily deals with multiple stakeholders that may represent conflicting and competing priorities.

An additional sector to highlight for water use and food demand is tourism. Literature has highlighted that there is a gender-water-tourism nexus and that women are particularly affected by water scarcity. Ethnographic research conducted in Labuan Bajo, Indonesia concluded that "tourism is out competing locals for access to water leading to women suffering in multiple ways" (Cole, 2017).

The paramount question is how could the balance be reached between local and national levels while ensuring food and nutrition security for all at a global level. For this issue to be addressed, water ownership needs to be clarified. One of the challenges is that, for instance, water basins can supply distanced fields, and thus, geographic ownership is not coherent if it fails to integrate the specifics of utilization. For example, literature (Peck, 2007) identified that in the case of the United States of America, there are several approaches adopted, such as Nebraska's' "Reasonable Use Doctrine" and the "Correlative Rights Doctrine"—where the right to use comes from owning the land above and Texas' "Rule of Capture" which grants unlimited extraction rights if the land is owned. This brings up the question of shared natural resources, a perspective that starts getting ground in policy approaches that involve both agricultural producers and human settlements. Global policy frameworks need to be coherent. However, studies (Canadian Agri-Food Policy Institute, 2017) have identified underreporting that veils the effects of subsidies is common practice.

The OECD (2016) Daegu's Multistakeholder Declaration on the OECD Principles on Water Governance brought together the public, private and nonprofit sectors and put forward the strategic consideration that water challenges are often governance and coordination challenges to mitigate fragmentation at national level.

In 2018, the UN Decade of Action on Nutrition has published a guidance document issued by the FAO and the WHO on how governments could address the priority areas set by the Second International Conference on Nutrition and the Declaration on Nutrition of Rome (2014). Recommendation 50 is on Water management for nutrition: "Implement policies and programmes using participatory approaches to improve water management in agriculture and food production." The guidance for Recommendation 50 provides a series of questions that are identified for countries and local authorities in order to map the current context:

- Are there national measures, strategies or programs to improve water management in agriculture and food production?
- Do the (sub-) national policies on water management include nutrition and food security objectives?
- Which temporal planning scenarios are applied in water resources management by both general/national water sector administration and the agriculture sector?
- What other forms of nonconventional water are used, and what is their proportion in light of the total supply?
- Are there any legal provisions for regulating and governing the use of wastewater irrigation in agriculture, such as regulations on quantity as well as microbiological and chemical quality of wastewater?

The UN Decade of Action on Nutrition is an opportunity to address the global challenge in governance for water—through the lens of healthy and sustainable diets for all.

2. Conclusions

This chapter provided evidence on the relevance of having a global approach to the concept of virtual water. The 2016 publication *Plates, Pyramids, and Planets: Developments in National Healthy and Sustainable Dietary Guidelines: A State of Play Assessment* (Fischer and Garnett, 2016) has identified water in several key points of the systemic approach to diets and sustainability. Water is a natural resource that links all the food and agricultural system actors—from seed production to end consumption and waste management.

End consumers should not be overseen. Households have three direct entry points to impact water: sustainable consumption of household water supply, prevention and reduction of household food waste, and rely on water management performed by others—in their jurisdiction. There is a large body of literature in support of diets with a high variety of micro- and macronutrients' and energy sources that also reduce the risk of noncommunicable diseases (Lonnie et al., 2018). However, from evidence provided by literature (FAO, 2018), the food and agricultural system encounters various challenges when addressing consumers' demands for a healthy and sustainable diet and there is long path to take through the ongoing global nutrition transition.

There is, on one hand, the policy perspective and recommendation to increase fresh fruits and vegetables consumption and, on the other hand, businesses, working with their supply chains to increase water efficiency and improve postharvest practices. Private sector codes of conduct have started to emerge to not only ensure longevity of the business but also to provide a growing body of consumers with knowledge on the private sector's efforts to increase, for instance, water efficiency.

One of the means through which diets can be linked with the supply chain behavior concerning water consumption is technology that can inform actors along the supply chain of their water impact as well as technology that can track consumers' behavior and calculate the estimated water impact. All of these options rely on data quality and availability.

In the case of food producers, drip irrigation is one of the solutions together with soil moisture sensors. For processors—operational optimizations that prevent and reduce wastes along the supply chains can be directly quantified in saved hydric resources. The same for retailers and food services—that can also optimize their water use for processing.

Among the instruments available for all food systems actors there are the food-based dietary guidelines that can be defined as a tool utilized "to establish a basis for public food and nutrition, health and agricultural policies and nutrition education programs to foster healthy eating habits and lifestyles" (FAO, 2016). Governments around the world are launching food-based dietary guidance for their populations and are showing interest also in educating consumers. Consumers' food and water literacy is key to drive change for water efficiency for food and security nutrition. However, when it comes to water, as embedded in the production, processing, and distribution cycle of a specific food item or for the ingredients of a meal, consumers may not be in the position to grasp the relevance of this impact as it is perceived as naturally linked with the food itself. In the same time, the information that the private sector may present to the consumers risks being discredited (Nestle, 2015).

Consumers are identified in the 2030 Agenda as a global multi-dimensional actor that should engage in actions that impact both the individual and the global food system. However, in order for the consumers to play an active role, their food literacy should be functional, interactive, and critical. Fig. 7.2 maps the main dimensions of consumers' diets that rely on water security.

FAO (AQUASTAT, 2018) estimates that depending on diet choices and other life-related activities, about 2000 to 5000 L of water is required to produce a person's daily food and meet their drinking and sanitation needs. The chapter argued that all water (blue, green, gray) becomes virtual as it is embedded in foodstuffs for end consumers—whether is traded intranationally or internationally and whether it is utilized for crops, livestock, fisheries, or nonwood forest products.

The 2030 Agenda SDG six needs additional investments of up to 1.7 trillion USD (Hutton and Varughese, 2016) in support of achieving universal and equitable access to safe and affordable drinking water for all. Concurrently, the global financing needs for water infrastructure range from USD 6.7 trillion by 2030 to USD 22.6 trillion by 2050 (OECD, 2018a,b).

Governance systems and policies with adequate following legislation are needed to address the real cost-accounting of water—from local to global level; in a way that reflects water security (availability, access, utilization, stability). The concept of a global food and agricultural system water requirement was put forward in support of water as a global resource that *feeds* the global population and will have to address food and nutrition security for future generations as well—from rural to peri-urban, and urban areas.

The 2030 Agenda, the UN Decade of Action on Nutrition and the Paris Agreement have prioritized water security. These three global policy processes have a local to

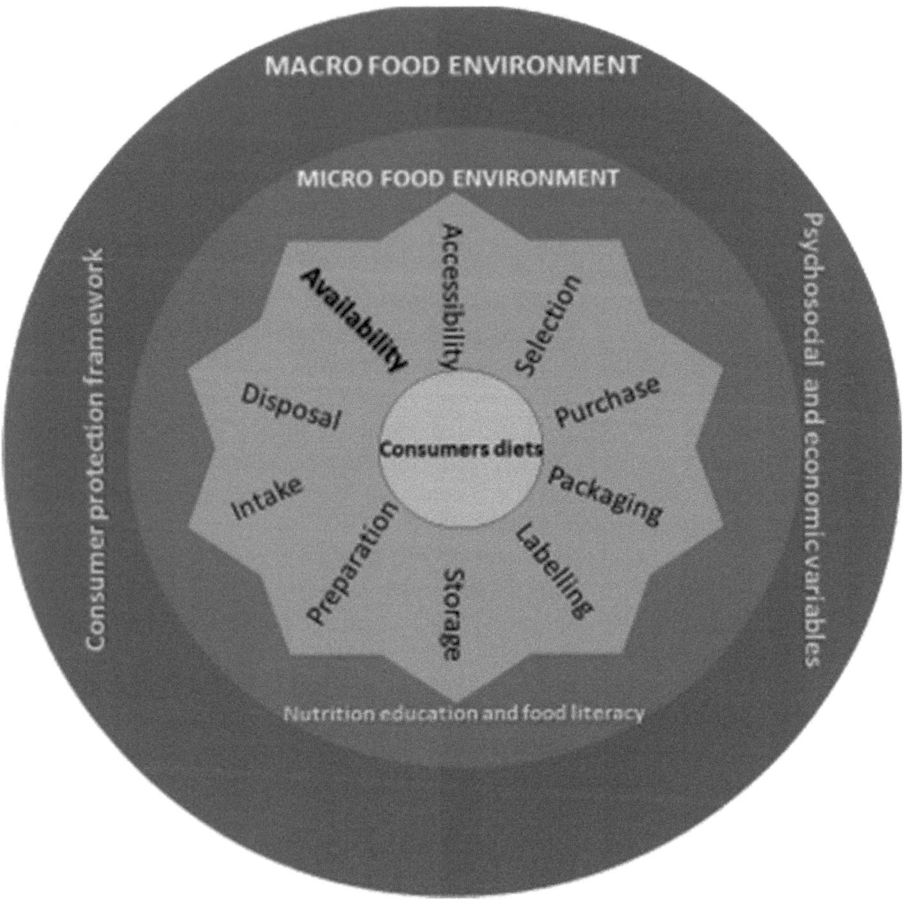

Figure 7.2 Main dimensions of consumers' diets that rely on water security (availability, accessibility, utilization, and stability).

global implementation approach. The SDGs have a monitoring system that relies on statistical information provided by countries. There is an opportunity to link these three processes, with their data sets, in order to provide enough and sufficient quality information to work toward a strengthened global approach to water security.

One step toward achieving this potential objective was completed with the launch in 2019 of the Water-Related Ecosystems platform that "monitors changes occurring in the spatial extent of open water bodies; identifies new or lost water bodies; and identifies where changes are occurring in seasonal water bodies" (www.sdg661.app). The platform was launched by UN Environment and Google with the European Commission Joint Research Center (JRC), the European Space Agency (ESA), the United States National Aeronautics and Space Administration (NASA), and the Group on Earth Observations (GEO).

References

Agricultural and food trade, 2017. European Commission. https://ec.europa.eu/agriculture/sites/agriculture/files/statistics/facts-figures/agricultural-food-trade.pdf.

Bigas, H., 2013. Water security & the Global Water Agenda. A UN-Water Analytical Brief: Institute for Water. Environment and Health.

Bredehoeft, J.D., 2002. The water budget myth revisited: why hydrogeologists model. GroundWater 40 (4), 340–345.

Canadian Agri-Food Policy Institute (CAPI), 2017. http://www.capi-icpa.ca/pdfs/2017/CAPI_Subsidies_Summary_2017_ENG.pdf.

Castañeda, A., Doan, D., Newhouse, D., Nguyen, M.C., Uematsu, H., Azevedo, J.P., 2016. Who Are the Poor in the Developing World? The World Bank.

Cazcarro, I., Duarte, R., Martín-Retortillo, M., Pinilla, V., Serrano, A., 2015. Water Scarcity and Agricultural Growth in Spain: From Curse to Blessing? https://doi.org/10.13140/2.1.2685.1843.

Chapagain, A., Hoekstra, A., Savenije, H., 2006. Water saving through international trade of agricultural products. Hydrol. Earth Syst. Sci. 10 https://doi.org/10.5194/hess-10-455-2006.

Chen, J., Shi, H., Sivakumar, B., Peart, M.R., 2016. Population, water, food, energy and dams. Renew. Sustain. Energy Rev. 56, 18–28. https://doi.org/10.1016/j.rser.2015.11.043.

Cole, S., 2017. Water worries: an intersectional feminist political ecology of tourism and water in Labuan Bajo, Indonesia. Ann. Tourism Res. 67, 14–24.

Dugan, P., Dey, M.M., Sugunan, V.V., 2006. Fisheries and water productivity in tropical river basins: enhancing food security and livelihoods by managing water for fish. Agric. Water Manag. 80 (1–3), 262–275. https://doi.org/10.1016/j.agwat.2005.07.017. ISSN 0378-3774. http://www.sciencedirect.com/science/article/pii/S0378377405003021.

FAO, 2011. Global Food Losses and Food Waste – Extent, Causes and Prevention. Rome.

FAO, 2014a. Food Wastage Footprint. Impacts on Natural Resources. Rome. http://www.fao.org/3/ar429e/ar429e.pdf.

FAO, 2014b. Transboundary Water Infographic. Rome. http://www.fao.org/nr/water/aquastat/infographics/Transboundary_eng.pdf.

FAO, 2015a. Irrigation Management. http://www.fao.org/3/a-i4591e.pdf.

FAO, 2015b. Towards a Water and Food Secure Future: Critical Perspectives for Policy-Makers. Food and Agriculture Organization of the United Nations, Rome and World Water Council, Marseille, 61. http://www.fao.org/nr/water/docs/FAO_WWC_white_paper_web.pdf.

FAO, 2015c. Water at a Glance: The Relationship between Water, Agriculture, Food Security and Poverty. Rome, 15. http://www.fao.org/nr/water/docs/waterataglance.pdf.

FAO, 2016. The State of Food and Agriculture. http://www.fao.org/3/a-i6030e.pdf.

FAO, 2017a. Does Improved Irrigation Technology Save Water? a Review of the Evidence Discussion Paper on Irrigation and Sustainable Water Resources Management in the Near East and North Africa.

FAO, 2017b. Water for Sustainable Food and Agriculture A Report Produced for the G20 Presidency of Germany. Rome. http://www.fao.org/3/a-i7959e.pdf.

FAO, 2019. FAO Framework for the Urban Food Agenda. Rome, p. 44. Licence: CC BY-NC-SA 3.0 IGO.

FAO/WHO, 2018. Proceedings of the FAO/WHO International Symposium on Sustainable Food Systems for Healthy Diets and Improved Nutrition. Rome.

Fischer, C.G., Garnett, T., 2016. Plates, Pyramids, and Planets: Developments in National Healthy and Sustainable Dietary Guidelines: A State of Play Assessment. http://www.fao.org/3/a-i5640e.pdf.

Gephart, J.A., Troell, M., Patrik, J., Henriksson, G., Beveridge, M.C.M., Verdegem, M., Metian, M., Mateos, L.D., Deutsch, L., 2017. The `seafood gap' in the food-water nexus literature—issues surrounding freshwater use in seafood production chains. Adv. Water Resour. 110, 505−514. https://doi.org/10.1016/j.advwatres.2017.03.025. ISSN 0309-1708. http://www.sciencedirect.com/science/article/pii/S0309170817302488.

Gleeson, T., Wada, Y., Bierkens, M.F.P., van Beek, L.P.H., 2012. Water balance of global aquifers revealed by groundwater footprint. Nature 488 (7410), 197−200.

Hoekstra, A.Y., Chapagain, A.K., 2007. The water footprints of Morocco and The Netherlands: global water use as a result of domestic consumption of agricultural commodities. Ecol. Econ 64 (1), 143−151. Elsevier.

Hoekstra, A.Y., Chapagain, A.K., 2008. Globalization of Water: Sharing the Planet's Freshwater Resources. Blackwell Publishing, Oxford, UK.

Hoekstra, A.Y., Mekonnen, M.M., 2012. The water footprint of humanity. Proc. Natl. Acad. Sci. U.S.A. 109 (9), 3232−3237.

Hutton, G., Varughese, M., 2016. The Costs of Meeting the 2030 Sustainable Development Goal Targets on Drinking Water, Sanitation, and Hygiene. World Bank, Washington, DC. © World Bank. https://openknowledge.worldbank.org/handle/10986/23681 License: CC BY 3.0 IGO.

Jackson, S., Barber, M., 2015. Recognizing Indigenous water cultures and rights in mine water management: the role of negotiated agreements. Aquatic Procedia 5, 81−89.

Keller, A., Weight, E., Taylor, S., 2014. Rapid Assessment of Water Availability and Appropriate Technologies, vol. 156. IWMI.

Kibler, K.M., Reinhart, D., Hawkins, C., 2018. Amir Mohaghegh Motlagh, James Wright, food waste and the food-energy-water nexus: a review of food waste management alternatives. Waste Manag. 74, 52−62. https://doi.org/10.1016/j.wasman.2018.01.014. ISSN 0956-053X. http://www.sciencedirect.com/science/article/pii/S0956053X18300151.

Konar, M., Hussein, Z., Hanasaki, N., Mauzerall, D., Rodriguez-Iturbe, I., 2013. Virtual water trade flows and savings under climate change. Hydrol. Earth Syst. Sci. Discuss. 10, 67−101. https://doi.org/10.5194/hessd-10-67-2013.

Kuishuang, Feng., Hubacek, Klaus., 2015. A Multi-Region Input-Output Analysis of Global Virtual Water Flows.

Lonnie, M., Hooker, E., Brunstrom, J.M., Corfe, B.M., Green, M.A., Watson, A.W., Williams, E.A., Stevenson, E.J., Penson, S., Johnstone, A.M., 2018. Protein for life: review of optimal protein intake, sustainable dietary sources and the effect on appetite in ageing adults. Nutrients 10 (3), 360. https://doi.org/10.3390/nu10030360.

Mejias Moreno, P., Lagana, M.H., 2019. Évaluation de la résilience climatique des petits exploitants agricoles au Mali. FAO, Rome, p. 52. Licence: CC BY-NC-SA 3.0 IGO.

Mekonnen, M., Pahlow, M., Aldaya, M., Zarate, E., Hoekstra, A., 2015. Sustainability, efficiency and equitability of water consumption and pollution in Latin America and the Caribbean. Sustainability 7 (2), 2086−2112.

Merrett, S., 2003. 'Virtual Water and Occam's Razor', Occasional Paper 62. SOAS Water Issues Study Group.

Nestle, M., 2015. Soda Politics: Taking on Big Soda (And Winning). Oxford University Press, New York, p. 525.

OECD, 2016. Water Governance in Cities, OECD Studies on Water. OECD Publishing, Paris. https://doi.org/10.1787/9789264251090-en.

OECD, 2018a. Financing Water. Investing in Sustainable Growth. OECD Publishing, Paris. https://www.oecd.org/water/Policy-Paper-Financing-Water-Investing-in-Sustainable-Growth.pdf.

OECD, 2018b. Reforming Water Policies in Agriculture Lessons from Past Reforms Guillaume.

Peck, J.C., 2007. Groundwater management in the high plains aquifer in the USA: legal problems and innovations, Chapter 14. In: Giardano, M., Vilholth, K. (Eds.), Agricultural

Groundwater Revolution: Opportunities and Threats to Development. CABI, Wallingford, UK, pp. 296–319.
Rockström, J., Steffen, W., Noone, K., Persson, Å., Chapin III, F.S., Lambin, E., Lenton, T.M., Scheffer, M., Folke, C., Schellnhuber, H., Nykvist, B., De Wit, C.A., Hughes, T., van der Leeuw, S., Rodhe, H., Sörlin, S., Snyder, P.K., Costanza, R., Svedin, U., Falkenmark, M., Karlberg, L., Corell, R.W., Fabry, V.J., Hansen, J., Walker, B., Liverman, D., Richardson, K., Crutzen, P., Foley, J., 2009. Planetary boundaries: exploring the safe operating space for humanity. Ecol. Soc. 14 (2), 32 [online] URL: www.ecologyandsociety.org/vol14/iss2/art32.
Ryan, C., Elsner, P., 2016. The potential for sand dams to increase the adaptive capacity of East African drylands to climate change. Reg. Environ. Change 16 (7), 208–2096. https://doi.org/10.1007/s10113-016-0938-y.
Turral, H., Burke, J., Faurès, J.M., 2011. Climate Change, Water and Food Security. FAO, Rome.
United Nations, Department of Economic and Social Affairs, Population Division, 2017. World Population Prospects: The 2017 Revision, Key Findings and Advance Tables. Working Paper No. ESA/P/WP/248.
United Nations Food and Agricultural Organization (UN FAO), 2018. AQUASTAT Database. http://www.fao.org/nr/water/aquastat/data/query/results.html.
Ward, S.M., Holden, N.M., White, E.P., Oldfield, T.L., 2016. The 'circular Economy' Applied to the Agriculture (Livestock Production) Sector – Discussion Paper Presented at Presented at Workshop on the Sustainability of the EU's Livestock Production Systems Hosted by European Commission, DG Agriculture and Rural Development, pp. 14–15. http://ec.europa.eu/information_society/newsroom/image/document/2016-48/ward_-_circular_economy_applied_to_the_livestock_production_sector___brussels__2_40231.pdf.
WWAP (UNESCO World Water Assessment Programme), 2019. The United Nations World Water Development Report 2019: Leaving No One behind. UNESCO, Paris.
World Bank, 2016. "High and Dry: Climate Change, Water, and the Economy". World Bank, Washington, DC. License: Creative Commons Attribution CC BY 3.0 IGO.
Zekster, I.S., Everett, L.G. (Eds.), 2004. Groundwater Resources of the World and Their Use, IHP-VI, Series on Groundwater No. 6. UNESCO (United Nations Educational, Scientific and Cultural Organization), p. 342.
Zhuo, La, Mekonnen, M.M., Hoekstra, A.Y., 2016a. The effect of inter-annual variability of consumption, production, trade and climate on crop-related green and blue water footprints and inter-regional virtual water trade: a study for China (1978–2008). Water Res. 94, 73–85. https://doi.org/10.1016/j.watres.2016.02.037. ISSN 0043-1354. http://www.sciencedirect.com/science/article/pii/S0043135416300963.
Zhuo, L., Mekonnen, M.M., Hoekstra, A.Y., 2016b. Water Footprint and Virtual Water Trade of China: Past and Future, Value of Water Research Report Series No. 69. UNESCO-IHE, Delft, the Netherlands.
Zhao, D., Klaus, H., Feng, K., Sun, L., Liu, J., 2019. Explaining virtual water trade: a spatial-temporal analysis of the comparative advantage of land, labor and water in China. Water Res. 153 https://doi.org/10.1016/j.watres.2019.01.025.

Further reading

Alcamo, J., Döll, P., Henrichs, T., Kaspar, F., Lehner, B., Rösch, T., Siebert, S., 2003. Development and testing of the WaterGAP 2 global model of water use and availability. Hydrol. Sci. J. 48, 317–337.

Allan, J.A., 2003. Virtual water – the water, food, and trade nexus useful concept or misleading metaphor? Int. Water Resour. Assoc. Water Int. 28 (1), 4–11. https://doi.org/10.1080/02508060.2003.9724812.

Allan, T., 2011. Virtual Water: Tackling the Threat to Our Planet's Most Precious Resource. I.B. Tauris & Co Ltd, London.

aus der Beek, T., Flörke, M., Lapola, D.M., Schaldach, R., Voß, F., Teichert, E., 2010. Modelling historical and current irrigation water demand on the continental scale: Europe. Adv. Geosci. 27, 79–85. https://doi.org/10.5194/adgeo-27-79-2010.

Comprehensive Assessment of Water Management in Agriculture, 2007. Water for Food, Water for Life: A Comprehensive Assessment of Water Management in Agriculture. Earthscan, and Colombo: International Water Management Institute, London.

FAO, 2012. Sustainable Diets and Biodiversity. Rome. http://www.fao.org/3/i3004e/i3004e.pdf.

FAO, 2013. Good Practices in Planning and Management of Commercial Poultry Production in South Asia. http://www.fao.org/3/a-y4991e.pdf.

FAO/IFAD. Forthcoming. Adapting Irrigation to Climate Change (AICCA) in West and Central Africa. Project Report Validated by Country Counterparts. Rome. http://www.fao.org/3/a-i7143e.pdf.

Flörke, M., Kynast, E., Bärlund, I., Eisner, S., Wimmer, F., Alcamo, J., 2013. Domestic and industrial water uses of the past 60 years as a mirror of socio-economic development: a global simulation study. Glob. Environ. Chang. 23, 144–156.

Gruère, C.A., Cadilhon, J.J., https://www.oecd-ilibrary.org/docserver/1826beee-en.pdf?expires=1551703352&id=id&accname=guest&checksum=A3C86B6DB5CFC3818040BF2306468648.

Gruère, G., Ashley, C., Cadilhon, J., 2018. Reforming Water Policies in Agriculture: Lessons from Past Reforms. OECD Food, Agriculture and Fisheries Papers, No. 113. OECD Publishing, Paris. https://doi.org/10.1787/1826beee-en.

Hoekstra, A.Y. (Ed.), February 2003. Virtual Water Trade – Proceedings of the International Expert Meeting on Virtual Water Trade. IHE Delft, the Netherlands.

Inyinbor Adejumoke, A., Adebesin Babatunde, O., Oluyori Abimbola, P., Adelani-Akande Tabitha, A., Dada Adewumi, O., Oreofe Toyin, A., 21, March 2018. Water Pollution: Effects, Prevention, and Climatic Impact, Water Challenges of an Urbanizing World. Matjaž Glavan, IntechOpen. https://doi.org/10.5772/intechopen.72018. Available from: https://www.intechopen.com/books/water-challenges-of-an-urbanizing-world/water-pollution-effects-prevention-and-climatic-impact.

Iyer, R.R., 2008. Water: a critique of three concepts. Econ. Pol. Wkly. 43 (1), 15–18.

Konikow, 2011. Contribution of global groundwater depletion since 1900 to sea-level rise. Geophys. Res. Lett. 38.

Margat, Van der Gun, 2013. Groundwater Around the World - A Geographic Synopsis. CRC Press/Balkema.

Mekonnen, M.M., et al., 2015. Sustainability, efficiency and equitability of water consumption and pollution in Latin America and the Caribbean. Sustainability 7 (2), 2086–2112.

OECD, 2017. Water Risk Hotspots for Agriculture, OECD Studies on Water. OECD Publishing, Paris. https://doi.org/10.1787/9789264279551-en.

Bihn, E.A., Schermann, M.A., Wszelaki, A.L., Wall, G.L., Amundson, S.K., 2014. On-Farm Decision Tree Project: Postharvest Water—v7 07/16/2014 1. www.gaps.cornell.edu.

Ray, C., McInnes, D., Sanderson, M., 2018. Virtual water: its implications on agriculture and trade. Water Int. 43 (6), 717–730. https://doi.org/10.1080/02508060.2018.1515564.

Sadoff, C.W., Hall, J.W., Grey, D., Aerts, J.C.J.H., Ait-Kadi, M., Brown, C., Cox, A., Dadson, S., Garrick, D., Kelman, J., McCornick, P., Ringler, C., Rosegrant, M., Whittington, D., Wiberg, D., 2015. Securing Water, Sustaining Growth: Report of the GWP/OECD Task Force on Water Security and Sustainable Growth. University of Oxford, UK, p. 180.

Shared global vision for Groundwater Governance 2030 Revised Edition, 2016. http://www.fao.org/3/a-i5508e.pdf.

Srinivasan, V., Konar, M., Sivapalan, M., 2017. A dynamic framework for water security. Water Secur. 1, 12–20. https://doi.org/10.1016/j.wasec.2017.03.001. ISSN 2468-3124. http://www.sciencedirect.com/science/article/pii/S2468312416300220.

Sun, S., Wang, Y., Engel, B.A., Wu, P., 2016. Effects of virtual water flow on regional water resources stress: a case study of grain in China. Sci. Total Environ. 550, 871–879. https://doi.org/10.1016/j.scitotenv.2016.01.016. ISSN 0048-9697. http://www.sciencedirect.com/science/article/pii/S0048969716300092.

Varis, O., Keskinen, M., Kummu, M., 2017. Four dimensions of water security with a case of the indirect role of water in global food security. Water Secur. 1, 36–45. https://doi.org/10.1016/j.wasec.2017.06.002. ISSN 2468-3124. http://www.sciencedirect.com/science/article/pii/S2468312416300256.

Wichelns, D., 2010a. An economic analysis of the virtual water concept in relation to the agri-food sector. In: Gestion durable des ressources en eau dans le secteur agricole. OECD Publishing, Paris. https://doi.org/10.1787/9789264083592-8-fr.

Wichelns, D., 5, January 2010b. Virtual Water: A Helpful Perspective, but Not a Sufficient Policy Criterion. Water Resources Management, pp. 2203–2219.

WWAP (United Nations World Water Assessment Programme)/UN-Water, 2018. The United Nations World Water Development Report 2018: Nature-Based Solutions for Water. UNESCO, Paris.

Yang, Zehnder, 2008. Globalization of Water Resources through Virtual Water Trade.

Yang, H., Wang, L., Abbaspour, K.C., Zehnder, A.J.B., 2006. 'Virtual water highway: water use efficiency in global food trade', Hydrology and. Earth Syst. Sci. Data 3, 1–26.

Zimmer and Renault, Virtual Water in Food Production and Global Trade – Review of Methodological Issues and Preliminary Results, World Water Council, FAO-AGLW.

Wastewater treatment and water reuse in the food industry

8

A.G. Tekerlekopoulou[1], *Ch.N. Economou*[2], *T.I. Tatoulis*[1], *C.S. Akratos*[3], *D.V. Vayenas*[2,4]

[1]Department of Environmental and Natural Resources Management, University of Patras, Agrinio, Greece; [2]Department of Chemical Engineering, University of Patras, Patras, Greece; [3]Department of Civil Engineering, Democritus University of Thrace, University Campus, Xanthi, Greece; [4]Institute of Chemical Engineering Sciences (ICE-HT), Patras, Greece

Chapter outline

1. Introduction 246
- 1.1 Food industry wastewaters 246
- 1.2 Characteristics of food industry wastewaters 247

2. Treatment methods 247
- 2.1 Biological methods 250
  - *2.1.1 Aerobic processes 250*
  - *2.1.2 Anaerobic processes 252*
  - *2.1.3 Microalgal-based processes 253*
- 2.2 Physicochemical methods 255
- 2.3 Electrochemical methods 257
- 2.4 Constructed wetlands 258
- 2.5 Hybrid systems 259

3. Case studies of food industry wastewater treatment 260
- 3.1 Biological trickling filters 261
  - *3.1.1 Biological trickling filters and olive mill wastewater 261*
  - *3.1.2 Biological trickling filters and table olive wastewater 264*
  - *3.1.3 Biological trickling filters and second cheese whey 265*
- 3.2 Constructed wetlands 265
  - *3.2.1 Constructed wetlands and olive mill wastewater 268*
  - *3.2.2 Constructed wetlands and table olive wastewater 268*
  - *3.2.3 Constructed wetlands and second cheese whey 268*
- 3.3 Electrocoagulation treatment 270

4. Reuse of treated food industry wastewaters 272

5. Conclusions 273

Acknowledgments 274

References 274

The Interaction of Food Industry and Environment. https://doi.org/10.1016/B978-0-12-816449-5.00008-4

1. Introduction

1.1 Food industry wastewaters

Food processing requires high volumes of water. Specifically, in the last decades this sector has become the third largest industrial user of water after the refined petroleum, primary metals and chemicals industries (Barbera and Gurnari, 2018). Water consumption in the food industry is connected with the required high quality of the final product and varies depending on the type and capacity of each production line, the equipment and processes applied to yield the final product, the cleaning equipment systems used, and the level of automation carried out (Jaszczyszyn et al., 2018; Valta et al., 2014). However, the food industry is not only a significant user of water, but inevitably results in the production of large quantities of wastewaters (Abdallh et al., 2016). Wastewaters derived from food industries are distinguished from municipal and industrial wastewaters as they are biodegradable and do not contain toxic chemicals (Emara et al., 2017). Food industry wastewaters are also characterized by significant values of Biological and Chemical Oxygen Demand concentrations (BOD and COD, respectively), caused by organic wastes entering the wastewater stream and the chemicals used in the various processing stages, high concentrations of dissolved (DS) and/or suspended solids (SS) (including oils, fats, grease), nutrients such as nitrogen (N, including ammonia) and phosphorous (P), minerals (e.g., salts), and varying pH values. These characteristics and the volume and rates of wastewater production all differ depending on the specific type (e.g., fruit, dairy, oil, vegetable, meat) and scale of production (small-scale local, large-scale or international producer). In addition, food industry wastewater production shows wide regional and seasonal variations (Onet, 2010).

The above-mentioned variations make the determination of suitable and cost-effective treatment technology difficult and as such the selection of a suitable treatment process must sometimes be made on a case-by-case basis. Nevertheless, there are certain standards that must be kept in mind. Treatment technologies should follow the principle of reducing the amounts of water and energy used as well as reducing the contamination level of produced wastewaters. In addition, the operating strategies should provide an easy "Return on Investment" regardless of the scale of operation (Klemes et al., 2008).

A wide range of technologies and techniques have been reported in the literature for the treatment of food industry wastewaters (De Gisi and Notarnicola, 2017; Abdallh et al., 2016). These include physicochemical systems, biological systems (anaerobic or aerobic), constructed wetlands, electrochemical methods, membrane bioreactors, advanced oxidation processes or hybrid systems including two or more of the above-mentioned methods. Some can be applied only to small units (e.g., due to more flexible production processes), while others can be applied only to larger units (e.g., because of their high costs). In addition, some of the above-mentioned methods have been examined only in laboratory-scale reactors and not in pilot- or industrial-scale plants.

This chapter provides a general overview of various types and characteristics of agri-food industrial wastewaters and describes their different treatment techniques.

Case studies on dairy, table olive and olive mill wastewaters treatment in Greece are presented with emphasis on biological methods, constructed wetlands and electrochemical processes that have been applied in pilot- or industrial-scale units. The quality of posttreatment wastewaters is demonstrated and their potential for reuse and recycling is discussed.

1.2 Characteristics of food industry wastewaters

Knowledge of the characteristics of food industry wastewaters is crucial to develop treatment systems that are technically feasible, economically viable, and comply with current environmental policy and regulations. Knowledge of wastewater characteristics also allows the design of a precise recycling-reuse-recovery plan of valuable substances for each wastewater. For example, wastewaters from vegetable and fruit processing contain fermentable sugars which could be utilized within a novel treatment method that combines high purification performance and recovery of valuable by-products (Bories, 2000).

The concentration and composition of food processing wastewaters varies from high-strength substrates (e.g., olive mill, winery, cheese wastewaters, second cheese whey with COD concentrations above 3000 mg/L), to low and medium strength (e.g., dairy effluents or wash water from sugar mills with COD concentrations up to 3000 mg/L) in terms of organic load (determined as COD and BOD), nutrients (nitrogen and phosphorus), suspended solids (TSS: Total Suspended Solids), etc. (Rajagopal et al., 2013). Table 8.1 presents reported characteristics of a variety of agri-food wastewaters and shows that many are potential environmental pollutants if disposed of untreated. Therefore, authorities set strict criteria for wastewater discharge (some countries have stricter requirements than others) which have, in turn, forced all food processing industries (small-, medium-, and large-scale) to find effective and cost-viable technologies to treat their effluents.

2. Treatment methods

Properties of food industry wastewater, economic viability and site conditions of the processing factory, are the most important factors that should be considered for the selection of the proper treatment process. Before the 1990s, lagoons and activated sludge systems (including aeration and settling tanks) were the most widely used food industry wastewater treatment processes worldwide. However, the strict environmental criteria for discharge of these wastewaters led to the development of new technologies offering improved efficiencies at lower cost.

Nowadays, one or more treatment steps are required for each wastewater depending on its strength. Generally, wastewater treatment technologies can be classified as preliminary treatment, primary treatment, secondary treatment and tertiary/advanced treatment. During preliminary treatment large, suspended material can be removed via screening (through coarse filters) or settlement. Primary treatments (sedimentation,

Table 8.1 Examples of food industry wastewaters and their characteristics.

Food industry wastewater	COD (mg/L)	BOD (mg/L)	TS/TSS (mg/L)	TKN/N–NH_3 (mg/L)	TP (mg/L)	pH	Reference
Dairy wastewater	650–3000	300–1400	–/250–2700	–/10–20 (N-NH_4)	10–130 (P_2O_5)	4–12	Cecconet et al. (2018)
	4013 ± 2593	–	–	202 ± 75 (TN)/ 137 ± 89	–		Shams et al. (2018)
	921–9004	483–6080	–/134–804	8–230 (TN)	9–111.5	5.5–5.8	Farizoglu & Uzuner (2011)
Second cheese whey	43,000 ± 2000	20,000 ± 1000	49,300/19,400	1100/–	0.34	4.5–6	Tatoulis et al. (2015); Tsolcha et al. (2018)
Poultry slaughterhouse wastewater	3000–4800	750–1890	300–950/–	–/16–165	16–32	7.0–7.6	Rajakumar et al., 2011, 2012
	2800 ± 537	1500 ± 137	–/1565 ± 374	290 ± 40/–	–	6.51 ± 0.07	Affes et al. (2017)
Swine	5608	1984	464 (SS)	476 (N-NH_4^+)	158.61	–	Li et al. (2013)
	1500–6500 (TOC)	2500–4500	–/3000–8000	1500–2000 (TN)/900–1500		7.9	De Prá et al. (2012)
Cattle	2900 ± 20			618 ± 0.9/ 498 ± 0.21	23.5 ± 0.01		De Mendonça (2017)

Fish processing	300–90,000	114–78,000	–/95–10,000	77–3000 (TKN)		3.8–10	Chowdhury et al. (2010)
Potato chips manufacturing wastewater	6000	5000	–/5000	250 (TN)	100		Rajagopal et al. (2013)
Olive mill wastewater	96,000 ± 2000			46 (TN)		5 ± 0.03	Jaouad et al. (2018)
	130,100	–	–/75,500	460			Rajagopal et al. (2013) Gonçalves et al. (2012)
Winery	18,000–21,000		–/150–200	310–410 (TN)	40–60		Rajagopal et al. (2013); Ganesh et al. (2010)
Sugar	6600		–/6100	10	2.07		Rajagopal et al. (2013)

filtration or centrifugation are widely used) can remove most suspended solids (SS) or oils and grease. According to the bibliography, SS and oils/grease can be removed with gravity separation (40%–90% and 50%–90%, respectively), coagulation (about 60% for TSS), dissolved air floatation (60%–95% and 60%–95%, respectively), or dissolved air floatation and chemical addition (80%–93% and 58%–99%, respectively) (Falletti et al., 2015). In addition, some COD and BOD content can be removed during primary treatment (50%–60%, De Gisi and Notarnicola, 2017). However, these pollutants are mainly removed by other processes, biological or physicochemical, considered as secondary treatment. A tertiary/advanced treatment step may be applied as a polishing stage before the wastewater is disposed of into a recipient water body. This treatment stage may comprise sand filtration, membrane filtration, activated carbon adsorption, reverse osmosis, constructed wetlands, etc. However, there is no specific categorization of the earlier-mentioned processes at each treatment stage. Therefore, depending on a wastewater's characteristics and strength, biological or physicochemical processes can be used either as primary/secondary treatment or as polishing steps.

Several methods for the treatment of wastewaters from the food industry have been examined extensively based on biological (under aerobic or anaerobic conditions) or physicochemical technologies. However, in cases of high-strength food industry wastewaters (i.e., that contain high amounts of organic matter), hybrid systems have been developed by combining aerobic and/or anaerobic processes with physicochemical processes, resulting in higher organic load removal efficiencies and substantial energy and water savings.

2.1 Biological methods

The organic load found in food industry wastewaters is considerable and complex, however it is mostly biodegradable (BOD/COD >0.5) (Farraji et al., 2015). Therefore, biological treatment is considered an appropriate method for such waste streams. Both aerobic and anaerobic processes are included in biological treatment. In these processes aerobic, anaerobic or facultative in nature microorganisms are responsible for removing contaminants from the wastewater. The most common literature-referred processes are: suspended growth processes (aerobic or anaerobic), attached growth processes (aerobic or anaerobic) and digestion (aerobic or anaerobic), as well as microalgal-based processes. Constructed wetlands are also an attractive solution for food industry wastewaters, especially in rural areas (Metcalf and Eddy, 2003).

2.1.1 Aerobic processes

In aerobic processes, organic load is oxidized (with oxygen present) into CO_2, H_2O, and new biomass. Aerobic methods can be categorized as either suspended or fixed film (attached growth or biofilm) processes. According to Gavrilescu and Macoveanu (1999), aerobic processes are the most effective and less costly treatment methods for food industry wastewaters, especially for low strength waste streams (concentrations up to 4000 mg COD/L) (Abdulgader et al., 2007). Suspended growth reactors include aerated lagoons, sequencing batch reactors (SBR, single-stage or two-stage systems), granular activated sludge processes (GAS) and aerobic jet loop reactors.

In almost all these suspended growth systems, COD removal depends significantly on the strength of the food industry wastewater. However, the bioprocess in these systems is also dependent on other operating conditions such as dissolved-oxygen concentration, hydraulic retention time (HRT), recycling ratio of sludge, mixed liquor-suspended solid (MLSS), temperature, pH, etc. High organic load, total nitrogen and phosphorus percentage removals (above 90%, 80% and 70%, Schwarzenbeck et al., 2005) were observed in all the above-mentioned suspended systems at an organic loading rate (OLR) up to 3 kg COD/m^3 d, while for higher OLR the COD removal efficiency decreased (Abdulgader et al., 2007). In these processes dissolved oxygen should be above the value of 2–3 mg/L, while MLSS should be around 3000 mg/L (Kushwaha et al., 2011; Mohseni-Bandpi and Bazari, 2004). On the other hand, lower COD removal was obtained for lagoons, SBR and GAS systems at higher organic loads (e.g., 70% for loadings above 4 kg COD/m^3 d) (Abdulgader et al., 2007). In these cases, the residual organic load exceeded significantly the legal value for direct discharge (150 mg/L) (Prazeres et al., 2013). All the above-mentioned systems have been applied in pilot and full-scale units for food industry wastewater treatment, however not as single step but in combination with other processes (including pretreatment or posttreatment steps). According to the literature, membrane bioreactors can be coupled to SBRs (membrane-sequencing batch reactor, MSBR). This system was found to achieve significant removal efficiencies of COD and total Kjeldahl nitrogen (TKN), above 90% and 60%, respectively, even at the high load of 3 kg COD/m^3 d (corresponding about to 1.34 kg BOD$_5$/m^3 d) (Sirianuntapiboon et al., 2005). Jet loop reactors have also been combined with membrane technologies. Jet loop membrane bioreactors (JLMBR) are considered advanced reactors achieving high COD removal efficiencies. These reactors present high oxygen transfer and mixing and are additionally small-sized. Değermenci et al. (2016) examined this type of reactor for pretreated olive mill wastewater treatment and achieved COD removal up to 93% at volumetric loads of 17.8 kg COD/m^3 day. Winery, brewery and cheese whey wastewaters have also been treated in JLMBRs, yielding COD, total nitrogen and PO_4^{3-} removal efficiencies higher than 93%, 99% and 65%–88%, respectively (Bolzonella et al., 2007; Farizoglu and Keskinler, 2006). However, it should be mentioned that sludge generation leads to settling problems while the efficiency of this configuration decreases with usage time.

Research on aerobic biological treatment of food industry wastewaters in the last 2 decades reveals that aggregation of cells either into flocs or biofilms and their attachment onto support materials allows the bioremoval of large amounts of biodegradable organic load even at very high concentrations (e.g., up to 43,000 mg d-COD/L for second cheese whey treatment) (Tatoulis et al., 2015). Attached growth systems also produce less sludge than suspended ones, resulting in high savings in sludge handling, storage and disposal costs (Vayenas, 2011). For this reason, research has concentrated on these fixed film processes. Specific surface area of these systems plays an important role in the process and varies depending on the support material used (Vayenas, 2011). The most commonly used systems are trickling filters, packed bed reactors, and rotating biological contactors. However, fluidized-bed attached systems and submerged systems are also used for food industry wastewaters. It should be noted that attached growth systems, such as single-step treatment, yield higher COD degradation rates than suspended ones. Packed and fluidized-bed reactors can achieve COD percentage removals of over 78% for high organic load food industry wastewaters up to 17 kg COD/m^3 d (Abdulgader et al., 2007) and over 90% for 6.5 kg COD/m^3 d (Tatoulis et al., 2015), however, they fail to completely remove the organic load. In most cases, the treated waters do not meet the stringent regulations for final disposal, and therefore, hybrid systems are usually examined. A literature review reveals that although there are many research works on aerobic attached growth systems for food industry wastewaters, only a few have been performed in pilot or full-scale units (Prazeres et al., 2012) as most of them have been conducted in laboratories under controlled conditions. Aerobic attached growth systems commonly face several difficulties in full-scale systems, such as start-up and sludge generation. However, bioaugmentation (the addition of extra microorganisms with high degradation capacity for specific wastewaters) and adequate bed cleaning during backwashing, may control long-term operation of these systems in full-scale units (Vayenas, 2011). A description of aerobic attached growth systems in pilot and full-scale applications is given in Section 3.1.

2.1.2 Anaerobic processes

Anaerobic treatment is an effective treatment method for high-strength wastewaters such as agricultural and food processing wastewaters. Anaerobic microorganisms, without oxygen, convert organic matter into CO_2 and methane. Anaerobic systems present some advantages over aerobic ones, such as lower sludge production (5%–20%), low area demand, high removal efficiencies even at significant loadings, and biogas recovery (De Gisi and Notarnicola, 2017). Nevertheless, they require high energy levels to maintain the needed temperature. Additionally, these systems are not easy to operate (making them difficult to apply in small industries) and they cannot tolerate shock loadings. Finally, the anaerobic sludge generated is more difficult to treat than aerobic sludge, demanding extra equipment and finances. The most common anaerobic treatment units for food industry wastewaters include anaerobic lagoons, anaerobic filters (AF), anaerobic fluidized-bed reactors, up-flow anaerobic sludge blankets (UASBs), expanded granular sludge beds (EGSB), and anaerobic SBRs (Alayu and Yirgu, 2018; Rajagopal et al., 2013). Many research teams have studied the above-mentioned anaerobic configurations for a variety of food industry wastewaters (Cuff et al., 2018; Hung et al., 2008). According to Rajagopal et al. (2013, Table 1), Cruz-Salomón et al. (2017, Table 1) and Alayu and Yirgu (2018), COD removal efficiencies range between 75% and 96% depending on the type and strength of wastewater studied. However, it is extremely difficult for anaerobic systems to produce effluents suitable for final disposal in water bodies according to legal permitted values, especially for nitrogen (Rajagopal et al., 2013). Usually, post-treatment steps are necessary for anaerobic systems (De Gisi and Notarnicola, 2017). The parallel production of hydrogen and/or methane in these systems could help reduce treatment costs. A biogas yield of 10 ± 5 m^3 can be produced when 18.5–40 kg of volatile solids (VS) is added into anaerobic systems, while 65%

VS removal is achieved (Rajagopal, 2013). Analysis of literature indicates that although full-scale applications have been tested (Austermann-Haun et al., 1999), there is more need to examine the parameters that influence the processes of anaerobic systems (such as clogging, packing materials, biomass washout, biogas quality, etc.) in order to improve their operation at lower hydraulic operating times and higher organic loadings.

2.1.3 Microalgal-based processes

Even though most biological treatment technologies applied to food industry wastewaters include the use of bacteria, microalgae (including cyanobacteria) are gaining importance as they are capable of assimilating significant amounts of nutrients (mainly N, P and COD) phototrophically, heterotrophically, or mixotrophically, and can convert them into high-value bioproducts (Economou et al., 2015; Koutra et al., 2018; Tsolcha et al., 2017). Pollutant removal by microalgae is affected by several factors such as pH, temperature, light intensity, and inoculum and nutrients concentration, which all differ according to the algal strains used, the reactor type (closed photobioreactors, open ponds) and the operation mode (batch, fed-batch, continuous) (Koutra et al., 2018). Nowadays, many researchers propose microalgae cultivation using food industry wastewaters as an alternative low-cost and environmentally friendly technology (Cheng et al., 2013; Di Caprio et al., 2018; Tsolcha et al., 2016, 2017). However, the efficiency of light utilization should be always considered when working with photosynthetic microorganisms. Therefore, an important factor for algal cultivation is the turbidity of the growth medium, and wastewaters should be pretreated or diluted to reduce their solid particle contents and thus allow sufficient light penetration into the culture. Various pretreatment methods such as sedimentation, filtration, and centrifugation (see Section 2.2), have been applied to reduce medium and wastewater turbidity (Lu et al., 2015) and these methods also aim to minimize the toxic effects of high ammonia concentrations on algal growth. Food industry wastewaters have been diluted with fresh water (Tsolcha et al., 2016), however this technique is economically unfeasible for industrial applications. Alternatively, other wastewaters characterized by low initial nutrient concentrations and solid content could be used to dilute food industry wastewaters (Koutra et al., 2018; Tsolcha et al., 2018). Additionally, algal growth is affected by ratios of C/N and N/P within the culture medium and this parameter should be considered when using wastewaters in biological treatment systems (Tsolcha et al., 2017).

The pre-treatment or dilution (with water or wastewater containing lower organic load, nutrients, color and turbidity) of food industry wastewaters has been seen to increase pollutant removal rates. For example, the application of an air-stripping technique prior to algal cultivation in piggery wastewater (in laboratory-scale units) produced removal rates of 50%–88% for TN, 48%–93% for TP (total phosphorus) and 10%–42% for COD (the raw wastewater resulted in algal aggregations when used directly as the growth medium, showing inhibitory growth of the algal strain) (Cheng et al., 2013). Alternatively, high removal rates (almost 100%) of NH_4^+-N and PO_4^{3-}-P were recorded by Markou (2015) when applying fed-batch cultivation mode to control nutrient limitations and ammonia inhibition of anaerobically digested poultry litter wastewater. Significant pollutant removal rates have also been reported in blends of winery and raisin (weaker than winery) wastewaters (i.e., 92.8% for COD, 78.1% for TN, and 99% OP) (Tsolcha et al., 2017), and in second cheese whey diluted with tap water. In the latter case, the recorded removal rates of COD, TN, total sugars and orthophosphates were 94.0%, 91.2%, 84.8%, and 83.8%, respectively (Tsolcha et al., 2018).

Although high percentages of nitrogen and phosphorous can be removed from food industry wastewaters by axenic cultures of microalgae species, as in the case of *Scenedesmus obliquus* in olive mill wastewaters (Di Caprio et al., 2018), *Chlorella zofingiensis* in piggery wastewaters (Zhu et al., 2013) and *Chlorella* sp. in dairy wastewater (Lu et al., 2015), COD removal efficiencies range from 10% to 80%, due to the presence of nonbiodegradable compounds in the medium (Di Caprio et al., 2018; Lu et al., 2015). For this reason, the growth of mixed microalgae-bacteria cultures in food wastewaters has also been studied in nonaseptic conditions to reduce the cost of algal cultivation and increase pollutant removal rates (Tsolcha et al., 2016, 2017). In these cultivation types, oxygen is provided to bacteria via algal photosynthesis and this process degrades the organic substances present in the medium, while the carbon dioxide produced from the oxidation of organic compounds can be used by the microalgae as a carbon source for their growth. Additionally, bacteria convert ammonia to nitrate which can be utilized more efficiently by microalgae (Hena et al., 2015). However, the bacterial population should remain fairly low in these cultivation systems to ensure low turbidity of the growth medium.

Microalgae consortia comprising microalgae, cyanobacteria and bacteria, have been successfully applied to treat various food industry wastewaters such as fish processing wastewater (Queiroz et al., 2013), dairy farm wastewater (Hena et al., 2015), second cheese whey (Tsolcha et al., 2016) and brewery effluents (Raposo et al., 2010), and in all cases the consortia achieved high removal percentages (>80%) of initial COD, ammonia and phosphorus concentrations in laboratory-scale photobioreactors. Furthermore, consortia of cyanobacteria have demonstrated higher removal efficiencies of COD and phenolic compounds (84.8% and 93.8%, respectively) in the treatment of olive mill wastewaters than pure cyanobacterial cultures (*Anabaena oryzae*, *Nostoc muscorum* and *Spirulina platensis*) (El Shimi and Mostafa, 2016). Similarly, high removal efficiencies of COD and N from swine wastewater have been reported in open ponds (working volume of 464 L) by De Godos et al. (2009) using a mixed microalgal population of *Oocytis* (dominant genus, 72%), *Ankistrodesmus*, *Chlorella*, *Selenastrum*, and *Microspora*. Nevertheless, in this case, *phosphorous* removal was less than 10% probably due to the absence of P precipitation (pH measured 8.5 during the experiments, whereas P precipitation occurs at pH 9–11). In an outdoor pilot-scale treatment unit (working volume 30 L) tested by Lu et al. (2015) for the cultivation of green algae *Chlorella* sp., the maximum values of TN and TP removal recorded from diluted dairy wastewater were 85% and 88%, respectively. However, TP was removed slowly during the early stages of cultivation, and the maximum removal percentage of COD was just 41%. According to current research, it appears that a combination of pretreated wastewater combined with a consortium of microalgae could further improve pollutant removal efficiencies either in laboratory or industrial-scale treatment applications. Finally, it is worth noting that the produced algal biomass in most of the aforementioned studies has been used as feedstock for biodiesel production.

2.2 Physicochemical methods

Physicochemical methods, such as coagulation-flocculation, chemical precipitation, ion exchange, neutralization, filtration, ultrafiltration, nanofiltration, reverse osmosis, ozonation, advanced oxidation processes (AOPs) and combined physicochemical systems, have been extensively applied in the treatment of food industry wastewaters. Nevertheless, they are more costly than biological treatment methods. All the above physicochemical processes have been applied at industrial scale with the exception of AOPs that are mostly applied in laboratory and pilot-scale treatments. Additionally, AOPs are mainly proposed as posttreatment processes. The removal efficiency of each method in terms of COD, BOD, TS/TSS, TN and TP depends on the initial physicochemical characteristics of each raw wastewater (Ioannou et al., 2015).

Of the various physicochemical methods, coagulation-flocculation is the simplest and most economical, and is usually applied in the first stage of solid-liquid separation to reduce colloids, suspended particles and color in wastewaters. One study on the application of coagulation-flocculation using $FeSO_4 \cdot 7H_2O$ (6.67 g/L) as the coagulant and an anionic polyelectrolyte (0.287 g/L) as the flocculant to treat olive mill wastewater resulted in a 97%, 72% and 40% reduction of TSS, COD and phenolic compounds concentrations, respectively (Papaphilippou et al., 2013). In this study, photo-Fenton (0.2 g/L Fe^{2+}, 5 g/L H_2O_2 and pH = 3) was also applied for 240 min as a polishing step and reduced the remaining COD by 85%. However, the treated olive mill wastewater exhibited phytotoxicity, probably due to the formation of toxic compounds after the oxidation process. It is worth noting that coagulation and flocculation have also been separately investigated for their potential to treat food industry wastewaters. For instance, different coagulants (inorganic and polymeric) and their dosages were studied for their ability to remove COD and turbidity from dairy wastewater by improving floc settling velocity (Loloei et al., 2014).

Filtration is another process in which solid matter is removed by a filter that retains any solid particles and permits wastewater to flow through (Makara and Kowalski, 2015). Membrane technology is based on the same principle as filtration and can be classified into microfiltration, ultrafiltration, nanofiltration and reverse osmosis. All four methods allow both water reuse and nutrient recovery in the same operational unit (Luo et al., 2011; Víctor-Ortega et al., 2016). Ultrafiltration and microfiltration separation methods are used to remove suspended solids as well as large particles at low transmembrane pressure (Luo et al., 2011), while nanofiltration uses membranes with smaller-sized pores (1–10 nm) to remove inorganic salts and small molecular weight organic substances at moderate pressure (Andrade et al., 2014). El-Abbassi et al. (2014) applied the ultrafiltration process using a polyethersulfone membrane to remove COD, color, TS and total phenolic content from olive mill wastewater and table olive wastewater. In this study, various initial pH values were tested to investigate the membrane performance concerning pollutant removal. It was observed that maximum COD removal (up to 85%) was recorded at acidic pH values for both types of wastewaters, while the reduction of phenolic compounds was facilitated at high pH values (pH = 12). Andrade et al. (2014) used nanofiltration to treat dairy wastewater. Three feed flow rates were investigated to determine the optimum operating conditions for maximum pollutants removal. However, the concentration of dissolved solids in the effluent was found to be high and this prevented the reuse of the treated wastewater. In another study, Luo et al. (2011) applied a two-stage ultrafiltration and nanofiltration processes to recycle dairy wastewater. Lipids and proteins were removed during the ultrafiltration stage by an Ultracel PLGC UF membrane and the filtrate obtained was concentrated using an NF270 membrane to receive lactose and inorganic salts in retentate. In other cases, reverse osmosis is applied at high pressure to force wastewater across a semipermeable membrane to remove inorganic salts with the further aim of water reuse. Using this technology water recovery could reach 90%–95% from dairy wastewaters as low concentrations of organic matter remain in the effluents (Vourch et al., 2008). Nevertheless, the main disadvantage of membrane technologies is that membrane fouling dramatically effects separation efficiency and, consequently, membrane processing cost. For this reason, a wastewater pretreatment step is necessary to avoid membrane flux reduction and ensure increased membrane life.

The ion exchange process involves the exchange of ions between a liquid and an adsorbent. Anion exchange resins and zeolites are typical ion exchangers that have been applied in the treatment of food industry wastewaters. For example, Víctor-Ortega et al. (2016) examined two continuous fixed-bed ion exchange resins for their ability to separate phenolic compounds from olive mill wastewater. The results obtained from column experiments showed that high phenols recovery (approximately 60%) could be achieved at neutral pH values. Kolakovic et al. (2013) used zeolites to remove COD, NH_4^+-N, NO_3^--N, PO_4^{3-} from dairy wastewater as zeolites are known for their high specific surface areas and adsorption capacities. The maximum removal efficiencies recorded were approximately 70% and 50% for NO_3^--N and COD, respectively, while the maximum removal percentages for PO_4^{3-} and NH_4^+-N were lower. Neutralization is another chemical reaction in which H^+ ions and OH^- ions react quantitatively with each other to form water and salts (Makara and Kowalski, 2015). The combination of different physicochemical methods such as mineralization (i.e., inorganic substances precipitate to the organic matrix) with phosphoric and sulfuric acids, neutralization using lime milk and filtration, for the treatment of pig manure significantly improved the removal efficiency of COD, TN and TP (95%, 80%, and 96%, respectively) in comparison with single methods. This treatment technique also eliminated odor emissions in the liquid and solid fractions. The solid fraction resulting from the above processes could be utilized as an organic fertilizer or as feedstock for biogas production (Makara and Kowalski, 2015).

AOPs, including Fenton oxidation, photocatalysis, ozonation and wet oxidation, have also been investigated for the treatment of food industry wastewaters as they promote the formation of highly reactive hydroxyl radicals (OH) that can oxidize a wide range of pollutants. The Fenton oxidation process is characterized by lower cost in comparison with other AOPs. Pollutant oxidation is performed using H_2O_2 reactants and mainly Fe^{2+} salts as a catalyst, while the concentration of reactants and pollutants, the pH value, temperature and the reaction time, all affect the efficiency of the process (Ochando-Pulido et al., 2013). Prazeres et al. (2013) applied the Fenton oxidation process to posttreat cheese whey wastewater following coagulation-flocculation and $Ca(OH)_2$ precipitation-aerobic digestion stages and recorded an 80% reduction of the initial COD value (300 mg/L). In another study, winery wastewater was treated using the wet oxidation process and the results showed that an oxidant of hydrogen peroxide (H_2O_2) was effective and removed up to 80% of organic matter (Domínguez et al., 2014).

Photocatalysis requires a photoactive semiconductor and UV light for the excitation of semiconductor electrons. Titanium dioxide and silver are the most widely used and effective semiconductors for this purpose (Kanjwal et al., 2015). The treatment of dairy wastewater by photocatalysis using silver-titanium dioxide nanofibers and silver-titanium dioxide nanoparticles in a batch-scale glass reactor resulted in maximum dairy effluent degradation of 95% and 75%, respectively, in a 2-h reaction time (Kanjwal et al., 2015).

Ochando-Pulido et al. (2013) proposed a sequence of Fenton oxidation, coagulation-flocculation, filtration using gravel and olive stones, and finally reserve osmosis for the treatment of effluents obtained from olive washing and olive mill wastewaters. The pH, EC, TSS and COD values of the effluents recorded in the laboratory and pilot-scale experiments were acceptable for water reuse applications. Therefore, a combination of physicochemical techniques to treat food industry wastewaters could improve effluent quality and increase water recycling and reuse.

2.3 Electrochemical methods

Electrochemical processes, such as electrocoagulation (EC), electrochemical oxidation, and electro-Fenton, have been widely applied as suitable methods for the treatment of various types of wastewaters, however their role in the treatment of food industry wastewaters has not been investigated extensively. Of these processes, electrocoagulation is possibly the most advantageous, since it is easy to use, is less expensive than other methods, requires a short reaction time, and generates only small volumes of sludge (Fernandes et al., 2015). Bazrafshan et al. (2012) studied the use of electrocoagulation as a second stage (following chemical coagulation) in the treatment of slaughterhouse wastewater. Both processes achieved high removal efficiencies of the tested pollutants (COD, BOD, TN, and TSS), however, electrocoagulation produced effluents that met with regulatory standards for effluents discharged into the environment. Tsioptsias et al. (2015) also tested electrocoagulation to posttreat biologically treated molasses wastewater for additional reduction of color, COD and nitrogen concentration. Furthermore, a combination of electrocoagulation and electrochemical oxidation in the treatment of dairy wastewaters proved that the COD, TS, and color of raw wastewater can be reduced to 60%, 100%, and 80%, respectively (Chakchouk et al., 2017). Finally, Iglesias et al. (2015) optimized the electro-Fenton process (H_2O_2 and Fe^{2+} are generated in situ resulting in the production of less sludge than Fenton oxidation) at laboratory-scale to generate OH radicals that reduce the organic matter of diluted winery wastewater. It was observed that COD removal reached 80% after 24 h at optimum conditions (iron concentration of 0.72 mM, 15 V, air flow of 1 L/min and pH 3.2.). Although all the above electrochemical processes have been proposed as effective for the rapid treatment of food industry wastewaters, their pilot-scale applications are currently limited. An example of a pilot-scale case study on the electrocoagulation of table olive wastewater is presented in Section 3.3.

2.4 Constructed wetlands

A review of the literature reveals that constructed wetlands have been tested for their effectiveness at treating various food industry wastewaters for several decades, especially for small agroindustries (De La Varga et al., 2016; Sultana et al., 2015). The technology of constructed wetlands combines the use of plants (macrophytes) with physicochemical and biochemical processes and is therefore considered a green technology as it uses natural energy sources. This method has many advantages over conventional biological and physicochemical methods, such as simple operation, low capital and operating costs and good tolerance to shock loadings (Licata et al., 2017; Skrzypiec and Gajewska, 2017; Wang et al., 2018). However, efficient and long-term operation of these systems presupposes the proper design of several operating parameters including: type of wetland, plant species used, surface area planted, pollutant loads, HRT, recirculation of treated wastewater, media types, and pretreatment steps (Stefanakis and Tsihrintzis, 2012; Wu et al., 2015). Three categories of constructed wetlands are distinguished:

(a) free water surface flow (FWS),
(b) subsurface flow (SSF) that includes two different groups: horizontal subsurface flow (HSF) and vertical subsurface flow (VSF), and
(c) hybrid wetlands that combine both VSF and HSF stages and achieve high organic removal efficiencies and significant nitrification rates.

The majority of food industry wastewaters contain significant quantities of suspended solids and these tend to clog the porous media of constructed wetlands when applied directly. To avoid clogging of the porous media, food industry wastewaters should be diluted with tap water or undergo pretreatment steps. Therefore, constructed wetlands are usually used as a tertiary treatment step. The choice of pretreatment step (biological, electrochemical oxidation, coagulation, sludge digestion, settling tanks, stabilization lagoons, septic tanks, or a combination of these, such as solid separation—anaerobic digestion—aerobic oxidation) depends mainly on the type of wastewater to be treated (De La Varga et al., 2016). Various scales (laboratory-, pilot-, and full-scale) of wetlands have been examined for food industry wastewater treatment with surface areas ranging between 0.25 and 7600 m^2 (Sultana et al., 2015; Skrzypiec and Gajewska, 2017; Wang et al., 2018). Studies show that HSF wetlands present similar efficiencies to FWS wetlands for the tertiary treatment of food wastewaters (organic matter: 28%−99%, TKN: 10%−99%, TP: 2%−99%, TSS: 76%−99%), however the Hydraulic retention times (HRTs) applied in HSF systems were lower, ranging from 2 to 60 days, whereas 4−120 days were applied in FWS wetlands (Sultana et al., 2015). VSF wetlands can tolerate up to 10 times higher pollutant loadings (organic matter: 10−6589 gr/(m^2 d), TKN: 0.6−575 g/(m^2 d), TP: 0.08−20 g/(m^2 d), TSS: 35−1836 g/(m^2 d)), than FWS and HSF systems (organic matter: 0.17−376 g/(m^2 d),

TKN: 0.007−2.7 g/(m^2 d), TP: 0.004−4.7 g/(m^2 d), TSS: 0.2−62.4 g/(m^2 d)) and also show efficient removal rates (organic matter: 24%−95%, TKN: 10%−99%, TP: 47%−95%, TSS: 21%−99%) (Herouvim et al., 2011; Sultana et al., 2015). Hybrid wetland systems are of major interest as they have been shown to successfully treat food wastewaters of very high strength (organic matter: 1.28−1500 g/(m^2 d), TKN: 0.3−1500 g/(m^2 d) for, TP: 0.06−40 g/m^2 d, TSS: 1.96−400 g/m^2 d) with notable removal efficiencies (83%−96%, 55%−92%, 52%−96% and 83%−99%, respectively) (Sultana et al., 2015; Wang et al., 2018). However, in many studies (laboratory-, pilot- and full-scale) the final effluents produced contained pollutant concentrations that remain above the permitted EU limits for safe disposal (Sultana et al., 2015). In California two full-scale HSF wetlands were tested to treat winery wastewaters. The first wetland (in a small-production winery near Glen Ellen) was able to completely remove COD (from about 8000 mg/L to 5 mg/L) while the second (serving a moderate-production winery near Hopland) showed COD removal between 49% and 79%, depending on the season (peak season and off season, respectively) (Masi et al., 2015). In the town of Bishoftu in Ethiopia, the Ada milk factory uses full-scale HSF wetlands to treat its wastewaters. However, treated effluents from this factory contain COD concentrations of about 360 mg/L which are higher than the EU permitted limit (Alayu and Yirgu, 2018). Generally, COD concentrations can be lowered by: including a pretreatment step (a simple settling tank for wastewaters with low initial organic loads or a biological system for higher loads), selecting the most appropriate wetland type, together with the optimum HRT-media-vegetation density combination, and finally a posttreatment step to further improve effluent quality (Sultana et al., 2015; Wu et al., 2015).

2.5 Hybrid systems

From the above descriptions of techniques used to treat food industry wastewaters (both experimental and full-scale application), it becomes apparent that it is currently unfeasible to reduce effluent pollutant concentrations to below legislated discharge limits (especially for undiluted high-strength wastewaters) using only a single treatment stage. For this reason, a combination of two or more stages is usually applied to treat food wastewaters in real, working, industrial units. These are termed hybrid systems.

A pilot study on hybrid systems by Alayu and Yirgu (2018) reported that for a cheese whey influent with 34,000−65,700 mg COD/L the first treatment step using an anaerobic rotating biological conductor achieved 46.3%−62.6% COD removal, while the second treatment step using an aerobic fixed fluidized-bed reactor removed COD by up to 95%.

To treat its wastewaters, the St. George Brewery (Addis Ababa, Ethiopia) uses a hybrid system comprising mechanical screening and an equalization tank as the pretreatment step, a full-scale up-flow anaerobic sludge blanket reactor for secondary treatment, and a reaeration treatment unit for posttreatment. This hybrid system is able to treat successfully $2 \cdot 10^5$ L/d (COD: 92.2%, TSS: 87.3%, TN: 11.3%, PO_4^{3-}: 6.61%) with a HRT of 17.5 h. The CH_4 produced during the process is burned by flare (Bula, 2014). Integrated anaerobic and aerobic reactors are also promising treatment systems for high-strength food wastewaters. Anaerobic reactors aim mainly to recover energy and useful by-products, while aerobic reactors provide organic load degradation. Liguri Food Specialties Spa in Chiusanico, Italy, produces products such as basil sauces, creamy sauces, mushrooms and other food products, and employs a hybrid system to treat the wastewaters (100 m^3/d) resulting from production processes. This hybrid system is described in detail by Ghimpusan et al. (2015). For pretreatment, the wastewater passes through screening, a flotation tank and an equalization tank, while secondary treatment comprises two aerobic biological reactors. Following treatment by the second biological reactor, COD removal reaches 80%–90%. An advanced oxidation method (ozonization) was further used for disinfection, while a filter press was applied for sludge treatment. The above system was applied for about 3 years, however, in some cases the produced effluent did not meet with legal requirements and for this reason membrane (MBR) filtration and ozonization were added into the system as posttreatment steps. Following MBR installation COD concentrations at the end of treatment had decreased to almost 163 mg/L (Italian Standards: $<$160 mg COD/L).

De Gisi and Notarnicola (2017) described various case studies of hybrid treatment systems for a variety of food wastewaters and concluded that different strategies should be applied according to the strength of the raw wastewater in question. Pretreatment steps including screening, settling, fat separation and coagulation/flocculation, are used in the majority of cases. The authors proposed that the removal of organic matter nutrients from low strength wastewaters could be performed by membrane technologies, while a combination of biological treatment and membrane technologies or chemical oxidation is suitable for medium and high-strength wastewaters.

Since food industry wastewaters exhibit a wide variety of characteristics (i.e., wide-ranged pollutant concentrations), their treatment is complex and usually involves the combination of different treatment technologies. Nevertheless, an ideal solution would be to develop one simple, low-cost, system which, with only minor modifications, would be able to effectively treat a great variety of food industry wastewaters. A combination of biological trickling filters and constructed wetlands seems to fit this purpose as it presents several advantages such as: high pollutant removal efficiencies even when treating high-strength wastewaters, tolerance to loading shocks, simplicity of operation, low capital and operation costs, and durability (Metcalf and Eddy, 2003; Vayenas, 2011).

3. Case studies of food industry wastewater treatment

This section presents case studies on the treatment of olive mill wastewaters, table olive processing wastewater and dairy wastewaters using primarily biological methods, constructed wetlands and the electrochemical process. All the case studies were carried out by one research group based in Greece (University of Patras) and concentrate on the treatment of wastewaters generated by small, local, agroindustrial businesses located in western Greece. These small agroindustries are somewhat typical of the Mediterranean landscape and economy and the wastewaters they produce are copious in quantity and potentially pose an environmental hazard if not treated before disposal. The research team focuses on creating and implementing systems (applying one or more treatment technologies) that are simple to operate, economic to install and run, environmentally friendly and compact in size, and as such are ideal for the treatment of wastewaters produced by small, low-capacity, agroindustrial units scattered across rural Greece and other countries in the Mediterranean or other regions.

3.1 Biological trickling filters

Biological trickling filters (BTFs) have frequently been used to treat and co-treat a variety of wastewaters originating from small agrofood industries (Michailides et al., 2011; Tatoulis, 2016; Tatoulis et al., 2015, 2016; 2017a, 2018; Tziotzios et al., 2007a,b). Their effectiveness at treating food industry wastewaters was first examined by conducting laboratory-scale experiments where the biofilters comprised Plexiglas tubes filled with hollow plastic tubes used as support media. These laboratory-scale experiments were used to optimize parameters of filter operation before constructing and operating pilot- and full-scale filters. Table 8.2 presents operational parameters for the earlier-mentioned experiments.

Laboratory-scale experiments were all conducted using the draw-fill mode with recirculation, to obtain a completely mixed flow pattern within the bioreactor and sufficient exploitation of the filter, while pilot- and full-scale filters were operated under draw-fill mode or continuous flow mode with recirculation (Michailides et al., 2011; Tatoulis, 2016; Tatoulis et al., 2015, 2016, 2017a, 2018; Tziotzios et al., 2007a,b). All the examined wastewaters were obtained from small, local, food industries: Olive mill wastewater was obtained from a three-phase olive mill near the town of Amfilochia in western Greece, secondary cheese whey from the Papathanasiou A.B.E.E. cheese factory situated near the city of Agrinio in western Greece, and black table olive processing wastewater was obtained from the Ilida A.B.E.E. olive processing plant, also located in Agrinio.

3.1.1 Biological trickling filters and olive mill wastewater

Table 8.2 presents the performance of various treatment stages of olive mill wastewater (OMWW). Tziotzios et al. (2007a,b), presented the first results of treating OMWW using trickling filters as attached growth biological reactors, and focusing on the removal of phenols. The effect of several operating parameters was examined in these experiments (Tziotzios et al., 2007a,b). Initial experiments showed that the optimum support media for OMWW treatment was plastic tubes, as the low porosity of fine gravel led to frequent filter clogging and the spatial restriction of biofilm growth (Tziotzios et al., 2007a). Another crucial parameter effecting filter performance is temperature, as ambient temperatures below 16°C were proved to slow down and reduce filter performance. Nevertheless, if the OMWW has a temperature of above 40°C when added to the filter, a full-scale BTF will maintain a rather stable performance (Tziotzios et al., 2007b).

Table 8.2 Operational parameters and pollutant influent and effluent concentrations recorded in experiments on olive mill wastewater, table olive wastewater and second cheese whey treatment using biological trickling filters.

Wastewater	**Olive mill wastewater**			**Table Olive wastewater**		**Second cheese whey**		
Reference	**Tziotzios et al. (2007a)**	**Tziotzios et al. (2007b)**	**Michailides et al. (2011)**	**Tatoulis et al. (2016); 2017a**	**Tatoulis (2016)**	**Tatoulis et al. (2015)**	**Tatoulis et al. (2015)**	**Tatoulis (2016)**
Parameter								
Experimental Scale	Laboratory	Pilot	Full	Laboratory	Full (Three biological trickling filters in parallel)	Laboratory	Pilot	Full (Three biological trickling filters in parallel)
Total height (cm)	160	160	300	160	210 (each)	160	113	210 (each)
Filters diameter (cm)	9	180	180	9	160 (each)	9	68	160 (each)
Support media depth (cm)	143	120	180	143	190	143	100	190
Support media type	Gravel and plastic material	Plastic material (small-sized/large-sized)	HDPE	Plastic material	HDPE	Plastic material	HDPE	HDPE
Support media diameter (mm)	5.75/16 (length 30.2 mm)	16 (length 30.2 mm)/ 26.5 (length 39.5 mm)	35 (length 38 mm)	16 (length 30.2 mm)	35 (length 38 mm)	16 (length 30.2 mm)	35 (length 38 mm)	35 (length 38 mm)

Support media specific area (m^2/m^3)	1123/500	500/301.9	188	500	188	500	188	188
Porosity	0.36/0.8	0.8/0.95	0.95	0.8	0.95	0.8	0.95	0.95
Reactor mode	Draw and fill	Continuous flow with recirculation	Continuous flow with recirculation	Draw-fill mode with recirculation	Continuous flow with recirculation	Draw-fill mode with recirculation	Draw-fill mode with recirculation	Continuous flow with recirculation
COD influent concentration (mg/L)	–	15,000	43,000	5500–15,000	2300–23,280	11,000–43,000	11,000–23,000	9750–25, 400
COD effluent concentration (mg/L)	–	4500	21,201	1350–3900	550–6900	<150	2300–6601	2200–9100
COD removal load (g/Ld)			21.79	1.24–4.09	2.1–10.92	14.53–26.3	2.66–6.41	1.9–27.9
Phenols influent concentration (mg/L)	500–2750	2900	9500	500–1150	–	–		–
Phenols effluent concentration (mg/L)	0	1305	4311	285–483	–	–		–

HDPE, High Density Polyethylene.

Figure 8.1 Full-scale biological trickling filters used to treat: (A) olive mill wastewater, (B) table olive wastewater, (C) second cheese whey, and (D) the portable container (dimensions 12 × 2.4 × 2.6 m) containing the biological filters.

An effective, stable, full-scale BTF treating OMWW was presented by Michailides et al. (2011) (Fig. 8.1A). Based on the pilot-scale experiments of Tziotzios et al. (2007a,b), Michailides et al. (2011) constructed and operated a full-scale BTF to treat the total wastewater effluent produced by a small olive oil mill in one season (approximately 30 m^3 of OMWW per day). The filter removed 50% of COD and phenol concentrations (Table 8.2) when working with a hydraulic retention time (HRT) of just 24 h.

3.1.2 Biological trickling filters and table olive wastewater

Table olive wastewater (TOWW) is another important by-product of olive processing industries and due to its high salt content (salinity: 58.5−84.0 g/L) (Tatoulis et al., 2016) it cannot be easily treated. Nevertheless, laboratory-scale BTF proved effective at treating TOWW, as removal values of approximately 70% organic matter and 50% phenol content were recorded (Tatoulis et al., 2016, Table 8.2). A full-scale unit that was built and operated at the local, black table olive processing plant of Ilida A.B.E.E. (producing approximately 18−35 m^3 of TOWW per day), showed COD removal efficiencies of between 61% and 82%, depending on the initial COD concentration (from 2300 to 23,280 mg/L) of the raw TOWW (Tatoulis, 2016, Table 8.2). Fig. 8.1B shows the full-scale BTF installed at the processing plant. Although the performance of these filters on TOWW treatment is higher that of suspended growth biological treatment methods (Tatoulis, 2016), the final effluent still contained high concentrations of organic matter and phenols and also retained its color. Therefore, additional post-treatment stages are imperative.

3.1.3 Biological trickling filters and second cheese whey

Second cheese whey (SCW) has extremely high organic matter contents making it a potential environmental hazard if released untreated. Nevertheless, the organics it contains can be biodegraded fairly easily and this makes it especially suitable for treatment using biological methods such as BTF and constructed wetlands.

Tatoulis et al. (2015) presented a series of experiments using BTF to treat SCW that included both laboratory- and pilot-scale applications. The laboratory-scale experiments resulted in complete removal of the organic matter load after 16, 26 and 42 h for initial COD concentrations of 10,000, 22,500 and 40,000 mg/L, respectively. It should be mentioned that these initial COD concentrations were among the highest reported in the literature (Tatoulis et al., 2015, Table 8.2). Pilot-scale filters that were then constructed and operated on-site (at the Papathanasiou cheese factory in western Greece), also succeeded in removing the majority of the organic matter load, as COD concentrations of approximately 2000 mg/L were recorded in the effluent (Tatoulis et al., 2015). The pilot-scale filters were operated continuously for 12 months and their performance was not found to be affected by fluctuations in ambient temperature (Tatoulis et al., 2015). Industrial-scale BTF were then installed and operated in the cheese factory (Fig. 8.1C,D). This industry, with the capacity to process over 15 tons of raw milk per day, produces 30−35 m^3 of dairy wastewaters daily and these contain both SCW and cleaning waters. The industrial-scale filter produced COD removal efficiencies of between 64% and 80%, despite daily fluctuations in initial COD concentrations measured in the influent wastewater. Although the industrial-scale biological filter used in this case study demonstrated significantly high COD removal rates from this high-strength food industry wastewater, organic matter content in the effluent remained above permitted EU limits for safe direct disposal into the environment. For this reason an additional posttreatment stage was proposed (Tatoulis, 2016).

3.2 Constructed wetlands

Constructed wetlands are established treatment solutions for food processing industry wastewaters and are used as a posttreatment stage to further remove pollutant loads from effluents produced by BTF during primary treatment (Table 8.3). Several experiments have been conducted using all constructed wetland types (i.e., FWS, HSF, and vertical flow (VF)), to post-treat olive mill wastewater, table olive wastewater and second cheese whey. These experiments included both pilot- and full-scale wetland units (Table 8.3).

Table 8.3 Operational parameters and pollutant influent and effluent concentrations recorded using constructed wetlands to treat various wastewaters. (OMWW: Olive mill wastewater, TOWW: table olive wastewater, SCW: second cheese whey, VF: vertical Flow, FWS: free water surface flow, HSF: horizontal subsurface flow, CW: constructed wetland, HRT: hydraulic retention time, TKN: total Kjeldahl nitrogen, TP: total phosphorus).

Wastewater	**OMWW**		**TOWW**	**SCW**	
Reference	**Herouvim et al. (2011)**	**Michailides et al. (2015)**	**Tatoulis et al. (2017a)**	**Sultana et al. (2015)**	**Tatoulis et al. (2017b)**
Parameter					
Experimental Scale	Pilot	Full	Pilot	Pilot	Pilot
CW type	VF	FWS	HSF	HSF	HSF
CW area (m^2)	0.37	2050	0.24	1.08	0.24
Support media type	Gravel	Soil	Gravel	Gravel	Plastic material -zeolite
Support media diameter (mm)	0.5–60	–	6	6	6
Vegetation	*Phragmites australis*				
Hydraulic load ($m^3/m^2/d$)	0.08–0.405	0.014	0.007–0.29	0.01–0.11	0.015–0.08
Porosity	–	–	0.29	0.25	0.17–0.46
HRT (d)	–	60	2–8	1–8	2–4
COD influent concentration (mg/L)	14,120	27,400	2750	1200–7200	456–5740
COD effluent concentration (mg/L)	3600	1644	825	108–648	0–760
COD removal load ($gr/m^2/d$)	4900	57	2005	2.2–362.6	14–494.4

Phenols influent concentration (mg/L)	2841	4800	200	–	–
Phenols effluent concentration (mg/L)	657	240	60	–	–
Phenols removal load ($gr/m^2/d$)	750	3.5	145.8	–	–
TKN influent concentration (mg/L)	506	770	–	–	NH_4^+-N 8.77–34.01
TKN effluent concentration (mg/L)	98	15.4	–	–	NH_4^+-N 1.46–11.57
TP influent concentration (mg/L)	95	105	–	–	PO_4^{3-}-P 0.2–29.66
TP effluent concentration (mg/L)	11	5.25	–	–	PO_4^{3-}-P 0.12–4.54

3.2.1 Constructed wetlands and olive mill wastewater

The most efficient type of constructed wetland used to treat OMWW was the VF wetland used by Herouvim et al. (2011) (Fig. 8.2A). In this study, the pilot-scale VF wetland removed extremely high pollutant loads (4900 g/d m^2, 130 g/d m^2, 15 g/d m^2 and 750 g/d m^2 for COD, TKN, OP and phenols, respectively) from pretreated OMWW, while its performance was not influenced significantly by ambient temperature. Nevertheless, pollutant levels (particularly phenols and color) in the posttreated effluent remained above legislation limits and the wastewater could not be directly discharged into the environment (Herouvim et al., 2011).

FWS wetlands were also used to treat pretreated OMWW by Michailides et al. (2015). In this experiment the full-scale FWS wetland (Fig. 8.2B) was constructed by vegetating half the olive mill's original evaporation lagoon. Although the data cover the start-up period only, the wetland removed approximately 80% of the initial COD and phenol content (Table 8.3). These high removal rates were attributed to biological degradation and the effect of vegetation, as photo-oxidation alone could only remove approximately 20% organic matter and phenols (Michailides et al., 2015). Generally speaking, the amounts of pollutants removed by FWS-type wetlands are extremely lower than those removed by VF-type wetlands (Table 8.3). However, the FWS-type can be built in such a way that it can retain the entire volume of OMWW produced by one olive mill in one production season, which will finally be evapotranspirated (Michailides et al., 2015). The wastewater treatment sludge produced from this process could become an alternative source of soil organic matter following stabilization. Research has investigated the compostability of wastewater treatment sludge from the food industry with different bulking agents (e.g., leaf, wheat straw, corncob) (Uçaroğlu and Alkan, 2016).

3.2.2 Constructed wetlands and table olive wastewater

Constructed wetlands of the HSF type have been used to treat TOWW, and proved to be rather effective, as Tatoulis et al. (2017a) reported 70% removal of organic matter and 80% removal of phenol loads from diluted TOWW in pilot-scale experiments (Fig. 8.2C). However, the pollutant removal loads recorded were slightly lower than those recorded in VF-type wetlands, which were used to treat OMWW, which is comparable with TOWW (Table 8.3). The extremely high removal rates recorded in this study were achieved with an HRT of less than four (4) days, which is one of the lowest times applied to constructed wetlands treatment, even for municipal wastewater (Tatoulis et al., 2017b).

3.2.3 Constructed wetlands and second cheese whey

Experiments by Sultana et al. (2015) showed that although BTF could successfully remove the majority of the organic matter load in SCW samples, COD concentrations in the filter effluent were above legislation limits for safe disposal. Therefore, a

Figure 8.2 Constructed wetlands for wastewater treatment: (A) pilot-scale vertical flow, (B) full-scale free water surface units used for olive mill wastewater treatment, (C) pilot-scale horizontal subsurface flow wetland used for table olive wastewater treatment, (B) pilot-scale horizontal subsurface flow wetland for second cheese whey treatment with gravel and zeolite substrate, and (E) pilot-scale horizontal subsurface flow wetland for second cheese whey treatment with plastic tubes and zeolite substrate.

posttreatment stage comprising constructed wetlands was applied to the SCW treatment system. These applications showed that the pilot-scale wetlands:

(a) could reduce and maintain an effluent COD level constantly below legislation limits when the influent concentration was below 3500 mg/L, and
(b) the wetlands could be operated with a minimum HRT of two (2) days and their performance was stable (Sultana et al., 2015).

Although constructed wetlands have been proved to be effective treatment technologies for a variety of food processing wastewaters with minimum construction and operating costs, they do have two major drawbacks:

(a) their high surface area requirements compared to other biological treatment methods, and
(b) their substrate material is prone to clogging.

To overcome these drawbacks Tatoulis et al. (2017b) used plastic tubes and zeolite as substrate material in HSF-type wetlands (Fig. 8.2D,E) that were used to treat SCW. This substrate combination increased porosity and hydraulic load three-fold when compared to a standard wetland substrate comprising gravel and zeolite. In addition, the wetland's pollutant removal efficiency was not affected, as COD removal rates were similar in both wetlands that contained plastic tubes and those that contained gravel (Tatoulis et al., 2017b). Furthermore, the use of a zeolite substrate increased nutrient removal efficiency (Table 8.3).

3.3 Electrocoagulation treatment

Treated effluents from BTF and constructed wetlands still contain significant loads of organic matter, phenols and color, thus, an additional posttreatment stage is imperative to lower pollutant concentrations below legislation limits and fully decolor final effluents. The process of electrocoagulation (EC) is a potential posttreatment technology, as it can fully mineralize and decolorize a variety of wastewaters in just a few minutes.

As mentioned in Section 2.3, the pilot application of EC for the treatment of food industry wastewaters is currently limited. One set of experiments was performed in a small table olive processing plant (Ilida A.B.E.E., western Greece) using a portable EC pilot unit (Fig. 8.3A) to investigate the removal percentage of COD and color from raw (not pretreated) TOWW. The portable unit comprises an electrolytic cell with 16 aluminum (eight anode and eight cathode) electrode plates (15 cm long and 40 cm wide) with interelectrode gaps of 2 cm (Fig. 8.3B). The pilot unit was operated for 100 min at current density 3 mA/cm^2 using a 2.6 V power supply. The working volume of the reactor was 0.2 m^3 and the experiments were conducted at ambient temperature. According to the results obtained, the maximum COD and color removal was 25% and 95%, respectively (Fig. 8.3C; unpublished data). Future research will focus on reactor operation in terms of current density and electrode types, and reaction time will be optimized for maximum EC efficiency.

Tatoulis et al. (2016) posttreated TOWW effluent (with COD concentrations of about 5100 mg/L) obtained from a biological filter with electrooxidation and achieved

Figure 8.3 The electrocoagulation process showing the: (A) pilot-scale unit, (B) electrode configuration, and (C) effluent from treated table olive wastewater.

full mineralization and decolorization in 240 min at 187.5 mA cm^2. The treatment time was reduced to 60 min at 250 mA cm^{-2}. However, when the same effluent passed through a constructed wetland prior to the electrooxidation process, COD concentrations fell below legislation limits and the effluent was fully decolorized after 45 min at 62.5 mA/cm^2 (initial COD concentration of the electrooxidation system was c. 450 mg/L) (Tatoulis et al., 2017a).

4. Reuse of treated food industry wastewaters

Due to the high demand for fresh, clean, water and concerns to reduce environmental pollution by minimizing wastewater discharge, the reuse of treated food processing wastewaters has become a significant issue. However, several factors prevent recycling these treated wastewaters within the food industry itself or public spaces (parks, etc.), such as microbiological risks and public health regulations. Current research shows that there is potential in reusing water within the food production system, depending on the type of wastewater, volume of waste, the available treatment technology and the effluent standards (Meneses et al., 2017).

The reuse of food industry wastewaters for agricultural irrigation could clearly reduce water demands for irrigation (especially in arid agricultural areas), and has therefore been extensively examined (Tarantino et al., 2017). Several studies show that many treated food wastewaters can be used for irrigation without significantly increasing potential health risks, and simultaneously provide significant amounts of valuable nutrients to soils and positively affect crop growth. For example, Prazeres et al. (2014) found that tomato plants irrigated with pretreated (by precipitation) cheese whey wastewater presented improved fruit quality. Sawalha et al. (2014) examined the possibility of using treated (filtered) OMWW to irrigate tomato crops and revealed that applying a 10% dilution of the filtered wastewater significantly increased plant nutrients, provided strong crop structure, longer, greener leaves and longer stems.

Some studies have explored (mainly in laboratory- and pilot-scale bioreactors) the use of treated food industry wastewater as growth medium for algal growth (Franchino et al., 2013; Tsolcha et al., 2016, 2017, 2018). Specifically, Franchino et al. (2013) examined the cultivation of three microalgal strains (*Neochloris oleoabundans*, *Chlorella vulgaris*, and *Scenedesmus obliquus*) in anaerobically digested cheese whey. All three strains were able to completely remove nitrogen and phosphate in 11 days while microalgal biomass production offers real opportunities for addressing issues such CO_2 sequestration, biofuel production and wastewater treatment. Tsolcha et al. (2016, 2017; 2018) studied the operation of a mixed cyanobacterial/algal system using aerobically treated wastewaters originating from cheese, raisin and wine industries. They conclude that agroindustrial wastewaters can be used as substrate for cyanobacterial/microalgal biomass growth and simultaneously produce lipids suitable for biodiesel production (i.e., 14.8% lipid content in dry biomass using SCW). However, as food industry wastewater characteristics vary not only between industries but also within each industry, optimization of all growth parameters is required before large-scale application.

One alternative reuse of treated food processing waters is their use as a potential carbon source for the treatment of industrial wastewaters. An advantage of using biological filters and constructed wetlands is their ability to co-treat food wastewaters with other industrial effluents (Sultana et al., 2015; Tatoulis et al., 2018). Tatoulis et al. (2018) tested two food wastewaters (i.e., second cheese whey and winery wastewater) as an organic matter source for the biological reduction of Cr(VI) in laboratory-scale biological filters with the same characteristics as those presented in Table 8.2. The biofilters achieved complete Cr(VI) biological reduction for all the initial COD

(1000–25,000 mg/L) and Cr(VI) (5–110 mg/L) concentrations tested, with extremely high Cr(VI) removal rates (8.4 mg Cr(VI)/d) (Tatoulis et al., 2018). In these experiments the Cr(VI) was removed by biological reduction and biosorption, as abiotic reduction and Cr(VI) absorption were negligible. Increased COD concentrations led to increased Cr(VI) removal rates, which suggests that a combined treatment system for food wastewaters and Cr(VI) is a viable solution. Although complete Cr(VI) removal was achieved, COD concentrations in the effluent were high, therefore a second biofilter was used as a posttreatment stage and removed over 97.0% and 90.5% of COD in the SCW and winery effluents, respectively (Tatoulis et al., 2018).

Sultana et al. (2015) investigated the ability of constructed wetlands to co-treat Cr(VI) with SCW in pilot-scale wetlands similar to those presented in Table 8.3. The cheese whey proved to facilitate biological Cr(VI) reduction and biosorption, as complete Cr(VI) removal was recorded for both planted and unplanted units (Sultana et al., 2015). In contrast to Cr(VI), organic matter removal was negatively affected by the presence of cheese whey, as COD removal rates were lower than those recorded in constructed wetlands treating SCW alone and a posttreatment stage was required to further remove organic matter (Sultana et al., 2016).

5. Conclusions

Large quantities of wastewaters are generated from food processing industries. The characteristics of these wastewaters vary significantly according to the raw materials and processes used; therefore different wastewater treatment technologies are required to ensure effluents comply with environmental legislation before being discharged. Before the 1990s, the basic practices for medium- and large-scale food processing industries to treat their wastewaters included aeration (activated sludge or lagoons) and settling tanks, while for small local units these techniques were not economically viable. However, due to stringent water quality standards for treated food wastewaters, new processes and technologies have been developed, offering improved treatment ability, lower operation cost, and environmental protection, even for small-capacity industries.

Obtaining pollutant concentrations under discharge limits is very difficult to achieve when treating food industry wastewaters with a single-step technology, especially for undiluted high-strength wastewaters. Therefore, hybrid systems combining two or more steps are usually applied in full-scale applications. Effective, simple and low-cost treatment systems need to be developed to, with minor modifications, treat a great variety of food wastewaters. The case studies presented in this chapter show that a hybrid system comprising a biological trickling filter and a constructed wetland could fulfill this role. These two treatment technologies can remove the majority of the pollutant loads even from high-strength wastewaters. This hybrid treatment system could also include a physicochemical polishing stage (e.g., electrooxidation) to further reduce pollutant concentrations below legislation limits and simultaneously remove color and other substances. Low and medium strength wastewaters could be successfully treated using a combination of a biological trickling filter and electrooxidation, or constructed wetlands and electrooxidation.

Acknowledgments

This study was financially supported by “INVALOR: Research Infrastructure for Waste Valorization and Sustainable Management” (MIS 5002495), which is implemented under the Action “Reinforcement of the Research and Innovation Infrastructure,” funded by the Operational Program “Competitiveness, Entrepreneurship and Innovation” (NSRF 2014-2020) and co-financed by Greece and the European Union (European Regional Development Fund).

References

Abdallh, M.N., Abdelhalim, W.S., Abdelhalim, H.S., 2016. Industrial wastewater treatment of food industry using best techniques. Int. J. Eng. Sci. Invent. 5 (8), 15–28.

Abdulgader, M., Yu, J., Williams, P., Zinatizadeh, A., 2007. A review of the performance of aerobic bioreactors for treatment of food processing wastewater. In: Proceedings of the International Conference on Environmental Management, Engineering, Planning and Economics Skiathos, June 24–28, pp. 1131–1136.

Affes, M., Aloui, F., Hadrich, F., Loukil, S., Sayadi, S., 2017. Effect of bacterial lipase on anaerobic co-digestion of slaughterhouse wastewater and grease in batch condition and continuous fixed-bed reactor. Lipids Health Dis. 16 (1), 195. https://doi.org/10.1186/s12944-017-0587-2.

Alayu, E., Yirgu, Z., 2018. Advanced technologies for the treatment of wastewaters from agro-processing industries and cogeneration of by-products: a case of slaughterhouse, dairy and beverage industries. Int. J. Environ. Sci. Technol. 15 (7), 1581–1596.

Andrade, L.H., Mendes, F.D.S., Espindola, J.C., Amaral, M.C.S., 2014. Nanofiltration as tertiary treatment for the reuse of dairy wastewater treated by membrane bioreactor. Separ. Purif. Technol. 126, 21–29.

Austermann-Haun, U., Meyer, H., Seyfried, C.F., Rosenwinkel, K.-H., 1999. Full scale experiences with anaerobic/aerobic treatment plants in the food and beverage industry. Water Sci. Technol. 40 (1), 305–312.

Barbera, M., Gurnari, G., 2018. Wastewater Treatment and Reuse in the Food Industry. Springer Briefs in Molecular Science. https://doi.org/10.1007/978-3-319-68442-0.

Bazrafshan, E., Mostafapoor, F.K., Soori, M.M., Mahvi, A.H., 2012. Application of combined chemical coagulation and electro-coagulation process for carwash wastewater treatment. Fresenius Environ. Bull. 21 (9 A), 2694–2701.

Bolzonella, D., Cecchi, F., Pavan, P., 2007. Treatment of food processing wastewater. In: Woodhead Publishing Series in Food Science, Technology and Nutrition, Handbook of Waste Management and Co-product Recovery in Food Processing, vol. 1. Woodhead Publishing Editor(s), Keith Waldron, pp. 573–596. https://doi.org/10.1533/9781845692520.5.573.

Bories, A., 2000. A Novel Bio-Physical Way for the Treatment and Volarisation of Wastewaters from Food Industries. Proceedings 1st World Conference on Biomass for Energy and Industry, Seville, Spain, pp. 5–9.

Bula, K., 2014. Treatment and Biogas Performance Efficiency of St. George Brewery Full Scale Wastewater after Treatment System. Published M.Sc. thesis. Addis Ababa University, Ethiopia.

Cecconet, D., Molognoni, D., Callegari, A., Capodaglio, A.G., 2018. Agro-food industry wastewater treatment with microbial fuel cells: energetic recovery issues. Int. J. Hydrog. Energy 43 (1), 500–511.

Chakchouk, I., Elloumi, N., Belaid, C., Mseddi, S., Chaari, L., Kallel, M., 2017. A combined electrocoagulation-electrooxidation treatment for dairy wastewater. Braz. J. Chem. Eng. 34 (1), 109–117.

Cheng, H., Tian, G., Liu, J., 2013. Enhancement of biomass productivity and nutrients removal from pretreated piggery wastewater by mixotrophic cultivation of *Desmodesmus* sp. CHX1. Desalination Water Treat. 51, 7004–7011.

Chowdhury, P., Viraraghavan, T., Srinivasan, A., 2010. Biological treatment processes for fish processing wastewater – a review. Bioresour. Technol. 101, 439–449.

Cruz-Salomón, A., Ríos-Valdovinos, E., Pola-Albores, F., Meza-Gordillo, R., Lagunas-Rivera, S., Ruíz-Valdiviezo, V.M., 2017. Anaerobic treatment of agro-industrial wastewaters for COD removal in expanded granular sludge bed bioreactor. Biofuel Res. J. 4 (4), 715–720.

Cuff, G., Turcios, A.E., Mohammad-Pajooh, E., Kujawski, O., Weichgrebe, D., Rosenwinke, K.H., 2018. High-rate anaerobic treatment of wastewater from soft drink industry: methods, performance and experiences. J. Environ. Manag. 220, 8–15. https://doi.org/10.1016/j.jenvman.2018.05.015.

De Gisi, S., Notarnicola, M., 2017. Industrial wastewater treatment. In: Abraham, M. (Ed.), Encyclopedia of Sustainable Technologies, Chapter: Module in Earth System and Environmental Sciences, first ed. Publisher: Elsevier Ltd, pp. 23–42. https://doi.org/10.1016/B978-0-12-409548-9.10167-8.

De Godos, I., Blanco, S., García-Encina, P.A., Becares, E., Muñoz, R., 2009. Long-term operation of high rate algal ponds for the bioremediation of piggery wastewaters at high loading rates. Bioresour. Technol. 100 (19), 4332–4339.

De La Varga, D., Soto, M., Arias, C.A., Oirschot van, D., Kilian, R., Pascual, A., Álvarez, J.A., 2016. Constructed wetlands for industrial wastewater treatment and removal of nutrients. In: del Rio, A.V., Gómez, J.L.C., Corral, A.M. (Eds.), Technologies for the Treatment and Recovery of Nutrients from Industrial Wastewater. Publisher: IG Global, pp. 202–230. https://doi.org/10.4018/978-1-5225-1037-6.ch008.

De Mendonça, H.V., Ometto, J.P.H.B., Otenio, M.H., dos Reis, A.J.D., Marques, I.P.R., 2017. Bioenergy recovery from cattle wastewater in an UASB-AF hybrid reactor. Water Sci. Technol. 76 (9), 2268–2279. https://doi.org/10.2166/wst.2017.325.

De Prá, M.C., Kunz, A., Bortoli, M., Perondi, T., Chini, A., 2012. Simultaneous removal of TOC and TSS in swine wastewater using the partial nitritation process. J. Chem. Technol. Biotechnol. 87 (12), 1641–1647. https://doi.org/10.1002/jctb.3803.

Değermenci, N., Cengiz, İ., Yildiz, E., Nuhoglu, A., 2016. Performance investigation of a jet loop membrane bioreactor for the treatment of an actual olive mill wastewater. J. Environ. Manag. 184 (2), 441–447.

Di Caprio, F., Altimaria, P., Iaquaniellob, G., Torob, L., Pagnanellia, F., 2018. T. obliquus mixotrophic cultivation in treated and untreated olive mill wastewater. Chem. Eng. 64, 625–630. https://doi.org/10.3303/CET1864105.

Domínguez, C.M., Quintanilla, A., Casas, J.A., Rodriguez, J.J., 2014. Treatment of real winery wastewater by wet oxidation at mild temperature. Separ. Purif. Technol. 129, 121–128.

Economou, C.N., Marinakis, N., Moustaka-Gouni, M., Kehayias, G., Aggelis, G., Vayenas, D.V., 2015. Lipid production by the filamentous cyanobacterium *Limnothrix* sp. growing in synthetic wastewater in suspended- and attached-growth photobioreactor systems. Ann. Microbiol. 65 (4), 1941–1948.

El Shimi, H.I., Mostafa, S.S., 2016. Phycoremediation of Olive Mill Wastewater (OMW) using cyanobacteria for sustainable biofertilizer and biofuel production. In: International Conference on Pollution Control & Sustainable Environment, 11, pp. 10259–10272. https://doi.org/10.4172/2375-4397.C1.002.

El-Abbassi, A., Kiai, H., Raiti, J., Hafidi, A., 2014. Application of ultrafiltration for olive processing wastewaters treatment. J. Clean. Prod. 65, 432–438.

Emara, M.M., Abd El-Razek, A.M.A., Sayed Ahmed, A.A.M., 2017. Industrial food processing wastewater treatment by modified moving bed biofilm reactor (MBBR). Int. J. Sci. Eng. Res. 8 (1), 929–934.

Falletti, L., Conte, L., Zaggia, A., Battistini, T., Garosi, D., 2015. Food industry wastewater treatment plant based on flotation and MBBR. Mod. Environ. Sci. Eng 1, 94–98. https://doi.org/10.15341/mese(2333-2581)/02.01.2015/006.

Farizoglu, B., Keskinler, B., 2006. Sludge characteristics and effect of crossflow membrane filtration on membrane fouling in a jet loop membrane bioreactor (JLMBR). J. Membr. Sci. 279 (1–2), 578–587.

Farizoglu, B., Uzuner, S., 2011. The investigation of dairy industry wastewater treatment in a biological high performance membrane system. Biochem. Eng. J. 57, 46–54.

Farraji, H., Zaman, N.Q., Aziz, H.A., Aqeel, M.A., Mojiri, A., Mohajeri, P., 2015. Enhancing BOD/COD ratio of POME treatment in SBR system. Appl. Mech. Mater. 802, 437–442. DOI: 10.4028/www.scientific.net/AMM.802.437.

Fernandes, A., Pacheco, M.J., Ciríaco, L., Lopes, A., 2015. Review on the electrochemical processes for the treatment of sanitary landfill leachates: present and future. Appl. Catal. B Environ. 176–177, 183–200.

Franchino, M., Comino, E., Bona, F., Riggio, V.A., 2013. Growth of three microalgae strains and nutrient removal from an agro-zootechnical digestate. Chemosphere 92, 738. https://doi.org/10.1016/j.chemosphere.2013.04.023.

Ganesh, R., Rajagopal, R., Torrijos, M., Thanikal, J.M., Ramanujam, R., 2010. Anaerobic treatment of winery wastewater in fixed bed reactors. Bioproc. Biosyst. Eng. 33, 619–628.

Gavrilescu, M., Macoveanu, M., 1999. Process engineering in biological aerobic wastewater treatment. Acta Biotechnol. 19 (2), 111–145.

Ghimpusan, M., Nechifor, G.D., Nechifor, A.C., Passeri, P., 2015. Performance of combination of treatment processes for food industry wastewater depuration. In: International Conference, IWWATV.

Gonçalves, M.R., Costa, J.C., Marques, I.P., Alves, M.M., 2012. Strategies for lipids and phenolics degradation in the anaerobic treatment of olive mill wastewater. Water Res. 46, 1684–1692.

Hena, S., Fatimah, S., Tabassum, S., 2015. Cultivation of algae consortium in a dairy farm wastewater for biodiesel production. Water Resour. Ind. 10, 1–14.

Herouvim, E., Akratos, C.S., Tekerlekopoulou, A., Vayenas, D.V., 2011. Treatment of olive mill wastewater in pilot-scale vertical flow constructed wetlands. Ecol. Eng. 37 (6), 931–939.

Hung, Y.-T., Kajitvichyanukul, P., Wang, L.K., 2008. Advances in anaerobic systems for organic pollution removal from food processing wastewater. In: Klemes, J., Smith, R., Kim, J.-K., CRC (Eds.), Handbook of Water and Energy Management in Food Processing, first ed. Woodhead Publishing Limited, pp. 755–775. https://doi.org/10.1533/9781845694678.5.755.

Iglesias, O., Meijide, J., Bocos, E., Sanromán, M.Á., Pazos, M., 2015. New approaches on heterogeneous electro-Fenton treatment of winery wastewater. Electrochim. Acta 169, 134–141.

Ioannou, L.A., Puma, G.L., Fatta-Kassinos, D., 2015. Treatment of winery wastewater by physicochemical, biological and advanced processes: a review. J. Hazard Mater. 286, 343–368.

Jaouad, Y., Villain-Gambier, M., Mandi, L., Marrot, B., Ouazzani, N., 2018. Key process parameters involved in the treatment of olive mill wastewater by membrane bioreactor. Environ. Technol. https://doi.org/10.1080/09593330.2018.1464064.

Jaszczyszyn, K., Góra, W., Dymaczewski, Z., Borowiak, R., 2018. A comparative analysis of selected wastewater pretreatment processes in food industry, E3S Web Conf. In: The First Conference of the International Water Association IWA for Young Scientist in Poland "Water, Wastewater and Energy in Smart Cities", vol. 30. https://doi.org/10.1051/e3sconf/20183002003.

Kanjwal, M.A., Barakat, N.A.M., Chronakis, I.S., 2015. Photocatalytic degradation of dairy effluent using $AgTiO_2$ nanostructures/polyurethane nanofiber membrane. Ceram. Int. 41 (8), 9615–9621.

Klemes, J., Smith, R., Kim, J., 2008. Handbook of Water and Energy Management in Food Processing, first ed. Woodhead Publishing Limited, CRC, pp. 1–1029.

Kolakovic, S., Stefanovic, D., Milicevic, D., Trajkovic, S., Milenkovic, S., Kolakovic, S., Andjelkovic, L., 2013. Effects of reactive filters based on modified zeolite in dairy industry wastewater treatment process. Chem. Ind. Chem. Eng. Q. 19 (4), 583–592.

Koutra, E., Economou, C.N., Tsafrakidou, P., Kornaros, M., 2018. Bio-based products from microalgae cultivated in digestates. Trends Biotechnol. 36 (8), 819–833.

Kushwaha, J.P., Srivastava, V.C., Mall, I.D., 2011. An overview of various technologies for the treatment of dairy wastewaters. Crit. Rev. Food Sci. Nutr. 51 (5), 442–452. https://doi.org/10.1080/10408391003663877.

Li, W.-Y., Peng, Z.-P., Yu, J.-H., Huang, J.-C., Xu, P.-Z., Yang, S.-H., 2013. Wastewater pollution characteristics from typical intensive pig farms in the Pearl River Delta and its ecological risk assessment. Huan. Jing. Ke. Xue. 34 (10), 3963–3968.

Licata, M., Tuttolomondo, T., Virga, G., Leto, C., La Bella, S., 2017. The use of constructed wetlands for the treatment of agro-industrial wastewater – a case study in a dairy-cattle farm in Sicily (Italy). Desalination Water Treat. 76, 300–310. https://doi.org/10.5004/dwt.2017.20720.

Loloei, M., Nekonam, G., Alidadi, H., Kor, Y., 2014. Study of the coagulation process in wastewater treatment of dairy industries. Int. J. Environ. Health Eng. 3 (1), 12.

Lu, W., Wang, Z., Wang, X., Yuan, Z., 2015. Cultivation of Chlorella sp. using raw diary wastewater for nutrient removal and biodiesel production: characteristics comparison of indoor bench-scale and outdoor pilot-scale cultures. Bioresour. Technol. 192, 382–388.

Luo, J., Ding, L., Qi, B., Jaffrin, M.Y., Wan, Y., 2011. A two-stage ultrafiltration and nanofiltration process for recycling dairy wastewater. Bioresour. Technol. 102 (16), 7437–7442.

Makara, A., Kowalski, Z., 2015. Pig manure treatment and purification by filtration. J. Environ. Manag. 161, 317–324.

Markou, G., 2015. Fed-batch cultivation of *Arthrospira* and *Chlorella* in ammonia-rich wastewater: optimization of nutrient removal and biomass production. Bioresour. Technol. 193, 35–41.

Masi, F., Rochereau, J., Troesch, S., Ruiz, I., Soto, M., 2015. Wineries wastewater treatment by constructed wetlands: a review. Water Sci. Technol. 71 (8), 1113–1127. https://doi.org/10.2166/wst.2015.061.

Meneses, Y.E., Stratton, J., Flores, R.A., 2017. Water reconditioning and reuse in the food processing industry: current situation and challenges. Trends Food Sci. Technol. 61, 72–79. https://doi.org/10.1016/j.tifs.2016.12.008.

Metcalf, Eddy, I., 2003. Wastewater Engineering: Treatment and Reuse, fourth ed. McGraw-Hill, New York.

Michailides, M., Panagopoulos, P., Akratos, C.S., Tekerlekopoulou, A.G., Vayenas, D.V., 2011. A full-scale system for aerobic biological treatment of olive mill wastewater. J. Chem. Technol. Biotechnol. 86 (6), 888–892.

Michailides, M., Tatoulis, T., Sultana, M.-Y., Tekerlekopoulou, A., Konstantinou, I., Akratos, C.S., Pavlou, S., Vayenas, D.V., 2015. Start-up of a free water surface constructed wetland for treating olive mill wastewater. [Izgradnja vestacke mocvare u cilju tretiranja otpadne vode nastale usled rada mlina za masline]. Hem. Ind. 69 (5), 577–583.

Mohseni-Bandpi, A., Bazari, H., 2004. Biological treatment of dairy wastewater by sequencing batch reactor. Iran. J. Environ. Health Sci. Eng. 1 (2), 65–69.

Ochando-Pulido, J.M., Hodaifa, G., Victor-Ortega, M.D., Rodriguez-Vives, S., Martinez-Ferez, A., 2013. Reuse of olive mill effluents from two-phase extraction process by integrated advanced oxidation and reverse osmosis treatment. J. Hazard Mater. 263, 158–167.

Onet, C., 2010. Characteristics of the Untreated Wastewater Produced by Food Industry, 4V. Chronicles of the University of Oradea, University of Oradea, Faculty of Environmental Protection, pp. 709–714.

Papaphilippou, P.C., Yiannapas, C., Politi, M., Daskalaki, V.M., Michael, C., Kalogerakis, N., Fatta-Kassinos, D., 2013. Sequential coagulation-flocculation, solvent extraction and photo-Fenton oxidation for the valorization and treatment of olive mill effluent. Chem. Eng. J. 224 (1), 82–88.

Prazeres, A.R., Carvalho, F., Rivas, J., 2013. Fenton-like application to pretreated cheese whey wastewater. J. Environ. Manag. 129, 199–205.

Prazeres, A.R., Carvalho, F., Rivas, J., Patanita, M., Dôres, J., 2014. Reuse of pretreated cheese whey wastewater for industrial tomato production (*Lycopersicon esculentum* Mill.). Agric. Water Manag. 140, 87–95.

Prazeres, A.R., Carvalho, F., Rivas, F.J., 2012. Cheese whey management: a review. J. Environ. Manag. 110 (48), 68.

Queiroz, M.I., Hornes, M.O., Gonçalves da Silva Manetti, A., Zepka, L.Q., Jacob-Lopes, E., 2013. Fish processing wastewater as a platform of the microalgal biorefineries. Biosyst. Eng. 115, 195–202.

Rajagopal, R., Saady, N.M.C., Torrijos, M., Thanikal, J.V., Hung, Y.-T., 2013. Sustainable agro-food industrial wastewater treatment using high rate anaerobic process. Water 5 (1), 292–311.

Rajakumar, R., Meenambal, T., Banu, J.R., Yeom, I.T., 2011. Treatment of poultry slaughterhouse wastewater in UAF under low upflow velocity. Int. J. Environ. Sci. Technol. 8 (1), 149–158.

Rajakumar, R., Meenambal, T., Saravanan, P.M., Ananthanarayanan, P., 2012. Treatment of poultry slaughterhouse wastewater in hybrid upflow anaerobic sludge blanket reactor packed with pleated polyvinylchloride rings. Bioresour. Technol. 103, 116–122.

Raposo, M.F.D.J., Oliveira, S.E., Castro, P.M., Bandarra, N.M., Morais, R.M., 2010. On the utilization of microalgae for brewery effluent treatment and possible applications of the produced biomass. J. Inst. Brew. 116 (3), 285–292.

Sawalha, H., Houshia, O., Hussein, A., Mahariq, Q., Khader, A., 2014. Scope of using treated olive mill wastewater in tomato production. Int. J. Agric. For. 4 (4), 304–309. https://doi.org/10.5923/j.ijaf.20140404.07.

Schwarzenbeck, N., Borges, J.M., Wilderer, P.A., 2005. Treatment of dairy effluents in an aerobic granular sludge sequencing batch reactor. Appl. Microbiol. Biotechnol. 66, 711–718.

Shams, D.F., Singhal, N., Elefsiniotis, P., 2018. Effect of feed characteristics and operational conditions on treatment of dairy farm wastewater in a coupled anoxic-upflow and aerobic system. Biochem. Eng. J. 133, 186–195. https://doi.org/10.1016/j.bej.2018.02.012.

Sirianuntapiboon, S., Jeeyachok, N., Larplai, R., 2005. Sequencing batch reactor biofilm system for treatment of milk industry wastewater. J. Environ. Manag. 76, 177–183.

Skrzypiec, K., Gajewska, M.H., 2017. The use of constructed wetlands for the treatment of industrial wastewater. J. Water Land Dev. 34 (1), 233–240. https://doi.org/10.1515/jwld-2017-0058.

Stefanakis, A.I., Tsihrintzis, V.A., 2012. Effects of loading, resting period, temperature, porous media, vegetation and aeration on performance of pilot-scale vertical flow constructed wetlands. Chem. Eng. J. 181–182, 416–430.

Sultana, M.-Y., Mourti, C., Tatoulis, T., Akratos, C.S., Tekerlekopoulou, A.G., Vayenas, D.V., 2016. Effect of hydraulic retention time, temperature, and organic load on a horizontal subsurface flow constructed wetland treating cheese whey wastewater. J. Chem. Technol. Biotechnol. 91, 726–732.

Sultana, M.-Y., Tatoulis, T.I., Akratos, C.S., Tekerlekopoulou, A.G., Vayenas, D.V., 2015. Effect of operational parameters on the performance of a horizontal subsurface flow constructed wetland treating secondary cheese whey and Cr(VI) wastewater. Int. J. Civ. Struct. Eng. 2 (1) [ISSN: 2372-3971].

Tarantino, E., Disciglio, G., Gatta, G., Libutti, A., Frabboni, L., Gagliardi, A., Tarantino, A., 2017. Agro-industrial treated wastewater reuse for crop irrigation: implication in soil fertility. Chem. Eng. Trans. 58, 679–684.

Tatoulis, T., 2016. Biological Treatment of Agro-Industrial Wastewaters. Ph.D. Thesis. University of Patras, School of Engineering, Department of Environmental and Natural Resources Management (in Greek with English summary).

Tatoulis, T.I., Tekerlekopoulou, A.G., Akratos, C.S., Pavlou, S., Vayenas, D.V., 2015. Aerobic biological treatment of second cheese whey in suspended and attached growth reactors. J. Chem. Technol. Biotechnol. 90, 2040–2049. https://doi.org/10.1002/jctb.4515.

Tatoulis, T.I., Zapantiotis, S., Frontistis, Z., Akratos, C.S., Tekerlekopoulou, A.G., Pavlou, S., Mantzavinos, D., Vayenas, D.V., 2016. A hybrid system comprising an aerobic biological process and electrochemical oxidation for the treatment of black table olive processing wastewaters. Int. Biodeterior. Biodegrad. 109, 104–112.

Tatoulis, T., Akratos, C.S., Tekerlekopoulou, A.G., Vayenas, D.V., Stefanakis, A.I., 2017b. A novel horizontal subsurface flow constructed wetland: reducing area requirements and clogging risk. Chemosphere 186, 257–268.

Tatoulis, T., Stefanakis, A., Frontistis, Z., Akratos, C.S., Tekerlekopoulou, A.G., Mantzavinos, D., Vayenas, D.V., 2017a. Treatment of table olive washing water using trickling filters, constructed wetlands and electrooxidation. Environ. Sci. Pollut. Control Ser. 24 (2), 1085–1092.

Tatoulis, T.I., Michailides, M.K., Tekerlekopoulou, A.G., Akratos, C.S., Pavlou, S., Vayenas, D.V., 2018. Simultaneous treatment of agro-industrial and industrial wastewaters: case studies of Cr(VI)/second cheese whey and Cr(VI)/winery effluents. Water (Switzerland) 10 (4) art. no. 382.

Tsioptsias, C., Petridis, D., Athanasakis, N., Lemonidis, I., Deligiannis, A., Samaras, P., 2015. Post-treatment of molasses wastewater by electrocoagulation and process optimization through response surface analysis. J. Environ. Manag. 164, 104–113.

Tsolcha, O.N., Tekerlekopoulou, A.G., Akratos, C.S., Bellou, S., Aggelis, G., Katsiapi, M., Moustaka-Gouni, M., Vayenas, D.V., 2016. Treatment of second cheese whey effluents using a *Choricystis*-based system with simultaneous lipid production. J. Chem. Technol. Biotechnol. 91 (8), 2349–2359.

Tsolcha, O.N., Tekerlekopoulou, A.G., Akratos, C.S., Aggelis, G., Genitsaris, S., Moustaka-Gouni, M., Vayenas, D.V., 2017. Biotreatment of raisin and winery wastewaters and simultaneous biodiesel production using a *Leptolyngbya*-based microbial consortium. J. Clean. Prod. 148, 185–193.

Tsolcha, O.N., Tekerlekopoulou, A.G., Akratos, C.S., Antonopoulou, G., Aggelis, G., Genitsaris, S., Moustaka-Gouni, M., Vayenas, D.V., 2018. A *Leptolyngbya*-based microbial consortium for agro-industrial wastewaters treatment and biodiesel production. Environ. Sci. Pollut. Control Ser. https://doi.org/10.1007/s11356-018-1989-z.

Tziotzios, G., Economou, C.N., Lyberatos, G., Vayenas, D.V., 2007a. Effect of the specific surface area and operating mode on biological phenol removal using packed bed reactors. Desalination 211 (1–3), 128–137.

Tziotzios, G., Michailakis, S., Vayenas, D.V., 2007b. Aerobic biological treatment of olive mill wastewater by olive pulp bacteria. Int. Biodeterior. Biodegrad. 60 (4), 209–214.

Uçaroğlu, S., Alkan, U., 2016. Composting of wastewater treatment sludge with different bulking agents. J. Air Waste Manag. Assoc. 66 (3), 288–295. https://doi.org/10.1080/10962247.2015.1131205.

Valta, K., Kosanovic, T., Malamis, D., Moustakas, K., Loizidou, M., 2014. Overview of water usage and wastewater management in the food and beverage industry. Desalination Water Treat. 53 (12), 3335–3347. https://doi.org/10.1080/19443994.2014.934100.

Vayenas, D.V., 2011. Attached growth biological systems in the treatment of potable water and wastewater. In: Moo-Young, M. (Ed.), Comprehensive Biotechnology 6, second ed. Elsevier, pp. 371–383.

Víctor-Ortega, M.D., Ochando-Pulido, J.M., Martínez-Ferez, A., 2016. Performance and modeling of continuous ion exchange processes for phenols recovery from olive mill wastewater. Process Saf. Environ. Protect. 100, 242–251.

Vourch, M., Balannec, B., Chaufer, B., Dorange, G., 2008. Treatment of dairy industry wastewater by reverse osmosis for water reuse. Desalination 219 (1–3), 190–202.

Wang, M., Zhang, D., Dong, J., Tan, S.K., 2018. Application of constructed wetlands for treating agricultural runoff and agro-industrial wastewater: a review. Hydrobiologia 805 (1). https://doi.org/10.1007/s10750-017-3315-z.

Wu, H., Zhang, J., Ngo, H.H., Guo, W., Hu, Z., Liang, S., Fan, J., Liu, H., 2015. A review on the sustainability of constructed wetlands for wastewater treatment: design and operation. Bioresour. Technol. 175, 594–601. https://doi.org/10.1016/j.biortech.2014.10.068.

Zhu, L., Wang, Z., Shu, Q., Takala, J., Hiltunen, E., Feng, P., Yuan, Z., 2013. Nutrient removal and biodiesel production by integration of freshwater algae cultivation with piggery wastewater treatment. Water Res. 47 (13), 4294–4302.

Control of odors in the food industry

9

Nastaein Qamaruz-Zaman, Nurashikin Yaacof, Fatin Farhana Kamarzaman
School of Civil Engineering, Universiti Sains Malaysia, Engineering Campus, Nibong Tebal, Penang, Malaysia

Chapter outline

1. Introduction 282
2. Example sources of odor in the food industry 283
2.1 Brewery industry 284
2.2 Baking industry 285
2.3 Palm oil industry 286
2.4 Poultry industry 288
3. Control of odor through odor management plan 289
3.1 Odor management plan documentation 289
3.2 Identification of odor sources from processing activities 290
3.2.1 The quantification of odor emission 292
3.2.2 International standards of odor assessment 293
3.3 Validation of odor exposure at the surroundings (ambient) 294
4. Odor control technologies 295
4.1 Adsorption 295
4.2 Absorption 296
4.3 Thermal oxidation 297
4.4 Catalytic oxidation 297
4.5 Nonthermal plasma injection 298
4.6 Biotechnologies 299
5. Modifications of biofilter technology for improved odor removal 301
5.1 Activated carbon assisted biofilters 301
5.2 Nonthermal plasma assisted biofilters 302
5.3 Multistaged high performance modules 302
5.4 Engineered biofilter media 303
6. The selection of odor control systems 304
6.1 Selection based on investment cost 304
6.2 Selection based on operational requirements 305
6.3 Maintenance requirement 305
6.4 Selection based on preceding environmental impacts 306

The Interaction of Food Industry and Environment. https://doi.org/10.1016/B978-0-12-816449-5.00009-6

7. Conclusions 307
References 307

1. Introduction

The food industry is defined by Collins Dictionary (Collins English Dictionary, 2018), as: "the industry surrounding the production of food." The food industry is also known as a *food system*, which has been defined as (GRACE Communications Foundation, 2018). "Food systems comprise all aspects of food production (the way the food is grown or raised; the way the food is harvested or slaughtered; and the way the food is processed, packaged, or otherwise prepared for consumer purchase) and food distribution (where and how the food is sold to consumers and how the food is transported)."

Both of the above definitions bear the same meaning and reflect the diversity of the food production sector. It is thus expected that odor emissions from food production activities, can be wide ranging, from a pleasant bakery smell to the unpleasant odor such as those from the poultry farms. In certain conditions, even the pleasant-smelling emissions like the bakery, can become an odor nuisance if being repeatedly exposed to the odor over a prolonged period of time (odor frequency). The severity of the odor nuisance is also governed by other factors such as intensity (odor strength), duration of the odor exposure, offensiveness which relates to its pleasant or unpleasant character (or odor hedonic tone) and the location (of sensitive receptors). These five factors, Frequency—Intensity—Duration—Offensiveness—Location, are more commonly known as "FIDOL" factors and are often considered when assessing odor impacts. People living close to industrial sites should not be subjected to odor pollution that may cause physical and mental symptoms, interference to the enjoyment of daily activities as well as downgrading aesthetics and property values (Oiamo et al., 2015; and Lowman et al., 2013).

The production of chili sauce in an industrial town east of Los Angeles, California, in the United States has caused complaints of spicy odor, with some experiencing headaches and sore throats. Chili grinding was found to be the smelliest phase of the production process leading the affected residents demanding for the closure of the plant (Carroll, 2014). Plant closures due intense odor nuisance is usually the last resort in addressing odor emissions after all means of odor treatment proved to be futile. Such examples include the closure of a pig slaughterhouse in Toronto which had been in operation close to a century, but the stench of pig no longer being tolerable in a rising residential area (Atkins, 2014). Oftentimes, the food processing industry is also a major income provider for the local community, for example, 1000 jobs in poultry production in a town in Texas, United States (Patrick, 2017), and employer to 200 workers for a chili sauce factory in a small town of 1400 people

(Drost et al., 2017). Unfortunately, the shutdown of a manufacturing facility, although effective in mitigating odor by ceasing the production of smell, is not only at cost to the plant owners, but also other stakeholders. The local agencies may be affected from the lost revenue from licensing fees and land rental, and more worrying is the loss of income for a large number of the plant workers. In order for a food industry to remain sustainable in the long run, a good rapport with the community is important, which means an investment in an effective odor management plan and treatment system is sometimes inevitable. Plant owners must be aware and ready to meet both the growing demands for food supply and also the environmental responsibility that is needed in food production.

The world population today is about 7.6 billion which is projected to increase to 9.7 billion people by 2050 (Population Reference Bureau, 2018). The population growth will consequently spur the global food demand anywhere between 59% and 98% by 2050 (Elferink and Schierhorn, 2016), and with it the expansion of the food industry. It is very likely that there will be more and more conflict between the food industry and the community due to industrial growth, encroachment of residential areas to the food industrial areas, and increasing community awareness for a clean and pleasant environment. In Ireland for example, the Environmental Protection Agency (EPA) received 788 complaints in relation to odor from licensed sites, with a significant percentage (40%) related to food and drink processing activities. Historically, waste management activities were the dominant source of odor complaints received by the EPA (Environmental Protection Agency, 2014).

The traditional isolation of industries to curtail the odor impact to the community, is no longer a viable option, due to the growing scarcity of land. Isolation also does not bring any changes toward odor problem, thus odor control measures such as biofilter, absorbers or scrubbers, nonthermal radiation technology and catalytic oxidation becomes necessary to assist food production companies to overcome their odor issues.

This chapter aims to highlight issues of odor pollution related to the food industry. A few examples of food production sector are discussed earlier in the chapter giving an overview about its food processing and the resultant odor problem with identification of its sources. Next, the chapter looks at ways to control the odor pollution through an odor management plan which includes odor detection and quantification as well as control technologies to curtail odor at its source. In the conclusion section, inventive modifications on biofiltration technique are showcased and also summarizes approaches that are most suitable to manage odor from the food industry.

2. Example sources of odor in the food industry

Odors in the food industry are generated under numerous conditions, for example in the handling of raw materials or as a result of process reactions. A variety of odorous substances can also be emitted from the microbial decomposition of organic compounds usually at the waste or wastewater treatment plants in the food industry. Hydrogen sulfide or mercaptans, as well as nitrogen compounds in the form of ammonia, amines, and skatole maybe be emitted from the treatment processes causing nuisance to the surroundings. There is an exhaustive list of food industries that emits or prone to causing odor, such as brewery, animal husbandries, baking industries and others. It is not possible to describe all of these odor-causing food industries, and thus only a few examples of plants processes and activities which are well-known emitters of odorous compounds are presented in this section.

2.1 Brewery industry

Beer is one of the most widely consumed alcoholic beverages, and the process in making beer is called brewing. It is mainly derived from cereal grains such as maize, wheat and rice but is most commonly from malted barley. There are essentially eight important stages in brewing process; milling, mashing, lautering, the boil, wort separation, and cooling, fermentation, maturation, and filtration, with some involving microbial reaction. Various microbes are involved in the different processes of brewery production with the *Saccharomyces* strains being largely responsible for the fermentation of cereal extracts which produces ethanol as the fermentation end product (Bokulich and Bamforth, 2013).

For the mashing step, malt or other grains are first milled to break down the complex carbohydrates into simpler sugars. The crushed malt produces grist, a fine sugar which is then mixed with water at an optimum temperature of 72−75°C and optimum pH of 5.6−5.8 to initiate the mashing process (Pires and Brányik, 2015). This mash is then pumped into the lauter where wort (sweet liquid) is separated from the grain husks. The aqueous solution will go through the filtration process to extract and separate the wort which will be boiled after that for about 90 min to develop color and flavor. After the cooling session, the wort will be aerated and chilled. At this stage, the fermented and cooled liquid must be clear from any malt or hop particle. The fermentation process will follow where the brewing yeast utilizes sugars and other nutrients to generate ethanol and carbon dioxide which will change the flavor and nutritional content of the beverage. This young or "green" beer will be left to mature for some time in the conditioning stage to further maximize its flavors and produce a smooth finish.

Odor and dust are the major public concern regarding air emission from the breweries (Olajire, 2012). Some of the steps as discussed above are sources for the odor emissions, with its character and pleasantness varies according to the processing steps. Brew kettle for wort boiling, vessels, distiller's grain and brewer's yeast dryers mostly emit a pleasant bread-like smell. On the contrary, unpleasant odor detected from breweries is mostly associated with the fermentation process, by-product storage tanks, silos and the wastewater treatment systems (Olajire, 2012). Not only that, the packaging and bottling operations may also lead to odor pollution which may arise due to poor ventilation in the packaging area.

With breweries, volatile organic compounds (VOCs) is of utmost concern, with ethanol being the highest contributor to the odor. Other substances that emit pungent smell include alcohols, aldehydes, dimethyl sulfide and organic acids. Most of it can be found in the fermentation process, product storage and from waste storage.

Sulfur-containing compounds like hydrogen sulfide and mercaptans causes pungent and rotten egg smell, a by-product from the reduction of inorganic sulfur compounds under anaerobic condition. Breweries adopt anaerobic wastewater treatments systems, which if left unmonitored may emit fugitive sulfide emissions. Sulfides and odor have also been reported from wort boiling, a process which utilizes high heat thus prone to emitting higher odor concentration (Needham and Freeman, 2009).

2.2 Baking industry

Baking is an industry that produces fresh and frozen pastries, bread, cookies and cakes. The baking steps are essentially similar across the various baking products, with a few small exceptions. As shown in Fig. 9.1, a ferment step is required when making breads, rolls, cookies and cracker with shortening usually added into the latter. Product form and chill is usually a required step for the cakes, pies and frozen products. The steps shown in Fig. 9.1 is only a guideline, as the steps involved in making each product is rather unique, and would require steps specific to the baking product.

Ingredients selection and scaling are the two crucial steps in baking process. Each types of breads, cakes or cookies had its own specific ingredients and measurement. Flour, water and yeast are often times, compulsory ingredients in baking. Flour will give structure to the bread or cookies while yeast will ensure an airy, nice and soft

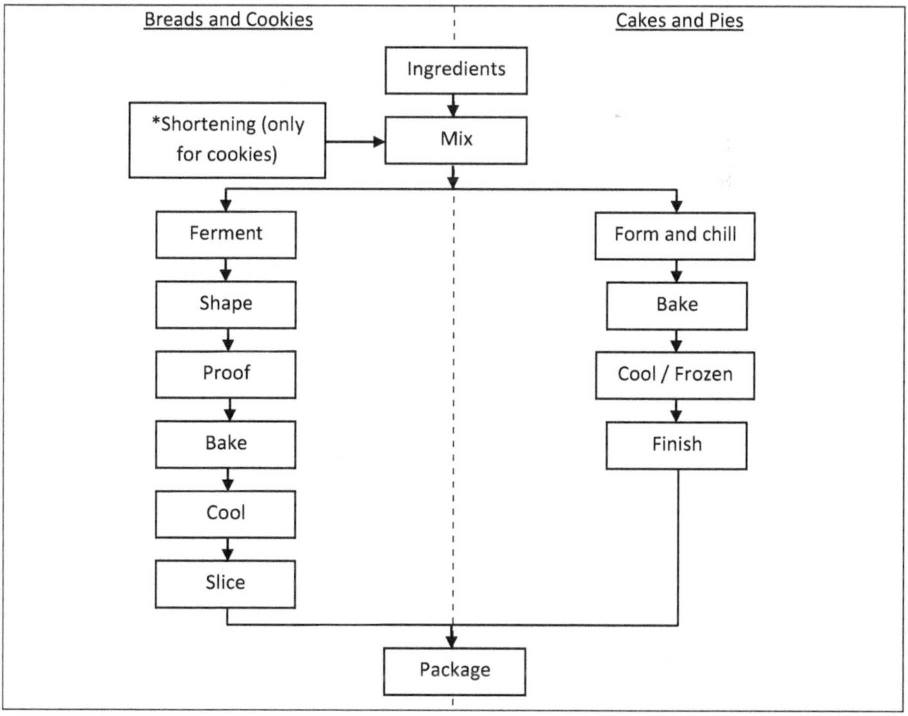

Figure 9.1 Typical processing in the baking industry.

product. Not only that, yeast will add the flavor to the dough during the fermentation process. In bread and cookie making, the added yeast during the fermentation step will react to rise the dough by converting sugars into carbon dioxide, alcohol and organic acids. The fermentation process in baking is similar to brewery fermentation, except that the former is performed under a dry and humid environment. Sufficient time is important in making sure the dough is properly fermented, where it could be as short as 15 min or as long as 12 h, perhaps more, depending on the amount of dough. The prepared dough is then shaped into the desired shapes and sizes which is later baked and packaged.

There are generally two main sources of odor in the baking industry; the fermentation step and the actual baking. Various ingredients are added for the bakery products, which reacts with the yeast during fermentation, releasing aroma compounds such as lactic acid and alcohol. Lactic acid is practically odorless although at times may present a weak unpleasant odor. Alcohol on the other hand usually emits a pungent smell.

In baking, two different reactions occur, the first being Maillard reaction and followed by sugar caramelization. During Maillard reaction 2-acetyl-1-pyrroline compound is generated while 2-acetyl tetrahydropyridine compound is a product of sugar caramelization reactions. Both compounds smelled of bread roasted, cracker-like aroma. Odor from the baking process are usually the ones causing concerns, as heat from the baking ovens are often channeled out through the oven exhausts into the surroundings. With baking industry, the acceptance of the local community toward its odor is rather subjective, as demonstrated by Sowka et al. (2011). The odor concentration from its biscuit production ranged from 1300 to 10,000 OU m^{-3} with an emission rate of 800–1000 OU s^{-3}, which can be considered as a relatively strong odor. Residents living close to a biscuit and wafers bakery plant described the odor as sweet, vanilla, and chocolate-like. The odor from the bakery plant, which was mainly producing wafers, was not bothersome given its rather pleasant nature which did not deteriorate their quality of life.

2.3 Palm oil industry

Palm oil production has increased over the years and has become a very large industry, especially in Malaysia and Indonesia (Rupani and Singh, 2010). The sustained growth of the palm oil in the global market is impressive and it is predicted that the average annual production will reach 15.4 million tons between 2016 and 2020 (Abdullah and Sulaiman, 2013). The use of palm oil is wide ranging from food and beverages to personal care products, cleaning products to cosmetics. The production of palm oil is carried out in a mill, with Malaysia having about 454 such mills (MPOB, 2018a), servicing 5.81 million hectares of the local palm oil plantation (MPOB, 2018b). The process (as shown in Fig. 9.2) starts with the receival of the fresh fruit bunches at the mill and later undergoes sterilization under steam temperature of 143°C and 4 bar absolute pressure for 50 min (Mahlia et al., 2011). The sterilization process softens the fresh fruit bunches to help ease the extraction of oil. Then, the small individual fruits which contained the mesocarp and kernel, are separated from the bunch stalks in a rotating divesting machine during the

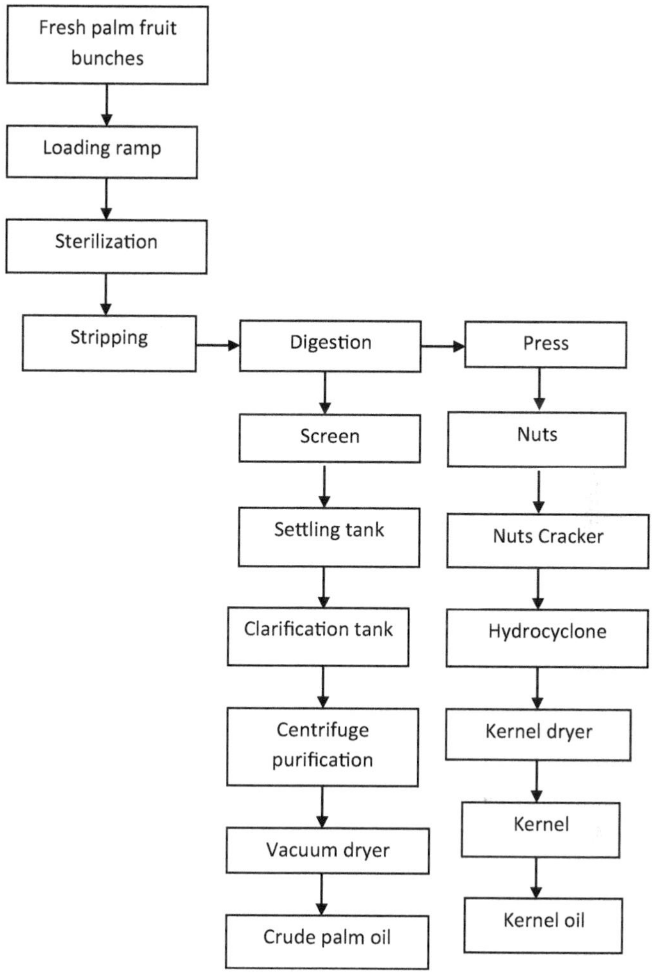

Figure 9.2 The processing of crude palm oil and palm kernel oil.

stripping process. The empty fruit bunches (EFB) will fall into the collector and are typically disposed as boiler fuel or plantation mulch (Abdullah and Sulaiman, 2013). After that, the sterilized fruits are fed into a digester where water at 80°C is added, followed by pressing to extract crude oil composing also some impurities such as cell debris, fibrous materials etc. These impurities are removed in the following step, clarification, whereby the extracted oil is diluted with water, settled and screened to separate between the oil and the debris. Further purification is carried out using centrifuge and vacuum condition to improve the quality of the oil before being cooled and stored. Palm oil mill also produces palm kernel oil, which is oil that is obtained from the kernel. The kernel oil extraction can be done via mechanical extraction, solvent extraction or a combination of both.

The production of palm oil as with many other industrial processing, are not without emissions, either as solid, liquid or air emissions. Palm oil mill effluent (POME) is a liquid waste that contains high colloidal suspension in the form of thick brownish viscous slurry. About 28% of POME is generated during each cycle of the palm oil production. As of 2016, most of the effluent is treated using a series of open lagoons comprising of pretreatment ponds, anaerobic, aerobic, facultative and polishing pond compared to only 21% adoption of biogas plants (Loh, 2017). Even though the open lagoons were removing the debris and other pollution in the POME, its uncovered effluent meant emissions from the anaerobic processes were released uncontrollably into the surroundings, including its unpleasant odor. Hydrogen sulfide (H_2S) and ammonia (NH_3) gases that generated throughout the POME treatment emits pungent and rancid smell (Abdullah and Sulaiman, 2013).

High concentrations of 8500, 6700 and 10,000 OU m^{-3} on average, has been reported for anaerobic, facultative and aerobic ponds, at palm oil mills respectively (Qamaruz Zaman et al., 2016). This "palm oil mill effluent" odor has been described as very strong with an intensity scale of five by people living near the palm oil mill and can at times detected more than 1 km away (Yaacof et al., 2015).

2.4 Poultry industry

Poultry refers to meat products related to chickens, ducks, turkey and geese. The global poultry consumption was forecasted to increase by 1 million metric tons in 2017 from 117 million metric tons the previous year due in part to the increased population and economy in the developing countries and their change in diets (Conway, 2017). Fig. 9.3 shows a typical poultry processing which begins with the chickens brought over from the breeder farm to the processing plant to a finished chilled poultry meat. Chickens are first slaughtered (stunning and bleeding stage), followed by the removal of feathers, oil gland, feet and head and internal organs. A food safety inspector then inspects the products against the local and international food safety demands to ensure consumer's safe consumption. The inspected poultry will be washed, rinsed, chilled, weighed and graded and proportioned according to the marketed products. The prepared products are packed and chilled while awaiting distribution.

An important element to the whole poultry industry and its processing as in Fig. 9.3, is the poultry supply or the chickens itself. Chickens are bred either as free range, fully-housed, or as organic chickens which are differentiated by their placement either indoors, outdoors or caged (Brulliard, 2017). Regardless of the breeding system, the odor emission from the litter had been seen as the primary source of odor problem (Dunlop et al., 2016). Litters are prone to be wet due to spillage of drinking water systems. Coupled with poor litter management practices and inefficient housing ventilation, further escalates the litter odor issue. Odor monitoring by Guo et al. (2007) during daytime operation of poultry barns in Saskatchewan, Canada found that poultry nursery room contributed most to odor (1159 OU m^{-3}) and was lowest (213 OU m^{-3}) at the finishing, farrowing and gestation room. Poor ventilation will increase the temperature and humidity in the building, consequently heightening the concentration of the odorous gases. Thus, good barn ventilation is important to reduce litter odor to acceptable levels.

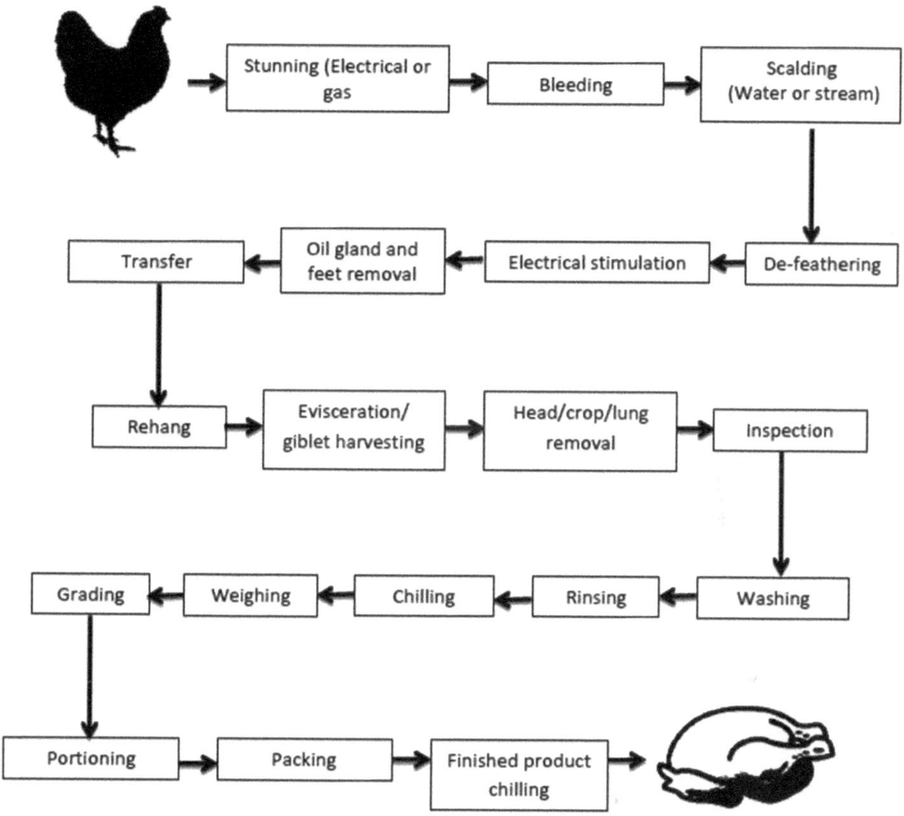

Figure 9.3 Poultry processing steps from raw chickens to finished chilled product (King et al., 2018).

Another odor source at poultry farms are the decomposition of carcasses and waste products including litter and feathers. The breakdown of sulfide in these materials produces H_2S, mercaptans and organic sulfide while ammonia and skatole are generated due to the breakdown of nitrogen compounds (Wlazlo et al., 2016). Volatile fatty acids, alcohols and aldehydes may also be produced.

3. Control of odor through odor management plan

3.1 Odor management plan documentation

To prevent or minimize odorous effects, a systematic approach toward an extensive odor management is required. Such aspiration can be achieved through the adoption of an odor management plan, a written document of the approach that is taken by the facility to manage odor and its nuisance impact that may arise from processing and other site activities. OMPs are described in a number of guidance documents from the United Kingdom "Guidance on the Assessment of Odor for Planning" (Bull et al., 2018) and New Zealand "Good Practice Guide for Assessing and Managing Odor" (Ministry for the Environment, 2016) among others.

Developing the odor management plan involves a process of describing the site and its general management, identifying the sources of odor at site by including the description of the whole process and its waste activities, as well as its surrounding areas such as the odor pathway and receptors characterization. Consideration must be given to the distance between the site under investigation and the neighbors, prevailing wind directions, and the mostly affected community. The obtained information would provide sufficient data for the selection of the most suitable odor control solution for the facility's needs. Because all odor issues and control are thought through beforehand in a holistic and systematic manner, odor incidences and possible crisis management are diverted, instead, is a well-planned execution.

3.2 *Identification of odor sources from processing activities*

Understanding the sources of odor, where it originates from, whether the gas emissions, dusts, liquids, or solids is important to help determine the most suitable odor control systems to be implemented. To aid with the identification of the possible sources, developing a simple diagram of the process flow will initiate the thought process. Fig. 9.4 shows a process flow diagram where the product manufacturing is

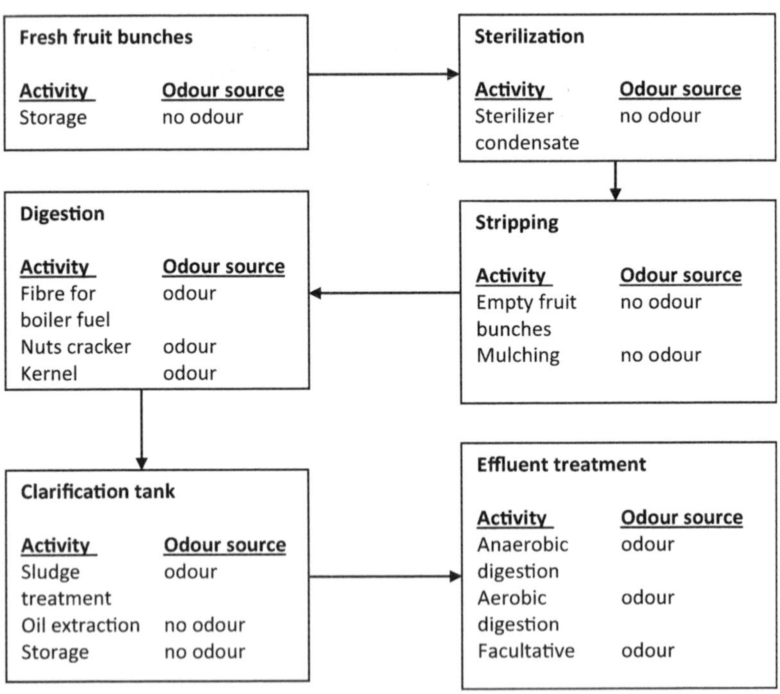

Figure 9.4 Example of a process flow diagram from palm oil milling with the identification of odor sources.

diminished into smaller processing sections, with each activities causing odor being listed down. The process flow diagram should take into consideration all conditions of the manufacturing facility, conditions that are considered normal or typical activities as well as upset conditions. Sudden leakages or spills, ineffective pollution control equipment, malfunctioned machineries etc. may result in unwanted release of odor, thus should be highlighted as a potential odor source.

The process flow diagram can be further complemented with a site plan as shown in Fig. 9.5 that details out the processing facility relative to the other amenities, buildings, storage etc. The marking of the material handling and storage areas, all buildings,

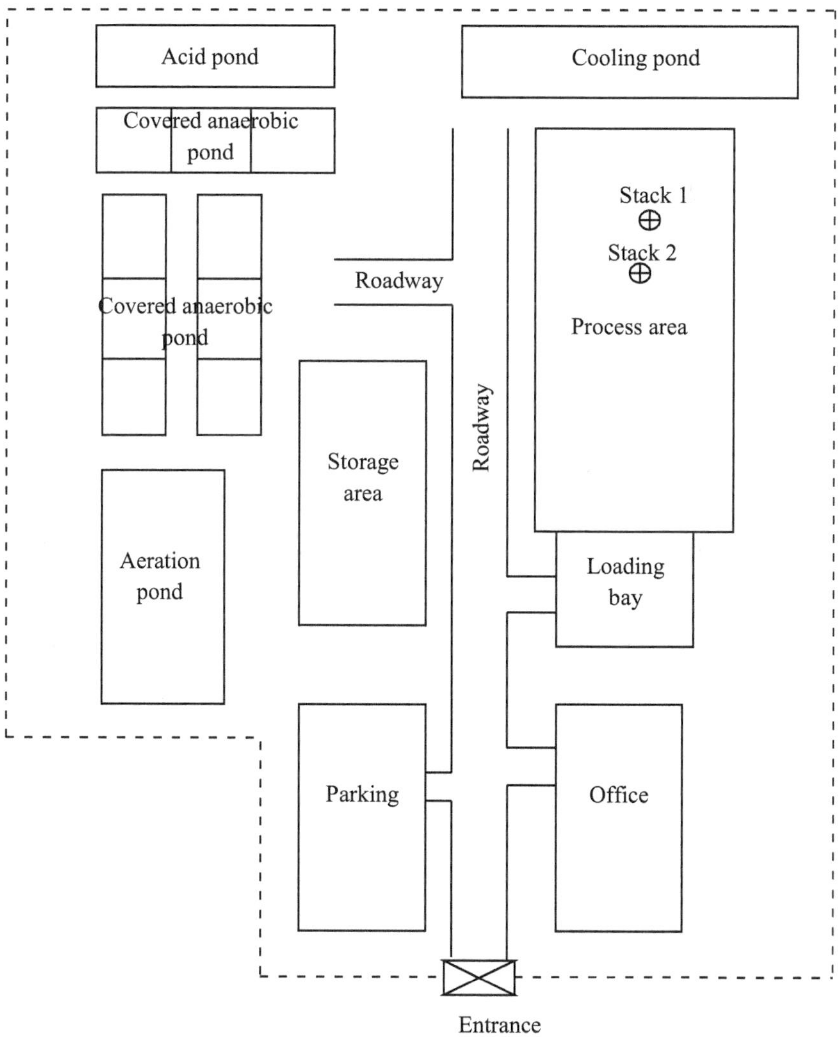

Figure 9.5 A site layout of a palm oil mill with clear indication of the processing area, other buildings, amenities and site boundary.

exhaust or stacks, waste and wastewater treatment plants and storage, roading, and the property boundary should be made clear on the site plan. The local wind conditions, its direction, speed and other prevailing meteorological conditions should also be included in the site details. A clear site overview would be useful in the management of odor emissions as well as for the planning of implementing new approaches in odor management, either by modifications of current site practices or the utilization of new odor control systems.

3.2.1 The quantification of odor emission

Odor is a mixture of odorous chemical compounds that invokes olfactory responses in individuals subjected to the smelly environment. Humans detects a combinations of odors rather than single odor molecules at a time, which then bind to receptors inside the nose and relaying messages to the brain (Williams, 2014). There are generally two spectra for quantifying odor, one that focuses on the sensorial techniques which uses human panelists for the detection of odor, and the other that measures the concentration of odorous chemical compounds as an indicator for the odor emission. The Japanese Offensive Odor Control Law (Law No. 91 of 1971 - Latest Amendment by Law No. 71 of 1995) for example specifies limits for 22 compounds including 0.02–0.2 ppm of hydrogen sulfide, 1–5 ppm ammonia, etc. in addition to the 10 to 21 odor index limit (Higuchi et al., 2007). The Netherlands Emission Guidelines (Nederlandse Emissie Richtlijn) for example, regulates a variable odor concentration threshold from 0.5 $ou_E m^{-3}$ to 25 $ou_E m^{-3}$ and the percentiles from 98th to 99.99th (Brancher et al., 2017).

The determination of odorous compounds can be done using physico-chemical methods, most popular being the gas chromatography mass spectrometry (GC–MS). Portable devices such as the Jerome 631-X or Drager tubes are more suitable for on-site odor analysis, but are typically limited to measuring only individual chemical compounds and within certain detection ranges. Although physico-chemical means allows for a more precise analysis of chemical compounds, its applicability as a surrogate for odor depiction is less suitable since environmental odors are constantly made up of various and many chemical compounds, thus making individual compounds as not being representative. Furthermore, highly sophisticated gas chromatography instruments are required with the ability for assessing specific compounds and at low concentration as well as a skilled GC operator.

For the sensorial technique, odor analysis is carried out by presenting the diluted odor sample to a panel of selected and screened human assessors via a dilution device known as an olfactometer. Normally around 4 to 10 odor panelists is involved in the sensorial assessment. A selection of average-sensitivity panels is recommended in order to produce more reproducible results while the gender of the panel does not bring any significant difference toward the test results. The odorous sample is presented to the human panel in increasing concentrations until the panel member is able to recognize an odor over a neutral reference air. The odor concentration in OU/m^3 resembles the dilution factors needed to achieve the detection threshold of a sample (Van Harreveld et al., 2013). As defined by EN 13725 (2003), the individual threshold estimate is defined by the two presentations in one dilution series, sorted on growing odor concentration, where a certain change in response from "false" to a consistently "true" response occurs. The individual threshold estimate is calculated as the geometric mean of the dilution factors of the two defined presentations.

The laboratory olfactometric test often begins with the collection of samples at the site. Point source such as stack emissions are sampled by drawing part of the odorous airflow into sampling bags (usually Nalophan or Tedlar material) using a depression pump. On the other hand, sampling for passive area sources such as effluent ponding systems requires the addition of a flux hood or a wind tunnel being placed over the emitting surface, to isolate from other background odor. A clean feed air e.g., nitrogen gas is fed into the flux hood at a fixed flowrate to imitate the wind action over a liquid or solid surface. Next, the odor samples are drawn into the Nalophan or Tedlar sample bags using a vacuum chamber. These collected samples are then sent to the odor laboratory for dynamic olfactometry analysis within 24–30 h of sample collection as specified by EN 13725 (2003).

In addition to odor concentration in OU/m^3 that quantifies the strength of odor, other common sensorial odor presentations includes hedonic tone, odor intensity and odor descriptors. Hedonic tone is a measure of the pleasantness or unpleasantness of the smell, based on a 7-point scale system. A negative value (e.g., −7) implies an unpleasant smell, a positive (+7) as a pleasant smell, whereas a 0 value indicates a nonodorous condition. Odor intensity is generally based on the same concept as hedonic tone, except that usually a 0–6 scale is used, with 0 being no odor which increases to a weak smell at scale 1 and extremely smelly at scale 6. Odor descriptors on the other hand, is a descriptive quantification of odor based on the individual's past experiences and recognition of the odor character for example sewery, pungent, solvent etc. Odor descriptors are useful for indicating the presence of possible odorous compounds e.g., hydrogen sulfide for rotten egg smells or ammonia from a pungent odor, and identifying potential sources of environmental odor. The odor wheels by Suffet and Rosenfeld (2007) and the Table *"Odor Descriptors for Commonly Encountered Compounds"* (UK Environment Agency, 2007) are some simplified versions of odor descriptors which has been developed for easy reference.

3.2.2 International standards of odor assessment

The sensorial techniques for odor assessment is able to portray the true representation of an odor. However, there is a concern with the subjectivity of the sensorial analysis, as well as the variability of the odor data. In order to address the concerns raised with the olfactometry test, standards and guidelines are established to ensure a reliable and quality assessment by guiding the sampling, testing and data analysis procedures. The EN 13725:2003 and the US ASTM E679-04 (ASTM, 2011) are among the internationally recognized standard methods for odor assessment which was developed to achieve a common basis for the evaluation of odor across various laboratories.

The European standard, *EN13725 Air Quality - Determination of Odor Concentration By Dynamic Olfactometry*, was approved by the European Committee for Standardization (CEN) on 6 December 2002 (Van Harreveld et al., 2013). Accuracy and repeatability has been one of the important element of EN 13725:2003. The standard guides users by providing details such as the selection and training of odor panelists, sampling and olfactometry equipment and its calibration requirement, methodologies as well as the procedures for data reporting.

The *ASTM* E679 *Standard Practice for Determination of Odor and Taste Threshold by a Forced-Choice Ascending Concentration Series Method of Limits* is the standard that had been developed by United State of America (USA) in 1979 and its procedure is based on a presentation method called 3-alternative forced-choice (3-AFC) or triangular forced-choice (TFC). The method is similar to the use of dynamic olfactometry in performing the dilutions of the odorous air and present it to the human assessors. Around 5 to 12 trained and experienced human assessors are needed to carry out the test.

3.3 Validation of odor exposure at the surroundings (ambient)

Odor incidences and its exposure to the surroundings can be validated via several odor assessment tools, tested either at the source of emission or at the receptors. Often times, a combination of several tools are necessary to give a reliable portrayal of an odor impact from an industry given that each tools have different strengths and limitations. This combination may be in terms of atmospheric modeling, sniff test or known also an in-field odor assessment as well as community survey. An example would be to corroborate the odor contours plot from the modeling with the locations of odor identified through surveys or sniff tests.

Atmospheric modeling is useful to forecast an odor exposure captured in long timeframe and over a vast coverage area. Examples of typical odor dispersion modeling includes ADMS, Aermod and Calpuff. One of the input data for the modeling is the odor emission rate, an odor concentration value that is obtained via literature or by sampling the odor at its point of emission and obtaining its concentration using a dynamic olfactometer. However, models are prediction of the real situation based on a set of input data, thus although reasonable assumptions and input data are used, there is still a considerable uncertainty in its prediction.

Field odor monitoring is done in-situ and in real time, which represents the odor exposure at specific locations under the prevailing conditions during the assessment. Trained odor panelists perform the field monitoring either according to the VDI 3940 (2006), sniff test (Bull et al., 2018) or portable olfactometers such as the Nasal Ranger and Scentroid SM100 Olfactometer. Both of the Nasal Ranger and Scentroid SM100 olfactometer constitutes a portable device which utilizes clean odorless air (either filtered or supplied from a portable air tank) and a dilution system based on several calibrated orifices.

The Scentroid SM100 Field Olfactometer quantifies the ambient odor strength using the same basic theory of laboratory based olfactometry. The equipment draws a sample of ambient air via venturi vacuum pump and dilutes it using carbon filtered air from a high pressure compressed air tank. The mixed air is sent to a disposable face mask through a flexible hose. Dilution ratio of clean air to sample air is controlled via the designated flow regulator valve, in which the assessor slowly increases the concentration of the mix until the odor of ambient air is detected and reported as Odor Unit (OU).

The Nasal Ranger Field Olfactometer is made up of a barrel with a nasal mask at the edge and two carbon filters attached to each opposite sides of the Nasal Ranger housing. The odorless air is mixed with the odorous ambient air at discrete volume ratios whereby the dilution ratio of clean air to sample air is controlled via the dilution to threshold (D/T) dial. Unlike the SM100 Olfactometer which had an external air supply, the Nasal Ranger makes use of the ambient air, but scrubbed odorless via the two carbon filters. The Scentroid SM100 Field Olfactometer specifies 15 dilution ratios whereas the Nasal Ranger Field Olfactometer contains six D/T positions, alternating with six blank positions for the user to inhale only odor-free filtered air.

With in-field odor assessments, although the odor observations is more "true" and representative of the actual exposure, this type of assessment does not cover a wide receptor location. The data obtained are snapshots rather than details that covers variable meteorological conditions over a typical year. Often times, the characteristics of odor are also noted during odor reporting which is useful in pin-pointing the source of odor emission.

The monitoring of odor at the site should be done in tandem with regular inspections over the actions taken for the avoidance, control and mitigation of odor as set out in the odor monitoring plan. Any elevated odor levels than that normally experienced during monitoring is an indication of an odor-related incidence. Such odor incidences are to be further investigated and undergoes appropriate remedial action to prevent recurrence. In order for the monitoring and its effect to work effectively, good recordkeeping, the delegation of responsibilities as wells as staff training and competency should be a vital practice for the industry.

4. Odor control technologies

Various odor components can be present in the air emitted from the industries, each with differing odor thresholds and concentration. Four main groups of odor components are nitrogen-containing compounds, sulfur compounds, aromatics and fatty acids. To treat these odorous emissions several technologies are available and can be grouped as a physical, biological and oxidation treatment system. A physical system includes adsorption and absorption whereas thermal, catalytic and nonthermal plasma (NTP) are examples of oxidation system. A biological process on the other hand relies on microorganism for the degradation of odorous components.

4.1 Adsorption

Adsorption is the process that occur when ions or molecules of gas or liquid adheres on a surface of an adsorbent. A molecular or atomic film will be formed, known as the adsorbate. Adsorption is a cost-effective process which is versatile and widely used (Qian et al., 2018) for the treatment of odorants and poison (Danish Environmental Protection Agency, 2002). The adsorption media can also be regenerated by removing the odorants by heating, soaking in chemical such as hydrogen peroxide (Guo Ping, 2014) or electron-fenton process (Trellu et al., 2018).

The capacity of the adsorbate that can attach onto the adsorbent depends on the surface area of adsorbent pores. The surface area is one of the important criteria for adsorbent selection and are widely studied to further increase the adsorption capacity. Cellulose impregnated adsorbent is a recent example whereby the utilization of cellulose as an adsorbent was widely recognized due to its sizable surface area ($1300\ m^2\ g^{-1}$) and total pore volume of $0.8\ cm^3\ g^{-1}$ (Suhas et al., 2016). Other types of adsorbent include alumina silica and metal hydroxides but activated carbon has been the commonly utilized adsorbent material thus far (Malik et al., 2002). Activated carbon has large surface area up to $3000\ m^2\ g^{-1}$ with highly developed porosity, variable characteristic of surface chemistry and high degree of surface reactivity. The carbons are traditionally sourced from coal and wood, but in recent years has seen the engineering of carbons from various sources including waste materials such as waste tires (Al-Rahbi and Williams, 2016), plastics (Yuliusman et al., 2017) and waste tea (Zhou et al., 2018).

The odor removal efficiency of activated carbon has been reported to be 99.9% (Corey and Zappa, 2018), which is able to reduce odor concentration to less than 100 OU. A standard activated carbon is efficient for low levels of hydrogen sulfide (1−2 ppm) and for most amine compounds found in municipal wastewater and agricultural processes (Thiesen et al., 2002). With higher hydrogen sulfide levels above 20 ppm, the use of impregnated, blended or catalytic carbon is more efficient and economical. During the operation of activated carbon, any solid particles or substance such as dust, mineral oil or large quantities of steam need to be removed prior, in order to enhance the adsorption capacity and more importantly to prevent any clogging of filter bed. The unwanted entrapment of this debris can reduce the adsorption capacity since certain metal can oxidize the adsorbent material and destruct the pore structure (Danish Environmental Protection Agency, 2002).

4.2 Absorption

Unlike adsorption, absorption is the process where the ions or molecules enter some bulk phase such as liquid or solid material. The resultant concentration of substance in the bulk material would usually increase as a result and would require removal either as liquid (wastewater) or solid (particulate) form. Absorption acts like a sponge, which is much like a scrubber system. There are essentially two types of scrubbers; a wet scrubber which uses scrubbing solutions, and a dry scrubber which requires a sorbent material.

In wet scrubbers, the system works by first receiving contaminated air from the industrial processes and passing it through a layer of steam. The steam traps the odorants in its droplets which are then collected in a filter or mist eliminator, while releasing clean treated air. In certain cases, acid, oxidizing agent or alkaline are added to the water (or steam) in order to create the optimum scrubber condition for the type of odor being treated. Gaseous emissions from fish and meat processing plants usually require strong oxidizers such as hypochlorites to remove its concentrated odorous compounds (Danish Environmental Protection Agency, 2002), while alkaline gases are needed for acidic emissions and vice versa (Biard et al., 2017). The absorption approach of scrubbers is often applicable to remove odorant with ammonia, hydrogen sulfide, chlorine, sulfur and fluoride compounds (Biard et al., 2017; Danish Environmental Protection Agency, 2002). A chemical wet scrubber reportedly have high removal efficiency up to 99% for hydrogen sulfide and ammonia while caustic wet scrubber was less effective for organic sulfur compounds removal between 20% and 70% (Corey and Zappa, 2018).

Scrubber is a technology that is simple and suitable for treating a large volume of air. Moreover, the technology is reusable after undergoing regeneration by gas washing. However, given its wet nature, poorly maintained scrubbers can be breeding grounds for bacteria and may cause the spread of unwanted diseases. The Legionnaires' outbreak in Norway is an example case of 10 deaths and 50 infections which was traced back to a few dirty local air scrubbers (Nygård et al., 2008).

4.3 Thermal oxidation

In thermal oxidation, the odorous air especially those containing VOC are demolished under high temperatures with the presence of oxygen by conversion to carbon dioxide and water. The odor treatment starts by diverting the air stream into the reaction chamber whereby the process is governed by the residence time, turbulence and temperature. An adequate residence time usually within 1−2 s is important to ensure complete oxidation of the odorous compound at or above 760°C. Turbulence within the chamber will equally distribute the heat to reach all parts of the injected air stream.

The treatment of contaminated air which contains combustible reactants is most suited for thermal oxidation as it can self-sustain the combustion. In comparison, air streams with low hydrocarbon contents although viable, would require the addition of external fuel such as natural gas, propane etc. As such, its application for low VOC laden emissions are more recommended for smaller industries (Higashi, 2004). Thermal oxidation is also not suitable for treating emissions with a highly variable flow due to the potentiality of poor mixing in the reaction chamber during increased flow conditions. Poor mixing conditions can cause reduce combustion rate by decreasing the temperature which in turn decreases the destruction efficiency.

The odor removal efficiencies from fish factories and meat processing from thermal oxidation can reach between 90%−99% with VOC being 99% reduced (Schiavon et al., 2017). A study by Janoško and Čery (2015), when comparing the treatment of odor from animal rendering plant in Krškany, Slovak Republic, found a marked difference between a thermal oxidizer and a bioscrubber system. Thermal oxidation of the malodor reached 99% removal resulting in 400 OU m^{-3} emissions, in comparison to only 76% removal with the bioscrubber. Odor was still a high 19,000 OU m^{-3} with the latter.

4.4 Catalytic oxidation

Catalytic oxidation is much like thermal oxidation, but with the aid of catalyst to initiate the oxidation of odorous compounds. Because the oxidation occurs on the surface of the catalyst, the same level of combustion can be achieved at lower temperatures e.g., 250–500°C which makes the operating costs small. Two groups of catalyst are available either homogeneous or heterogeneous usually found in different types including metal catalyst, platinum and vanadium pentoxide as well as ceramic porous material. The selection of catalysts is dependent on the odorous compound to be treated. For example, the oxidation of ammonia to nitric oxide is catalyzed by platinum under the Ostwald process, while the oxidation of hydrogen sulfide to sulfur is catalyzed by vanadium pentoxide under the Claus process (Lloyd, 2011).

The oxidation proceeds in two steps; first being the oxygen in the catalyst reacting with odorous compound e.g., adsorbed VOCs which will be reduced to metal oxide, followed by the re-oxidization of the reduced metal oxides by the gaseous oxygen present in the air flow (Song et al., 2001). During the oxidation process, the catalyst also undergoes various transformations. The catalyst will change itself during the process by interacting with other reactants, altering the reactions rates and then return to its original form after each cycle. The recurrence of the catalysts is known as "reborn" whereby the capability of the catalyst is slowly diminished during every use.

In the catalytic oxidation of pungent ammonia, the reaction usually occurs in aqueous condition for example, using catalytically activated oxygen and an alkaline condition. Ammonia will oxidize into nitrous acid molecule followed by the splitting of nitrous acid into smaller ion, nitrite ion and proton. The increased proton in turn decreases the pH level and consequently elevating the ammonium ion content. A homogeneous reaction between nitrous ion and ammonium ion will occur, which will generate nonodorous molecular nitrogen as the final product (Lee, 2003).

Odor removal efficiencies of 98% with reductions in ammonia emission have been reported when using a UOP catalyst for the treatment of emissions from the roasting stage of coffee manufacturing (Noordally et al., 2008). Phenol which smells like disinfectant and acetic acid with vinegar-like odor has been shown to be successfully removed by heterogeneous Ru/carbon catalysts. The Sibunit-4, PZ65, and PZ68 catalysts developed by Dobrynkin et al. (2012) only required 160–200°C operating temperature, compared to the typical 200–240°C requirement, to reach high degrees of mineralization (99%–99.5%), making the catalytic oxidation process practical for odor control uses.

4.5 Nonthermal plasma injection

Similar to catalytic oxidation, NTP injection relies on oxidation process to change or convert one compound to another. This technology is an advanced oxidation which is highly suitable for flue gas treatment. A large amount of active species can be catered to a vast temperature range for the selective catalytic reduction (An et al., 2017). In NTP injection, clean air is drawn into the system and transformed into active oxygen. The oxidation reaction will then transpire to oxidize the odor molecules in the air stream into odorless components undetectable by human nose.

Plasma is made up of ionized gas and is the fourth state of matter along with gas, liquid and solid. A mixture of balanced electrons, neutral particles and positive ions makes up the plasma which is able to produce its own radiation in the electromagnetic form. NTP performs at very high temperature compared to thermal and catalytic oxidation, in the range of 5000 to 10,000°C. The excessively hot condition is needed to ionize the ions since cold plasma results in lesser ionization. Despite the high temperature, the neutral particles and positive ions remain at a cooler condition while only the electrons exist in the hot conditions.

The NTP injection was effective in treating odor from the exhaust of cooking and frying process of a food industry (Holzer et al., 2018). The emission contained 2-methylthiophene, 2-methylpyrazine, 2-acetylthiazole, nonanal and trans-2-nonenal in the ppm range. The oxidation of these target compounds as multicomponent mixture in nonthermal gas plasma was more energy efficient when using a dry mixture of argon and oxygen. In addition, the selectivity of the oxidation toward carbon dioxide was significantly higher in humid air.

In the treatment of malodor from swine livestock production, a low-energy NTP was found to be a realistic solution for treating large odorous air volumes at a reasonable cost. Anderson et al. (2013) installed a pilot NTP system at the ventilation line of a pig house, which was tested under both negative and positive corona discharge modes. The system successfully lowered indole, 3-methyl-1H-indole methanol, 4-methylphenol, and 4-ethylphenol content for both positive and negatives modes. High removal efficiency above 90% was achieved when the system was operated under negative discharge from 15 to 45 kV voltages and at 135 m^3/h air flow.

4.6 Biotechnologies

Biotechnologies for the treatment of odor includes biofilter, membrane bioreactors and biotrickling filters. The biofiltration process involves the combined action of adsorption, absorption and microbial degradation. The treatment starts with the movement of the contaminated air through the filter bed in which the odorous compounds will either be adsorbed onto the media or absorbed into the water layer that exists on the media particles. The microorganism within the filter bed then biologically degrade the compounds into nonodorous end products such as carbon dioxide, water, mineral salts, nitrogen, sulfur, and others (Schiavon et al., 2017). The materials for the bed media may consist of natural materials like composts, coarse soil, gravel, or wood charcoal (Singh et al., 2017; Gómez-Borraz et al., 2017) or synthetic types like perlite (Rene et al., 2009), polyurethane foam (Filho et al., 2010), and zeolites (Zhang et al., 2012). Fungi, bacteria, and actinomycetes are some of the microorganisms that are naturally available or immobilized onto the filter media and would usually require some adaptation time for microbial growth before biofilter operation (Vikrant et al., 2018). Being a biological process, biofilters are vulnerable to unsteady pollutant concentration and flow rate which could negatively influence the microbial adaptation time (Elias et al., 2010; Hoimark, 2012). It has to be cautioned, that sudden flow loadings and concentration is not favorable as these fluctuations can affect the stability of the microbial population and consequently the biofilter performance (Schiavon et al., 2017).

There are essentially two types of biofilters; an open bed biofilter and a closed system as shown in Fig. 9.6. The open biofilter type is typically built as a concrete box with the upper part being in contact with the atmosphere. The design is simple, and thus suitable for various applications with limited costs and low odor requirements.

However, due to its open configuration, the biofilter is subjected to the weather conditions. The bed may get too wet after excessive rain and perhaps also plant growth or in periods of sunny days, may become too dry. The closed system on the other hand is encased in a steel or plastic container and is better protected against external weather conditions. The technology is more technically sophisticated and allows better control over the fluid flow, composition and temperature and efficient monitoring of biofilter and bed conditions.

Parameters such as moisture, pH, nutrients and temperature are important to ensure an effective odor removal by this biological treatment. A moisture content of 40%–60% will help maintain buffer capacity of the bed and maintain a relatively neutral pH range between 6 and 8 (Chan, 2006) for the survival and metabolism of the resident microorganisms. Microbial growth also require nutrients in the form of nitrogen, phosphorous, and trace elements, with preference toward a constant mesophilic temperature range of about 35°C (Wani et al., 1997).

(A)

(B)

Figure 9.6 Two types of biofilters in operation at wastewater treatment plant (A) an open biofilter bed and (B) a closed biofilter system.

Melse and Hol (2017) demonstrated that up to 74% ammonia and 62% odor can be removed from poultry and swine farmhouse using a simple biofilter with 0.25 m deep wood-chip-based media. The biofilter have potential for emission reduction at animal houses, but homogeneous moistening of the biobed along with a higher media depth is needed to prevent breakthrough of air at dry spots. An animal rendering plant had its odor, ammonia and hydrogen sulfide removed by more than 99% resulting in a low odor of 480 OU m^{-3} and less than 0.2 ppm of its odorous compounds (Almarcha et al., 2014). This high removal efficiency was enabled through a more sophisticated biofilter system. The advanced biofilter system in the study by Almarcha et al. (2014) were developed from a combination of inorganic and an organic media and inoculated with specific, genetically selected, natural microorganisms.

5. Modifications of biofilter technology for improved odor removal

The contaminant removal efficiencies and the costs involved with the acquirement of the technology are often the two determinants when deciding upon the most suitable investment. Biofilters has always been one of the preferred solutions in odor treatment due to its low operation and maintenance requirement (Sheridan et al., 2003). In addition, biofiltration is a sustainable treatment technology that uses natural biological process to treat several contaminants simultaneously. No additional chemicals are needed in the system thus avoids generating environmentally harmful by-product (Zhang et al., 2002). Much has evolved in the biofiltration system to meet the ever-challenging demands of odor treatment. With the increased complexity in product making, including the food industry, coupled with stringent environmental requirements, modifications to the biofilter system is necessary for an improved odor control.

5.1 *Activated carbon assisted biofilters*

The field application of biofilters are limited by the ability of biological processes to respond to variable loading patterns and shock —loads. Most air streams emit varying concentrations of the contaminants, with its characteristics differing depending to the processes, productions, shutdowns, and shifts. Microorganisms are particularly sensitive to the contaminant loading, which may lead to microbial starvation due to the change in feed and eventually death under a prolonged upset period. The incorporation of a nonbiological step, usually adsorption has been found to successfully ease the variable air flow loadings.

Activated carbons when placed before a biofilter enhanced the removal efficiency of styrene above 96% with an elimination capacity of 93.8 $g\ m^{-3}\ hr^{-1}$ after minimizing the inlet loading fluctuations (Li et al., 2012). The activated carbon works by temporarily absorbing and storing the contaminants during periods of over loadings and then slowly desorbing the contaminants into the biofilter during periods of low loadings. The dampening effect on the shock loads prevents microbial starvation thus providing a more stable system design. Such benefits has also been observed when treating *n*-hexane and benzene as well as toluene (Dorado et al., 2012) when the biofilters were preceded with activated carbons.

Alternatively, activated carbon has also been incorporated after the biofilter system, but as a polishing step, especially for meeting strict odor emission levels. The study by Gilani et al. (2007) demonstrated the possibility of achieving low odor of 100 D/T, along with hydrogen sulfide below 1 ppm when using the combined biofilter and AC system. Although the carbon stage maintained an excellent odor control, it was workable only for a short time. Gilani et al. (2007) found that the high humidity in the air stream affected the performance of the activated carbon, however, can be solved by an effective dehumidification system. Biofilters with activated carbon polishing designed to achieve 99% reduction of H2S and a 95% reduction of the odor concentration, has been reported to be a reliable investment and as one of the most preferred technology for odor control (Kraakman et al., 2014).

5.2 Nonthermal plasma assisted biofilters

Another means to counter the unsteady loading rates is the addition of NTP onto the biofilter system. NTP does not require replacement or regeneration of the filter material, unlike the activated carbon, which makes it an interesting alternative.

Schiavon et al. (2017) found that more than 93% of toluene, p-xylene, ethylbenzene and benzene were removed using the hybrid technology compared to only 60% toluene, 87% p-xylene, 76% ethylbenzene and 28% benzene reduced when using biofilter alone. NTP kept the concentrations at the biofilter inlet constant by regulating based on demand. The flow demand is controlled by its electrical operation parameters such as high voltage amplitude and frequency, and thus does not require an acclimation period. However, the dielectric barrier discharge produced carbon monoxide, which has a poisoning effect on the microorganisms with a reported 25% decrease in the microbial cell number of the continuing biofilter system.

On the contrary, an earlier investigation by Wei et al. (2013) found that the ozone produced by the NTP actually spurred the activity of microorganisms toward the removal of dimethyl sulfide, of up to 96%. The dimethyl sulfide was first oxidized into simpler compounds such as methanol and carbonyl sulfide (intermediary organic products) by the NTP followed by degradation to sulfate, carbon dioxide, and water vapors in the biofilter.

5.3 Multistaged high performance modules

Biofilter media after some time in operation will deteriorate and accumulate biomass, a case usually more severe when treating highly odorous air and for low cost organic materials. The decomposed and compacted bed is prone to causing pressure buildup, uneven distribution of water and airflow which significantly affects the biological activity and efficiency of biofilters.

Design innovations such as those by Estrada et al. (2013) and Lin et al. (2013) are some of the examples in the improvement of the conventional open bed and/or closed biofilter system. The step-feed biofilter design by Estrada et al. (2013) divides the packed column biofilter into three separated modules, each having its own set of feeding and irrigation channels. This new configuration reached toluene elimination capacity of 80 g m^{-3} h^{-1}, similar to a simple biofilter, but maintained pressure drop below 300 Pa m_{bed}-^{1} bed much longer (over 60 days), compared just 14 days with the conventional design. The improved performance observed with the step-feed biofilter is attributed to the more homogeneous biomass distribution, lower overall pressure drops and an efficient irrigation of the filter bed due to independent media and input air and water feeds.

Lin et al. (2013) focused its modular biofilter design into the separation of its microbial community to cater to a mixed foul odor such as nitrogen-containing groups, sulfur-containing groups, and short-chain fatty acids. The modular reactor by Lin et al. (2013) successfully removed 97% of 96 ppm input ammonia, 98% of 220 ppm input amines, and 100% of other odorous substances. Like Estrada et al. (2013), this modular bioreactor design also maintained a low pressure drop (<45 mm H_2O m−1) over an extended period which helps maintain the inoculated microflora of the system (>107 CFU g $^{-1}$ −GAC).

There is also a multistage biofilter configuration which is composed of a biotrickling filter and a biofilter, as demonstrated by Toretta et al. (2013). In this design, a highly odorous air emission is first treated by the biotrickling filter followed by subsequent polishing by the biofilter. The multistage set-up was highly effective in meeting high odor removals e.g., less than 300 OU m^{-3} and 81% VOC removal (Toretta et al., 2013) as well as low concentrations of 1.07 mg Nm^{-3} for benzene, 0.16 mg Nm^{-3} for toluene, 0.22 mg Nm^{-3} for ethylbenzene and 0.07 mg Nm^{-3} for xylene (mix), averaging at 96.1% removal efficiency (Raboni et al., 2017). A multistage biofilter is thus recommended when compliant for most stringent toxicological standard and threshold odor concentrations are of most importance.

5.4 Engineered biofilter media

Biofilter media provides the support for the attachment of microorganisms during the biofilter process, therefore must be able to provide the right environment and nutritional requirements for their continued growth. Materials having high surface area, porosity and water holding capacity are the characteristics that will warrant an optimal pollutant removal and at the same time have a sufficient long useful life.

The PVA/nitrocellulose(NC)/GAC/KNO_3 composite bead developed by Chan and Ho (2011), reportedly had a porous spherical particle with a diameter of 2.4−6.0 mm and a density of 1.125 g/cm^3. In addition, The PVA/nitrocellulose/GAC/KNO_3 composite bead had superior moisture-holding capacity which is favorable for the adaptation of microorganism and promotes the secretion of exocellular enzymes to break down odorous compounds.

Expanded schist is an example of inorganic media which has demonstrated excellent removals of odorous hydrogen sulfide (Dumont et al., 2012; Shareefdeen,

2015). The material is obtained from the thermal expansion of schist which resulted in its granular form and large surface area. Expanded schists is effective in hydrogen sulfide removal due to its strong mechanical behavior which is able to withstand acidic conditions for a long time (Jaber et al., 2016). It was discovered that the application of expanded schist is best operated with the addition of UP20 topped as a layer over the expanded schist to provide synthetic nutrients, making it a stable biofiltration process for various industrial applications (Courtois et al., 2014).

Another alternative for biofilter media lies in the utilization of waste products. The application of waste and residual materials provides a more sustainable option for its disposal while at the same time saving resources and cleaning the environment. Sugarcane bagasse is an agricultural residue from industrial sugar extraction process, and has been proven for the removal of dimethyl sulfide (DMS) (Fernández et al., 2013). A maximum DMS removal of 98% was achieved when the sugarcane bagasse biofilter was inoculated with *Hyphomicrobium* VS and operated at 180 s (empty bed residence time). Other innovative examples of waste utilization as biofilter media includes concrete waste for treating hydrogen sulfide gas (Jaber et al., 2017) and domestic waste with pruning waste compost for treating a-pinene vapors (Cabeza et al., 2013).

6. The selection of odor control systems

When deciding on the odor control technology to be implemented at an industrial facility, the effectiveness of the odor control system and the investment costs are usually two of the most considered determinants. The system should be able to suit with the air flow volume and the inlet concentrations. A high cost treatment system may seem daunting in the first instance but could prove to be a rewarding choice especially if compared to a less expensive, less efficient system or one which would require hefty maintenance and operational input. Not only are the removal efficiencies and costs being important deciding factors, the potential of environmental impact from the implementation of the technological choice is also of similar significance and should receive due recognition. The consideration for the solid, water and air discharges as well as their treatment and disposal, types of packing bed, chemicals or cleaning agents, logistics and also water and energy requirements, are examples of environmental impacts associated with the treatment of industrial malodor.

6.1 Selection based on investment cost

Biotrickling filters and biofiltration systems are higher in investment costs, ranging between (12–80 \$ m^{-3} h^{-1}) in comparison to the physical/chemical technologies such as chemical scrubbers and activated carbon filters (6–38 \$ m^{-3} h^{-1}) (Kraakman et al., 2014). The costs for the installation of biofilters are attributed to the size of the reactor for an optimum contact time as well as the supply of the packing material. Some biofilters require a large footprint which can mean an additional cost for sites with limited land availability.

The price estimation for the one-stage chemical scrubber could increase if a two-stage or multistage configuration is opted instead, especially if stricter removal efficiencies beyond 95% are required. A two-stage chemical scrubber is also often used when a mixture of odorous air need to be treated, which would otherwise suffice using one-stage scrubbing if solely for treating hydrogen sulfide.

6.2 Selection based on operational requirements

When looking at the context of system operation, it seems that activated carbon filtration is the cheapest technology when treating less problematic emissions, typically having hydrogen sulfide lower than 7.5 ppm and at low airflow of 7500 $m^3 h^{-1}$. On the other hand, activated carbon would be very expensive for higher hydrogen sulfide loadings, especially above 20 ppm (Kraakman et al., 2014). This is because at high H_2S loadings, the carbon filter would be saturated at a faster rate, subsequently diminishing the performance of the odor removal. Replacement of the carbon material is thus required to restore the efficiency of the activated carbon filtration system.

At elevated hydrogen sulfide content and higher airflows above 75,000 $m^3 h^{-1}$, biological treatment systems e.g., biotricking filters, biofilters and their hybrids becomes less expensive as the system works well with increased airflows. Shammay et al. (2018), when comparing 19 odor abatement systems in Australia, found that biotrickling filters were most consistent for hydrogen sulfide removals, while methyl mercaptan was mostly removed when using activated carbon filters. The poor performance of the activated carbon for other sulfur compounds was probably related to an improper treatment design leading to the selection of activated carbon material or retention times not properly suited to the inlet air. Other volatile sulfur compounds including carbon disulfide, dimethyl sulfide, dimethyl disulfide, and dimethyl trisulfide were treated most effectively by biofiltration systems.

In terms of volatile organic compounds, activated carbon filters generally exhibited low VOC removal, in comparison to the biological treatment approaches. The study by Shammay et al. (2018), found that biofilters were superior for the removal of toluene, xylene, and 1,4-dichloro benzene. Clearly, under conditions where a mixture of odorous compounds is expected (not solely hydrogen sulfide) which are usually the case with most industries, a biological system may be more beneficial compared to physical systems. It may be more effective too in certain conditions to employ a combination or two-stage systems in order to provide an additional removal of the odorous compounds.

6.3 Maintenance requirement

In chemical scrubbing, a high consumption of chemicals is needed such as both the caustic and hypochlorite dosing. Wet scrubber on one hand experiences gradual fouling of its packing surface, more so when dealing with airflow at 3–10 pH range. Deposits could also develop when hard water is used. Therefore, regular, properly scheduled maintenance is a must to prevent fouling incidences.

With biofilters, the lifespan of the media can be affected by the acidification of the biofilter bed which ultimately reduces the performance of the system. An average hydrogen sulfide concentration of 5 ppm usually permits a lifespan of five years, whereas a shorter lifespan of two years is the case when treating inlet H_2S of 40 ppm (Kraakman et al., 2014). A relatively short lifespan with the materials (either biofilter media or activated carbon filters) is not favorable in the industry as it adds costs for media disposal and its replacement. When no longer usable, a large volume of these spent materials would require special handling, transportation and disposal. Longer lasting packing materials are available with some are engineered to last beyond 10 years of application, but is considerably more expensive.

6.4 Selection based on preceding environmental impacts

Alfonsín et al. (2015) in their life cycle analysis of different odor treatment systems suggested that the environmental implication from the odor control technologies actually outweighs the negative impact from the odor emissions, if left untreated. Odor control technologies pose risks to the environment for example from the production of glass fiber for biotrickling filters and their disposal in landfills as well as the use of chemicals in activated carbons. The freshwater and marine, terrestrial and human are at risks of toxicity conditions due to the bed material disposal in landfills, energy use, glass fiber production, and large amounts of chemicals usage.

Although biofilters do not exhibit toxicity effects, the technology is also an environmental concern generally due to the secondary emissions of bioaerosol from biofilter operations especially at the outlets of the treatment system. Bioaerosols are formed when the microbial colonies are sheared from the packing media surface by the passing inlet stream and released from the bioreactor which usually aggravates with increasing gas flow rate. Yang et al. (2018) reported that most of the bacteria in bioaerosols released from the biofilters were assigned to the *Pseudomonas* sp., which was the main desulphurizing bacterial genus present on the surface of packing media PM after 81 days. The bacterial species are usually different between the inlet and outlet, with the latter usually having a greater number (about twice) of mesophilic bacteria (Schlegelmilch et al., 2005).

When the Cumulative Energy Demand (CED) of various odor control technologies are taken into consideration, the biofiltration systems accounted for the least CED impacts contributed mainly from minimal usage of energy particularly for the pumping of water and air for regulating moisture and pressure drops (Alfonsín et al., 2015). CED is the direct or indirect consumption of energy in the whole life cycle of a product or process, such as energy used in withdrawing, producing and disposing of raw and supplementary materials (VDI 4600, 1997). Alfonsín et al. (2015) also found that for activated carbon filters and chemical scrubbers, the main energy demands for these systems were contributed by the production of the packing material and the chemicals, respectively.

7. Conclusions

Odor emissions from the food and beverage industry may originate from the processing plant itself such as the fermentation stage with the brewery, or the exhaust from the baking process in the baking industries. In addition, the wastewater generated from the industrial processing presents another source of odor emission especially when anaerobically treated effluents are left uncovered. Some of these air emissions can be controlled via better management at the facility while for others, odor control systems must be installed to ensure the avoidance of odor nuisance to the surrounding community. This chapter has highlighted the possibilities of using a physical, biological or oxidation process for industrial applications and further recommends modifications to the biofiltration system to improve its odor removal capabilities. Research has shown that waste materials such as sugarcane bagasse, composts and concrete waste can be applied in a biofilter system with marked significance in removing odor and sulfur compounds. Further testing at pilot scale in terms of long-term applications are needed to realize the utilization of waste derived biofilter media application in the industries, including food processing, which presents a sustainable approach in both waste disposal and odor remediation.

References

Abdullah, N., Sulaiman, F., 2013. The Oil Palm Wastes in Malaysia. Biomass Now: Sustainable Growth and Use.

Al-Rahbi, A.S., Williams, P.T., 2016. Production of activated carbons from waste tyres for low temperature NOx control. Waste Manag. 49, 188–195.

Alfonsín, C., Lebrero, R., Estrada, J.M., Munoz, R., Kraakman, N.J.R., Feijoo, G., Moreira, M.T., 2015. Selection of odor removal technologies in wastewater treatment plants: a guideline based on Life Cycle Assessment. J. Environ. Manag. 149, 77–84.

Almarcha, D., Almarcha, M., Nadal, S., Poulssen, A., 2014. Assessment of odor and VOC depuration efficiency of advanced biofilters in rendering, sludge composting and waste water treatment plants. Chem. Eng. Transact. 40, 223–228.

An, J., Lou, J., Meng, Q., Zhang, Z., Ma, X., Li, J., 2017. Non-thermal plasma injection-CeO_2-WO_3/TiO_2 catalytic method for high-efficiency oxidation of elemental mercury in coal-fired flue gas. Chem. Eng. J. 325, 708–714.

Andersen, K.B., Beukes, J., Feilberg, A., 2013. Non-thermal plasma for odor reduction from pig houses. Chem. Eng. J. 223, 638–646.

ASTM E679-04, 2011. Standard Practice for Determination of Odor and Taste Thresholds by a Forced-Choice Ascending Concentration Series Method of Limits. American Society for Testing and Materials. ASTM International, West Conshohocken, PA, USA.

Atkins, E., May 30, 2014. End of a chapter in Hogtown history after Toronto's last pig plant shuts its doors. Globe Mail. Retrieved from: https://www.theglobeandmail.com.

Biard, P.-F., Couvert, A., Renner, C., 2017. Intensification of volatile organic compound absorption in a compact wet scrubber at co-current flow. Chemosphere 173, 612–621.

Bokulich, N.A., Bamforth, C.W., 2013. The microbiology of malting and brewing. Microbiol. Mol. Biol. Rev. 77 (2), 157–172.

Brancher, M., Griffiths, D., Franco, D., de Melo Lisbo, H., 2017. A review of odor impact criteria in selected countries around the world. Chemosphere 168, 1531–1570.

Brulliard, K., June 15, 2017. Many People Think a Cage-free Life is Better for Hens. It's Not that Simple. The Washington Post. Retrieved from: https://www.washingtonpost.com.

Bull, M., McIntyre, A., Hall, D., Allison, G., Redmore, J., Pullen, J., Caird, L., Stoaling, M., Fain, R., 2018. IAQM Guidance on the Assessment of Odor for Planning – Version 1.1. Institute of Air Quality Management, London. Retrieved from: www.iaqm.co.uk/text/guidance/odor-guidance-2018.

Cabeza, I.O., López, R., Giraldez, I., Stuetz, R.M., Díaz, M.J., 2013. Biofiltration of a-pinene vapours using municipal solid waste (MSW) – pruning residues (P) composts as packing materials. Chem. Eng. J. 233, 149–158.

Carroll, R., April 10, 2014. Sriracha Hot Sauce Production Declared Public Nuisance by California City. The Guardian. Retrieved from: https://www.theguardian.com.

Chan, A.A., 2006. Biofiltration of Odorous Gas Emissions. (Doctoral Dissertation, Luleå University of Technology).

Chan, W.-C., Ho, S.-J., 2011. A new type of poly(vinyl alcohol)/nitrocellulose/granular activated carbon/KNO_3 composite bead used as a biofilter material. J. Polym. Res. 18, 1871–1877.

Collins English Dictionary, 2018. HarperCollins Publishers. Retrieved from: https://www.collinsdictionary.com.

Conway, A., 2017. World Poultry Market Growth Remains Stagnant in 2016–17. Poultry-Trends 2017. Retrieved from: http://www.poultrytrends.com.

Corey, K., Zappa, L., 2018. Odor Control "ABC's" How to Compare and Evaluate Odor Control Technologies. Retrieved from: http://www.gesodorcontrol.com/abcs.pdf.

Courtois, A., Andrès, Y., Dumon, E., 2014. H_2S biofiltration using expanded schist as packing material: influence of packed bed configurations at constant EBRT. J. Chem. Technol. Biotechnol. 90 (1), 50–56.

Danish Environmental Protection Agency, 2002. Environmental Guidelines No. 9. Industrial Odor Control. Retrieved from: https://eng.mst.dk/trade/industry/industrial-air-pollution/regulation-of-industrial-air-pollution/.

Dobrynkin, N.M., Batygina, M.V., Noskov, A.S., Besson, M., Gallezot, P., 2012. Wet air oxidation of organic acids and pphenol for odor control processes. Chem. Eng. Transact. 30, 277–282.

Dorado, A.D., Lafuente, J., Gabriel, D., Gamisans, X., 2012. Interaction between sorption and biodegradation in a biofilter packed with activated carbon. Water Sci. Technol. 66 (8), 1743–1750.

Drost, E.A., Mcguire, S.J.J., Kwong, K., Zhang, Y., Abousaidi, N., Tamekuni, N., Gandara, C., Shinn, J., Sanchez, P., Gray, T., 2017. Huy Fong foods' Sriracha: Irwindale turns up the heat. J. Case Res. Inq. 3, 47–89.

Dumont, E., Guzman, L.M.A., Susa, M.S.R., Andrès, Y., 2012. H_2S biofiltration using expanded schist as packing material: performance evaluation and packed-bed tortuosity assessment. J. Chem. Technol. Biotechnol. 87, 725–731.

Dunlop, M.W., Blackall, P.J., Stuetz, R.M., 2016. Odor emissions from poultry litter- a review litter properties, odor formation and odorant emissions from porous materials. J. Environ. Manag. 177, 306–319.

Elferink, M., Schierhorn, F., 2016. Global Demand for Food Is Rising. Can We Meet it? Harvard Business Review. Retrieved from: https://hbr.org/2016/04/global-demand-for-food-is-rising-can-we-meet-it.

Elias, A., Barona, A., Gallastegui, G., Berastegi, G.I., 2010. Preliminary acclimation strategies for successful start-up in conventional biofilters. J. Air Waste Manag. Assoc. 60 (8), 959–967.

EN13725, 2003. Air Quality-Determination of Odor Concentration by Dynamic Olfactometry. Committee for European Normalization (CEN), Brussels, Belgium.

Environmental Protection Agency, 2014. EPA Licensed Sites 2014 Report on Air Emissions. Office of Environmental Enforcement. Retrieved from: http://www.epa.ie/pubs/reports/enforcement/.

Estrada, J.M., Quijano, G., Lebrero, R., Munoz, R., 2013. Step-feed biofiltration: a low cost alternative configuration for off-gas treatment. Water Res. 47, 4312–4321.

Fernández, H.T., Rico, I.R., Prida, J.J.d. l., Van Langenhove, H., 2013. Dimethyl sulfide biofiltration using immobilized *Hyphomicrobium* VS and *Thiobacillus thioparus* TK-m in sugarcane bagasse. Environ. Technol. 34 (2), 257–262.

Filho, J.L.R.P., Sader, L.T., Damianovic, M.H.R.Z., Foresti, E., Silva, E.L., 2010. Performance evaluation of packing materials in the removal of hydrogen sulfide in gas phase biofilters: polyurethane foam, sugarcane bagasse, and coconut fibre. Chem. Eng. J. 158 (3), 441–450.

Gilani, C.A., Haug, R.T., Redd, K.R., Wojslaw, J.A., 2007. Prototype testing of an Integrated bio-scrubber and carbon system at the hyperion treatment plant. Water Pract. 1 (2), 1–11.

Gómez-Borraz, T.L., González-Sáncheza, A., Bonilla-Blancas, W., Revah, S., Noyola, A., 2017. Characterization of the biofiltration of methane emissions from municipal anaerobic effluents. Process Biochem. 63, 204–213.

GRACE Communications Foundation, 2018. Local and Regional Food Systems. What is a "Food System?". Retrieved from: http://www.sustainabletable.org/254/local-regional-food-systems.

Guo, H., Dehod, W., Agnew, J., Feddes, J.R., Lague, C., Pang, S., 2007. Daytime odor emission variations from various poultry barns. Am. Soc. Agric. Biol. Eng. 50 (4), 1365–1372.

Guo Ping, W., 2014. Discussion on activated carbon regeneration method. Appl. Mech. Mater. 641–642, 1127–1130.

Higashi, R.M., 2004. Whitepaper: Analysis of Volatile Organic Tropospheric Ozone Precursors from Biogenic Sources. Center for Health and Environment and Crocker Nuclear Laboratory, UC Davis, CA.

Higuchi, T., Ukita, M., Sekine, M., Imai, T., 2007. A case study and recent improvements in odor management in Japan. Water Pract. 1 (2), 1–7.

Hoimark, R.L.L., 2012. Energy Loss during Air Flow through Biofilter Media – Impact of Entrance Region Minor Loss, Biomass Accumulation and Media Characteristics. Aalborg University. Masters dissertation.

Holzer, F., Kopinke, F.D., Roland, U., 2018. Non-thermal plasma treatment for the elimination of odorous compounds from exhaust air from cooking processes. Chem. Eng. J. 334, 1988–1995.

Jaber, M.B., Couvert, A., Amrane, A., Rouxel, F., Cloirec, P.L., Dumont, E., 2016. Biofiltration of H_2S in air—experimental comparisons of original packing materials and modelling. Biochem. Eng. J. 112, 153–160.

Jaber, M.B., Couvert, A., Amrane, A., Cloirec, P.L., Dumont, E., 2017. Removal of hydrogen sulfide in air using cellular concrete waste: biotic and abiotic filtrations. Chem. Eng. J. 319, 268–278.

Janoško, I., Čery, M., 2015. Degradation of animal malodor. Res. Agric. Eng. 61, 60–66.

King, T., Osmond-McLeod, M.J., Duffy, L.L., 2018. Nanotechnology in the food sector and potential applications for the poultry industry. Trends Food Sci. Technol. 72, 62–73.

Kraakman, N.J.R., Estrada, J.M., Lebrero, R., Cesca, J., Muñoz, R., 2014. Evaluating odor control technologies using reliability and sustainability criteria – a case study for water treatment plants. Water Sci. Technol. 69 (7), 1426–1433.

Lee, D.K., 2003. Mechanism and kinetics of the catalytic oxidation of aqueous ammonia to molecular nitrogen. Environ. Sci. Technol. 37, 5745–5749.

Li, L., Lian, J., Han, Y., Liu, J., 2012. A biofilter integrated with gas membrane separation unit for the treatment of fluctuating styrene loads. Bioresour. Technol. 111, 76–83.

Lin, Y.-H., Chen, Y.-P., Ho, K.-L., Lee, T.-Y., Tseng, C.P., 2013. Large-scale modular biofiltration system for effective odor removal in a composting facility. J. Environ. Sci. Health A 48 (11), 1420–1430.

Lloyd, L., 2011. Handbook of Industrial Catalysts, Fundamental and Applied Catalysis. Springer Science+Business Media.

Loh, S.K., 2017. The potential of the Malaysian oil palm biomass as a renewable energy source. Energy Convers. Manag. 141, 285–298.

Lowman, A., McDonald, M., Wing, A., Muhammad, S.N., 2013. Land application of treated sewage sludge: community health and environmental justice. Environ. Health Perspect. 121, 537–542.

Mahlia, T., Abdulmuin, M., Alamsyah, T., Mukhlishien, D., 2011. An alternative energy source from palm wastes industry for Malaysia and Indonesia. Energy Convers. Manag. 42, 2109–2118.

Malaysian Palm Oil Board, 2018a. Number and Capacities of Palm Oil Sectors in Operation as at December 2017 (Tonnes/Year). Retrieved from: http://bepi.mpob.gov.my/index.php/en/statistics/sectoral-status/179-sectoral-status-2017/803-number-a-capacities-of-palm-oil-sectors-2017.html.

Malaysian Palm Oil Board, 2018b. Overview of the Malaysian Oil Palm Industry 2017. Retrieved from: http://bepi.mpob.gov.my/images/overview/Overview_of_Industry_2017.pdf.

Malik, D.J., Strelko Jr., V., Streat, M., Puziy, A.M., 2002. Characterisation of novel modified active carbons and marine algal biomass for the selective adsorption of lead. Water Resour. 36 (6), 1527–1538.

Melse, R.W., Hol, J.M.G., 2017. Biofiltration of exhaust air from animal houses: evaluation of removal efficiencies and practical experiences with biobeds at three field sites. Biosyst. Eng. 159, 59–69.

Ministry for the Environment, 2016. Good Practice Guide for Assessing and Managing Odor. Retrieved from: http://www.mfe.govt.nz/sites/default/files/media/Air/good-practice-guide-odor.pdf.

Needham, C.E., Freeman, T.J., 2009. Case studies in the use of source specific odor modelling guidelines. In: Water NZ Conference and Expo 2009. Retrieved from: https://www.waternz.org.nz/Attachment?Action=Download&Attachment_id=1274.

Noordally, E., Richmond, J.R., Drumm, K.J., 2008. Catalytic oxidation processes for odor and VOC control. Stud. Environ. Sci. 61, 459–467.

Nygård, K., Werner-Johansen, O., Rønsen, S., Caugant, D.A., Simonsen, Ø., Kanestrøm, A., Ask, E., Ringstad, J., Ødegård, R., Jensen, T., Krogh, T., Høiby, E.A., Ragnhildstveit, E., Aaberge, I.S., Aavitsland, P., 2008. An outbreak of legionnaires disease caused by long-distance spread from an industrial air scrubber in Sarpsborg, Norway. Clin. Infect. Dis. 46 (1), 61–69.

Oiamo, T.H., Luginaah, I.N., Baxter, J., 2015. Cumulative effects of noise and odor annoyances on environmental and health related quality of life. Soc. Sci. Med. 146, 191–203.

Olajire, A.A., 2012. The brewing industry and environmental challenges. J. Clean. Prod. https://doi.org/10.1016/j.jclepro.2012.03.003.

Patrick, W., August 1, 2017. Neighbors Complain of Odors from Local Chicken Plant. Palestine Herald-Press. Retrieved from: https://www.palestineherald.com.

Pires, E., Brányik, T., 2015. Biochemistry of Beer Fermentation, SpringerBriefs in Biochemistry and Molecular Biology. Springer International Publishing.

Qamaruz Zaman, N., Yusup, Y., Yaacof, N., 2016. Verification of receptor exposure to palm oil mill odor using in-field olfactometry with odor characteristic. Chem. Eng. Transact. 54, 271–276.

Qian, H., Lin, Y.-L., Xua, B., Wang, L.-P., Gao, Z.-C., Gao, N.-Y., 2018. Adsorption of haloforms onto GACs: effects of adsorbent properties and adsorption mechanisms. Chem. Eng. J. 349, 849–859.

Raboni, M., Torretta, V., Viotti, P., 2017. Treatment of airborne BTEX by a two-stage biotrickling filter and biofilter, exploiting selected bacterial and fungal consortia. Int. J. Environ. Sci. Technol. 14 (1), 19–28.

Population Reference Bureau, 2018. PRB Projects 2.3 Billion More People Living on Earth by 2050. Retrieved from: https://www.prb.org/2018-world-population-data-sheet-with-focus-on-changing-age-structures/.

Rene, E.R., Veiga, M.C., Kennes, C., 2009. Performance of a biofilter for the removal of high concentrations of styrene under steady and non-steady conditions. J. Hazard Mater. 168 (1), 282–290.

Rupani, P., Singh, R., 2010. Review of current palm oil mill effluent (POME) treatment methods: vermicomposting as a sustainable practice. World Appl. Sci. 11 (1), 70–81.

Schiavon, M., Schiorlin, M., Torretta, V., Brandenburg, R., Ragazzi, M., 2017. Non-thermal plasma assisting the biofiltration of volatile organic compounds. J. Clean. Prod. 148, 498–508.

Schlegelmilch, M., Herold, T., Streese, J., Hensel, A., Stegmann, R., 2005. The potential to reduce emissions of airborne microorganisms by means of biological waste gas treatment systems. Waste Manag. 25, 955–964.

Shammay, A., Evanson, I., Stuetz, R.M., 2018. A comparison of removal performance of volatile organic and sulfurous compounds between odor abatement systems. Water Sci. Technol. 77 (11), 2657–2667.

Shareefdeen, Z., 2015. Hydrogen sulfide (H2S) removal using schist packings in industrial biofilter applications. Korean J. Chem. Eng. 32 (1), 15–19.

Sheridan, B.A., Curran, T.P., Dodd, V.A., 2003. Biofiltration of n-butyric acid for the control of odor. Bioresour. Technol. 89 (2), 199–205.

Singh, K., Giri, B.S., Sahi, A., Geed, S.R., Kureel, M.K., Singh, S., Dubey, S.K., Rai, B.N., Kumar, S., Upadhyay, S.N., Singh, R.S., 2017. Biofiltration of xylene using wood charcoal as the biofilter media under transient and high loading conditions. Bioresour. Technol. 242, 351–358.

Song, L., Akoh, C.C., Reynolds, A.E., 2001. Recovery of used frying oils with adsorbent combinations: refrying and frequent oil replenishment. Food Res. Int. 34 (2), 159–166.

Sowka, I., Skretowicz, M., Szklarczyk, M., Zwozdziak, J., 2011. Evaluation of nuisance of odor from food industry. Environ. Protect. Eng. 37 (1), 6–12.

Suffet, I.H., Rosenfeld, P., 2007. The anatomy of odor wheels for odors of drinking water, wastewater, compost and the urban environment. Water Sci. Technol. 55 (5), 335–344.

Suhas, V.K.G., Carrott, P.J.M., Singh, R., Chaudhary, M., Kushwaha, S., 2016. Cellulose: a review as natural, modified and activated carbon adsorbent. Bioresour. Technol. 216, 1066–1076.

Thiesen, P.H., Beneke, K., Lagaly, G., 2002. Silylation of a crystalline silicic acid: an MAS NMR and porosity study. J. Mater. Chem. 12, 3010–3015.

Toretta, V., Raboni, N., Copelli, S., Caruson, P., 2013. Application of Multi-Stage Bioiflter Pilot Plants to Remove Odor and VOCs from Industrial Activities Air Emissions, Energy and Sustainability IV. WIT Press.

Trellu, C., Oturan, N., Keita, F.K., Fourdrin, C., Pechaud, Y., Oturan, M.A., 2018. Regeneration of activated carbon fiber by the electro-fenton process. Environ. Sci. Technol. 52 (13), 7450–7457.

UK Environment Agency, 2007. Using Science to Create a Better Place. Review of Odor Character and Thresholds. Science Report SC030170/SR2. Environment Agency, Bristol, United Kingdom.

Van Harreveld, T., Guillot, J., Baas, N., Bilsen, I., Bree, F., Both, R., Franzen-reuter, I., Mannebeck, D., Nicolas, J., Ogink, N., Arne, O., Claus, R., Andrea, R., Robert, S., Selena, S., Hugo, V.B., Van Boheemen, A., Van Elst, T., Peter, W., September 2013. Progress of EN13725 revision, the standard used to determine odor concentration and odor treatment efficiency. In: Proceedings of the Biotechniques for Air Pollution Control and Bioenergy, Nimes, France, pp. 10–13.

VDI 3940, 2006. Vdi 3940 Part 1. Measurement of Odor Impact by Field Inspection – Measurement of the Impact Frequency of Recognizable Odors Grid Measurement, VDI (Verein Deutsche Ingenieure), Handbuch Reinhaltung der Luft.

VDI 4600, 1997. VDI-Richtilinie 4600: Cumulative Energy Demand, Terms, Definitions, Methods of Calculation, VDI (Verein Deutscher Ingenieure), Duseldorf.

Vikrant, K., Kailasa, S.K., Tsang, D.C.W., Lee, S.S., Kumar, P., Giri, B.S., Singh, R.S., Kim, K.-H., 2018. Biofiltration of hydrogen sulfide: trends and challenges. J. Clean. Prod. 187, 131–147.

Wani, A.H., Branion, R.M.R., Lau, A.K., 1997. Biofiltration: a promising and cost-effective control technology for odors, VOCs and air toxics. J. Environ. Sci. Health 32 (7), 2027–2055.

Wei, Z.S., Li, H.Q., He, J.C., Ye, Q.H., Huang, Q.R., Luo, Y.W., 2013. Removal of dimethyl sulfide by the combination of non-thermal plasma and biological process. Bioresour. Technol. 146, 451–456.

Williams, S.C.P., March 20, 2014. Human nose can detect a trillion smells. Science. Retrieved from: https://www.sciencemag.org/news/2014/03/human-nose-can-detect-trillion-smells.

Wlazlo, L., Nowakowicz-Debek, B., Kapica, J., Kwiecien, M., Pawlak, H., 2016. Removal of ammonia from poultry manure by aluminosilicates. J. Environ. Manag. 183, 722–725.

Yaacof, N., Qamaruz Zaman, N., Yusop, Y., 2015. Odor intensity assessment at different area of a palm oil mill using olfactometry method. Appl. Mech. Mater. 802, 472–477.

Yang, K., Li, L., Ding, W., Liu, J., Xue, S., 2018. A Full-scale thermophilic biofilter in the treatment of sludge drying exhaust: performance, microbial characteristics and bioaerosol emission: a full-scale thermophilic biofilter for sludge drying exhaust treatment. J. Chem. Technol. Biotechnol. 93 (8), 2216–2225.

Yuliusman, N., Sanal, A., Bernama, A., Haris, F., Ramadhan, I.T., 2017. Preparation of activated carbon from waste plastics polyethylene terephthalate as adsorbent in natural gas storage. IOP Conf. Ser. Mater. Sci. Eng. 176, 1–6.

Zhang, Q., Feddes, J., Edeogu, I., Nyachoti, M., House, J., Small, D., Clark, G., 2002. Odor Production, Evaluation and Control. Final Report Submitted to Manitoba Livestock Manure Management Initiative Inc. Retrieved from: https://www.researchgate.net/publication/237329737_Odor_Production_Evaluation_And_Control/download.

Zhang, W., Qu, Z., Li, X., Wang, Y., Ma, D., Wu, J., 2012. Comparison of dynamic adsorption/desorption characteristics of toluene on different porous materials. J. Environ. Sci. 24 (3), 520–528.

Zhou, J., Luo, A., Zhao, Y., 2018. Preparation and characterisation of activated carbon from waste tea by physical activation using steam. J. Air Waste Manag. Assoc. https://doi.org/10.1080/10962247.2018.1460282.

Innovation management and sustainability in the food industry: concepts and models

10

Barbara Bigliardi[1], *Charis Galanakis*[2,3]
[1]Department of Engineering and Architecture, University of Parma, Parma, Italy; [2]Department of Research & Innovation, Galanakis Laboratories, Chania, Greece; [3]Food Waste Recovery Group, ISEKI Food Association, Vienna, Austria

Chapter outline

1. Introduction 315
2. How to classify innovation in the food industry 317
3. Models of innovation in the food industry 321
3.1 Food incremental innovation: the low-fat margarine and me-too food products 323
3.2 Food market breakthrough: the soy-based food 324
3.3 Food technological breakthrough: 3D food printing, the digital gastronomy and foodomics 324
3.4 Food radical innovation: functional foods and foodomics 326
4. Sustainability driven foods innovation: the case of food waste recovery 327
4.1 Other sustainability-driven foods innovation cases 329
5. Concluding remarks 331
References 333
Further reading 340

1. Introduction

Until the end of the 20th century, the prevalent policy was mainly to increase food production, without improving the efficiency of the food systems. This fact increased generation of food lost or wasted along supply chains. In the 21st century, escalating demands for processed foods have required identification of concrete opportunities to prevent depletion of natural resources, restrict energy demands, minimize economic costs as well as reduce food losses and wastes. Besides, recent changes in the legislative frameworks and environmental concerns have stimulated industry to reconsider their management policy and, in some cases, to face the concept of "recovery" as an opportunity. The food industry is today increasing attention toward sustainability, which has been developed into a trendy word characterizing a frame of advances and

The Interaction of Food Industry and Environment. https://doi.org/10.1016/B978-0-12-816449-5.00010-2

modernization in the years to come. However, sustainability is neither easy to specify nor to implement. In theory, it reflects the principle that we must meet the needs of the present without compromising the ability of future generations to meet their own needs. Responding to this goal, sustainability requires the maximum utilization of all raw materials produced and integration of activities throughout all the production-to-consumption stages. Efforts begin at the production stage, with activities to reduce postharvest losses and increase use of waste, and continue through food manufacturing and beyond, to ensure that energy, water, and other resources are used most efficiently, and environmental impacts are minimized. Refrigeration of fresh produce is an example of an action that reduces loss and increases the edible life of the product.

In order to achieve sustainability in the food industry, research and development (R&D) activities based on the creation of new technologies are required. For the implementation of these activities, knowledge and innovation management is a prerequisite. But innovation should be first defined prior exploring its management. And despite its frequent usage, it is neither easy to specify nor to identify. What does innovation truly represent? It is a new idea that when implemented leads to a more effective process, product, service, or technology. Innovation provides better solutions that meet advanced, unaddressed, or existing market needs. It can be considered as a breakthrough that provides a different way of thinking, consuming, or living. However, the key element of innovation is the precondition noted herein—"when implemented." Without implementation, "innovation" turns back to the "idea" status. Today, there is an extensive dialogue about the need of food industries to introduce innovations in the market in order to survive competition. Innovation in the food industry has been discussed not just as an opportunity but also as a precondition for assuring the sustainability of the food sector. It is an important instrument to stand out from the competition and satisfy consumer demands. Ultimately, it is the tool to success in a hard economic environment, the carrier to penetrate new markets, and the key to establish new products or processes (Galanakis, 2016b).

Innovation efforts in the food industry typically target to reduce costs, eliminate waste production and environmental impact. Eco-innovation has been defined "new products and processes creating value for enterprises and clients and reducing (negative) environmental effects" (Fussler and James, 1996). In recent years, eco-innovations, European strategic policy and specifically the generation of mutual benefits for companies and affected stakeholders have received considerable academic attention (Rennings, 2000; Romani-Dias et al., 2018; Segarra-Oña et al., 2011; Díaz-Garcia et al., 2015; Marin et al., 2015). Indeed, climate awareness and external societal pressure encourage companies to develop innovative technologies in order to remain sustainable, avoid environmental deterioration or enable savings that mitigate climate change (Rennings, 2000; Calza et al., 2017). A new knowledge opportunity through open innovation to achieve environmental improvements is thus an emerging field of interest for the food industry.

On the other hand, it's not new that in the innovation literature, the food industry is traditionally regarded as a sector with low research intensity (e.g., Garcia Martinez and Briz, 2000; Christensen et al., 1996; Galanakis, 2016a). Paraphrasing Denning (2011), "the food industry has plenty of apples, but no Apple." With this sentence, Denning (2011) means that "there is no company like Apple showing transformational innovation with exponential growth." Although researchers continuously develop innovative products and technologies, their applications in the food industry face several obstacles. The latest fact concerns more the introduction of the innovations within the food industry, as well as the reaction of related consumers, and less the technological adequacy of the innovative techniques. Indeed, there is a gap between R&D strategy developers and technical R&D associates, as well as a lack of interpretation between the information received by food technologists and consumer scientists.

However, innovations in the form of new products, processes or services (but not only), are recognized as a key factor for companies belonging to the food industry to satisfy consumer expectations, thus standing out from competitors (Menrad, 2004). As a consequence, innovation is more and more widely investigated also within this traditional industry. Innovation developed in the food industry usually aims at developing new or improved products, has to follow nutritional directions and is subject to food additive regulations (Bigliardi and Galati, 2013a). It can focus on a specific food technology area (e.g., food qualities or consumer needs), or on more than one area (e.g., process engineering and product formulation). Moreover, it can follow different implementation strategies: in particular, the innovation literature identifies the two generally accepted strategies labeled as "technology push" and "market pull" (e.g., Cooper et al., 2001). Recently, Galati et al. (2016) explored these concepts in the food industry and under an open innovation perspective.

Despite the huge amount of research on the topic of innovation in the food industry, to date no classification of innovation in this context exists. We believe that such a classification could be useful for both scholars and managers operating in the food industry in order to identify the key factors that drive innovation in this environment, especially within the urgent need for sustainability and reduced environmental impact of our times. Thus, the main aim of this chapter is to identify such a classification and in particular an innovation model suitable for the food industry. For this purpose, we first present the extant literature on innovation in general and in the food industry in particular, with particular emphasis on the innovation models available in the research literature. Afterward, we propose a classification of innovation according to the Chandy and Tellis' model (1998) and provide examples for each kind of innovation. The chapter concludes with comments on recent trends.

2. How to classify innovation in the food industry

The food industry has high relevance for both the economic output and the employment in the European Union, and consequently it is recognized to be one of the most important industrial branches. For several years this economic sector has been seen as a low-research intensity industry, but the technical and economic changes in the production and processing of food, as well as in society (Menrad, 2004), forced the industry to become more and more innovative. Consequently, innovation is an instrument (if not "the" instrument) for food companies to stand out from competitors and fulfill consumer expectations (Menrad, 2004). A broad range of theories and approaches to assess consumer response to innovations have been developed in the past. Many of these approaches have in common that information about the innovative product and its attributes lead to perceptions or attitudes toward the innovative product. Positive attitudes in turn lead to positive responses and subsequently to more consumers buying the product. Modern insights of consumer behavior show that the picture is not as straightforward. Implicit, often unconscious, gut feelings and the influence of the environment are also influencing consumer choice even without the awareness of the consumers themselves (Fischer and Reinders, 2016). Innovation is a multi-dimensional concept. To make a product innovative, one or more of the following requisites are needed: a new feature, changed features and new technology, a new target user, a new concept, new marketing mix, new answer to an existing need or to a latent need, a cost efficiency and new organization/management (E.C.R. Europe, 1999). In the economic literature, numerous classifications of innovation have been developed and applied. Two of the main approaches proposed to classify innovation are based on the type of innovation and on the nature of innovation (Baregheh et al., 2012).

The first way to classify innovation (that is according to the type of innovation) distinguishes innovation according to the outcome of the innovation process. Knight (1967) for instance proposed a four-category classification: (1) organizational structure, (2) production process, (3) people, and (4) product/service. Clarysse et al. (1998) distinguished between technological and nontechnological innovations. Furthermore, they divided these two categories into product and process innovations, and organizational and market innovations respectively. Francis and Bessant (2005) identified four kinds of innovation (that is, position, process, product and paradigm innovation), while Oke et al., (2007) identified three type of innovation, namely: product, service, and process innovation.

Many classifications based on the type of innovation have been proposed or have been applied also in the food context. Earle (1997), distinguished between (1) technological, (2) social and (3) cultural innovations, stressing that innovation has to combine these kinds of innovations in order to produce a food product that satisfies the nutritional, personal and social needs and customer requirements. Social innovation is a new idea that meet social needs and create new relationships. Food redistribution is an example of successful social innovation that tackles food waste and food poverty (Easteal and WRAP, 2013). Cultural innovations are internal changes that depend (and are limited) upon the recombination of already existing elements in culture. Earle (1997) proposed as example of cultural innovation those food products from one country that are adapted for consumers in other countries, by blending with the traditional ethnic food profile and absorbing any religious or cultural needs. The same author highlighted how these innovations are interrelated. That is, technological innovations can lead to social innovations and vice versa: for example, the innovations of gas, electric and then microwave heating on the one hand revolutionized kitchens and cooking methods and on the other hand changed the food industry. The majority of the research follows the classification in product and process innovation (Avermaete, 2002;

Avermaete et al., 2004; Menrad, 2004; Brewin et al., 2009; Capitanio et al., 2010). A product innovation is any good that is perceived by someone as new, that results in a new or improved product and that offers consumer an advantage and a higher utility compared to already existing products (Sabisch, 1991). Product innovations usually focus on market or demand-oriented issues, and in the majority of the case, they focus in application fields known to the food company that innovate. The improvement of useful properties of the product, the increasing of quality, as well as the changing of design are only some examples of important product innovation attributes (Pleschak and Sabisch, 1996). Functional foods are a typical and successful example of product innovation in the food industry (Bigliardi and Galati, 2013a). In a similar way, a process innovation is an adaptation of existing production lines as well the installation of an entirely new infrastructure and the implementation of new technologies that allow the creation of new products (Bigliardi and Dormio, 2009). Process innovations are usually more wide-ranging (that is, they do not refer necessarily to a specific field), and their main aims are the improvement of product quality and the improvement of the production process in terms of time, cost and flexibility. To cite only some, high pressure technology or ultrafiltration techniques are typical examples of process innovations introduced in the food industry (Menrad and Feigl, 2007). However, Weindlmaier (2001) and Brewin, Monchuk, DC & Partridge (2009) highlighted that often product and process innovations go hand in hand, thus making difficult a clear distinction between them. Potato chips provide an example of the interdependence of product and process innovations: the development of the extruder technique (the process innovation) was a prerequisite to produce potato chips (the product innovation). As far as nontechnological innovation is concerned, organizational innovation refers to changes in marketing, purchases, sales, administration, management, staff policy, while market innovation includes the exploitation of new territorial markets and the penetration of new market segments within existing markets. Grunert et al. (1997) added a third category of innovation to the product/process orientation, considering also the market orientation. A market innovation can be defined as the "exploitation of new territorial markets and the penetration of new market segments within existing markets" (Avermate et al., 2003). High levels of market innovation usually characterize R&D intense food products, such as for example functional foods (Bigliardi and Galati, 2013a). In addition to product, process and market innovations, Gellynck and Vermeire (2009) proposed also packaging innovation: food packaging was simply a container to hold food, while in recent years it can play an active role in food quality. An example of packaging innovation is represented by active packages (Risch, 2009). More recently, Baregheh et al. (2012) adopted in their study the classification originally proposed by Francis and Bessant (2005) identifying position and paradigm innovation in addition to the classic product and process one. Specifically, position innovation refers to changes in how a specific product or process is perceived symbolically and how they are used, while they refer to paradigm innovation as any "changes in the underlying mental models which frame what organization does." To understand what is a position innovation in the food industry, the glucose based drink Lucozade can be considered: it originally aimed at helping children and invalids in convalescence and was thus associated with sickness. Smith Kline Beecham abandoned the original concept of sickness and relaunched the product as a health drink aimed at the growing fitness market. In such a way, its new aim was the enhancement of performance aid to healthy exercise (Rao, 2010). Recent examples of paradigm innovation in the food industry include the repositioning of drinks like coffee and fruit as premium designer products (Rao, 2010). Finally, Bigliardi & Galati (2013a), stressing that innovations may occur throughout all parts of the food system, proposed the following classification of food innovations: (1) new food ingredients and materials (from the early purification of vegetable oils and raw sugar to the present-day development of low calorie fat substitutes and sweetener), (2) innovations in fresh foods (including new types or new varieties of fresh foods), as well as innovation in process (e.g., atmospheric and temperature control during storage and transport), new methods of food preservation (from the traditional methods, heat, cold, chemical, dehydration and mechanical, to the more innovative methods including the use of electric fields, magnetic fields, ultrasound, pulsed light and irradiation), or new types of packaging), (3) new food process techniques (such as the realization of complex food formulations combining ingredients that are designed to give specific flavors, structures and nutrition), (4) innovations in food quality (e.g., the ready-to-eat poultry products by Moyra Mac's (CSIRO), (5) new packaging methods (e.g., from the early cellophane packs to the PET bottle), and (6) new distribution or retailing methods (e.g., McDonald's fast foods).

Moreover, even if usually it is not considered within a specific classification, another type of innovation to be considered also within the food context is business model innovation. Paraphrasing Matzler et al. (2013), "business model innovation occurs when a company increases simultaneously customer value and creates a new value creation and revenue model that allows the company to capture some of the value created in a new way." Such a definition emphasizes the system level nature of business model innovation, which contrarily to previous types of innovation involves the firm and its environment as a whole. Nespresso is a convincing example of such kind of innovation (Matzler et al., 2013).

A second classification of innovation refers to its nature that is to its degree of novelty: radical and incremental innovations can be identified according to this classification. An innovation is radical when it results in something new and is associated with radical or fundamental changes. Thus, radical innovations represent a novelty for both the company and the market, and often require complex changes in different company's fields, in particular in terms of intensive R&D activities. High financial expenditures and a high market risk characterize these innovations (Kotler and Bliemel, 1999). Consequently, radical innovations are rare and applied only discontinuously in the food industry: only 2.2% of total product launches (Costa and Jongen, 2006) in front of the high number of incremental innovative products (77% of total product launches) (Avermaete et al., 2004). Conversely, an innovation is incremental when it results in something improved and is associated with an increment or "add-ons" to a previous innovation (Bessant and Tidd, 2007). Most of the innovation developed in the food industry is incremental, mainly because they are quicker to be realized and require lower financial expenditures, because the food industry is low-tech, and because there are few barriers to market entry. Therefore, competitors, who produce the so-called "me-too products," often copy food product characteristics (Tetra Pak,

2004). Indeed, according to Sabisch (1991), an incremental innovation can be further divided into "quasi-new products" and "me-too products." The former refers to products that already exist on the market, but that present at least one different characteristic feature with respect to the existing products. On the contrary, "me-too products" are imitation products that represent a novelty for the company but not for the market. Companies develop me-too products by observing trends in the marketplace and then imitating a product, often a brand leader. "Me-too products" are the vast majority and are introduced by retailers (private label) or by minor niche players (Winger and Wall, 2006).

Finally, according to E.C.R. Europe (1999), food innovations can be classified according to their degree of novelty (high – medium – low) also in a different way. Specifically, "true new product," "new EAN codes" and "not new product" can be identified. These categories can be further split in the following subcategories:

- *high degree of innovativeness*: "classically innovative products" (that is, breakthrough products that appear to the consumer to bring true innovation to a category, or that create new categories) and "equity transfer products" (that is, products that are new to the category but recognized by the consumer, such as for example Mars ice cream);
- *medium degree of innovativeness*: "line extension products" (that is, a new version of a product within the same category, such as the first yogurt with coconut flavor), and "me-too products" (that is, substantially the same products as existing ones, such as a retailer's private label or the second coconut flavor yogurt);
- *low degree of innovativeness*: "seasonal temporary products" (that is products with a short life cycle, such as Easter eggs or bonus packs) and "conversion/substitution products" (that is, products that replace ones already marketed without adding new value for the consumer, such as a food product packaging size raised from 4.5 to 5 kg). This latter subcategory represents products that are not recognized as an innovation at all.

Even if these two main classifications have been proposed separately, they are often considered in an integrated way: that is, a product or a process innovation can be further classified as radical or incremental (e.g., Oke et al., 2007). Moreover, innovation classification is often subjective: what may be new and radical for one person, may be old for another.

3. Models of innovation in the food industry

In addition to the classification of innovation, in the research literature a number of contributions are available proposing theoretical models of innovation. Popadiuk and Chou (2006), for example, proposed a comparison of four different models, namely: the Abernathy and Clark's model (1985), the Henderson and Clark's (1990) model, the Tushman, Anderson and O'Reilly's model (1997) and the Chandy and Tellis' model (1998). All these models are based on a specific classification of innovation and combine technology and market perspectives.

The model proposed by Abernathy and Clark (1985) differentiates between the preservation or destruction of market and technology knowledge and capability, thus classifying innovations in four groups: regular innovation, niche innovation, revolutionary innovation and architectural innovation. In a similar way, Tushman, Anderson, and O'Reilly's model (1997) considers technology cycles and innovations streams, in combination with market knowledge and technology. The classification that derives from these dimensions is as follows: architectural innovation, incremental products, service or process innovation, major product or service innovation, and major process innovation. In addition, the same authors propose a fifth kind of innovation (namely, the generational innovation), as an intermediate phase. Henderson and Clark's model (1990) proposes a classification based on the concepts of knowledge of a product's components and architectural knowledge. From such a combination, four kinds of innovation arise: incremental innovation, radical innovation, architectural innovation and modular innovation. Finally, the model by Chandy and Tellis (1998) suggests the two dimensions technology and markets as base for most definitions of innovations, and as a result four types of innovation arise: incremental innovation, market breakthrough, technological breakthrough, and radical innovation.

As generally stressed by food researchers, most of the innovations introduced in the food industry are market driven (Menrad and Feigl, 2007), and consequently the role of the consumer is fundamental (Bigliardi and Galati, 2013b; Knudsen, 2007). For this reason, we assume that the model that better fit the characteristics and specific issues of this industry is the one proposed by Chandy and Tellis (1998). In this model, the dimension "technology" decides if a technology involved in a product development is new or different from previous ones, while the dimension "markets" determines if a new product satisfy key consumer requirements better than existing ones. These dimensions represent the two most important sources of innovation (Henzing, 2010). Considering two levels (low and high) for each dimension, the earlier-mentioned four kinds of innovation arise. More specifically, an incremental innovation occurs when both the newness of technology and the consumer need fulfillment are low. This kind of innovation does not create a temporary monopoly position for the company, and usually has a low degree of novelty (Menrad and Feigl, 2007). In other words, it involves minor changes in the technology adopted in the new product, and provides relatively low incremental customer benefits in terms needs satisfaction. When the newness of technology is low but consumer fulfillment is high, an innovation labeled as market breakthrough occurs. This kind of innovation is based on core technology, thus similar to existing products, but provides higher customer benefits in terms of needs satisfaction. Conversely, in the opposite situation, that is when the newness of technology is high and the consumer need fulfillment is low, the innovation represents a technological breakthrough: it adopts a different technology, but does not provide superior customer benefits. Finally, in contrast to the previous three, a radical innovation is associated with the combination of high newness of technology and high customer need fulfillment: it involves new technology and provides greater customer needs fulfillment.

The analysis of the extant research on the issue of innovation in the food industry provides examples for each of this kind of innovation. In the following of the section, such a model of classification, depicted in Fig. 10.1, is adopted and adapted to the food industry. Moreover, and examples of each type of innovation are provided.

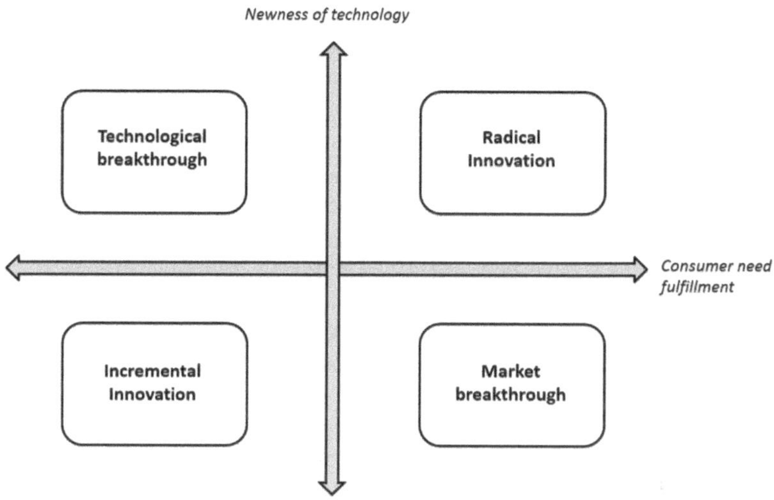

Figure 10.1 The Chandy and Tellis' model (authors' own creation).

3.1 *Food incremental innovation: the low-fat margarine and me-too food products*

Researchers agree in stating that food companies are more focused on incremental innovations as opposed to radical ones. Incremental food innovation refers to changes in food products that do not offer consumers superior benefits, and that, in general, do not create significant additional demand for the company's products. Being the majority of food innovation incremental, numerous examples could be provided. According to the model proposed in Fig. 10.1, both the newness of technology and the customer need fulfillment are low. Menrad and Feigl (2007), for example, cited the low fat margarine as example on incremental innovation, and specifically as example of a "quasi-new product." Margarine may contain trans fat, which raises bas cholesterol, lowers the good one and increases health diseases risk, while low fat margarine contains (as different feature with respect to the existing products) less saturated and trans fat that regular margarine, thus representing a healthier product for the consumer (American Health Association, 2008). With such an innovation, the food industry follows one of the primary trends in food product innovation and contextually one of the major challenges the food industry is facing, that is reducing fat and cholesterol content (Ma and Boye, 2013).

Similarly, numerous examples of food "me-too products" exist, ranging from imitation products introduced by retailers (with their own private label) to imitation products introduced by minor niche players. Examples refer to food products (e.g., a retailer's private label or the second coconut flavor yogurt) but also to other kind of innovation according to the classifications previously described (e.g., McDonald's copied White Castle, inventor of the fast-food burger joint) (White, 2012). The reasons that lead a food company at introducing a me-too product is simple: it sees a competitor dominating a market and wants a piece of the action. Doing so, followers have lower R&D costs, and less risk of failure because the product has already been tested on the market. However, also barriers to this kind of innovation exist: the main refers to the fact that the consumers are often perfectly happy with the present product and do not want to make the change (confirming the low customer need fulfillment provided by incremental innovation). Thus, they have to be shown that there are new benefits, so that they do not create resistance to the me-too product as well as to the brand (Earle and Earle, 2009). Three main reasons for the high mortality rate among me-too products have been identified referring to their distribution channels (private label items are only distributed within the parent retailer), a lack of resources (niche players often lack the sales and marketing resources needed), and their positioning in the market (they are not first to market) (E.C.R. Europe, 1999).

3.2 Food market breakthrough: the soy-based food

Soy-based foods provide high customer benefits in terms of needs satisfaction, and specifically in terms of both real and perceived health-related needs (Drake and Gerard, 2003). Their functions may vary from emulsification to fat or water absorption. When a market breakthrough occurs, it generally creates a new market that the innovator can dominate, at least for some time. The positive health properties recognized to soy-based foods, combined with the health claim approval, lead in fact to the creation of a large market potential for these foods (Drake and Gerard, 2003). Conversely, in terms of technology, the newness of technology is low: usually, the technology used in the production process of soy-based products is similar to that used for the preparation of other products: however, the raw material is natural, without artificial substances added. That is why we classify soy-based foods as market breakthrough.

Soy-based foods come in dozens of forms: some types of soy-based foods are used as ingredients in commercially produced food products, while other kinds are sold as ready-to-eat products. Moreover, these foods can take the place of meat, cheese, eggs, milk, and other animal products in recipes. The Food and Drug Administration (FDA) recognized the health-promoting properties of soy-based foods (e.g., properties allowing the reduction of health-related chronic diseases) and consequently in the 90s allowed the food companies to use health claims on soy-based foods with at least 6.25 g of soy protein/serving (Chang et al., 2012).

3.3 Food technological breakthrough: 3D food printing, the digital gastronomy and foodomics

Food industries have used mass customization to provide customers with individualized information as well as deliver natural, sustainable and environmentally friendly products. At the same time data technologies (e.g., artificial intelligence) and internet of things have already affected the way of food production. In the next decade, the food industry will be profoundly changed by numerous disruptive innovations. We live in a generation where a bundle of technologies will enable the development of personalized health, food and nutrition. The fusion of food and technology will become an intrinsic part of our food production system and diets, leading to the creation of a new economic growth model. Disruptive technologies are a revolutionary type of innovation that can displace an established technology and shakes up the industry or create a completely new industry. Nowadays, this type of innovation is growing fast in all fields including food science and technology. Food companies often rely their sustainability on established (but obsolete) technologies, whereas other food industry producers take the opportunity and risk to produce disruptive innovations in order to become leaders and game changers in the market. Disruptive technology is linked to innovation in products, services or even business models that offer alternative solutions to the market and in many cases target nontraditional consumers.

The mass-producing nature of the food industry has changed the way food is produced by companies, distributed to consumers, as well as the way it is stored and consumed. In addition to the quite traditional approach to food represented by the so-called "slow food movement," we assisted also the development of new technologies aimed at helping to compensate the negative effects of mass production. In this direction, digital gastronomy is seen as a typical example of technological revolution with respect to traditional cooking process (Zoran and Coehlo, 2011). For this reason, we classify digital gastronomy and 3D food printing as technological breakthrough: they adopt a completely new technology, but without providing superior consumer benefits. In digital gastronomy, ingredients can be determined according to online information on nutritional content, personal and social preference. Digital gastronomy is defined by Zoran and Coelho (2011) as the application of digital technologies to cooking in order to create tastes, forms and new food transformations. As they stressed, "it represents an unconventional approach to reimagining the way we eat and get our food in the future."

3D food printing integrates 3D printing and digital gastronomy and revolutionizes the food industry allowing realizing food products with customized shape, color, taste, taste. In addition, nutritional aspects can also be taken into consideration (Sun et al., 2015). With respect to traditional food preparation processes, 3D food printing allows to meet the new demand of food consumers such as, for example, consumer-designed birthday cakes. As described by Huang et al. (2013), it is "a digitally controlled, robotic construction process which can build up complex 3D food products layer by layer." For this reason, it is also known as Food Layered Manufacture (Wegrzyn et al., 2012). Nanotek Instruments, Inc., patented the first-generation of food printer more than 10 years ago, but no prototype was built. Different food printing technologies are available and potentially applicable to food printing: selective laser and hot air sintering, hot-melt extrusion, blinder jetting, inkjet printing (Sun et al., 2015). Currently, applications of 3D food printing are still primitive and only a few examples are available (e.g., the BurritoBot 3D printing or the 3D scan allowing Japanese consumers to order chocolates with their faces shape for Valentine's Day). Different benefits might be reached with 3D food printing: efficient delivery of high-quality food products, personalized food items in terms of shapes, flavors but also nutrition, and so on (Sun et al., 2015).

3.4 Food radical innovation: functional foods and foodomics

Functional foods can bring specific components with healthy benefits beyond those of basic nutrition. The term "functional food" (hereafter, FF) has been defined several times in the literature (Roberfroid, 2002). Bigliardi and Galati (2013a) after reviewing 39 definitions of FFs available in the literature, highlighted that FFs definition encapsulates all or in part the following three main concepts: the concept of health benefits (almost all definitions), the technological process at the basis of the FFs, and their nutritional function. Functional food must be scientifically evaluated prior to commercialization, and once approved, the use of a health claim as a marketing tool significantly improves the competitive position of the food company. Besides, the actual framework for the development of functional foods is diverse in different countries, influencing greatly the innovation policies of food companies. European legislation, does not consider FFs as specific food categories (Coppens et al., 2006; Stanton et al., 2005): they are positioned in a transitional zone between food and pharmaceuticals, in a kind of "gray zone." Legal requirements for the commercialization of these products have profoundly affected the market of functional foods, especially in Europe. A success factor for the marketing of FF is their price premium with respect to "conventional" food products: the literature, however, indicate that consumers are only willing to accept limited price premium for such products (Richardson, 1996). Since the 90s, several multinational food companies (like e.g., Nestle, Danone, Unilever, and others) have developed and introduced FFs into the market, based on the statement "helps the body protect itself" (Young, 1996).

An FF is often a novel food that has resulted from various technological applications (Ronteltap et al., 2007): it requires high newness of technology (e.g., nutrigenomics) and provides greater customer needs fulfillment (in term of health needs). In addition, the design of a functional product means high development costs and important challenges sometimes difficult to achieve by a particular food company. For these reasons numerous FFs, and in particular the probiotic yogurt LC1 of Nestlè, are proposed as examples of radical innovation (Menrad and Feigl, 2007). It was introduced in the early 1990s by researchers and developers at the Nestlé Research Center in Lausanne, Switzerland (Wittkopp, 2004). Two main concepts are at the basis of Nestlé's strategy: innovation (that is develop new products) and renovation (that is renovate existing products). Nestlè's researchers cultivated, due to its probiotic characteristics, the Lactobacillus bacterial strain and brought it to the market as the first ever probiotic yogurt. In doing so, they created a new segment of the yogurt market, and positioned it as an FF. Since its introduction, LC1 had seized a 15% share of the French bio-yogurt market and accounted for 20% of the company's European trade in yogurts and fermented milks (Young, 1996). The product development and market introduction of Nestle's LC1 yogurt is estimated around 50 million US dollars (Grasso et al., 2014). While in the France market LC1 did not reach the success that Nestlè has hoped (consumers did not recognize the better taste in comparison to other yogurt and too scientific was needed to understand its benefits), it has been a successful innovation for the German market (German consumers preferred its taste and this kind of yogurt was new to the market). After its introduction on the market, a "Nutrition" division was established in Nestlé.

Foodomics, defined as "a discipline that studies the Food and Nutrition domains through the application of advanced omics technologies to improve consumer's well-being, health, and confidence," is also an example of radical innovation in the food industry. Indeed, it may help researchers in facing the daunting tasks to meet consumers' needs in terms of food traceability, sustainability, quality, safety and integrity, thus promoting human health and well-being. Over the last years, food and nutrition sciences have changed classical analytical methods with very sophisticated omic tools that create an enormous amount of data. The latest can only be analyzed using bioinformatics. Work published so far has shown the power of foodomics using a couple of approaches like transcriptomics and metabolomics. Integrating and connecting data sets from different expression levels such as gene, transcript, protein, and metabolite is now the major challenge. Nevertheless, a lot of work still have to be conducted in order to transfer this methodology from research laboratories to the food industry. This is happening because most of the foodomics techniques are complex and costly as well as needing highly trained and interdisciplinary personnel (having skills in chemistry, analysis, statistics, biology, and bioinformatics) to obtain robust data. Governments have developed large "omics" centers, sometimes using joint ventures with private industries, whereas some companies have invested money and provide services to laboratories of the public sector. Initiatives of sharing databases and bioinformatics tools have been proposed, too (Sébédio and Malpuech-Brugère, 2016).

4. Sustainability driven foods innovation: the case of food waste recovery

To this line, there is a new type of innovation in the food industry that might not fall in any of the aforementioned four categories. In most cases, these innovations neither show novel technological advances nor meet new consumer needs. However, they are driven by the urgent need for sustainability in the food industry and the modern bioeconomy frame. The so-called "food waste recovery" trend is a typical example of this category. This kind of innovations is nowadays really hype within food science field, although their industrial development has not been fully implemented yet (Galanakis and Schieber, 2014).

Food waste recovery uses underutilized food-based materials (e.g., by-products and removed materials during processing, peel, pulps, wastewaters etc) as a source to extract high added-value micromolecules (e.g., antioxidants, polyphenols, carotenoids) and macromolecules (e.g., dietary fiber, pectin, β-glucan, proteins) prior formulating them in new products, usually additives in foodstuff (Galanakis, 2011, 2012, 2013, 2015a, 2018a,b,f,g,h; Galanakis et al., 2010a, b, c, d, e, 2018a,d; Ananey-Obiri et al., 2018; Galanakis 2019b, c) and cosmetics (Galanakis et al., 2018b, c). Added value can be related to the shelf life (applications as natural antioxidants) or color of the final product (colorants), health benefits (health-promoting substances), formulation, or other product feature (Galanakis et al., 2015a, 2016a).

At this case, innovation concerns basically the utilization of the new source (by-products) and less the application of new technologies or the development of novel food products. In particular, by-products may have decreased quantities of the target compounds compared to the initial source (e.g., the fruit peel might have less antioxidants than the whole fruit), whereas recovery procedures are normally conducted using conventional technologies such as freeze or spray drying, mechanical pressing, ethanol precipitation, isoelectric solubilization, biotechnological approaches, solvent extraction, hydrodistillation, membranes, adsorption on resins, chromatography, and extrusion (Cassano et al., 2016; Deng et al., 2014; Galanakis, 2012, 2017a,b, 2018c, 2018d; Galanakis et al., 2012, 2013a, b, 2014a, 2015c, b, 2016b; Heng et al., 2015, 2016; Patsioura et al., 2011; Rahmanian et al., 2014; Zinoviadou et al., 2015). These technologies have been used in the food industry for a long time. Most of the suggested ones could anyhow be developed using the initial source (e.g., the fruit) instead of its processing by-product. In fact, this might be an additional obstacle.

For instance, olive polyphenols are well-known for their antioxidant activities and health-promoting properties. According to Commission Regulation (EU) 432/2012, "olive oil polyphenols contribute to the protection of blood lipids from oxidative stress. The claim may be used only for olive oil, containing at least 5 mg of hydroxytyrosol and its derivatives (e.g., oleuropein complex and tyrosol) *per* 20 g of olive oil. In order to bear the claim information shall be given to the consumer that the beneficial effect is obtained with a daily intake of 20 g of olive oil." This claim allows olive oil producers to create an innovative product with the health claim on the label if the above requirements are met (Souilem et al., 2017; Galanakis and Kotsiou, 2017; Seçmeler and Galanakis, 2019; Galanakis, 2017b). This is a "food radical innovation" similar to the case of functional food. What if a food industry recovers polyphenols from olive mill wastewater (the by-product of olive oil production) and used it as an additive in another food product or even olive oil? Does this procedure allow the utilization of the health claim? The answer is no (at least within EU) due to the strict regulation and/or the lack of scientific evidence about the health claims induced by polyphenols recovered from olive mill wastewater.

So why not selling olive oils rich in polyphenols instead of foods fortified with polyphenols recovered from olive mill wastewater? These questions indicate that food waste recovery is not certainly driven by new market needs. Of course, there are objections on this claim. For example, there are markets that prefer other kind of products instead of olive oils, e.g., Northern European markets prefer to consume sparkling waters with tastes. However, as a general rule, food waste recovery applications are not driven by market needs. For example, consumers in general prefer natural products but are not always eager to pay additionally for them. Natural antioxidants from food processing by-products are more expensive and less effective compared to the synthetic ones. Primary synthetic antioxidants such as butylated hydroxy anisole and butylated hydroxy toluene have been used in the food industry (especially in oil, fats, meat and bakery products) over the last 50 years (Rojas and Brewer, 2008;

Nanditha and Prabhasankar, 2009). Despite the increasing concern about their safety, these compounds are still preferred by the food industry.

Considering all of the above examples, is food waste recovery a real innovation or more particularly do new sources count enough to consider these approaches "innovations" that should additionally be categorized? This is certainly a debate of our times. Unfortunately, we do not have time to debate on this issue as we run out of resources. In the next decades, increasing population, urbanization and rising incomes will bring a sharp growth in the food industry and alter forever food supply chains. Pressures in natural resources to feed world population and waste disposal costs are already global priorities, whereas consumers' consciousness about environmental issues and legislative pressures increase the requirements of news methods to recover valuable compounds from food waste, rather than its disposal. This is why food waste recovery is a new type of innovation, the sustainability-driven one. This approach is enhanced by more needs of the food industry. For example, to prevent stifling of innovation (from the mounting safety concerns and strict regulations) and simultaneously address emerging wellness issues, a new direction for developing, acquiring and implementing the vast potential of scientific breakthroughs is needed. The establishment of a new label (similar to organic foods) or reduced tax to relevant products could reveal the potentiality of recovering high added-value ingredients from food by-products and reutilizing them in foods (Galanakis et al., 2015a).

4.1 Other sustainability-driven foods innovation cases

As noted above, the sustainability of a product or a process is assessed according to environmental, social, and economic dimensions. Sustainability challenges occur at all stages in the food system from production through processing, distribution, and retailing to consumption and waste disposal. Consequently, the promotion of organic and local foods is not the only way; there is another possibility that may foster continuing the production hegemony, emphasizing biotechnology and technological panaceas. Increasing research is being focused on the development of healthy, quality, and safe food products adapted to consumers' needs and more environmentally friendly processes, that is, processes consuming energy more efficiently, generating less waste, and emitting less greenhouse-effect gases, among other features (Betoret et al., 2016). For instance, energy is typically consumed by changing the temperature (increasing or decreasing) between the different processes. In order to recover part of the consumed energy, a known strategy is to preheat the untreated products with regenerated heat during product's cooling. To improve energy saving efficiency, many other techniques have been proposed either in the literature or already implemented in the processing chain of food products. The so-called emerging (mainly nonthermal) processing technologies have been gaining interest among food researchers, due to their lower impact on nutritional and sensory properties of the products, compared to the conventional thermal techniques (Galanakis, 2013; Barba et al., 2016). Emerging technologies promise shortening of processing and residence times, accelerated heat and mass transfer, control of Maillard reactions, improvement of product quality, enhancement of functionality, protection from environmental stresses, and extended preservation. The most popular emerging or reconsidered technologies examined in the broad field of food science are (Galanakis, 2013; Kovacevic et al., 2018; Roselló-Soto et al., 2015a,b; Ucak et al., 2018, 2019), radio-frequency drying, electro-osmotic dewatering, pressurized extraction, cold plasma treatment, high-hydrostatic pressure, ultrasound-assisted extraction, laser ablation, high voltage electrical discharge, pulsed electric field, pulsed fluid bed agglomeration, nanoencapsulation and others. Improved process efficiency in the food industry is considered in terms of cost minimization via reducing processing steps and restricting overall energy consumption (Barba et al., 2015).

The emerging nonthermal technologies promise to overcome the obstacles of conventional techniques with regard to the energy requirements, but in some cases, this has not been achievable yet. In addition, they have disadvantages that concern mainly the product safety, especially in the case of the most advanced techniques. For instance, the cells membrane permeability of cold plasma or nanotechnology and their effect on biological matrices still remain uninvestigated. Nevertheless, the market and consumers demand for tailored-made processes and products, respectively, will lead producers to adapt sooner or later the aforementioned technologies (Galanakis, 2013). This kind of technologies has both technological breakthrough and radical innovation elements. However, they can also be considered as sustainability-driven innovations, as the main reason behind their development is the reduction of processing cost via reducing energy demands.

Another example of sustainability-driven innovations includes packaging materials. Food packaging is an essential part of the food industry since it protects products from surrounding stress challenges, often of detrimental nature. Most of the currently used packaging plastics are produced from fossil recourses because of the low cost and their advantageous properties such as lightness, flexibility and seal ability. However, increased use of petroleum-based packaging has led to major ecological problems due to their lack of biodegradability (Zinoviadou et al., 2016). Today, plastics cover up to 50% of primary food packaging thanks to their positive features such as lightweight, flexibility, strength, transparency, impermeability, ease of sterilization, and so on. This massive consumption is accompanied by a consistent waste generation with an important environmental footprint. Biobased products are wholly or partly derived from materials of biological origin, excluding materials embedded in geological formations and/or fossilized. These products could reduce dependency on fossil fuels, developing a more sustainable economy. They provide additional product functionalities, less resource intensive production and efficient use of all-natural resources. Biobased industries use as many renewable raw materials as possible for products such as chemicals, materials and energy. This is vital if further depletion of nonrenewable raw materials and an increase in CO_2 emissions is to be avoided. To this line, biobased and biodegradable packaging (containing raw materials originating from agricultural and marine renewable sources) has been gaining lots of interest recently, driven by the need for sustainability (Galanakis, 2018e). However, for food packaging application, the prolongation of food shelf life is required to save money, material consumption and waste, both for food and plastic point of view and not many biopolymers could be employed for this scope. Therefore, sustainability-driven innovation at this case leads to the development of new biobased packaging materials.

Modern sanitation is another example of sustainability-driven innovation in the food industry. Food sanitation is the exercise of subsequent measures to avoid the microbial contamination of food processing. Cleaning and sanitizing procedures should actively destroy the vegetative cells of hazardous microorganisms in the production line, and also eliminate undesirable microorganisms without adversely affecting the quality of food products and the environment. Besides, the attachment of food pathogenic bacteria to food-contact can lead to possible hygienic complications. Decontaminating procedures were typically performed with different chemical and physicochemical techniques, but some of them possess numerous drawbacks, e.g., adverse effects on the quality of the products, remaining chemicals, negative impact on the environment, etc. These facts led to development of alternative sanitizers effective against food spoilage and pathogenic bacteria, yet harmless to humans and the environment. Electrolyzed Water is an interesting option on this direction, as it concerns ecologically, friendly, cleaning and biocidal solution, as well as a clean-in-place sanitizing agent of the industrial equipment. In addition, cleaning and sanitization in food processing surfaces are significant measures, but disinfectants and other cleaning material alone are not competent to eradicate the anchored biofilm and do not prevent the microbial populations to colonize the surfaces. The prevention of microbial biofilms could be attained through the use of biosurfactants (Galanakis, 2018e).

5. Concluding remarks

One of the keys to any successful business is being able to come up with new innovations to keep operations, products and services fresh. Innovations play an important role in food production keeping firms safe from their competitors in the market. The process of innovation should be a key aspect of any food organization's strategy. The Chandy and Tellis' model (1998) emphasizes two main concepts to be considered when classifying innovation in the food context: the market, and in particular the consumer, and the technology. The earlier-mentioned model, adapted to the food industry and including the main examples described in the chapter, is depicted in Fig. 10.2.

An understanding of the consumer's needs is vital if opportunities for innovation are to be identified (Cooper et al., 2001). Consumers are becoming more and more heterogeneous and creative, thus making difficult for food companies to understand and predict their food choice. Moreover, the growing importance of values such as quality of life, health and well-being is also influencing the way consumers perceive and evaluate foods, playing a significant role in determining their choice. Also, the management of the technology, both at the process and product level, is fundamental in order to achieve innovation success. The level reached by technology in recent years has allowed to food companies to develop products that only few decades ago were beyond imagination. We refer for example to the development made both in the fields

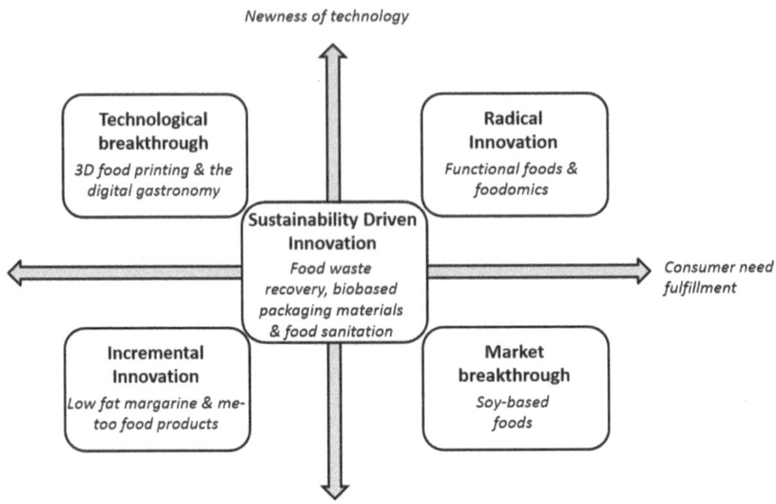

Figure 10.2 The Chandy and Tellis' model for the food industry (authors' own creation).

of biotechnology or chemistry, but also in terms of new process for "constructing" food (e.g., Zoran and Coelho, 2011). All these technological progresses have allowed, for example, the fulfillment of the new consumers' health-related trends.

Additionally, we assisted to a recent stream of literature investigating another link between consumer and technology (but not at the product or process level as previously discussed). We refer to crowdsourcing, defined as "a type of participative online activity in which an individual, an institution, a non-profit organization, or company proposes to a group of individuals of varying knowledge, heterogeneity, and number, via a flexible open call, the voluntary undertaking of a task." (Estellés-Arolas and González-Ladrón-de-Guevara, 2012). The literature shows a growing number of recent examples of its application in the food industry: for example, the cases of Starbucks, Unilever, and Coca Cola (Arcese et al., 2015). These crowdsourcing applications facilitate the identification of food consumers' requirements, being the consumer the central player, and highlight the role of (information and communication) technology as enabling factor.

In the past the food industry was traditionally focused on the cost minimization of production line, giving little attention to real consumer needs. Thereafter, food chain was reversed, and the supply based need for food was switched to a demand-based concept. At this case the consumers tell producers what they want to eat. Following this trend, modern innovations in food companies is conducted by several drivers, e.g., to improve the quality of the products and increase their range, develop flexible processes, provide better services, reduce labor costs, etc. The food industry faces today challenges that cannot be dealt with by the sector's present state in organization, technology, cooperation and communication. In addition, the growing awareness of environmental sustainability has fully reached business reality in the food industry. Companies and consumers are looking for alternatives to mitigate pressing environmental demands resulting from continuous population and economic growth. Modern food companies must deal with an increasingly competitive scenario in which innovation is regarded as a survival requirement in most markets. Thus, major challenges include also assuring food security and safety while respecting society's environmental and social concerns toward sustainability. In a world with increasing population, urbanization and requirements on higher value diets, the challenges should be addressed with limitations and reductions in the availability of resources. Indeed, in many cases challenges were obtained because of responsibilities placed by society on the food sector. Although not enough elucidated yet, the increasing demand for sustainability within the food industry will bring soon more innovations of this kind in the market. Some examples were already provided in this chapter.

In addition, we briefly presented the concept of "business model innovation" in the food industry. In our opinion, this type of innovation should be taken into consideration more and more also in the food context, in order to create a company's value distinct from that of its competitors. This implies the redefinition of consumers' value, the development of a new way to structure the value creation process and the definition of the activities needed to capture that value. In such a context (information and communication) technology may offer unprecedented opportunities to rearrange value creation and value appropriation activities in new and different ways (Matzler et al., 2013). Consequently, food companies should consider business model innovation as a concrete opportunity to build new and sustainable competitive advantage. Last but not least, we proposed a new kind of innovation in the food industry: the sustainability-driven one.

References

Abernathy, W.J., Clark, K.B., 1985. Innovation: mapping the winds of creative destruction. Res. Pol. 14 (1), 3–22.

American Heart Association, 2008. In: Grundy, S.M., Winston, M. (Eds.), Low-fat, Low-Cholesterol Cookbook: Heart-Healthy, Easy-To-Make Recipes that Taste Great. Times Books.

Ananey-Obiri, D., Matthews, L., Azahrani, M.H., Ibrahim, S.A., Galanakis, C.M., Tahergorabi, R., 2018. Application of protein-based edible coatings for fat uptake reduction in deep-fat fried foods with an emphasis on muscle food proteins. Trends Food Sci. Technol. 80, 167–174.

Arcese, G., Flammini, S., Luchetti, M.C., Martucci, O., 2015. Evidence and experience of open sustainability innovation practices in the food sector. Sustainability 7, 8067–8090.

Avermaete, T., 2002. Systems of innovation: the case of small food firms in the EU. In: DRUID PhD Winter Conference, pp. 17–19.

Avermaete, T., Viaene, J., Morgan, E.J., Crawford, N., 2003. Determinants of innovation in small food firms. Eur. J. Innov. Manag. 6 (1), 8–17.

Avermaete, T., Viaene, J., Morgan, E.J., Pitts, E., Crawford, N., Mahon, D., 2004. Determinants of product and process innovation in small food manufacturing firms. Trends Food Sci. Technol. 15 (10), 474–483.

Barba, F.J., Galanakis, C.M., Esteve, M.J., Frigola, A., Vorobiev, E., 2015. Potential use of pulsed electric technologies and ultrasounds to improve the recovery of high-added value compounds from blackberries. J. Food Eng. 167, 38–44.

Barba, F.J., Orlien, V., Mota, M.J., Lopes, R.P., Pereira, S.A., Saraiva, J.A., 2016. Chapter 7. Implementation of emerging technologies. In: Galanakis, C.M. (Ed.), Innovation Strategies for the Food Industry: Tools for Implementation. Elsevier Inc., Waltham, pp. 117–148.

Baregheh, A., Rowley, J., Sambrook, S., Davies, D., 2012. Innovation in food sector SMEs. J. Small Bus. Enterp. Dev. 19 (2), 300–321.

Bessant, J., Tidd, J., 2007. Innovation and Entrepreneurship. Whiley, Chichester.

Betoret, E., Calabuig-Jimenez, L., Betoret, N., Barrer, C., Segui, L., Fito, P., 2016. Chapter 8. Sustainable innovation in food science and engineering. In: Galanakis, C.M. (Ed.), Innovation Strategies for the Food Industry: Tools for Implementation. Elsevier Inc., Waltham, pp. 149–166.

Bigliardi, B., Dormio, A.I., 2009. An empirical investigation of innovation determinants in food machinery enterprises. Eur. J. Innov. Manag. 12 (2), 223–242.

Bigliardi, B., Galati, F., 2013a. Innovation trends in the food industry: the case of functional foods. Trends Food Sci. Technol. 31 (2), 118–129.

Bigliardi, B., Galati, F., 2013b. Models of adoption of open innovation within the food industry. Trends Food Sci. Technol. 30 (1), 16–26.

Brewin, D.G., Monchuk, D.C., Partridge, M.D., 2009. Examining the adoption of product and process innovations in the Canadian food processing industry. Can. J. Agric. Econ./Revue Canadienne d'agroeconomie 57 (1), 75–97.

Calza, F., Parmentola, A., Tutore, I., 2017. Green innovation development: a multiple case study analysis. In: Loué, C., Ben Slimane, S. (Eds.), 12th European Conference on Innovation and Entrepreneurship ECIE 2017, pp. 116–124. Paris, France.

Capitanio, F., Coppola, A., Pascucci, S., 2010. Product and process innovation in the Italian food industry. Agribusiness 26 (4), 503–518.

Cassano, A., Conidi, C., Muñoz, R.C., Galanakis, C.M., 2016. Chapter 7: recovery of polyphenols from olive mill wastewaters by membrane operations. In: Figoli, A., Cassano, A., Basile, A. (Eds.), Membrane Technologies for Biorefining. Elsevier Ltd., Waltham, pp. 163–188.

Chandy, R.K., Tellis, G.J., 1998. Organizing for radical product innovation: the overlooked role of willingness to cannibalize. J. Mark. Res. 35 (4), 474–487.

Chang, J.B., Moon, W., Balasubramanian, S.K., 2012. Consumer valuation of health attributes for soy-based food: a choice modeling approach. Food Policy 37 (3), 335–342.

Christensen, J.L., Rama, R., Von Tunzelmann, N., 1996. Study on Innovation in the European Food Products and Beverages Industry, European Innovation Monitoring System, vol. 35. EIMS Publication. European Commission, Directorate General XIII, Luxembourg.

Clarysse, B., Van Dierdonck, R., Gabriëls, W., Lambrechts, J., Uytterhaegen, M., 1998. Strategische verschillen tussen innovatieve KMOs: een kijkje in de zwarte doos. Publication No. 5. IWT, Brussels.

Commission Regulation (EU), May 25, 2012. No 432/2012 of 16 May 2012 establishing a list of ermitted health claims made on foods, other than those referring to the reduction of disease risk and to children's development and health. Off. J. Eur. Union L 136/1, 1–40.

Coppens, P., Fernandes Da Silva, M., Pettman, S., 2006. European regulations on nutraceuticals, dietary supplements and functional foods: a framework based on safety. Toxicology 221, 59–74.

Cooper, R.G., Edgett, S.J., Kleinschmidt, E.J., 2001. Portfolio Management for New Products. Basic Books.

Costa, A.I., Jongen, W.M.F., 2006. New insights into consumer-led food product development. Trends Food Sci. Technol. 17 (8), 457–465.
CSIRO, January19, 2019. Food safety and quality in the food industry. Available from: http://www.csiro.au/en/Research/FNF/Areas/Food-innovation-centre/Our-expertise/Food-safety.
Deng, Q., Zinoviadou, K.G., Galanakis, C.M., Orlien, V., Grimi, N., Vorobiev, E., Lebovka, N., Barba, F.J., 2014. The effects of conventional and non-conventional processing on glucosinolates and its derived forms, isothiocyanates: extraction, degradation and applications. Food Eng. Rev. 7, 357–381.
Denning, S., 2011. Why No Apple in the Food Industry? Available from: http://www.forbes.com/sites/stevedenning/2011/08/07/why-no-apple-in-the-food-industry/#37135d9d3f4c.
Díaz-García, C., González-Moreno, Á., Sáez-Martínez, F.J., 2015. Eco-innovation: insights from a literature review. Innovation 17 (1), 6–23.
Drake, M.A., Gerard, P.D., 2003. 'Consumer attitudes and acceptability of soy-fortified yogurts'. J. Food Sci. 68 (3), 1118–1122.
Earle, M.D., 1997. Innovation in the food industry. Trends Food Sci. Technol. 8, 166–175.
Earle, M.D., Earle, R.L., 2009. Creating New Foods. The Product Developer's Guide, the Web Edition, 2009. The New Zealand Institute of Food Science & Technology (Inc.).
Easteal, S., WRAP, 2013. 'How Can Social Innovation Help Reduce Food Waste?', WP4, T4.1 June 2013.
Estellés-Arolas, E., González-Ladrón-de-Guevara, F., 2012. Towards an integrated crowdsourcing definition. J. Inf. Sci. 38 (2), 189–200.
Europe, E.C.R., 1999. Efficient Product Introductions: The Development of Value-Creating Relationships. ECR Europe. Available from: http://ecr-all.org/content/ecropedia_element.php?ID=11914.
Fischer, A.R.H., Reinders, M.J., 2016. Chapter 14. Consumer acceptance of novel foods. In: Galanakis, C.M. (Ed.), Innovation Strategies for the Food Industry: Tools for Implementation. Elsevier Inc., Waltham, pp. 271–292.
Francis, D., Bessant, J., 2005. Targeting innovation and implications for capability development. Technovation 25 (3), 171–183.
Fussler, C., James, P., 1996. A Breakthrough Discipline for Innovation and Sustainability. Pitman Publishing, London, UK.
Galanakis, C.M., 2011. Olive fruit and dietary fibers: components, recovery and applications. Trends Food Sci. Technol. 22 (4), 175–184.
Galanakis, C.M., 2012. Recovery of high added-value components from food wastes: conventional, emerging technologies and commercialized applications. Trends Food Sci. Technol. 26, 68–87.
Galanakis, C.M., 2013. Emerging technologies for the production of nutraceuticals from agricultural by-products: a viewpoint of opportunities and challenges. Food Bioprod. Process. 91, 575–579.
Galanakis, C.M., 2015a. Separation of functional macromolecules and micromolecules: from ultrafiltration to the border of nanofiltration. Trends Food Sci. Technol. 42, 44–63.
Galanakis, C.M., 2016a. Preface. In: Galanakis, C.M. (Ed.), Innovation Strategies for the Food Industry: Tools for Implementation. Elsevier Inc., Waltham, pp. xvii–xviii.
Galanakis, C.M., 2016b. In: Galanakis, C.M. (Ed.), 'Chapter 1. Preface' in Innovation Strategies For the Food Industry: Tools for Implementation. Elsevier Inc., Waltham, pp. xvii–xviii.
Galanakis, C.M., 2017a. Modeling in food and bioproducts processing using Boltzmann entropy equation: a viewpoint of future perspectives. Food Bioprod. Process. 106, 102–107.
Galanakis, C.M., 2017b. Olive Mill Waste: Recent Advances for the Sustainable Management. Elsevier-Academic Press. ISBN: 9780128053140.

Galanakis, C.M., 2018a. Phenols recovered from olive mill wastewater as additives in meat products. Trends Food Sci. Technol. 79, 98–105.
Galanakis, C.M., 2018b. Preface. In: Galanakis, C.M. (Ed.), Sustainable Food Systems from Agriculture to Industry: Improving Production and Processing. Elsevier Inc., Waltham.
Galanakis, C.M., 2018c. Polyphenols: Properties, Recovery and Applications. Elsevier-Academic Press.
Galanakis, C.M., 2018d. Sustainable Meat Production and Processing. Elsevier-Academic Press. ISBN: 9780128148747.
Galanakis, C.M., 2018e. Sustainable Food Systems from Agriculture to Industry: Improving Production and Processing. Elsevier-Academic Press.
Galanakis, C.M., 2019a. Proteins: Sustainable Source, Processing and Applications. Elsevier Inc.
Galanakis, C.M., 2019b. The Role of Alternative and Innovative Food Ingredients and Products in Consumer Wellness. Elsevier Inc.
Galanakis, C.M., Kotsiou, K., 2017. Chapter 10. Recovery of bioactive compounds from olive mill waste. In: Galanakis, C.M. (Ed.), Olive Mill Waste: Recent Advances for the Sustainable Management. Elsevier Inc., Waltham, pp. 205–230.
Galanakis, C.M., Schieber, A., 2014. Editorial. Special Issue on Recovery and utilization of valuable compounds from food processing by-products. Food Res. Int. 65, 299–230.
Galanakis, C.M., Tornberg, E., Gekas, V., 2010a. A study of the recovery of the dietary fibres from olive mill wastewater and the gelling ability of the soluble fibre fraction. LWT-Food Sci. Technol. 43 (7), 1009–1017.
Galanakis, C.M., Tornberg, E., Gekas, V., 2010b. Clarification of high-added value products from olive mill wastewater. J. Food Eng. 99 (2), 190–197.
Galanakis, C.M., Tornberg, E., Gekas, V., 2010c. Dietary fiber suspensions from olive mill wastewater as potential fat replacements in meatballs. LWT-Food Sci. Technol. 43 (7), 1018–1025.
Galanakis, C.M., Tornberg, E., Gekas, V., 2010d. Recovery and preservation of phenols from olive waste in ethanolic extracts. J. Chem. Technol. Biotechnol. 85 (8), 1148–1155.
Galanakis, C.M., Tornberg, E., Gekas, V., 2010e. The effect of heat processing on the functional properties of pectin contained in olive mill wastewater. LWT-Food Sci. Technol. 43 (7), 1001–1008.
Galanakis, C.M., Fountoulis, G., Gekas, V., 2012. Nanofiltration of brackish groundwater by using a polypiperazine membrane. Desalination 286, 277–284.
Galanakis, C.M., Goulas, V., Tsakona, S., Manganaris, G.A., Gekas, V., 2013a. A knowledge base for the recovery of natural phenols with different solvents. Int. J. Food Prop. 16, 382–396.
Galanakis, C.M., Markouli, E., Gekas, V., 2013b. Fractionation and recovery of different phenolic classes from winery sludge via membrane filtration. Separ. Purif. Technol. 107, 245–251.
Galanakis, C.M., Chasiotis, S., Botsaris, G., Gekas, V., 2014a. Separation and recovery of proteins and sugars from Halloumi cheese whey. Food Res. Int. 65, 477–483.
Galanakis, C.M., Barba, F.J., Prasad, K.N., 2015a. In: Galanakis, C.M. (Ed.), 'Chapter 14: Cost and Safety Issues of Emerging Technologies against Conventional Techniques', Food Waste Recovery: Processing Technologies and Techniques. Elsevier Inc., Waltham, pp. 321–336.
Galanakis, C.M., Kotanidis, A., Dianellou, M., Gekas, V., 2015b. Phenolic content and antioxidant capacity of Cypriot Wines. Czech J. Food Sci. 33 (2), 126–136.

Galanakis, C.M., Patsioura, A., Gekas, V., 2015c. Enzyme kinetics modeling as a tool to optimize food biotechnology applications: a pragmatic approach based on amylolytic enzymes. Food Sci. Technol. 55 (12), 1758–1770.

Galanakis, C.M., Cvejic, J., Verardo, V., Segura-Carretero, A., 2016a. Chapter 11. Food use for social innovation by optimizing food waste recovery strategies. In: Galanakis, C.M. (Ed.), Innovation Strategies for the Food Industry: Tools for Implementation. Elsevier Inc., Waltham, pp. 211–236.

Galanakis, C.M., Muñoz, R.C., Cassano, A., Conidi, C., 2016b. Chapter 8: recovery of high-added value compounds from food waste by membrane technology. In: Figoli, A., Cassano, A., Basile, A. (Eds.), Membrane Technologies for Biorefining. Elsevier Ltd, Waltham, pp. 189–216.

Galanakis, C.M., Tsatalas, P., Charalambous, Z., Galanakis, I.M., 2018a. Control of microbial growth in bakery products fortified with polyphenols recovered from olive mill wastewater. Environ. Technol. Innov. 10, 1–15.

Galanakis, C.M., Tsatalas, P., Galanakis, I.M., 2018b. Implementation of phenols recovered from olive mill wastewater as UV booster in cosmetics. Ind. Crops Prod. 111, 30–37.

Galanakis, C.M., Tsatalas, P., Galanakis, I.M., 2018c. Phenols from olive mill wastewater and other natural antioxidants as UV filters in sunscreens. Environ. Technol. Innov. 9, 160–168.

Galanakis, C.M., Tsatalas, P., Charalambous, Z., Galanakis, I.M., 2018d. Polyphenols recovered from olive mill wastewater as natural preservatives in extra virgin olive oils and refined olive kernel oils. Environ. Technol. Innov. 10, 62–70.

Galati, F., Bigliardi, B., Petroni, A., 2016. Open innovation in food firms: implementation strategies, drivers and enabling factors. Int. J. Innov. Manag. 20 (3), 1650042.

Garcia Martinez, M., Briz, J., 2000. Innovation in the Spanish food & drink industry. Int. Food Agribus. Manag. Rev. 3, 155–176.

Gellynck, X., Vermeire, B., 2009. The contribution of regional networks to innovation and challenges for regional policy. Int. J. Urban Reg. Res. 33 (3), 719–737.

Grasso, S., Brunton, N.P., Lyng, J.G., Lalor, F., Monahan, F.J., 2014. 'Healthy processed meat products–regulatory, reformulation and consumer challenges'. Trends Food Sci. Technol. 39 (1), 4–17.

Grunert, K.G., Harmsen, H., Meulenberg, M., Kuiper, E., Ottowitz, T., Declerck, F., Göransson, G., 1997. A Framework for Analysing Innovation in the Food Sector. Springer, US, pp. 1–37.

Henderson, R.M., Clark, K.B., 1990. Architectural innovation: the reconfiguration of existing product technologies and the failure of established firms. Adm. Sci. Q. 35 (1), 9–30.

Heng, W.W., Xiong, L.W., Ramanan, R.N., Hong, T.L., Kong, K.W., Galanakis, C.M., Prasad, K.N., 2015. Two level factorial design for the optimization of phenolics and flavonoids recovery from palm kernel by-product. Ind. Crops Prod. 63, 238–248.

Heng, W.W., Yang, B., Prasad, K.N., Ramanan, R.N., Galanakis, C.M., Sun, J., Ismail, A., Ti, T.B., 2016. Extraction of phytochemicals using hydrotropic solvents. Separ. Sci. Technol. 51, 1151–1165.

Henzing, C., 2010. Product Innovation in the Dutch Food and Beverage Industry. Wageningen Academic Publishers.

Huang, S.H., Liu, P., Mokasdar, A., Hou, L., 2013. Additive manufacturing and its societal impact: a literature review. Int. J. Adv. Manuf. Technol. 67 (5–8), 1191–1203.

Knight, K.E., 1967. A descriptive model of the intra-firm innovation process. J. Bus. 40 (4), 478–496.

Knudsen, M.P., 2007. The relative importance of interfirm relationships and knowledge transfer for new product development success. J. Prod. Innov. Manag. 24, 117–138.

Kotler, P., Bliemel, F., 1999. Marketing Management. C.E. Poeschel Verlag, Stuttgart.

Kovacevic, D.B., Barba, F.J., Granato, D., Galanakis, C.M., Herceg, Z., Dragovic-Uzelac, V., Prutnik, P., 2018. Pressurized hot water extraction (PHWE) for the green recovery of bioactive compounds from steviol glycosides from Stevis rebaudiana Bertoni Leaves. Food Chem. 254, 150–157.

Ma, Z., Boye, J.I., 2013. Advances in the design and production of reduced-fat and reduced-cholesterol salad dressing and mayonnaise: a review. Food Bioprocess Technol. 6 (3), 648–670.

Marin, G., Marzucchi, A., Zoboli, R., 2015. SMEs and barriers to Eco-innovation in the EU: exploring different firm profiles. J. Evol. Econ. 25 (3), 671–705.

Matzler, K., Bailom, F., Friedrich von den Eichen, S., Kohler, T., 2013. Business model innovation: coffee triumphs for Nespresso. J. Bus. Strat. 34 (2), 30–37.

Menrad, K., 2004. Innovations in the food industry in Germany. Res. Pol. 33, 845–878.

Menrad, K., Feigl, S., 2007. 'Innovations in Traditional Food Products in Small and Medium-Sized Companies of the Food Industry. Available from: http://www.wz-straubing.de/fachhochschuleweihenstephan/download/literature%20review_truefood.pdf.

Nanditha, B., Prabhasankar, P., 2009. Antioxidants in bakery products. Crit. Rev. Food Sci. Nutr. 49, 1–27.

Oke, A., Burke, G., Myers, A., 2007. Innovation types and performance in growing UK SMEs. Int. J. Oper. Prod. Manag. 27 (7), 735–753.

Pak, T., 2004. Company Magazine Number 89, Lund, Sweden.

Patsioura, A., Galanakis, C.M., Gekas, V., 2011. Ultrafiltration optimization for the recovery of β-glucan from oat mill waste. J. Membr. Sci. 373 (1–2), 53–63.

Pleschak, F., Sabisch, H., 1996. Innovationsmanagement. Schäffer-Pöschel, Stuttgart.

Popadiuk, S., Choo, C.W., 2006. Innovation and knowledge creation: how are these concepts related? Int. J. Inf. Manag. 26 (4), 302–312.

Rahmanian, N., Jafari, S.M., Galanakis, C.M., 2014. Recovery and removal of phenolic compounds from olive mill wastewater. J. Am. Oil Chem. Soc. 91, 1–18.

Rao, S.R., 2010. Types of Innovation. Available from: http://www.citeman.com/11298-types-of-innovation.html.

Rennings, K., 2000. 'Redefining innovation—eco-innovation research and the contribution from ecological economics'. Ecol. Econ. 32 (2), 319–332.

Richardson, D., 1996. Probiotics and product innovation. Nutr. Food Sci. 96 (4), 27–33.

Risch, S.J., 2009. Food packaging history and innovations. J. Agric. Food Chem. 57 (18), 8089–8092.

Robertfroid, M.B., 2002. Global view on functional foods: European perspectives. Br. J. Nutr. 88, S133–S138.

Rojas, C.M., Brewer, S., 2008. Effect of natural antioxidants on oxidative stability of frozen, vacuum-packaged beef and pork. J. Food Qual. 31, 173–188.

Romani-Dias, M., dos Santos Barbosa, A., Walchhutter, S., 2018. Green Innovation: complexidades e revelações a partir da literatura internacional. Revista de Gestão Ambiental e Sustentabilidade 7 (2), 359–377.

Ronteltap, A., Van Trijp, J.C.M., Renes, R.J., Frewer, L.J., 2007. Consumer acceptance of technology-based food innovations: lessons for the future of nutrigenomics. Appetite 49 (1), 1–17.

Roselló-Soto, E., Barba, F.J., Parniakov, O., Galanakis, C.M., Grimi, N., Lebovka, N., Vorobiev, E., 2015a. High voltage electrical discharges, pulsed electric field and

ultrasounds assisted extraction of protein and phenolic compounds from olive kernel. Food Bioprocess Technol. 8 (4), 885–894.
Roselló-Soto, E., Galanakis, C.M., Brnčić, M., Orlien, V., Trujillo, F.J., Mawson, R., Knoerzer, K., Barba, F.J., 2015b. Clean recovery of antioxidant compounds from plant foods, by-products and algae assisted by ultrasounds processing. Modeling approaches to optimize processing conditions. Trends Food Sci. Technol. 42 (2), 134–149.
Sabisch, H., 1991. Produktinnovationen. C.E. Poeschel, Stuttgart.
Sébédio, J.-L., Malpuech-Brugère, C., 2016. Chapter 13. Implementation of foodomics in the food industry. In: Galanakis, C.M. (Ed.), Innovation Strategies for the Food Industry: Tools for Implementation. Elsevier Inc., Waltham, pp. 251–270.
Segarra-Oña, M.V., Peiró-Signes, A., Albors-Garrigós, J., Miret-Pastor, P., 2011. Impact of innovative practices in environmentally focused firms: moderating factors. Int. J. Environ. Res. 5 (2), 425–434.
Seçmeler, Ö., Galanakis, C.M., 2019. Chapter 8. Olive fruit and olive oil. In: Galanakis, C.M. (Ed.), Innovations in Traditional Foods. Elsevier Inc., Waltham.
Souilem, S., El-Abbassi, A., Kiai, H., Hafidi, A., Sayadi, S., Galanakis, C.M., 2017. Chapter 1. Olive oil production, environmental effects and sustainability challenges. In: Galanakis, C.M. (Ed.), Olive Mill Waste: Recent Advances for the Sustainable Management. Elsevier Inc., Waltham, pp. 1–28.
Stanton, C., Ross, R.P., Fitzgerald, G.F., Van Sinderen, D., 2005. Fermented functional foods based on probiotics and their biogenic metabolites. Curr. Opin. Biotechnol. 16, 198–203.
Sun, J., Zhou, W., Huang, D., Fuh, J.Y., Hong, G.S., 2015. An overview of 3D printing technologies for food fabrication. Food Bioprocess Technol. 8 (8), 1605–1615.
Tushman, M.L., Anderson, P.C., O'Reilly, C., 1997. Technological cycles, innovation streams, and ambidextrous organizations: organizational renewal through innovation streams and strategic change. In: Tushman, M.L., Anderson, P. (Eds.), Managing Strategic Innovation and Change: A Collection of Readings. Oxford University Press, New York.
Ucak, I., Gokoglu, N., Toepfl, S., Galanakis, C.M., 2018. Inhibitory effects of high pressure processing on Photobacterium phosphoreum and Morganella psychrotolerans in vacuum packed herring (Clupea harengus). J. Food Saf. 38, 6. https://doi.org/10.1111/jfs.12519.
Ucak, I., Gokoglu, N., Kiessling, M., Toepfl, S., Galanakis, C.M., 2019. Inhibitory effects of high pressure treatment on microbial growth and biogenic amine formation in marinated herring (Clupea harengus) inoculated with Morganella psychrotolerans. LWT – Food Sci. Technol. 99, 50–56.
Wegrzyn, T.F., Golding, M., Archer, R.H., 2012. Food Layered Manufacture: a new process for constructing solid foods. Trends Food Sci. Technol. 27 (2), 66–72.
Weindlmaier, H., 2001. General aspects of food innovation management. In: R Treillon (Ed.), Food Innovation Management from Idea to Success. ENSIA, pp. 3–33.
White, T., 2012. Copycats and 'Me Too' Products. Available from: http://www.tutor2u.net/business/blog/copycats-and-me-too-products.
Winger, R., Wall, G., 2006. Food product innovation: a background paper. Agric. Food Eng. Work. Doc. 2.
Wittkopp, A., 2004. Produktinnovation und performance. Eine empirische analyse des deutschen Ernährungsgewerbes. In: European University Studies, vol. 3069. Peter Lang Europäischer Verlag der Wissenschaften, Frankfurt/M.
Young, J., 1996. Functional Foods: Strategies for Successful Product Development, FT Management Report. Pearson Professional Publishers, London.

Zinoviadou, K.G., Galanakis, C.M., Brnčić, M., Grimi, N., Boussetta, N., Mota, M.J., Saraiva, J.A., Patra, A., Tiwari, B., Barba, F.J., 2015. Fruit juice sonication: implications on food safety, physicochemical and nutritional properties. Food Res. Int. 77, 743–752.

Zinoviadou, K.G., Gougouli, M., Biliaderis, C.G., 2016. Chapter 9. Innovative biobased materials for packaging sustainability. In: Galanakis, C.M. (Ed.), Innovation Strategies for the Food Industry: Tools for Implementation. Elsevier Inc., Waltham, pp. 167–190.

Zoran, A., Coelho, M., 2011. Cornucopia: the concept of digital gastronomy. Leonardo 44 (5), 425–431.

Further reading

Dietary Fiber: Properties, 2019. In: Galanakis, C.M. (Ed.), Recovery & Applications. Elsevier Inc.

Galanakis, C.M., 2015b. Chapter 3: development of a universal recovery strategy. In: Galanakis, C.M. (Ed.), Food Waste Recovery: Processing Technologies and Techniques. Elsevier Inc., Waltham.

Galanakis, C.M., 2015h. Food Waste Recovery: Processing Technologies & Techniques. Elsevier-Academic Press. ISBN: 9780128003510.

Galanakis, C.M., 2016b. 'High Value-Added Compounds from Food Waste', Reference Module in Food Sciences. Elsevier, pp. 1–8. https://doi.org/10.1016/B978-0-08-100596-5.03510-1.

Galanakis, C.M., 2019d. Separation of Functional Molecules in Food by Membrane Technology. Elsevier-Academic Press.

Galanakis, C.M., Martínez-Saez, N., del Castillo, M.D., Barba, F.J., Mitropoulou, V.S., 2015b. Chapter 15: patented and commercialized applications. In: Galanakis, C.M. (Ed.), Food Waste Recovery: Processing Technologies and Industrial Techniques. Elsevier-Academic Press.

Prokopov, T., Goranova, Z., Slavov, A., Baeva, M., Galanakis, C.M., 2015. Effects of white cabbage outer leaves powder on sponge cake quality. Int. Agrophys. 29 (4), 493–500.

Food waste valorization opportunities for different food industries

11

Charis Galanakis [1,2]
[1]Department of Research & Innovation, Galanakis Laboratories, Chania, Greece; [2]Food Waste Recovery Group, ISEKI Food Association, Vienna, Austria

Chapter outline

1. **Introduction 341**
2. **The Universal Recovery Strategy 343**
3. **Innovation challenges and commercialization aspects of food waste recovery 344**
4. **Cereal processing by-products 347**
5. **Coffee processing by-products 358**
6. **Meat processing co-products and by-products 366**
7. **Olive mill processing by-products 370**
8. **Grape processing by-products 377**
   8.1 Fruit processing by-products 387
9. **Conclusion 402**

References 403
Further reading 421

1. Introduction

Until the end of the 20th century, increases in food production without improving the efficiency of the food systems was the prevalent policy around the globe. Nowadays, increasing demands for processed foods have required certain strategies to minimize energy demands and economic costs as well as to reduce food losses and waste. The quantities of the latest are huge and the numbers are shocking. FAO published in 2011 a report listing food losses and food waste around the world and referring that the amount of lost or wasted food produced for human consumption accounts 1.3 billion tones globally each year. European Commission has also referred to this significant issue as in 2010 denoted that 90 million tons of food waste are generated within EU annually. In addition, this number is expected to grow, as food waste would reach 126 Mtn in

The Interaction of Food Industry and Environment. https://doi.org/10.1016/B978-0-12-816449-5.00011-4

2020 within EU (from about 89 Mtn in 2006), without additional prevention measures. From 2006 up to 2020, food waste tonnages are expected to be 3.7 million in EU27 when population increases by nearly 21 million (Otles et al., 2015).

All these tonnages of waste are generated in different stages of food chain. Food supply chains begin from the primary agricultural phase, proceed with manufacturing and retail, and end with household consumption. During this life cycle, food is lost or wasted because of technological, economic or societal reasons. Food and Agricultural Organization (FAO) 2014 global voluntary definitional framework defined "food loss" as the decrease in quantity or quality of food, caused mainly by food production and supply system functioning or its institutional and legal framework. "Waste" is defined as "any substance or object, which the holder discards or intends or is required to discard" according to the EU Commission Council Directive 2008/98/EC. On the other hand, food waste is defined as "edible material intended for human consumption that is discarded, lost, degraded or consumed by pests as food travels from harvest to consumer" in the Foresight Project report prepared by The Government Office for Science (GO-Science) (Foresight Project, 2011). In any case, within the sustainability and bioeconomy frame of our times, all the food-related substrates that are included in "food loss" term (lost within the food supply chain), should be considered as by-products of potential reuse or valorization. Those substrates that are not able to be valorized, are considered as food waste and are finally discharged. This point of view generates two simple approaches.

$$\text{Food Losses} - \text{Reutilized By-products} = \text{Food Waste} \tag{11.1}$$

$$\text{Food Waste} = \text{Wasted By-products} \tag{11.2}$$

Considering the side streams that are generated within the production line of food industries, once not being reutilized, they become wasted by-products. Olive oil production process is the perfect example. This process has one product (olive oil) and two wasted by-products (olive kernel and olive mill wastewater). As the olive kernel is utilized for many years as a substrate for the production of olive kernel oil, its status turns from wasted to reutilized by-products. Olive mill wastewater (OMW) is lately explored for its valorization as a substrate for the production of valuable polyphenols. If this approach becomes an established practice, OMW will also turn from a wasted to a reutilized by-product (Rahmanian et al., 2014).

It is also important to state that when being fresh, food processing by-products do not lack of nutritional characteristics compared to fresh foods and possess similar organoleptic attributes. This allows us considering not only the valorization of these biomasses (e.g., for compost, biofuels, etc.), but also the extraction of valuable compounds from them in order to be recycled in the food chain, e.g., for the animal feeding or fortification of foods, etc. The current examples in literature and market have been increased rapidly over the last years. Therefore, water soluble (polyphenols) and insoluble (carotenoids) antioxidants could be recovered from fruit peels and other processing by-products (e.g., OMW) prior being reutilized as bioactive agents in natural foods, beverages and even cosmetics (Galanakis et al., 2015c, 2018a,b,c,d; Heng et al., 2015, 2016;

Galanakis, 2018r). Dietary fiber such as pectin from apple and olive pomace (Galanakis, 2011, 2012, Galanakis et al., 2010a,b,c,d,e; Roselló-Soto et al., 2015a) or β-glucan from oat mill waste (Patsioura et al., 2011) could be utilized as gelling agent in confectionary or as fat replacement in meat. Proteins and saccharides from cheese processing whey could be used for the development of nutritional supplements and soft drinks (Galanakis et al., 2014), whereas different protein hydrolyzates from meat processing by-products could be utilized as edible coatings for fat uptake reduction in deep-fat fried foods (Ananey-Obiri et al., 2018).

The potential of wasted by-products to create new markets has been underestimated from the food industry. The term "Food Waste Recovery" was first proposed and established by the Special Interest Group 5 of ISEKI Food Association (FA) in 2012. It was used as a short term for the extraction of valuable compounds from wasted by-products and their reutilization within food chain. Besides, this is the main difference between the widely used term "Food Waste Valorization" and "Food Waste Recovery". The second targets to upgrade compounds and ingredients lost within food waste streams by creating high added-value products (e.g., functional foods). Food Waste Recovery Group of IFA has developed a number of teaching tools (e.g., a reference module, books, a webinar series, an e-learning course, courses in different universities, etc.), research efforts (Deng et al., 2015; Zinoviadou et al., 2015; Barba et al., 2015; Roselló-Soto et al., 2015a,b) and other actions (e.g., open innovation network) aiming at helping food industries to recover valuable compounds from wasted by-products, valorize their streams and improve their sustainability.

2. The Universal Recovery Strategy

Food wasted by-products generally contain lower concentrations of valuable compounds compared to the initial sources, fruits or vegetables. This fact results in higher processing cost, lower recovery yield and ultimately lower revenues. Being already processed materials, they are also susceptible to microbial growth and require both preservation and fast treatment. Fig. 11.1 illustrates the "Universal Recovery Strategy" (Galanakis, 2015j) that provides a holistic strategy for the development of an economically feasible, sustainable and safe recovery of high added-value compounds from wasted by-products.

The first level includes the determination of macroscopic characteristics, e.g., the adjustment of phases' (water, oils and solids) content could be an effective way to fit in the different compositions of the initial substrates. The second one is the determination of microstructure characteristics in order to get an overview of waste matrix prior designing the recovery process. The third level is the determination of compounds groups that allows having a clearer view of main macro- and micromolecules content prior designing separation processes. The fourth step of the characterization is the identification of particular target macro- and micromolecules. The fifth level includes the determination of microbial and enzyme load, whereas the final step of the recovery strategy is the so-called "5-Stage Universal Recovery Process" that is illustrated in Fig. 11.2 (adapted by Galanakis, 2012).

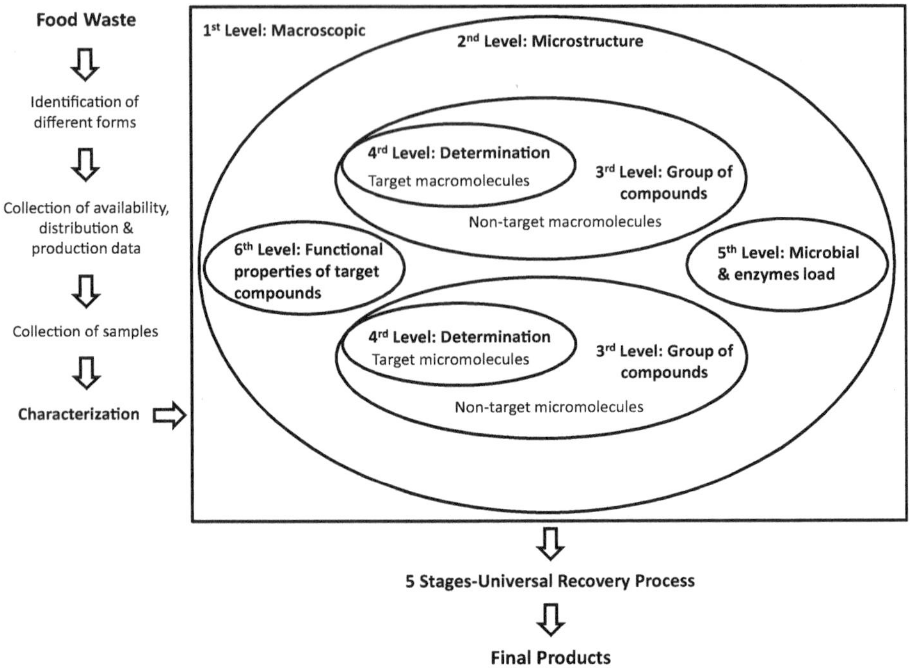

Figure 11.1 The Universal Recovery Strategy.
Adapted from Galanakis, C.M., 2015. Preface. In: Galanakis, C.M. (Eds.), Food Waste Recovery: Processing Technologies and Industrial Techniques.

This processing scheme moves from the macroscopic to the macromolecular level, prior the clarification and encapsulation of target compounds in a final product. For this purpose, different conventional and emerging technologies have been applied (Galanakis, 2013, 2016f; 2017f,g,h,i,q; 2018f,g,h; 2019a,b,c,d,e,f; Galanakis et al., 2013, 2015l; Kovacevic et al., 2018; Ucak et al., 2018, 2019; Zinoviadou and Galanakis, 2017; Skendi and Galanakis, 2018; Cassano and Galanakis, 2018; Incinur and Galanakis, 2018; Socaci et al., 2019). The ultimate goal of this approach is the selection of proper separation and extraction techniques in order to obtain as many different streams and compounds as possible.

3. Innovation challenges and commercialization aspects of food waste recovery

The industrial recovery of recovering valuable compounds from food by-products is a story that started few years ago. Nowadays, at least 50 companies around the world recover valuable compounds from food waste and sell them as clean-label ingredients for processed food products (e.g., natural preservatives to maintain shelf-life requirements, functional compounds), without impacting flavor or texture (Galanakis and Schieber, 2014). These commercially available applications inspired the scientific

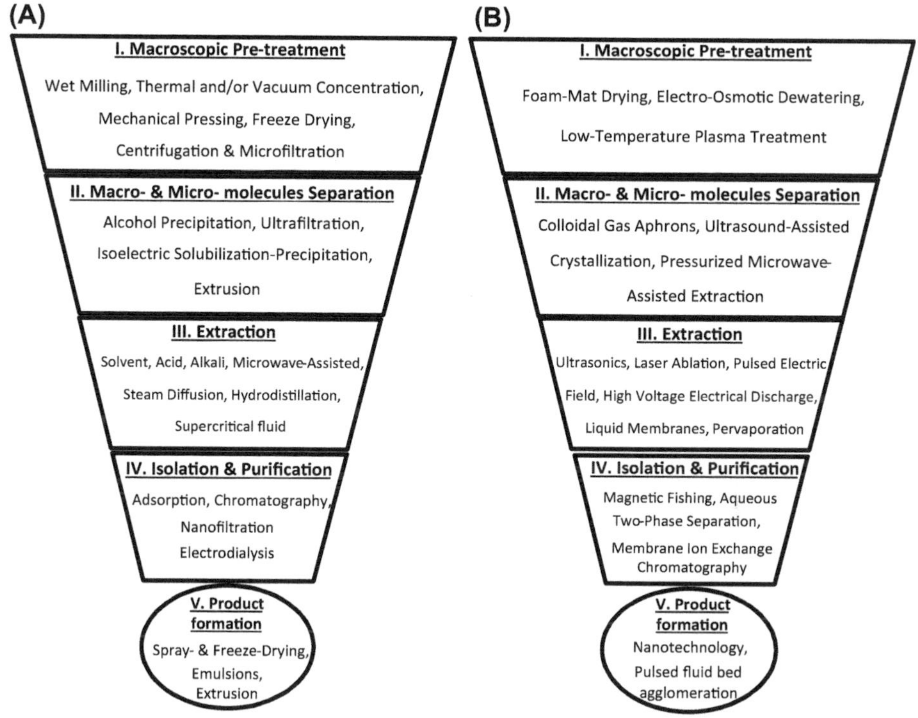

Figure 11.2 The "5-Stage Universal Recovery Process".
Adapted by Galanakis, C.M., 2012. Recovery of high added-value components from food wastes: conventional, emerging technologies and commercialized applications. Trends Food Sci. Technol 26 68–87.

community to intensify its efforts for the valorization of all kind of food by-products for recovery purposes. However, the industrialization of such processes and the implementation of the "Universal Recovery Strategy" is not an easy task. It typically includes laboratory research, transfer to pilot plan and full-scale production, protection of intellectual properties, development of definite applications and commercialization problems. For example, waste emerges seasonally and often in large quantities, whereas it is prone to microbial spoilage. Proper management of collection process, cooling of the material and addition of preservatives can provide solutions. Other problems such as the broad content of wasted by-products may be modified by adding a pretreatment step. Besides, transition of batch to continuous processes is usually accompanied with extension of mixing and heating time, heavier handling, increased air incorporation and higher degree of scrutiny. Subsequently, process cost is increased, as industrially recovered compounds are used in food formulations in higher concentrations compared to laboratory-recovered compounds. Therefore, solving these issues are necessary in order to ensure the sustainability of the process, the economic benefit and the perpetual establishment of the derived products in the market (Galanakis et al., 2016k).

A commercially feasible product can be generated only if a certain degree of flexibility and alternative solutions can be adapted in the developing methodology, e.g., less extraction steps are cheaper and scale up easier. However, they generate cruder products with lower concentrations of target compounds; thus it is recommended to use nonthermal technologies, green solvents and safer materials (possessing GRAS-status) to improve efficiency and clarification of the final product. The development of precise applications for well-defined markets is also very important (Galanakis, 2012). For example, natural antioxidants like polyphenols to be used in fresh, mom-approved and kid-friendly foods with clean ingredient lines and a minimum of three-month refrigerated shelf-life. Besides, modern new products should aim at the fulfillment of consumer needs and the realization of consumer value rather than at the development of products or enabling technologies per se. This means that the developed products should meet the high expectations of the consumers in an increasingly competitive market, e.g., development of "green" "organic" and "Protected Designation of Origin" products. Unlike the needs to delight the consumer and minimize environmental impact, developers should also ensure that the final product and process meet particular specifications, e.g., provide clean-label ingredients without impacting flavor or texture. Currently, the manufacturer's label typically provides only limited information about the origin and composition of the used extract in the final product formulation. A clearer label of the products containing recovered compounds would enable nutritionists and/or pharmacists to be more confident when recommending these products.

In addition, legal authorities regulate the way in which companies can advertise health benefits. However, those regulations (driven by the need of protecting consumers from dubious claims) create implications for the food industry. This is mainly because the demonstration of health benefits is very costly (constituting a high cost especially to small and medium-sized companies) and the risk of claims rejection by the corresponding authorities is too high. So far, only a small number of compounds and products have been approved, whereas the overwhelming majority of health claims has been declined by regulatory bodies (Galanakis et al., 2015c). The legislation challenges regulating health beneficial dietary products are lying to the nature of the products, which have the characteristics of both food and biologically active ingredients. For instance, EFSA of EU has approved health claims for only a small number of compounds (e.g., hydroxytyrosol in olive oil) (Galanakis and Kotsiou, 2017; Secmeler and Galanakis, 2019). This means that if you recover hydroxytyrosol from OMW in order to fortify foods (e.g., bakery, meat products or even oils), it is not allowed to add the health claim on the label (Galanakis et al., 2016k, Galanakis, 2019g).

Considering all the above innovation challenges, it can be concluded that food waste recovery may be an interesting but difficult to implement opportunity for food companies (compared to food waste valorization for nonfood applications). Decision upon recovery or valorization strategy should always be taken account in relation to the substrate and industry's goals. Below, a number of by-products cases from different food industries is explored and potential opportunities is highlighted.

4. Cereal processing by-products

Cereals comprise the most important source of food and feed for direct human consumption and meat production, respectively. It is a source of several phytochemicals such as carbohydrates, proteins, lipids, vitamins, mainly of B-complex and vitamin E, ferulic acid inorganic and trace elements, whereas most of them have well-established beneficial effects to human health (Table 11.1). For instance, sorghum bran is a unique dietary source of 3-deoxyanthocyanidins, which is a flavonoid with strong cytotoxic activities. Wheat and rye bran contain arabinoxylans that contribute to a reduction of the blood glucose, while oat bran is a starting raw material for the extraction of β-glucan. γ-Oryzanols found in rice bran and its oil have 10-fold higher antioxidant properties than tocopherols. Corn bran is one of the best sources of ferulic acid compared to other cereals, fruits and vegetables. Cereal (e.g., wheat, oat) brans can be used as food ingredients with no supplementary costs in baked products as natural sources of fibers to increase the dietary content of foods. They are the most common source of dietary fibers in baking. Wheat germ has also a great potential to be used as a functional food ingredient due to the rich composition in bioactive compounds which could improve the nutritional and healthy value when is added into a wide range of food products.

Cereals are manufactured using dry milling (wheat and rye), pearling (rice, oat, barley), wet milling (corn, wheat) and malting (barley, corn, wheat). All these processes generate by-products of different chemical composition. The current handling of cereal processing by-products include management practices that either degrade the substrate or lead to diminution of their pollution load, where like in rest cases, the major challenge is to incorporate ingredients or whole processing by-products into foods and nutraceutical formulations. Taking into account that the vast majority of cereal products consumed worldwide are made from refined grain, there is a need to increase cereal by-products' based functional foods and nutraceuticals market in order to assure the delivery of health beneficial compounds to final consumers. Economically, the valorization of cereal by-products streams to obtain high added value compounds is primarily focused on pharmaceutical and cosmetic applications, followed by food and feed uses, elaboration of biomaterials, chemicals and combustibles, and energy production.

Today, many ingredients derived from cereal processing by-product have already reached the market as dietary supplements. For instance, Oat Fiber Plus Tablets (Now Foods) which contain 3 g dietary fiber per recommended daily amount composed of oat bran, brown rice bran and apple pectin powder. Stabilized Rice Bran (NutraBio) has been merchandized as a source of natural B-complex vitamins, water-soluble polysaccharides, high quality protein and antioxidants. Gama Oryzanol (Swanson Health Products) is another similar product containing 60 mg of γ-oryzanol derived from rice bran oil. PROMITOR Soluble Fiber (Tate and Lyle) claims to help in reduction of calories in reduced-sugar and reduced-fat products.

Other compounds (e.g., proteins) of cereal processing by-products have advanced nutritional value and appropriate functional properties that prompted researchers, to

Table 11.1 Authorized health claims, their conditions of use and applicable restrictions for cereal by-products or bioactives from cereal by-products.

Functional ingredients	Claim	Usage conditions of the claim/ Restrictions of use/Reasons for non-authorization
Wheat bran fiber	Wheat bran fiber contributes to an acceleration of intestinal transit	Only for food which is high in that fiber as referred to in the claim HIGH FIBER as listed in the Annex to Regulation (EC) No 1924/2006.
Wheat bran fiber	Wheat bran fiber contributes to an increase in fecal bulk	Only for food which is high in that fiber as referred to in the claim HIGH FIBER as listed in the Annex to Regulation (EC) No 1924/2006.
Oat grain fiber	Oat grain fiber contributes to an increase in fecal bulk	Only for food which is high in that fiber as referred to in the claim HIGH FIBER as listed in the Annex to Regulation (EC) No 1924/2006.
Oat beta-glucan	Oat beta-glucan has been shown to lower/reduce blood cholesterol. High cholesterol is a risk factor in the development of coronary heart disease	Information shall be given to the consumer that the beneficial effect is obtained with a daily intake of 3 g of oat beta-glucan.
Beta-glucans from oats and barley	Consumption of beta-glucans from oats or barley as part of a meal contributes to the reduction of the blood glucose rise after that meal	The claim may be used only for food which contains at least 4 g of beta-glucans from oats or barley for each 30 g of available carbohydrates in a quantified portion as part of the meal.
Beta-glucans	Beta-glucans contribute to the maintenance of normal blood cholesterol levels	The claim may be used only for food which contains at least 1 g of beta-glucans from oats, oat bran, barley, barley bran, or from mixtures of these sources per quantified portion.
Barley grain fiber	Barley grain fiber contributes to an increase in fecal bulk	The claim may be used only for food which is high in that fiber as referred to in the claim HIGH FIBER as listed in the Annex to Regulation (EC) No 1924/2006.

Table 11.1 Authorized health claims, their conditions of use and applicable restrictions for cereal by-products or bioactives from cereal by-products.—cont'd

Functional ingredients	Claim	Usage conditions of the claim/ Restrictions of use/Reasons for non-authorization
Barley beta-glucans	Barley beta-glucans has been shown to lower/reduce blood cholesterol. High cholesterol is a risk factor in the development of coronary heart disease	Information shall be given to the consumer that the beneficial effect is obtained with a daily intake of 3 g of barley beta-glucan.
Barley beta-glucans	Barley beta-glucans has been shown to lower/reduce blood cholesterol. High cholesterol is a risk factor in the development of coronary heart disease	Information shall be given to the consumer that the beneficial effect is obtained with daily intake of 3 g of barley beta-glucan.
Rye fiber	Rye fiber contributes to normal bowel function	The claim may be used only for food which is high in that fiber as referred to in the claim HIGH FIBER as listed in the Annex to Regulation (EC) No 1924/2006.
α-linolenic acid (ALA) and linoleic acid (LA), essential fatty acids	Essential fatty acids are needed for normal growth and development of children	Information to the consumer that the beneficial effect is obtained with a daily intake of 2 g of α-linolenic acid (ALA) and a daily intake of 10 g of linoleic acid (LA).
Non-digestible carbohydrates	Consumption of foods/drinks containing < name of all used non-digestible carbohydrates > instead of sugars induces a lower blood glucose rise after their consumption compared to sugar-containing foods/drinks.	In order to bear the claim, sugars should be replaced in foods or drinks by non-digestible carbohydrates, which are carbohydrates neither digested nor absorbed in the small intestine, so that foods or drinks contain reduced amounts of sugars.

Continued

Table 11.1 Authorized health claims, their conditions of use and applicable restrictions for cereal by-products or bioactives from cereal by-products.—cont'd

Functional ingredients	Claim	Usage conditions of the claim/ Restrictions of use/Reasons for non-authorization
Plant sterols: Sterols extracted from plants, free or esterified with food-grade fatty acids.	Plant sterols have been shown to lower/reduce blood cholesterol. High cholesterol is a risk factor in the development of coronary heart disease	Information to the consumer that the beneficial effect is obtained with a daily intake of 1.5–3 g plant sterols. Reference to the magnitude of the effect may only be made for foods within the following categories: yellow fat spreads, dairy products, mayonnaise and salad dressings.
Plant sterols/Plant stanol esters	Plant sterols and plant stanol esters have been shown to lower/reduce blood cholesterol. High cholesterol is a risk factor in the development of coronary heart disease	Information to the consumer that the beneficial effect is obtained with a daily intake of 1.5–3 g plant sterols/stanols. Reference to the magnitude of the effect may only be made for foods within the following categories: yellow fat spreads, dairy products, mayonnaise and salad dressings.
Plant sterols and plant stanols	Plant sterols/stanols contribute to the maintenance of normal blood cholesterol levels	In order to bear the claim information shall be given to the consumer that the beneficial effect is obtained with a daily intake of at least 0.8 g of plant sterols/stanols.
Plant stanol esters	Plant stanol esters have been shown to lower/reduce blood cholesterol. High cholesterol is a risk factor in the development of coronary heart disease	Information to the consumer that the beneficial effect is obtained with a daily intake of 1.5–3 g plant stanols. Reference to the magnitude of the effect may only be made for foods within the following categories: yellow fat spreads, dairy products, mayonnaise and salad dressings.

Modified from Hadnađev, T.D., Hadnađev, M., Pojić, M., 2018. Chapter 2. The healthy components of cereal by-products and their functional properties. In Galanakis, C.M. (Eds.), Sustainable Recovery and Reutilization of Cereal Processing By-Products. Elsevier-Woodprint.

propose their recovery and their usage in food systems. However, since most of these proteins are enclosed in aleurone cells and enveloped by a polysaccharide matrix, their extraction is very difficult. Alkaline extraction coupled to isoelectric precipitation has been the classic extraction method, but there are concerns about its detrimental effects on the nutritive value and functional quality of proteins. Many other chemical, biotechnological and physical methodologies have been assayed for improving the extractability of proteins, but currently, the proteins are not extensively used, probably because the methods proposed for their extraction conflict with technical, economic, environmental and legislative issues. These include low yields and co-generation of other by-products, as well as the difficulty of introducing new additives into the food or feed markets.

Another example is Distillers' Dried Grains with Solubles (DDGS), which is the major by-product of bioethanol and distillery plants, owing a share of the market as animal feed due to its high protein, water-soluble vitamins and minerals content. The heterogeneous nature of DDGS allows its usage as a starting raw material within a biomass-based biorefining strategy, for the production of targeted valuable compounds with potential economic benefits (Skendi et al., 2018). Over the last decades, the utilization of some commonly used renewable materials, such as corn, sugar beet, wheat and potato as source of glucose for the production of energy commodities and chemicals has been partially industrialized. In addition, manufacturing of by-products in biorefineries are expected to be the technology of 2020s. Given the fact that prices of fossil fuel feedstocks have fluctuated, and the prices of biomass resources have slowly and steadily decreased gives the possibility to produce the existing bulk chemicals from biomass rather than fossils an attractive option. To this line, numerous methods and technologies using wheat bran and straw have been developed during the last years, resulting in high recovery of sugars and value added products. Table 11.2 lists the current state-of-the-art options for the valorization of wheat bran and straw within the biorefinery concept. Besides, multiple biofuels production from wheat bran increase the efficiency of material and energy can apparently be more economical process for biomass utilization as well as the technological and nutritional functionality of bran can be improved. Current pretreatment techniques and materials include acid hydrolysis, alkaline wet oxidation, steam explosion, ammonia fiber expansion, organic solvents to solubilize lignin and hemicellulose, ionic liquids, sulfite, and ozone. None of these deconstruction methods is universally advantageous over the others, and they vary in their outcomes depending on the type of feedstock, downstream process configurations, and a variety of other factors (Ucar et al., 2018).

The transformation of cereal processing by-products to alternative fuels, energy and other useful chemicals is a topic of great interest. Eco-innovative industrial ecology concepts are nowadays aiming at "zero waste economy", where wasted by-products are used as raw materials for new products development. Among the different technologies, membranes have a great potential due to their intrinsic properties and key advantages. Therefore, process flow-sheets based on a combination of membrane systems in a sequential design have been proposed in order to valorize oat, wheat, rice and corn by-products. These hybrid processes offer a synergetic effect, with overall important benefits in product quality, plant compactness, environmental impact and energetic aspects (Cassano and Galanakis, 2018).

Table 11.2 The state of the art of wheat (bran-straw) based biorefinery concept.

Material	Processing and technology	Products	Application	References
WS	Hydrothermal treatment Enzymatic treatment (Starch Hydrolysis)	Rhamnose, fucose, arabinose, xylose, mannose, galactose, glucose, vanillin p-Coumaric acid, ferulic acid	Energy, Biofuel	Merali et al. (2016)
WS	Liquefaction, fermentation, distillation and purification using molecular sieves	Ethanol and biomethane	Transport fuel	Murphy et al. (2008)
WS	Enzymatic hydrolysis, acid hydrolysis, SSF, Alkaline hydrolysis, steam explosion	reducing sugar (glucose), lignin, bioethanol	Biofuel	Sun et al. (2002)
WS	Dilute acid Steam explosion Alkaline peroxide Wet oxidation (alkaline) Lime, enzymatic hydrolysis and fermentation	Natural waxes, energy and chemical products	Biofuel, cosmetics, personal care products, polishes and coating	Talebnia et al. (2010)
WS	Alkali Extraction	sugars (glucose, xylose, arabinose), protein, lignin, cellulose	Monomer, protein, lignin	Maes et al. (2002)
WS	Enzymatic hydrolysis, fermentation, liquefaction, hydrothermal pretreatment, fermentation, acid hydrolysis	Bioethanol, biohydrogen and biogas	Biofuel, transport sector, food production	Kaparaju et al. (2009)
WB	Liquid Hot water Extraction Simultaneous Saccharification and Fermentation	Starch, lignin, sugar Glucose, xylose, arabinose, organic acids and furfural derivatives	Food flavorings, energy, biofuel	Wood et al. (2016)

WB	Liquefaction and saccharification	Starch, lignin, fermentable sugar; glucose, xylose, arabinose	Chemical process	Tirpanalan et al. (2015)
	Hydrothermal and enzymatic treatment	Lactic acid, hydroxymethylfurfural furfural (hmf), levulinic acid, acetic acid		
WB	Continuous fermentation	Lactic acid	Acidulant, preservative, flavor, monomer for PLA-production	González-Vara et al. (1996)
WB	Pretreatment, Extraction, Fermentation	Acetone, butanol, ethanol	Biofuel, additive to gasoline	Liu et al. (2010)
WB	Pretreatment, Enzymes, Fermentation	Bioethanol	Biofuel	Cripwell et al. (2015)
WB	Fermentation	IA and FA	Platform Chemicals	Jiménez-Quero et al. (2016)
WB	Extraction	Ferulic acid	Antioxidant, raw material for vanillin production	Buranov and Mazza (2009)
WB	Batch fermentation	Succinic acid	Acidulant, flavoring agent or as preservative	Song and Lee (2006)
WB	Starch Hydrolysis Hydrothermal pretreatment	Starch, HMF, acetic acid, lignin, fenolic acids	Biofuel	
WB	Starch Hydrolysis, Pretreatment of hemicellulose, enzymatic liquefaction and saccharification	Starch, lactic acid, succinic acid, ABE (acetone, butanol, ethanol), furfural and HMF, xylitol, ferulic acid, vanilin, amino acid	Food sector, cosmetic, fragrance, pharmaceutical, food and feed additives, antioxidant and raw material for vanillin production	Apprich et al. (2014)

Continued

Table 11.2 The state of the art of wheat (bran-straw) based biorefinery concept.—cont'd

Material	Processing and technology	Products	Application	References
WB	Enzymatic hydrolysis (liquefaction and saccharification)	Bioethanol	Biofuel	Juodeikiene et al. (2011)
WB	Starch hydrolysis, Acid hydrolysis, Heat pretreatment, Enzymatic hydrolysis	Bioethanol, arabinose, xylose, glucose	Converted sugar and biofuel	Palmarola-Adrados et al. (2005)
WB	Dry milling, Starch hydrolysis, Acid hydrolysis, enzymatic hydrolysis, liquefaction-saccharification	Bioethanol, furfural, HMF	wheat milling industry	
WB	Starch hydrolysis, Microwave pretreatment, enzymatic hydrolysis	Ethanol, DDGS, furfural, glucose, HMF	Animal fodder, biofuel	Linde et al. (2008)
WB	Starch hydrolysis, Alkaline hydrogen peroxide extraction	Arabinoxlyane, cellulose	Biofuel	Maes et al. (2001)
WB	Solid-state fermentation, Alkaline and acid pretreatment, enzyme hydrolysis	Arabinose, cellulose, xylan	Biofuel	Brijwani et al. (2010)
WB	Fermentation	Short-chain fatty acids (predominantly acetate)	The prebiotic food industry	

WB	Alkali extraction, hot water extraction	Ferulic acid and three forms of diferulic acid, arabinoxylan	Potential food and nonfood uses	Mandalaria et al. (2005)
WB	Saccharification, heat pretreatment, enzyme hydrolysis	Protein rich residues, melanoidins, furfural, 5-HMF, bioethanol	Animal feeding, biofuel	
WB	Enzymatic hydrolysis	Feruloyl oligosaccharides	Inexpensive renewable resources	Yuan et al. (2005)
WB	Starch hydrolysis, SSF	l(+) lactic acid	Inexpensive renewable resources	Naveena et al. (2005A)
WB	Acid pretreatment, Enzyme hydrolysis.	Bioethanol	Energy biofuel	Favaro et al. (2013)
WB	SSF with using L. amylophilus GV6	L(+) lactic	Food additives, inflammatory drugs, pharmaceutical, textile and tanning industries, green solvent	Naveena et al. (2005B)
WB	endoxylanase treatment	Arabinoxylans, oligosaccharides, protein, ferulic acid	Biofuel	Benamrouche et al. (2002)
WB	Alkaline extraction, enzyme and thermal treatments	Arabinoxylans, fats, ferulic, phytic and uronic acid	Biofuel	Aguedo et al. (2014)
WB	Acid and alkali hydrolysis	Xylose, arabinose, galactose, glucose, mannose and uronic acids	Pharmacology	Prisenžn; áková et al. (2010)
WB	Batch hydrolysis of wheat bran hydrolysate (AXH), Hydrolysis of AXH in a packed bed reactor	Pentose syrups	Biofuel, chemicals and polymers	Smaali et al. (2009)

Continued

Table 11.2 The state of the art of wheat (bran-straw) based biorefinery concept.—cont'd

Material	Processing and technology	Products	Application	References
WB	Dilute sulfuric acid pretreatment, enzymatic hydrolysis, fermentation	Acetone, butanol and ethanol	Butanol	Liu et al. (2010)
WB	Enzymatic hydrolysis, ethanol organosolv treatment	Lignin, minerals, glycerol, fatty acids	Supply of materials in biorefinery, animal feed	Reisinger et al. (2014)
WB	Acid pretreatment phosphoric acid, fermentation with edible fungi, Enzymatic hydrolysis	Ethanol, DDGS	Biofuel, dietary supplements, animal feed	Nair et al. (2014)
WB	Protein extraction, LHW hydrolysis, HP enzymatic hydrolysis	Proteins, fermentable sugars and lignin	Biofuel, supply of materials in biorefinery	Celiktas et al. (2014).
WB	Wet alkaline extraction, heat or acid precipitation	Proteins,	Food applications	Balandran-Quintana et al. (2015)
WB	Liquefaction, exogenous amylases for enzymatic hydrolysis, fermentation	Sugars, starch bioethanol	Biofuel	Cripwell et al. (2015)
WB	Alkali pretreatment, enzymatic hydrolysis, microbial fermentation using glucose.	Glucose, xylose, fermentable sugars, microbial fermentation using glucose utilizing mutant strain R. Eutropha (PHB)	Bioplastics	Annamalai and Sivakumar (2016)
WB	Ultrasound treatment	Arabinoxylans	Food, medicine, cosmetic	Li et al. (2016)

WB	Alkali extraction, GC and GC-MS Analyses	Fatty acids, acylglycerols, alkylresorcinols, steroid compounds	Phytochemicals	Prinsen et al. (2014)
WB	Acid treatments, microwave treatment	Arabinose, monosaccharide, xylose	Base material in the production of vitamin B2, human diabetes	Aguedo et al. (2013)
WB	Ultrasound-assisted enzymatic extraction	Arabinoxylans	Stimulate bowel movement, reduce blood sugar, inhibit elevated cholesterol, possess antitumor activity and increase immunopotentiating activity in mice	Wang et al. (2013)
WB	Alkali-pretreatment by T. virens, SSF by microbial hydrolytic enzymes, enzyme extraction	Hydrolytic enzymes CMCase, FPase, PGase and xylase	Sugar, fine chemicals and biofuels	El-Shishtawy et al. (2015)
WB	Microfluidization, ultrasonic-assisted extraction and pressurized liquid extraction	Arabinoxylans, ethanol	Biofuel	Mustafa et al. (2007)
WB	Extraction	Protein	Food additives	Roberts et al. (1985)

WB, Wheat Bran; *WS*, Wheat Straw.

Adapted from Ucar, R.C., Sngul, A., Celictas, M.S., 2018. Chapter 4. Biorefineries for wheat bran: sustainable utilization and market potential. In: "Sustainable Recovery and Reutilization of Cereal Processing By-Products". (Eds.), Galanakis, C.M. Elsevier Inc.: Waltham.

5. Coffee processing by-products

Coffee is among the most consumed beverages worldwide, and as more than 70 countries produce it (global production of 145 million of 60 kg bags in 2015/2016), it is the second largest commodity/trading volume after oil. The paradox with coffee crop is that while billions of coffee cups are daily consumed worldwide, the main parts of the coffee plant biomass is considered to be waste. In particular, approximately 90% of the edible parts of the cherry are discarded during its conversion into coffee brew. Thus, the coffee industry generates huge amounts of residue during production, including coffee silverskin, spent coffee grounds, coffee pulp, coffee husk, coffee cut-stems and wastewater. These residues represent a relevant environmental burden due to their production in high amounts and their richness in phytotoxic and/or antinutrient compounds (e.g., caffeine, tannins, and polyphenols) that can limit their use directly for soil and feed applications. The problem is getting even worse day by day, increasing the concerns that global warming will affect coffee cultivation severely, if not causing it to disappear (Galanakis, 2017f).

The great majority of coffee processing by-products is still dispatched for land fields or used for energy production. This handling has in many cases negative ecologic and economic impact, leading modern coffee industries to develop more profitable valorization strategies that allow not only the extraction of valuable compounds, but also their recycling via the generation of new products (Galanakis, 2017f). To this line, coffee cherries and their derivatives have a potential use as superfoods because of their composition on several phytochemicals with synergic biological effects and nutrients. Indeed, coffee by-products are very rich in antioxidant compounds like caffeine, chlorogenic acids, trigonelline, and diterpenes, as well as in polysaccharides, proteins and minerals (Mussato et al., 2011a). They also present components with technological interest, which may be used as natural colorants, aromas, texturizers among others. Most these compounds have not been subjected to the roasting process that the coffee beans undergo, thereby their levels and quality remains relatively unchanged.

For the recovery of bioactive compounds from coffee processing by-products, different technologies have been applied (Table 11.3). For example, caffeine is known to be recovered using liquid-liquid extraction following solid-liquid extraction, application of supercritical fluid extraction, and ultrafiltration for concentration and purification. In case of supercritical fluid extraction, higher operating pressure (300 bar) and the addition of a polar cosolvent such as ethanol will increase the yield. For the recovery of diterpens from spent coffee grounds spent coffee grounds, lower temperature and pressure (140 bar, 40°C) and no cosolvents are preferred. Total polyphenols can be optimally obtained after prior drying and size reduction of raw material using sequential extractions, and the following conditions: hydroethanolic solvent (50%–60% of ethanol), moderate extraction temperature (60°C) and 30–60 min of duration. Targeting particularly chlorogenic acids, the application of organic solvents or acidified water, higher extraction temperatures and enzyme pretreatment are preferable. For oil and lipid compounds recovery from spent coffee grounds, the highest yields are achieved by using Soxhlet extraction with *n*-hexane

Table 11.3 Overview of the employed conventional extraction techniques for recovery of polyphenolic compounds from coffee by-products.

Coffee by-product	Pretreatment	Extraction solvent	Extraction parameters	Polyphenolic yield	Reference
Spent cofee grounds (SCG)					
Arabica Robusta Commercial mixture	Sterilization (autoclave), 134°C, 40 min	Water 100%	Mixing(600 rpm) at room temperature; 60 min	14.97 22.56 mg GAE/g 11.04	Sousa et al. (2015)
Filter Espresso Plunger Mocha	Oven-drying to constant weight, 2 h, 102°C Defatting - Soxhlet extraction, petroleum ether, 3 h, 60°C	Water 100%	Filter coffeemaker extraction - 6 min, 90°C	>95% of original coffee brews TPC ~70% ~40% ~5%	Bravo et al. (2012)
40% Arabica +60% Robusta 70% Arabica +30% Robusta Regular Arabica coffee capsules Decaffeinated Arabica coffee capsules	Drying (forced-air food dehydrator), 40°C, 12–15h	Water 100%	60°C, 30 min constant stirring	19.62 17.43 mg GAE/g 7.43 6.33	Panusa et al. (2013)
Espresso SCG	Oven-drying, 45°C	Ethanol/water (50:50, v/v)	orbital shaker at 300 rpm; 5h	13.0 ± 5.6 mg GAE/g	Páscoa et al. (2013)
Blends of Arabica and Robusta varieties	Sterilization, 121°C, 20 min, powdering, extraction by Soxhlet	Isopropanol/water, 60/40 (v/v)	Extraction in a glass column, 27°C, evaporation in a rotary evaporator	10.2 mg GAE/g	Murthy and Naidu (2012)

Continued

Table 11.3 Overview of the employed conventional extraction techniques for recovery of polyphenolic compounds from coffee by-products.—cont'd

Coffee by-product	Pretreatment	Extraction solvent	Extraction parameters	Polyphenolic yield	Reference
40% Arabica +60% Robusta 70% Arabica +30% Robusta Regular Arabica coffee capsules Decaffeinated Arabica coffee capsules	Drying (forced-air food dehydrator), 40°C, 12–15 h	Ethanol/water (50:50, v/v)	Three stage extraction, 60°C, 30 min constant stirring	35.52 31.92 mg GAE/g 17.45 17.07	Panusa et al. (2013)
40% Arabica +60% Robusta 70% Arabica +30% Robusta Regular Arabica coffee capsules Decaffeinated Arabica coffee capsules	Drying (forced-air food dehydrator), 40°C, 12–15h	Ethanol/water (60:40, v/v)	Three stage extraction, 60°C, 30 min constant stirring	28.26 23.90 mg GAE/g 12.58 11.83	Panusa et al. (2013)
Arabica filter coffee	Freeze drying	Ethanol/water (60:40, v/v)	Shaking on an orbital shaker for 30 min (100 oscillation) at room temperature	19.2 mg/100 g dry matter	Jimenez-Zamora et al. (2015)
Espresso coffee capsules	Partial drying (air-oven)	100% ethanol 80% ethanol 60% ethanol 40% ethanol 20% ethanol 0% ethanol (water)	Batch extraction, temperature 20–60°C Extraction duration 30–180 min Liquid to solid ratio 10–50 mL/g	7.96–15.40 mg GAE/g	Zuorro and Lavecchia (2011)
Arabica Robusta	Extraction with distilled water in a column, 92°C, 6 h Drying (hot air-oven), 60°C, 3 h, defatted by Soxhlet, hexane	Methanol	Extraction duration: 8h	63.2 mg/g GAE 48.1 mg/g GAE	Ramalakshmi et al. (2009)

Not specified	Drying, 60°C until 5% of moisture	0%–100% methanol	Batch extraction by magnetic stirring 60–65°C Extraction duration 30–90 min Liquid to solid ratio 10–30 mL/g	6.0–18.2 mg GAE/g	Musatto et al. (2011)
Guatemala Arabica filter coffee SCG	Drying (oven), 102°C, 2 h Defatting (Soxhlet extraction), petroleum ether, 3 h, 60°C	Distilled water	Solvent extraction at 80°C for 10 min Filter coffee maker extraction for 6 min at 90°C Soxhlet extraction for 1h Soxhlet extraction for 3h	17.44 13.94 mg GAE/g DM 10.20 13.58	Bravo et al. (2013)
Guatemala Arabica filter coffee SCG	Drying (oven), 102°C, 2h Defatting (Soxhlet extraction), petroleum ether, 3 h, 60°C	pH 4.5 pH 7.0 pH 9.5	Filter coffeemaker extraction - 6 min, 90°C	8.17 13.94 mg GAE/g DM 19.01	Bravo et al. (2013)
Guatemala Arabica filter coffee SCG	Drying (oven), 102°C, 2 h Defatting (Soxhlet extraction), petroleum ether, 3 h, 60°C	Ethanol 0%–100% Methanol 0%–100%	Filter coffeemaker extraction - 6 min, 90°C	2.65–17.48 mg GAE/g DM 7.37–16.03 mg GAE/g DM	Bravo et al. (2013)
Cherry husk					
Blends of Arabica and Robusta varieties	Sterilization, 121°C, 20 min, powdering, extraction by Soxhlet	Isopropanol/water, 60/40 (v/v)	Extraction in a glass column, 27°C, evaporation in a rotary evaporator	12.2 mg GAE/g	Murthy and Naidu (2012)
Coffee pulp					
Arabica	Drying, 60°C until the moisture content ~13%	Ethanol Methanol Distilled water	Maceration for 3 days at room temperature	13.21 g GAE/100 g crude extract 9.09 20.02	Jaisan et al. (2015)

Continued

Table 11.3 Overview of the employed conventional extraction techniques for recovery of polyphenolic compounds from coffee by-products.—cont'd

Coffee by-product	Pretreatment	Extraction solvent	Extraction parameters	Polyphenolic yield	Reference
Blends of Arabica and Robusta varieties	Sterilization, 121°C, 20 min, powdering, extraction by Soxhlet	Isopropanol/ water, 60/ 40 (v/v)	Extraction in a glass column, 27°C, evaporation in a rotary evaporator	14.8 mg GAE/g	Murthy and Naidu (2012)
Coffee silverskin					
Commercial coffee blend - ~40% of arabica and ~60% of robusta coffee beans	Roasting and grinding	100% ethanol 75% ethanol 50% ethanol 25% ethanol 100% water	different temperatures (25, 30, 40, 50, and 60°C) 30, 60, 90, and 180 min Extractions performed on a heating plate with constant stirring (600 rpm)	4–6.5 10.5–14.5 9.5–16.5 mg GAE/g 9–14 7–12	Costa et al. (2014)
Not specified	*Not specified*	Ethanol, 20%–90%	60°C, solid/liquid ratio (1/10 to 1/40 g CS/mL solvent) Extraction time 30–90 min	5.26–13.53 mg/ g	Ballesteros et al. (2011)
Blends of Arabica and Robusta varieties	Sterilization, 121°C, 20 min, powdering, extraction by Soxhlet	Isopropanol/ water, 60/ 40 (v/v)	Extraction in a glass column, 27°C, evaporation in a rotary evaporator	13.2 mg GAE/g	Murthy and Naidu (2012)
Arabica coffee	Collecting by aspiration	Ethanol/ water (60: 40, v/v)	Shaking on an orbital shaker for 30 min (100 oscillation) at room temperature	17.3 mg/100 g dry matter	Jimenez-Zamora et al. (2015)
Whole coffee fruit					
Arabica Robusta	Grinding, sieving (0.5 mm)	Ethanol/ water (50: 50, v/v)	Vortexinng for 10 min in a tube, repetition 5-fold	33–50 16–84 mg/g	Mullen et al. (2013)

Adapted from Belscak-Cvitanovic, A., Komes, D., 2017. Chapter 4. Extraction and formulation of bioactive compounds. In: Galanakis, C.M. (Ed.), Handbook of Coffee Processing By-Products: Sustainable Applications. Elsevier-Academic Press.

or a mixture of polar and nonpolar solvents (e.g., *n*-hexane and isopropanol), after prolonging the extraction time and grinding the sample. On the other hand, Soxhlet extraction using ethanol as a solvent can be employed to provide polyphenolic antioxidants and caffeine, and in comparison to other conventional techniques it provides high recovery yields (Belscak-Cvitanovic and Komes, 2017). Solvents like ionic liquids and deep eutectic solvents are expected to play an important role in the near future, along with other technologies such as pervaporation, nanofiltration, sorption (adsorption and ion exchange), hydrothermal liquefaction and pulsed electric fields.

Following the extraction process and isolation processes (third and fourth step of 5-Stages Universal Recovery Process, Fig. 11.2), the recovered bioactive compounds are available in liquid extracts needing further concentration, drying or preferably encapsulation (fifth step). Among the different encapsulation techniques, hydrogel formation by ionic or covalent crosslinking of natural biopolymers enable the simultaneous entrapment of hydrophilic compounds and provide diverse encapsulates (from nano-to macro-sized scale). For this purpose, spray drying is the popular choice for coffee oil and cherry pulp anthocyanins. Nanoencapsulation using microemulsification or liposome entrapment can improve the functionality of formulated delivery systems and facilitate their potential food implementations.

Recovered bioactives could find a market in pharmaceutical, food or cosmetic industries, as novel food products based on coffee wastes are being very welcomed by western consumers due to their well-known health promoting and sensorial properties. The number of commercial products based on coffee-by-products has increased in the last few years with the aim to achieve a sustainable production and consumption of the coffee cherry (Table 11.4). Besides, the increasing number of patents claiming solutions for their extraction and implementation in various foods, medicinal and agricultural applications reveal their promising future potential for these purposes. Nevertheless, systematic studies are needed for a better understanding of the health promoting properties of compounds forming different anatomic parts of the cherry and inedible parts of the plant, too.

Cosmetics (e.g., skin care formulations) conclude another potential market for compounds recovered from coffee silverskin, spent coffee grounds, coffee husks, and defective coffee beans. Besides, considering that life-expectancy has increased in developed countries, raising new concerns in skin and body appearance, the demand for cosmetics with natural ingredients is stronger than ever, being considered a challenge to the worldwide economy and society. In particular, the field of cosmetics may benefit from coffee bioactives as they can fulfill a real skin function and activity. These compounds can be added to cosmetic formulations (e.g., sunscreens or antiaging formulations), targeting their photo protection and antiwrinkles effects. However, it is necessary to assure the presence and biological stability of compounds in the final products. In addition, other requirement for antioxidants functionality is the permeation in the *stratum corneum* and maintenance of adequate concentrations in the epidermis and dermis to have positive outcomes (Chen et al., 2012).

Considering potential biotechnological applications, coffee processing by-products appear as an interesting alternative substrate to be used in bacteria, filamentous fungi and yeast cultivation. The composition of the coffee processing by-products including

Table 11.4 Patented methodologies leading to commercial applications from coffee processing by-products (Galanakis et al., 2015).

Source	Patents	Applicant/ Company	Title/Treatment steps	Products/Brand	Potential/ Commercialized	Inventors/ Reference
	Application Number			names	Applications	
Coffee Silverskin	US 7927460	Ito En, Ltd., (Tokyo, Japan)	Silver skin-containing paper and method for producing the same	Functional silverskin containing-paper	Paper industry	Sato et al. (2011)
	WO 2013/ 004873	Consejo Superior de Investigaciones Cientificas/ CIAL (Madrid, Spain)	Application of products of coffee silverskin antiaging cosmetics and functional food	Bioactive silversink extract	Cosmetics, Nutrition and Health	del Castillo et al. (2013)
Spent Coffee Grounds	WO 2006/ 036208	Ajinomoto General Foods, Inc. (Tokyo, Japan), Kraft Foods Global Brands LLC (Northfield, Illinois, United States)	Mannooligosaccharide composition for body fat reduction	Mannooligosaccharides	Functional Food Ingredient	Asano et al. (2006)

	US2009/ 0005342 (PCT/ JP2006/ 301025)	Ajinomoto General Foods, Inc. (Japan)	Composition Having Blood Pressure Reducing and/or Elevation Suppressing Effect and Food and Drink Containing the Same	Mannologosaccharides	Functional Food Ingredient	Takao et al. (2009)
	PCT/ ES2014/ 070062	Consejo Superior de Investigaciones Cientificas/ CIAL (Madrid, Spain)	Healthy bakery products with high level of dietary antioxidant fiber	Antioxidant insoluble dietary fiber	Functional Food Ingredient	del Castillo et al. (2013)
	PT 105346	University of Minho. CEB - Center of Biological Engineering (Braga, Portugal)	Distilled beverage from spent coffee grounds and respective production method	Distellied beverage	Aroma	Mussatto et al. (2013)

water-soluble sugars, proteins, lignin, cellulose, hemicellulose, oils, etc., allowing its microbial conversion into products such as ethanol, acids (lactic, citric, gibberellic and gallic), aromatic volatile compounds (terpenes, higher alcohols and esters), pigments (carotenoids), microbial biomass (edible mushroom), and enzymes. However, although several examples of coffee by-products' reutilization use have already been reported, further studies are needed to maximize their potential.

Coffee processing by-products have also been proposed for environmental applications (e.g., adsorption of heavy metals, dye removal, regnation of K and Fe, etc.). For instance, considering the different raw residues as biosorbents for the removal of various heavy metals, spent coffee ground is the most efficient one for the removal of cadmium ions. Furthermore, coffee processing by-products have been applied for dyes removal from aqueous solutions and showed interesting adsorption capacities. In order to improve performance, physical and/or a chemical activation steps are required to improve the textural and surface chemistry of the elaborated adsorbents. Spent coffee grounds can also be used for the production of biochar (via pyrolysis), which is applied as soil amendment to improve agriculture due to its persistence in soil and nutrient-retention properties. More significantly, biochar's carbon sequestration can be used to actively remove carbon dioxide from the atmosphere, with the environmental implication for mitigation of climate change. Besides, the possibility of using coffee husks and spent coffee grounds for energy applications (e.g., biofuels, biodiesel, bioethanol, etc.) through thermochemical processes (pyrolysis, gasification, combustion, hydrothermal treatment, etc) has been proposed. The energy recovery from biomass is an ecological route to produce energy from renewable sources, reduce waste, produce cleaner-burning fuels, protect the environment, reduce fossil fuels consumption and dependence, decrease fuel cost and lower greenhouse gases emissions. Moreover, it has s good impact on economic, social, and agricultural development and ensures a regular supply of energy (Galanakis, 2017f).

Considering all the aforementioned alternatives, the integral valorization of by-products as raw materials to obtain antioxidants, biofuels, bioenergy, bioproducts and biofertilizers can be obtained using the biorefinery concept. However, it is important to conduct a techno-economic analysis including logistics and supply chain strategies to achieve the needed scales for biorefining these residues. For instance, in countries like Colombia, most of the coffee crops are located in mountains with difficult access and the size of the field usually (based on the ownership) is 1 to 2 hectares.

6. Meat processing co-products and by-products

Meat industry is one of the less environmental friendly sectors of food production and thus increasing attention toward sustainability requires maximizing the conversion of raw materials into consumer products. Therefore, the development of animal production systems for the optimal use of the available resources and their efficient application, the establishment of animal welfare standards and environmental sustainability are important. To make the meat producing sector sustainable, efforts begin to improve welfare, then to reduce energy consumption of meat processing, as well as to valorize co- and by-products. Additional challenges for the meat industry are addressing consumer demand for healthier meat products (e.g., low sodium) and meeting the so clean-label trends (e.g., free from synthetic additives, allergens, GMOs).

To give an example of sustainable meat production, an effective commercial animal welfare assessment system should have three parts:

(1) third party independent audits,
(2) internal farm audits by the producer, and
(3) audits by the corporate staffs who are buying the animal products.

Many welfare issues can be assessed at the abattoir. In addition, on-farm visits are required for evaluating compliance with animal housing specifications, pain relief after surgeries, and environmental enrichment.

Regarding meat co- and by-products, it is first important to clarify their difference. The term "meat by-product" is used in the United States to define the parts of the carcass other than dressed meat, and includes also materials that have the prospects of being reutilized, recycled or reused for several purposes, e.g., human nutrition and energy production. On the other hand, the European Union regulations specifically define animal by-products as "*any part of the animal carcass or any material of animal origin not intended for human consumption*". Definition excludes these materials from the human food chain and includes parts of the animal which may present a safety risk, inedible material, or material which the operator has decided to direct to purposes other than human consumption. Three categories of animal by-products are defined in the regulations, according to the level of risk presented:

- Category 1 is the highest risk category and including specific risk materials associated with transmissible diseases such as the bovine spongiform encephalopathy (BSE).
- Category 2 is considered high risk and includes materials such as infected or contaminated carcasses.
- Category 3 is considered low risk, and includes for example inedible carcass materials, such as hides and skins, free from infection. Frequently many edible products of the fifth quarter originating from healthy animals and suitable for human consumption are, for operational or commercial reasons, directed into category 3, excluding such items from the food chain.

Slaughter by-products (animal parts that are not intended for human consumption) have been used for a variety of applications like fertilizers, as well as in the chemical, medical or pharmaceutical industry and over the last years for energy generation in the form of biodiesel. On the other hand, meat processing co-products constitute a great source for the recovery of bioactive peptides and proteins, as well as good ingredients for pet foods. Large volumes are generated per slaughtered animal, the majority of which are directed to uses other than human consumption. Table 11.5 compares the direct use of fifth quarter products with their valorization as substrates for the recovery of low and high added-value compounds (Mullen et al., 2015).

Blood proteins are among the functional ingredients that can be recovered from meat processing co-products, whereas examples of already marketed product exist. Membrane technology constitutes an efficient and ecological process for the separation

Table 11.5 Comparison between direct uses of fifth quarter and high added-value products obtained with novel extraction techniques.

ABP	Direct preparation	Uses	Revalorizing techniques	High added-value products	References
Liver	Frozen, fresh or refrigerated	Braised, broiled, fried, patty and sausage	Enzymatic hydrolysis	Antioxidant peptides	Di Bernardini, Rai et al. (2011)
Heart	Whole or sliced Frozen, fresh or refrigerated	Braised, cooked, luncheon meat, patty, loaf	Isoelectric solubilization/ precipitation Phosphate buffer washing	High value protein with low ash, fat and cholesterol Myofibrillar concentrate as texturizing	Dewitt et al. (2002) Ionescu et al. (2007)
Skin	Fresh, refrigerated	Gelatin	Collagen recovery Enzymatic hydrolysis and chromatographic purification Collagen hydrolysis	Barrier membrane, drug delivery, fibroblast scaffolds, bioengineered tissues Antioxidant peptides and liver protectors Antioxidant activity, antimicrobial properties antihypertensive, biomimetic tissue	Kew et al. (2011) Lee et al. (2012); Di Bernardini, Harnedy et al. (2011); Ichimura et al. (2009); Zeugolis, Paul et al. (2008)

Blood	Fresh or refrigerated	Black pudding, sausages, blood and barley loaf	Enzymatic hydrolysis Chemical hydrolysis Ethanol precipitation High pressure treatment Subcritical water hydrolysis	Antioxidant, antibacterial, antihypertensive or iron-binding peptides Predigested peptides for animal and pet food Purified protein as food ingredient Peptides and biopreserved blood Amino acid and peptides production	Sun et al. (2011); Nedjar-Arroume et al. (2008); Lee and Song (2009) Alvarez et al. (2012) Álvarez et al. (2009) Toldrà et al. (2011) Rogalinski et al. (2005)
Bone	Frozen, fresh or refrigerated	Gelatin, soup, jellied products	Subcritical water Alkaline extraction	Hydroxyapatite and collagen New kind of sausages	Barakat et al. (2009) Boles et al. (2000)
Lung	Frozen, fresh or refrigerated	Blood preparations, pet food	Isoelectric solubilization/ precipitation and membrane filtration	Protein concentrates with good functional properties	Darine et al. (2010)
Feathers and hair	Incineration, rendering	Feather or hair meal	Keratinolytic bacteria fermentation and enzymatic hydrolysis	Keratinolytic protease production, culture medium, soil assessment, separation membranes	Chaturvedi et al. (2014); Gousterova et al. (2012); Sueyoshi et al. (2011)

Adapted fromMullen, A.M., Alvarez, C., Pojic, M., Hadnadev, T.D., Papageorgiou, M., 2015. Chapter 2. Classification and target compounds. In: Galanakis, C.M. (Eds.), Food Waste Recovery: Processing Technologies and Industrial Techniques.

(concentration, purification, and fractionation) of valuable molecules from co-products like blood. Certainly, these technologies can be easily adapted to the first four stages of the "5-Stages Universal Recovery Process" (Fig. 11.2). However, to increase membranes performance, a pretreatment (e.g., via enzymatic hydrolysis) of the co-products is needed. On the other hand, the scientists are currently working on the development of new membrane modules in order to enhance the selectivity of the existing ones, while the maintenance of the membrane process performance (in terms of productivity) is a current challenge based on the strong effect of fouling.

Nevertheless, despite the technological advances that have been achieved in the field of the valorization of animal blood and in its application as nutritional ingredients and/or techno-functional additives for the food industry, blood proteins are currently underutilized as food ingredients. Reasons include challenges associated with scale-up as well as technical, operational, commercial and regulatory ones. In addition, one of the key challenges relates to consumer acceptance as foods containing meat co-products may not be viewed as edible in some contexts (trial by curious consumers is not sufficient). Thus, interventions by a range of stakeholders are required to enhance capability, opportunity and motivation of consumers to alter their behavior to incorporate such products into their diets.

7. Olive mill processing by-products

The practice of cultivating olives to produce of olive oil has a long history in Mediterranean countries. Together with tradition, the production of olive oil comprises an important asset for the agriculture and food sector of these countries, not only in spite of culture but also in spite of wealth. About 750 million olive trees are cultivated and approximately 2.95 million tons of olive oil is produced annually. Most olive oil (98%) is produced in the Mediterranean region, mainly between October and February. Likewise, olive oil industry grows constantly in specific parts of the world. On the other hand, olive oil production is accompanied with the generation of huge amounts (e.g., an annual generation of more than 30 million m^3) of waste materials that in most cases are discharged without a treatment, leaving a congested environmental footprint. The uncontrolled disposal of OMW on soil may cause phytotoxic and antimicrobial effects, increasing soil hydrophobicity, decreasing water retention and infiltration rate, as well as affecting acidity, salinity, N immobilization, microbial activity, nutrient leaching, lipids concentration, organic acids and naturally occurred phenols. In surface waters it may decrease the dissolved oxygen content, increase the organic matter and K, Fe, Zn, and Mn contents.

Olive mill waste (e.g., olive tree leaves, OMW, exhausted olive kernel, etc.) are undesirable for the olive oil industry not only with regard to environmental impact, but also in terms of sustainability and disposal costs. To this line, olive mill waste has been considered as a substrate for treatment over the last decades. During the last decade, a large number of studies dealing with OMW treatment and/or valorization methods have been published (Table 11.6). Indeed, the proposed treatment methodologies, literature and references are endless. However, none of the proposed methods

Table 11.6 Treatment technologies of olive mill wastewater.

Category	Methodology	Description	Results	Notes	References
Physical	Sedimentation, filtration, flotation, centrifugation	Total solids removal	Total or partial removal of solids, 70% COD removal, 30% oil recovery	Common pretreatment methodology	(Velioglu et al., 1987; Georgacakis and Dalis, 1993)
	Micro-, ultra-, nanofiltration, reverse osmosis	Separation of compounds (existing in the same phase) according to their molecular weight	99% COD removal, but membrane fouling	They are applied in series, but the cost is high due to membrane fouling	(Paraskeva et al., 2007; Russo, 2007; Stoller, 2006; Turano et al., 2002)
	Evaporation				Masi et al. (2015)
	Sedimentation (settling)				(De Martino et al., 2011)
	Centrifugation/ultrafiltration				Turano et al. (2002)
	Adsorption/desorption	Nonionic XAD4, XAD16, and XAD7HP resins were used	A final concentration of 378 g/L in gallic acid equivalents was reached	The recovered phenolic compounds were concentrated through vacuum evaporation	Zagklis et al. (2015)
Thermal	Evaporation, distillation	Water removal and waste reduction	20%–80% COD removal, needing additional treatment	High energy demands	(Rozzi and Malpei, 1996; Tsagaraki et al., 2007)

Continued

Table 11.6 Treatment technologies of olive mill wastewater.—cont'd

Category	Methodology	Description	Results	Notes	References
	Combustion, pyrolysis	Decomposition, waste elimination	Toxic gases production and high cost	Very high energy demands	(Rozzi and Malpei, 1996; Caputo et al., 2003)
	Solar distillation				Potoglou et al. (2004)
	Combined	Membrane distillation			(El-Abbassi et al., 2013; El-Abbassi et al., 2012)
Physicochemical	Neutralization, precipitation, adsorption	Addition of $FeCl_3$, $Ca(OH)_2/MgO$, Na_2SiO_3, adsorption on activated carbon	30%–50% COD removal	80%–95% COD removal by combining precipitation with adsorption	(Sarika et al., 2005; Kestioglu et al., 2005; Adhoum and Monser, 2004)
	Oxidation and advanced oxidation processes	Ozonolysis, wet oxidation, O_3/H_2O_2 photolysis, photocatalysis	40%–60% COD removal using simple oxidation practices	85% COD removal by combining methods	(Chatzisymeon et al., 2009; Chatzisymeon et al., 2009)
	Chemical oxidation (Fenton reaction)	Ferric chloride catalyst was used for the activation of H_2O_2	86% COD removal	The produced water can be used for irrigation or can be directly discharged	Nieto et al. (2010)

	Wet hydrogen peroxide catalytic oxidation		System operating at 50°C reduced considerably the COD, color and total phenolic contents		Azabou et al. (2010)
	Electro-Fenton				(Kaplan et al., 2011; Kilic et al., 2013)
	Ozonation		Decrease of the total phenolic content (almost 50%) after 300 min of ozonation		Siorou et al. (2015)
	Lime treatment				Aktas et al. (2001)
	Electrocoagulation				Hanafi et al. (2010)
	Cloud point extraction (CPE)				(Gortzi et al., 2008; El-Abbassi et al., 2014)
	Combined	Ferric chloride coagulation, lime precipitation, electrocoagulation and Fenton's reagent			Gursoy-Haksevenler et al. (2014)
Biological	Anaerobic processes	Dilution, nutrients addition and alkalinity regulation	60%–80% COD removal for 2–5 digestion days	90% COD removal for 25 digestion days	(Dalis et al., 1996; Azbar et al, 2008)
	Aerobic processes	OMW co-composting with sesame bark			Hachicha et al. (2009)

Continued

Table 11.6 Treatment technologies of olive mill wastewater.—cont'd

Category	Methodology	Description	Results	Notes	References
		Biofilm, activated sludge	5%–75% COD removal for few days of digestion	80% COD for longer period	(El Hajjouji et al., 2007; Velioglu et al., 1992)
	Mixing and digestion	Together with other agricultural wastes	75%–90% COD removal	Nutrients and pH adjustment by combining wastes	(Azbar et al., 2008; Marques, 2001)
	Enzymatic	Oxidative transformation of phenols by *Trametes trogii* laccases	laccases were capable of efficiently removing phenolic compounds		
Combined	Oxidation and biological processes	2–3 sequential methods	75% phenols 80%–99% COD removal	High cost due to processes combination	Bressan et al. (2004)
	Evaporation - condensation followed by biological treatment			The process generated a concentrate containing high concentrations of valuable chemicals, including polyphenols	Masi et al. (2015)

Adapted from Souilem, S., El-Abbassi, A., Kiai, H., Hafidi, A., Sayadi, S., Galanakis, C.M., 2017. Chapter 1. Olive oil production sector: environmental effects and sustainability challenges. In: Galanakis, C.M. (Ed.), Olive Mill Waste: Recent Advances for the Sustainable Management. Elsevier Inc.: Waltham.

can be considered as a unique solution in terms of effectiveness, cost, environmental and economic impact. Despite the overabundance of scientific studies aimed at improving the efficiency of extraction systems and increase olive oil's quality, the industrial sector has shown little predispositions toward the possibility of transforming research findings into innovative plants. As a consequence, the entire olive milling process has changed very little over the last decades. The last big revolution in the technology field was the introduction of the horizontal centrifugal decanters. Other proposed technologies for olive oil extraction (e.g., pulsed electric field) have not found industrial applications.

Today, olive oil industries remain unsustainable and many problems still exist. For instance, consumers demand extra virgin olive oils of high organoleptic and nutritional quality, but the final price of the product varies often from one season to another, leaving a narrow margin for profit to the industries. Local authorities demand from production units to reduce their environmental impact. Under these stressing conditions, even low-cost treatment solutions may collapse financially olive oil industries. Subsequently, most of the aforementioned treatment solutions have been rejected in practice due to industries' denial that claim to close down production and society's tolerance that delay the enforcement of environmental legislations implementation. This fact remains a huge problem for the olive oil industries, as they cannot overpass environmental legislations forever.

The urgent need for sustainability within olive oil industry has turned the interest of researchers and professionals to investigate the management of olive mill waste with another perspective, looking at bioeconomy opportunities. The majority of olive oil producer countries in the Mediterranean area suffer from desertification, so the water and organic matter reuse would be beneficial to improve soil fertility and control the erosion processes. Olive mill waste contains high added-value ingredients and components (e.g., water, organic compounds and a wide range of nutrients) that could be recovered and recycled for different purposes, e.g., as fertilizer and/or biopesticides. However, phenols of olive mill waste are too valuable to be diminished or discharged to the environment. The recovery of phenols accompanied with their reutilization in different products and markets should come first. Besides, the prospect of recovering ingredients from olive mill waste has started many years ago with the valorization of olive kernel for the extraction of olive kernel oil. Today, it is considered an established commodity similarly to olive fruit, while technologists and environmentalists focus on the extraction of polyphenols from OMW (Galanakis, 2017h).

The selection of the most suitable olive mill waste management strategy depends on social, agricultural or industrial environment parameters. Solutions for the valorization of olive mill waste should consider aspects such as total amount of effluents, investment costs, available land, industrial or agronomic environment, local laws and most important needs. The valorization of olive mill waste is a challenging opportunity for the sustainable and competitive development of olive oil industry. However, there are several key points that have to be addressed for a more effective and rational reuse of olive oil by-products. Waste-to-energy process kills two birds with one stone as it is considered to be a sustainable waste management approach and a renewable energy technology. Nowadays, there are currently numerous studies and applications discussing energy recovery practices using olive mill waste, revealing that the olive oil industry can serve as a major supplier of the solid biofuels industry. Low/medium residues of olive oil production are considered for heat and power generation or for the production of biofuels (ethanol) or other bioproducts, i.e., xylitol, oligosaccharides and antioxidants. High moisture residues are examined for agricultural uses and the production of bioenergy and biofuels, previous extraction of polyphenols. Other advanced applications such as animal feed, the production of biosorbents or as ingredients in construction materials have also been proposed.

In many olive-producer countries (e.g., Italy), OMW could be applied as amendment on the agricultural soils under specific requirements. Such application has raised interest not only because of its relatively low cost, but also for its potential to enhance soil fertility. The application of olive mill waste as amendment may modify trace elements solubility and bioavailability in the soil, provide essential nutrients and organic matter, change soil structure and, therefore, improve the habitat function in soil. This would stimulate the development of soil microbial communities and promote the establishment of a healthy and self-sustaining vegetation cover, which would reactivate the nutrients cycles (usually inhibited in contaminated soils) and also protect the soil preventing the dispersion of the contaminants in the ecosystem. The ultimate goal is to reduce in situ the toxicity and the dissemination of contaminants, restore soil capacity and positively influence the overall ecosystem functioning. Nevertheless, the application form (if they are applied raw or transformed) needs to be carefully considered in order to avoid any adverse effect related mainly to their concentration of phenolic substances, which may cause modifications of soil pH and redox conditions. The latest could cause mobilization of certain elements (like As or Mn) and toxicity (Akratos et al., 2017).

OMW has high content in phosphorous, potassium and organic matter and thus could affect positively soil's physicochemical characteristics and improve crop productivity. However, this application should be conducted with rationality since the high concentration of polyphenols has a well-known bacteriostatic effect on microorganisms and phytotoxic effect on cultivations. In addition, careful management is required due to wastewater's acidic nature and high concentration in salts (Regni et al., 2017). Similar to OMW, olive pomace can be spread on agricultural soils, even though it is preferable to use after composting (Amirante, 2003; Proietti et al., 2015). The application of OMW and olive pomace compost is an answer to limit loss of soil fertility when soil organic matter content falls below 1%. This is very important where the environmental conditions (hot and arid) make particularly hard to maintain soil fertility with the dangerous consequences of desertification (IOOC, 2004).

Composting has been deemed a rather attractive valorization method for olive mill waste, as it could transform it into an excellent quality final product and could increase the income and viability of small-scale olive mill plants. In addition, co-composting of several other agro-industrial by-products makes this strategy an ideal method for the valorization of olive oil processing by-products while simultaneously provides environmental benefits. However, at this case, different operational parameters (e.g., bulking agents, aeration strategies, C/N ratios) should be adjusted. Composting should last at least 90 days, whereas suitable bulking agents should be selected to achieve porosity values between 35% and 50% and a final C/N ratio between 10 and 20. Aeration type has no significant effect on the composting process, as long as oxygen concentrations are between 15% and 20%. Besides, genotoxic and cytotoxic effects should be evaluated before composts are used on agricultural crops. However, in order to increase the sustainability of these facilities added-value chemicals like phenolic compounds, marketable products like waste oil and bioenergy through anaerobic digestion need to be extracted and produced prior to co-composting.

The recovery of polyphenols from OMW and wet olive pomace could be conducted with several technologies, most of them being already patented. The dominant scenario is to pretreat the initial material and convert oleuropein to hydroxytyrosol, prior extracting polyphenols with solvent and/or other technologies. Purification and product formation allow obtaining purer extracts with advanced antioxidant properties. The application of membrane processes for this purpose is summarized in Table 11.7. The beneficial health claims of the extracts are still under debate in Europe, but in other regions (e.g., USA, Asia) have been accepted. Potential health claim proof of these products will open the door in new opportunities and bring revolution in the olive oil industry. The development of certain applications for the recovered phenols is as much important as the development of the recovery process itself (Galanakis and Kotsiou, 2017).

Polyphenols recovered from OMW have been used to formulate and produce new functional foods and ingredients with biologically activities able to replace food additives of synthetic origin. Indeed, in the near future, the possibility of totally or partially replacing chemical preservatives employed today with natural additives derived from the by-products of the food industry will be increased. Another interesting application is cosmetics industry. Life-expectancy has increased in developed countries, raising new concerns in skin appearance. Particularly, face wrinkles are one of the major features of aging leading to new researches for active ingredients that are promising to slowdown the different symptoms. Presently, the best therapy against aging is to limit the DNA damage with antioxidants, vitamins, and minerals. OMW is extremely rich in such molecules (e.g., antioxidants, fatty acids and minerals) and to this line may become a source of possible active ingredients for cosmetic products with different claims, such as anti-aging or hydration (Galanakis, 2017h).

8. Grape processing by-products

Most grapes harvested worldwide are used to make wine, whereas winemaking has been one of the most important industries of many countries. Along with the increasing worldwide trend for wine consumption, an increasing volume of accompanied by-products, such as grape pomace, seeds, skins, stems, leaves, and wine lees, arises. These by-products conclude an interesting source for further exploitation since they are rich in polyphenols, tannins, procyanidins, anthocyanins, stilbenes, dietary fibers, tartarates, essential oils, proteins, tocopherols etc. To this line, they have the potential to be used as a source of ingredients that could be used in different applications in

Table 11.7 Summary of membrane techniques for OMW treatment and phenolic compounds recovery (Galanakis and Kotsiou, 2017).

Technique description	Membrane characteristics	Effect on COD	Phenolic compounds recovery	Reference
Centrifugation, ultrafiltration	Polysulphone flat-sheet, MWCO of 17 kDa	90%	0.2–0.3 g/L in the UF permeate	Turano et al. (2002)
Combination of centrifugation, microfiltration and ultrafiltration for COD followed by nanofiltration and reverse osmosis for phenols recovery	MF: Ceramic membrane, separation limits 0.1–1.4 μm; UF: Spiral-wound polymeric (polysulphone, polyethersulphone polyamide or regenerated cellulose acetate) membrane, MWCO 1–20 kDa; NF: Spiral-shaped polymeric composite (polyamide or nylon) membrane, MWCO 150–250 Da; RO: Spiral-shaped polymeric (composite polyamide) membrane	After RO COD lower than 100 ppm of O_2	30 g/L in the RO concentrate	Patent WO 2005/123603 (Pizzichini and Russo, 2005)
Substratums and filters of resins	Substratums: turf, sand, and sawdust; Filters of resins: cationic, anionic, mixed type or polyvinylpolypyrrolidone	Not reported	40.8% recovery	Patent WO 2005/003037 (Castanas et al., 2005)
Combination of rough filtration, microfiltration, ultrafiltration nanofiltration and reverse osmosis	7 RF modules 4.76–0.033 mm; 2 MF modules, MWCO 120 kDa; 2 UF membranes, MWCO 120-20 kDa and 20-1 kDa; NF: MWCO 1000-350 kDa; RO: MWCO 350 kDa	Lower than $100 mgO_2/L$	>1 g/L of hydroxytyrosol and >0.6 g/L of tyrosol in the RO extract	Patent US 2006/0070953 (Villanova et al., 2006)

Microfiltraion, ultrafiltration and reverse osmosis	MF: ceramic membrane, 0.45 μm, 0.8 μm, 500 Kd UF:Ceramic 1 kDa and polymeric 6, 20 and 80 kDa membranes; RO: Polymeric Hydronautics (composite polyamide) membrane	Not reported	464.870 ppm of low MW phenolic in the RO concentrate	Russo (2007)
Ultrafiltration, nanofiltration and reverse osmosis	UF: Zirconium oxide membrane, 100 nm; NF: Spiral-wound polymeric membrane, MWCO 200 Da; RO: Spiral-wound polymeric membrane, MWCO 100 Da	21% and 97% reduction after UF and NF respectively	9.962 g/L in the NF retentate (feed concentration 0.725 g/L); 6.782 g/L in the RO concentrate (feed concentration1.018 g/L)	Paraskeva et al. (2007)
Ultrafiltration and nanofiltration	UF: GE Osmonics GM4040F membrane (composite thin-film Desal-5); NF: GE Osmonics DK4040F membrane (composite thin-film Desal-5)	50% and 77% reduction after UF and NF respectively	1.64 mg/L in the UF permeate (initial feedstock concentration 3.1 mg/L) 16.0 mg/L in the NF residue	Stoller (2008)
Supercritical fluid extraction, nanofiltration and reverse osmosis individually or in integrated mode	NF: Desal DK type membrane (supplied by GE Osmonics), MWCO 250 Da; RO: Dow Filmtec SW 30 membrane	Not reported	Recovery of hydroxytyrosol of about 70% at the NF step.	Patent US 2008/0179246 (De Magalhaes Nunes Da Ponte et al., 2008)

Continued

Table 11.7 Summary of membrane techniques for OMW treatment and phenolic compounds recovery (Galanakis and Kotsiou, 2017).—cont'd

Technique description	Membrane characteristics	Effect on COD	Phenolic compounds recovery	Reference
Membrane distillation	Polytetrafluoroethylene (PTFE) membrane with mean pore size of 198.96 nm or Polyvinylidene fluoride (PVDF) membrane with mean pore size: 283.15 nm	Not reported	Concentration factor of 1.72 and 1.4 using PTFE and PVDF membranes respectively	El-Abbassi et al. (2009).
MF and NF as pretreatments, osmotic distillation (OD) or vacuum membrane distillation (VMD) for polyphenol recovery.	MF: tubular Al_2O_3 membrane (pore sizes 200 nm); NF: Nadir N30F spiral-wound membrane made of polyethersulfone, MWCO 578 Da OD: polyethylene membrane, containing microporous polypropylene hollow-fibers (mean pore size about 30 nm)	Not reported	78% of the initial phenols were recovered in the permeate stream of the MF 160.32 ppm in the NF permeate 0.5 g/L after OD	Garcia-Castello et al. (2010)
Microfiltration, ultrafiltration and reverse osmosis	MF: tubular polypropylene membrane, MWCO between 0.1 and 0.3 μm; UF: spiral membranes of polyamide and traces of polysulfone, MWCO of 7 kDa RO: spiral thin-film membrane (TFM) composed of Durasan and polysulfone, MWCO ~ 100 Da	Final reduction of 98%	19.3 g/L total phenols in the RO concentrate (initial OMW concentration 4.9 g/L)	Servili et al. (2011)
A sequence of two ultrafiltration processes followed by an nanofiltration process	UF: a hollow fiber membrane with nominal pore size of 0.02 μm and a flat-sheet composite fluoropolymer membrane with MWCO of 1000 Da; NF: spiral-wound membrane module, MWCO of 90 Da	Not reported	960 mg/L in the NF retantate (feed concentration: 625 mg/L; initial concentration: 1409 mg/L)	Cassano et al. (2013)

Sequential sieving, UF, NF, RO, Resin adsorption/desorption	Sieving: Stainless steel filters with final pore size of 0.125 mm UF: ceramic (zirconia) with a pore diameter 100 nm NF: Spiral wound, polymeric RO: Spiral wound, polymeric Resins: Nonionic resins (XAD4, XAD16, and XAD7HP)	6.47 g/L COD level in the final effluent	165 g in the UF concentrate, 74 g in the NF concentrate and 25 g of free simple phenols in the RO concentrate (initial 265 g of phenols) 28 g/L after rotary evaporation of RO concentrate or 378 g/L when the RO concentrate was treated with resins and vacuum evaporation	Zagklis and Paraskeva (2014) and Zagklis et al. (2015)
MF	MF: ceramic membrane (MWCO ~ 100,000 Da, membrane porosity 0.1 μm) UF: a polyethersulfone membrane (MWCO of 5000 Da) NF: polyamide membrane (MWCO of 200 Da)	Not reported	2456 μg/mL to 5284 μg/mL in MF fraction; 1404 μg/mL to 3065 μg/mL in UF fraction (55%–65% recovery); 373 μg/mL to 1583 μg/mL in NF fraction (15%–33% recovery)	
MF, UF, Distillation	MF: ceramic membrane, pore size 0.2 μm UF: organic membrane, MWCO 8KD	48.44 g/L COD level in the UF filtrate	7.2 g/L hydroxytyrosol in the UF permeate (initial concentration 0.23 g/L)	Hamza and Sayadi (2015)

Continued

Table 11.7 Summary of membrane techniques for OMW treatment and phenolic compounds recovery (Galanakis and Kotsiou, 2017).—cont'd

Technique description	Membrane characteristics	Effect on COD	Phenolic compounds recovery	Reference
MF	MF: polyvinyl-fluoride membrane, pore size 200 nm UF: ceramic membrane, MWCO 20 kD NF: DMS DK membrane (composite), MWCO 150–300 Da	34.8% reduction after MF; 44.9% reduction after UF; In the NF step, COD was reduced from 33 to 6.9 g/L	2.3 g/L in the MF retentate; 2.0 g/L in the UF retentate; 2.4 g/L in the UF retentate	Abdel-Shafy et al. (2015)

functional foods, dietary supplements, pharmaceutical and cosmetic products as well as for agricultural and oenological purposes. Table 11.8 revises the main conventional and nonconventional uses of grape processing by-products (Spigno et al., 2017). For instance, vine-shoots (which are currently left on the vineyards or even burnt) have an interesting composition of phenolic, volatile and mineral compounds. They could be used for different agricultural purposes as for example compost, soil amendments and more recently as foliar fertilizer to grapevines. Besides, grape marc, skins and seeds, constitute the most promising winemaking by-products to become oenological additives for improving the chemical composition and sensory attributes of wines. Red color (anthocyanincs and polymeric pigments) and phenolic compounds enhancement appear as the most straight forward application with recognized efficacy.

Conventional extraction methods such as solvent extraction, matrix solid-phase dispersion and enzymes assisted extraction can be applied for the recovery of biologically active compounds, whereas efficiency mainly depends on the choice of solvent. The main advantage of classic solvent extractions compared to other methods, is the low processing cost and ease of operation. However, they are slow and not selective, and in some cases solvents are toxic. The most important factor for the solvent selections is the polarity of the targeted compound. Therefore, the molecular affinity between solvent and solute, mass transfer, the use of co-solvent, environmental safety, toxicity to humans and financial feasibility are the factors which should be considered in selection of the extraction solvent. Many studies suggest that the use of ethanol/water solvent mixtures is preferred in terms of their similar extraction efficiencies, but most importantly because they are food-grade.

Membrane processes like microfiltration, ultrafiltration and nanofiltration can be used alone or in combination with other technologies for the recovery of the aforementioned compounds in different stages, e.g., to remove solids, separate macro- from micromolecules and clarify target compounds from impurities. The major disadvantage of membrane technologies is fouling, thus it is preferable to use them in combination with other extraction technologies, e.g., to perform target separations. The selection of membrane material is of critical importance for each application, as it governs the performance of the process. Nonpolar membranes like composite fluoro polymer allow the fractionation of compounds and the enhancement of concentrate and permeate streams with different phenolic classes. The latest could be used in several applications. For instance, concentrates rich in hydroxycinnamic acid derivatives could be used to fortify foods with antioxidants and permeates rich in flavonols and anthocyanins could be applied as flavorings and colorants (Galanakis and Schieber, 2014; Galanakis et al., 2015c). The composition of the extracts and the interactions of the dissolved compounds affect the antioxidant properties of the final product; thus separation should be carefully designed. Besides, several emerging techniques (ultrasounds, microwaves, electrotechnologies, high-pressure methods, etc.) have also been proposed to replace conventional ones in the purpose of saving energy, time and solvent consumption, while simultaneously conserving bioactive properties of the molecules. It is obvious that there is no optimal technique. Each one can be adapted to the specific uses and

Table 11.8 Main conventional and nonconventional uses of grape processing by-products (Spigno et al., 2017).

Residue	Conventional uses	Alternative/Emerging/Potential uses	References
Vine shoots	Spreading or burning in the field	Antioxidant/bio-stimulating extracts	Harris-Valle et al. (2007); Romain et al. (2012); Rajiha et al. (2015); Sánchez-Gómez et al. (2016z,b)
		Woody flavorings	Delgado de la Torre et al. (2014); Sánchez-Gómez et al. (2016z,b)
		Removal of toxic compounds	Karaoğlu et al. (2010)
		Biomass for energy production/ Lignocellulosic fractionation	Argun & Onaran (2015); Buratti et al. (2015); Cotana et al. (2015); Duca et al. (2016); San José et al. (2013); Spinelli et al. (2012)
		Composting	Rubio et al. (2013); Wang & Schuchardt (2010)
Stalks	Land-spreading; burning; landfill; animal feed; compost	Fermentation substrate; Composting; Biomass for energy production/lignocellulosic fractionation; antioxidant extracts; activated carbon; removal of toxic compounds	See Spigno et al. (2013) for detailed references

Pomace/skins	Distillation; coloring agents; land-spreading; incineration; (animal feeding)	Antioxidant extracts	Amendola et al. (2010); Lavelli et al. (2016); Pedroza et al. (2015); Spigno et al. (2015); Spigno et al. (2013b);
		Antioxidant fiber for food/feed applications	Brenes et al. (2016); Kara et al. (2016); Ke et al. (2015); Zhu et al. (2015)
		Fermentation substrate	Koutrotsios et al. (2014); Mateo & Maicas (2015); Mokochinski et al. (2015); Romo Sánchez et al. (2015)
		Composting/soil amendments	Achmon et al. (2016); Burg et al. (2014); Masowa et al. (2016); Zhang and Sun (2016)
		Biomass for energy production/lignocellulosic fractionation	Apolinar-Valienteet al. (2015):, Da Ros et al. (2016); Fabbri et al. (2015); Hsieh (2013); Martinez et al. (2015); Schönnenbeck et al. (2016)
Seeds	Oil recovery; burning	Antioxidant fiber for food/feed applications	Aghamirzaei et al. (2015); Correddu et al. (2015); Gül et al. (2013); Hajati et al., 2015; Kim et al. (2015); Mironeas et al. (2016); Özvural and Vural (2011)
		Antioxidant extracts	Ignea et al. (2013); Kakakei&Shahbazi (2016); Razavi et al. (2013); Yaragalla et al. (2016)
		Protein extracts	Vincenzi et al. (2013)

Continued

Table 11.8 Main conventional and nonconventional uses of grape processing by-products (Spigno et al., 2017).—cont'd

Residue	Conventional uses	Alternative/Emerging/Potential uses	References
Lees	Distillation; tartaric acid; coloring agents; nutritional supplements; incineration; landfill; land-spreading	Antioxidant extracts	Delgado De La Torre et al. (2015); Dimou et al. (2015); Giacobbo et al. (2015); Tao et al. (2014)
		Food application	Sharma et al. (2015)
		Wine production	Fia et al. (2016)
		Nutrients extracts/Fermentation substrate	Dimou et al. (2015); Pérez-Bibbins et al. (2015)
		Biochar	Zhu et al. (2016)
Filter cake	Tartaric acid; landfill; land-spreading; incineration	Composting (not much investigated)	De Mello Prado et al. (2013); EP Minerals (2013)

targets of the industry. However, an optimization process of the changeable parameters is required to obtain the highest quantity and quality of the desired components.

In order to commercialize the recovery of these beneficial ingredients from grape by-products, intellectual property rights of processes and ideas must be protected through patenting. In particular, there are many attempts to utilize them in food (confectionery products, food additives such as colors, flavors, preservatives, condiments), beverages (wine, juice), dietary supplements, pharmaceutical preparations and cosmetics (skin preparations, toothpaste). A vast number of patents already exist in WIPO and EPO databases (Table 11.9). Most of them deal with the recovery of polyphenolic compounds (procyanidin oligomers, anthocyanins, resveratrol, etc.) usually in a powdered form. The latest allows easier application mainly in functional food or dietary supplements. Besides polyphenols, many patents deal with isolation of grape seed oil, which is largely commercialized. However, grape seed extract in the form of a dietary supplement is probably the most commercially utilized grape by-product. Patents dealing with health benefits of grape-byproducts disclosed prevention of stress-induced injuries, neuronal and cardiovascular diseases, bacterial infections, inflammation, cancer etc. Likewise, many patents include other plants and extracts in combination with grape by-products, proposing synergistic actions of phytochemicals, which is already widely documented in literature. Still, many of these patents are not exploited and commercialized, especially the ones carrying health claims, due to many concerns arising from innovation such as safety and profitability (Vulic et al., 2017).

Overall, the research on winemaking by-products suggests that a wide variety of food applications is possible. Winemaking-derived ingredients could be used to replace synthetic antioxidants in various food matrices. Applications as colorants seem to be promising, though further improvements could be obtained from a better knowledge on the effects of food matrix on anthocyanin stability. Grape pomace and their components can exert antimicrobial activity against both spoilage and pathogen populations in food products, contributing to extended shelf-life. Their incorporation may help reducing the quantity of synthetic preservatives substances added to certain foods. Bioactive compounds from grape processing by-products could also be used as active ingredients for skin care products. Their potential emollient, antiwrinkle or antioxidant activity is known, whereas several cosmetic brands already use grape on their formulations. The main concern that surrounds the cosmetic use of grape processing by-products is based on their toxicity. At this case, their bioavailability and toxicity need to be carefully assessed by in vitro and in vivo assays. In addition the stability of antioxidants has to be improved (Galanakis, 2017g).

8.1 Fruit processing by-products

Fruit processing by-products have increased in recent years due to the high global demand and consumption of natural products, as a consequence of human desire for a healthy living. Table 11.10 lists fruit and vegetable by-products, as well as corresponded functional ingredients for recovery (Mullen et al., 2015). For instance, strawberry is a seasonal fruit that is unstable in the market. High perishability and

Table 11.9 Patent survey disclosing various grape by-products (Vulic et al., 2017).

Grape by-product	Patent application number	Applicant	Title	Outcome	Potential applicaton	Reference
Grape pomace (marc)	US6190716	Galbreath, JSO	Method for preparing a grape derived product	Method for preparation of readily digestible food supplement rich in resveratrol, ellagic acid, and other phenolic compounds and antioxidants	Ingredient of functional food, food supplement	Galbreath (2001)
	CN104287048	Xihua University	Preparation method of grape pomace essence and fruit and vegetable powder combined blending agent	Red powder containing fruit, vegetable and grape pomace extract	Ready to use food, food ingredient or dietary supplement	Xing et al. (2015)
	US6544581	Canandaigua Wine Company, Inc.	Process for extraction, purification and enrichment of polyphenolic substances from whole grapes, grape seeds and grape pomace	Method for producing free flowing powder containing monomeric and oligomeric procyanidins, and anthocyanins, flavanols and cinnamic acids in case of red grapes	Ingredient and/or colorant in foods, beverages, dietary supplements, nutraceutical products	Shrikhande et al. (2003)

	EP1288288	Balice Distillati SRL	Process and plant for calcium tartrate extraction from winemaking by-products	Plant for isolation of calcium tartarate from winemaking by-products with improved performances	Food preservative and acidity regulator	
	CN104798854	Northwest A&F University	Method for making grape pomace dietary fiber nutritious biscuits	Process for making confectionery products with improved processing performance, product crisp degree and nutritional characteristics	Functional biscuits, cookies and crackers	Wang et al. (2015)
	CN102268358	Shaanxi University of Science and Technology	Production process for brewing grape fruit vinegar by using grape pomace	Process for making vinegar utilizing grape pomace	Condiment, preservative, Flavoring agent, food and beverage ingredient	He et al. (2011)
	US20060121137	Hartle, DK, Greenspan, P, Hargrove, JL, Hofacre, CL and Bralley, EE	Muscadine grape pomace in the treatment of intestinal inflammation	Demonstrated reduction or prevention of colon inflammation in animals by administration of muscadine grape pomace through feed	Antiinflammatory supplements for food and feed	Hartle et al. (2006)

Continued

Table 11.9 Patent survey disclosing various grape by-products (Vulic et al., 2017).—cont'd

Grape by-product	Patent application number	Applicant	Title	Outcome	Potential applicaton	Reference
Grape skin	US20050002961	Capri Sun AG	Concentrate comprising green tea, grape skin extract and grape extract, the production thereof and use of the same	Method for producing liquid or solid concentrate with high antioxidant effect and reduced bitterness	For cold or hot beverages, supplementary food to other foodstuffs, such as mueslis, dairy products etc.	Wild and Sass (2006)
	US20160067298	Clark Atlanta University	Muscadine grape skin extract as treatment for bone metastatic cancer	Decrease in Snail expression, CatL and STAT-3 activity in prostate and breast cancer cells and formation of mature osteoclasts	Treating bone metastaic disease resulting from prostate or breast cancer	Odero-Marah (2016)
	US20110311680	Mercian Corporation	Soft drink, grape skin extract, and methods of preparation of same	The method for preparation of grape skin extract rich in precursors of 3-mercaptohexan-1-ol, and producing soft drinks with furity aroma and afteraroma	Food additives (falvours or flavor enhancing compounds), beverages	Takase et al. (2011)

	CN102675904	Yang Qingsong	Grape skin pigment extracting method	Method for preparaion of red grape skin pigment powder	Dyes Food additives (coloring)	Yang (2012)
	CN104448902	Qingdao Zhongren Zhiye Biotechnology Co., Ltd.	Grape skin pigment extraction technology	Method for preparaion of red grape skin pigment powder	Dyes Food additives (coloring)	Tan (2015)
	CN1380093	Xinjiang Medical University	Alcohol extraction method for extracting grape skin extract and its grape liver-nourishing capsule	Method for producing capsules with grape skin extract which have demonstrated protective action against chemical hepatic injury	Dietary supplements	Ma et al. (2002)
	CN1039841	Zhang Guangjin	Method of making vinegar using grape skin	Profitable and faster method for producing vinegar from grape skin. Residue can be used as feed	Condiment, acidity regulator Feed	Zhang and Zhao (1990)
	CN102286061	Not avalable in English	Method for extracting and purifying oleanolic acid from grape skin	Method for producing oleanolic acid from grape skins with improved efficicency, cost effectivness and purity of the final product	Dietray supplements Pharmacological preparations	Not avalable in English

Continued

Table 11.9 Patent survey disclosing various grape by-products (Vulic et al., 2017).—cont'd

Grape by-product	Patent application number	Applicant	Title	Outcome	Potential applicaton	Reference
	KR1020130064595	Taegu Technopark Foodwell Co., Ltd.	Antiobesity composition based on grape skin extracts as an effective component and a manufacturing method of the same	Method for preparation of grape skin extract with antiobesity effect	Dietary supplement Functional food ingredient	Sung et al. (2013)
Grape seeds	US7226627	Eckert Peter, Heinen Winfrid, Knaudt Carola	Grapeseed, cold-pressed grape oil, crushed grape and grape flour	Method for producing col-pressed grape seed oil with increased polyphenol content	Ingredient of food, food supplement, beverages, animal food, medicament and cosmetics	Eckert et al. (2007)
	US6291517	Dry Creek Nutrition, Inc	Method for preventing or reducing stress-induced gastric injury using grape seed proanthocyanidin extract	ability of grapeseed extract to protect against chronic and acute stress-induced gastrointestinal mucosal injury	Recommendation for preferred daily dose to prevent injury: 100–200 mg of GSPE, in two equally divided doses taken within 15–30 min after meals	Bagchi et al. (2003)

	US5484594	Tecnofarmaci S.p.A.; Indena S.p.A.	Process for preparing grapeseed extracts enriched in procyanidol oligomers	Process for the preparation of grape seed enriched in procyanidol oligomers, flavanol monomers, (+)-catechin	Procyanidin oligomers based dietary suppelements	Frangi et al. (2007)
	WO2015184291	Shaklee Corporation	Chardonnay grape seed extract	Process for extracting grape seeds, to obtain isolated procyanidin extract with low degree of polymerization and other biologically useful components	Dietary supplement or functional food, natural flavorant, prebiotic nutraceutical, dietary supplement	Ianiro and Fisher (2016)
	WO2015050883	ELC Management LLC	Topical preparations comprising grape seed, folic acid, biotin, *Bifidobacterium longum* and *Echinacea purpurea*	Formulation for topical skin care preparation, with optimal antioxidant activity utilizing grape seed	Ingredient for skin care preparations, hair care preparations, or color cosmetics	Lee et al. (2016)
	AU2016100373 A4	K D Health Food Australia Pty Ltd	"K.D. BIO-BEAUTY"	Formulation of skin serum comprising hyaluronic acid, nano placenta, grape seed extract and collagen serum	Skin care preparation	Lin (2016)

Continued

Table 11.9 Patent survey disclosing various grape by-products (Vulic et al., 2017).—cont'd

Grape by-product	Patent application number	Applicant	Title	Outcome	Potential applicaton	Reference
	WO2014141265	Nofar, Gil Cogan, Uri	Inhibition of neurodegenerative disease by grape seed extract, green tea extract and probiotic bacteria	Demonstrated synergism between green tea and grape seed extract and *Lactobacillus plantarum* in improving mental abilities	Dietray supplements for neurodegenerative diseases	Nofar and Cogan, 2016
	CN105343353	Wuhu Nokan Biotechnology Co LTD	Application of soft capsule containing sheep placenta, *Radix Salviae miltiorrhizae* and grape seeds in preparing cough medicine	Pharmaceutical preparation utilizing grape seeds that reduces the cough variant asthma	Dietary supplement with antitussive effect	Xu (2016)
	CN105342912	Shi Gengwei	Chinese toothpaste and a method for preparing a matrix using grape seed friction modifiers	Formulation of herbal toothpaste with improved friction effect, antibacterial, antiinflammatory, antipyretic activity and reduced cost	Toothpaste	Shi (2016)

Grape leaves	US6268002	Michael, Patrick T.	Method for making grape leaves with meat	Device for making grape leaves with meat with improved performances	Food	Michael (2001)
	KR1020110039756	Youngdong University Industry-Academia Cooperation Group	Powder containing high content of polyphenolic substances using grape leaves, and a producing method thereof	Drying method for producing powder rich in polyphenols from grape leaves	Dietary supplements Functional food Cosmetics	Chang et al. (2011)
	RU02341279	Ehsperester Anke; Shehfer Ehkkkhard; Zakher Fritts	Filmed tablet containing red grape leaves extract	Method for producing filmed tablet of improved stability including dried extract of red grape leaves	Dietary supplements	Ehsperester et al. (2008)
	CN101913995	Zhang Bo	Method for extracting resveratrol from grape leaves	Simple, low cost method for production of resveratrol crystals isolation from grape leaves	Dietray supplements Functional food ingredients	Zhang et al. (2010)
	KR1020090125988	Industry-Academic Cooperation Foundation, Dankook University Ko, Jung Moon	Massive extraction of quercetin glycoside derived from kyoho grape leaves using column chromatography	A method for extraction quercetin glycoside from grape leaves at industrial level	Dietary supplements Functional food Pharmaceutical preparations Cosmetics	Ko et al. (2009)

Continued

Table 11.9 Patent survey disclosing various grape by-products (Vulic et al., 2017).—cont'd

Grape by-product	Patent application number	Applicant	Title	Outcome	Potential applicaton	Reference
	RU02367464	Masuda Kendzhi; Matsumoto Kazuki; Okada Minoru; Takakhashi Koichi	Composition for treating chronic venous insufficiencies, containing aqueous extract of red grape leaves and antiinflammatory agent	Novel topical formulation for preventing or alleviating mild/ moderate chronic venous insufficiency of legs	Pharmaceutical products or medicated cosmetic	Masuda et al. (2009)
Grape stem (stalk)	KR1020090092405	Okcheon Nonghyup	Method for extracting resveratrol using grape stems, capable of using the by-product of grape as a material	Method for recovery of resveratrol from grape stems	Dietray supplements	Shin et al. (2009)
	WO2005048719	Wisconsin Alumni Research Foundation; Folts, John D.; Shanmuganayagam, Dhanansayan; Hutson, Paul R.	Polyphenol-containing stem and vine extracts and methods of use	Method for preparing grape stem and vine extracts and dietary supplements, nutraceutical and pharmaceutical compositions and methods of using the extracts to prevent or treat coronary artery disease	Dietary supplements Nutraceuticals Pharmaceutical compositions	Folts et al. (2005)

	US20110185629	Bettini Charles Randall; Bettini Deborah	Methods and systems for fuel generation	A method of generating a hybrid liquid hydrocarbon fuel distilling the grape stems into ethanol and mixing the ethanol with gasoline or diesel	Energy production	Bettini and Bettini (2011)
	KR1020120116115	Korea Research Institute Of Chemical Technology Kookmin University Industry Academy Cooperation Foundation	Anticancer drug composition including resveratrol derivative compound separated from grape stem shell	A method for preparation of grape stem extracts rich in resveratrol derivative and dietary supplement with it, in liquid form, with demonstrated anticancer effect	Dietary supplements Beverages	Choi et al. (2012)
Wine lees (dregs)	CN105349297	Sino French Joint Venture Dynasty Winery Ltd	Method for preparing wine from grape juice precipitation solution	Process for storage and processing of grape juice precipitate	Wine and organic fertlizer production	Zhang et al. (2016)
	JP2000135071	Sunstar Inc; University of Kyoto	Food or pharmaceutical compositions containing the wine lees extract	Method for obtaining wine lee extract for safe human use with no side effect, having demonstrated blood vessel-strengthening activity	Functional food Pharmaceutical preparations Dietary supplements	Mizutani et al. (2000)

Continued

Table 11.9 Patent survey disclosing various grape by-products (Vulic et al., 2017).—cont'd

Grape by-product	Patent application number	Applicant	Title	Outcome	Potential applicaton	Reference
	GB756854	Permutit Co Ltd.	Improvements in the recovery of cream of tartar from wine lees	Method for productuion of cream of tartar from wine lee solution	Food additive (stabilizing agent, thickening agent) Leavening agent Purgative Cleaning agent	Arden (1956)
	KR1020150096062	Bobsoonibobsooni	Manufacturing method of pickled yacon using wine lees	Manufacturing method for preparation of pickled yacon using wine lees	Food	Hwang and Jonghwang (2015)
	KR1020000049540	Technical Service Institute National Tax Service	Distilled liquor using wine lees	Method for re-saccharifying and fermenting the remaining sugar and unresolved starch in wine lees, distilling, deodorizing, and refining to produce a distilled liquor	Beverage	Kwon et al. (2000)
	US20020102287	Shanbrom Technologies, LLC	Antimicrobial lees	Method for recovery of pigments from wine lees; Demonstrated antioxidant and antimicrobial effects of wine lees	Demulcent formulations Dietary supplements Antimicrobial agents	Shanbrom (2002)

	WO1998024878	Oenodev S.A.R.L.; Ducournau, Patrick	Method for making wine on lees in barrel or vat and implementing device	A method for making wine on lees in barrel or vat and an implementing device	Beverage	Ducournau (1997)
	CN101845036	Xinjiang Hairuisheng Bio-Engineering Co., Ltd.	Method for extracting procyanidin from wine lees	Method for isolation of procyanidines from wine lees that overcomes the defects of large solvent consumption, long extracting time and high production costs	Dietray supplements Functional food	Li and Li (2010)

Table 11.10 Fruit and vegetable by-products and corresponded functional ingredients for recovery.

Fruits and vegetables	By-product	Target ingredient	References
Mandarin	Peel	Flavanone glycoside - narirutin	Kim et al. (2004)
	Peel	Essential oil (limonene and limonene/γ-terpinene)	Lota et al. (2000)
	Leaf	Essential oil (sabinene/linalool, linalool/γ-terpinene and methyl N-methylanthranilate)	Lota et al. (2000)
Orange	Peel	Flavanone glycoside - hesperidin	Di Mauro et al. (1999)
		Apocarotenoid	Chedea et al. (2010)
		Essential oil (limonene)	Farhat et al. (2011)
		Cellulose	Bicu and Mustata (2011)
Lemon	By-product	Pectin	Masmoudi et al. (2008)
Apple	Pomace	Pectin	Wang et al. (2007)
	Skin	Phenols	Schieber et al. (2001)
Peach	Pomace	Pectin	Pagan et al.(1999)
Apricot	Kernel	Protein	Sharma et al. (2010)
Grape	Pomace	Dietary fiber	Schieber et al. (2001)
	Skin	Phenols	Pinelo et al. (2006)
	Wine lees	Food preservative – calcium tartate	Braga et al. (2002)
		Pigment enocyanin	Braga et al. (2002)
Black currant	Seed residue after oil extraction	Phenols	Bakowska-Barczak et al. (2009)
Banana	Bracts (leaves below calyx)	Anthocyanin pigments	Pazmiño-Durán et al. (2001)
Kiwi	Pomace	Soluble and insoluble dietary fiber	Martin-Cabrejas et al. (1995)

Table 11.10 Fruit and vegetable by-products and corresponded functional ingredients for recovery.—cont'd

Fruits and vegetables	By-product	Target ingredient	References
Pear	Pomace	Soluble and insoluble dietary fiber	Martin-Cabrejas et al. (1995)
Pineapple	Core, peel, crown and extended stem	Proteolytic enzyme -bromelain	Umesh Hebbar et al. (2008)
Mango	Seed kernels	Phenolic compounds	Abdalla et al. (2007)
	Peel	Polyphenols, carotenoids, vitamins, enzymes and dietary fibers	Ajila et al. (2010)
Carrot	Peel	β-carotene	Chantaro et al. (2008)
		Phenols	Chantaro et al. (2008)
	Pomace	Carotenoids	Zhang and Hamauzu (2004)
Tomato	Pomace	Lycopene	Lavecchia and Zuorro (2008)
	Peel	Carotenoids (lycopene, lutein, β-carotene, and cis-β-carotene)	Knoblich et al. (2005)
	Seeds	Lycopene	Knoblich et al. (2005)
		Dietary fiber	Knoblich et al. (2005)
Cauliflower	Floret and curd	Pectin	Femenia et al. (1997)
Broccoli	Leaves or stalks	Glucosinolates, phenolic acids, flavonoids, vitamin C	Domínguez-Perles et al. (2010)

Adapted from Mullen, A.M., Alvarez, C., Pojic, M., Hadnadev, T.D., Papageorgiou, M., 2015. Chapter 2. Classification and target compounds. In: Galanakis, C.M. (Eds.), Food Waste Recovery: Processing Technologies and Industrial Techniques.

stringent quality criteria can lead to substantial surpluses with highly adverse impacts on social and economic conditions on production areas, and also on the environment. As with other fruits, strawberry surpluses can be used to obtain new products exploiting the rich chemical composition. In addition, the chemical composition of strawberry purée is ideal for a number of biotransformations potentially providing new products with an increased added value and healthy properties. Converting glucose in strawberry purée into gluconic acid is possible despite the natural tendency of *Gluconobacter* species to use the acid for further biotransformation. Besides, this bioprocess can be accomplished without using any excess of fructose to provide sweetness to the resulting product. The final product can be used to obtain a wide range of beverages and condiments by mixing it with vinegar in variable proportions.

Apricot (*Prunus armeniaca* L.) generates great amounts of waste that can reach thousands of tons of residues per year. The main by-product of the apricot is the stone, whereas the most valued part of the stone is the kernel. Apricot kernels have been used to extract essential oil that is very appreciated by the pharmaceutical and cosmetic industries, due to its antimicrobial, antiinflammatory, and antioxidant properties. Oil extraction from apricot kernel is typically conducted using hydrodistillation that is a cheap and scalable technology. However, more innovative techniques are required in order to reduce extraction times and increase extraction yields. Other less exploited by-products, such as thinned apricots or wastewaters concentrates, require deeper investigations for the better understanding of their potential.

Passion fruit is one of the well-known tropical fruits that can be consumed fresh and processed into several products. Due to the very large portion of by-products from peel and seed (more than 60% of whole fruit weight), utilizing of these industrial by-products is necessary and can add much value to fruit's supply chain. Current by-product valorizing schemes that have been already commercialized including an extraction of seed oil and further fractionation of unsaponifiable fraction for cosmetic uses. In addition, seed extract containing functional phenolic compounds, piceatannol and scirpusin B is also in the market.

Guava is another example, being a good source of dietary fiber and natural antioxidant compounds. Fruits like guava are being processed to fulfill the high demand but the by-products also need to be utilized. Although there are many research approaches conducted on the valorization of fruit by-products, the challenge resides on the practical implementation from lab scale to pilot scale as well as to the development of applications for the recovered products. To this line, valorization of fruit and vegetable by-products should be extended to consumable products, targeted both food and nonfood applications.

9. Conclusion

Nowadays, food processing by-products account as a source of high added-value compounds that could be recovered and reutilized to fortify different products such as cosmetics and foods. In this chapter, food waste related and newly introduced definitions were provided, prior explaining how the *Universal Recovery Strategy* could be applied to overcome innovation obstacles in the field and improve sustainability of food systems. Current trends of food waste recovery in the food industry were discussed, giving emphasis on different cases of processing by-products, namely derived materials from cereal, coffee, meat, olive oil, grape and fruit processing. The prospects in the field of food waste recovery are enormous and the different possibilities have just now started developing in commercialized applications. Future food industries will try to innovate by valorizing all of their processing by-products and recycling target ingredients in the food chain.

References

Abdalla, A.E.M., Darwish, S.M., Ayad, E.H.E., El-Hamahmy, R.M., 2007. Egyptian mango by-roduct 1. Compositional quality of mango seed kernel. Food Chem. 103, 1134–1140.

Abdel-Shafy, H.I., Schories, G., Mohamed-Mansour, M.S., Bordei, V., 2015. Integrated membranes for the recovery and concentration of antioxidant from olive mill wastewater. Desalination Water Treat. 56, 305–314.

Achmon, Y., Harrold, D.R., Claypool, J.T., Stapleton, J.J., VanderGheynst, J.S., Simmons, C.W., 2016. Assessment of tomato and wine processing solid wastes as soil amendments for biosolarization. Waste Manag. 48, 156–164.

Adhoum, N., Monser, L., 2004. Decolourization and removal of phenolic compounds from olive mill wastewater by electrocoagulation. Chem. Eng. Process. Process Intensification 43 (10), 1281–1287. https://doi.org/10.1016/j.cep.2003.12.001.

Aghamirzaei, M., Peighambardoust, S.H., Azadmard-Damirchi, S., Majzoobi, M., 2015. Effects of grape seed powder as a functional ingredient on flour physicochemical characteristics and dough rheological properties. J. Agric. Sci. Technol. 17 (2), 365–373.

Aguedo, M., Vanderghem, C., Goffin, D., Richel, A., Paquot, M., 2013. Fast and high yield recovery of arabinose from destarched wheat bran. Ind. Crops Prod. 43, 318–325.

Aguedo, M., Fougnies, C., Dermience, M., Richel, A., 2014. Extraction by three processes of arabinoxylans from wheat bran and characterization of the fractions obtained. Carbohydr. Polym. 105, 317–324.

Ajila, C.M., Aalami, M., Leelavathi, K., Rao, U.J.S.P., 2010. Mango peel powder: a potential source of antioxidant and dietary fiber in macaroni preparations. Innov. Food Sci. Emerg. Technol. 11, 219–224.

Akratos, C.S., Tekerlekopoulou, A.G., Vasiliadou, I.A., Vayenas, D.V., 2017. Chapter 8. Co-composting of olive mill waste for the production of soil amendments. In: Galanakis, C.M. (Ed.), Olive Mill Waste: Recent Advances for the Sustainable Management. Elsevier Inc., Waltham.

Aktas, E.S., Imre, S., Ersoy, L., 2001. Characterization and lime treatment of olive mill wastewater. Water Res. 35 (9), 2336–2340. https://doi.org/10.1016/s0043-1354(00)00490-5.

Álvarez, C., Bances, M., Rendueles, M., Díaz, M., 2009. Functional properties of isolated porcine blood proteins. Int. J. Food Sci. Technol. 44 (4), 807–814.

Alvarez, C., Rendueles, M., Diaz, M., 2012. Production of porcine hemoglobin peptides at moderate temperature and medium pressure under a nitrogen stream. Functional and antioxidant properties. J. Agric. Food Chem. 60 (22), 5636–5643.

Amendola, D., De Faveri, D.M., Spigno, G., 2010. Grape marc phenolics: extraction kinetics, quality and stability of extracts. J. Food Eng. 97, 384–392.

Amirante, P., 2003. I sottoprodotti della filiera olivicola-olearia. In: Fiorino, P. (Ed.), Olea, Trattato di olivicoltura. Edagricole, Bologna, pp. 291–303.

Ananey-Obiri, D., Matthews, L., Azahrani, M.H., Ibrahim, S.A., Galanakis, C.M., Tahergorabi, R., 2018. Application of protein-based edible coatings for fat uptake reduction in deep-fat fried foods with an emphasis on muscle food proteins. Trends Food Sci. Technol. 80, 167–174.

Annamalai, N., Sivakumar, N., 2016. Production of polyhydroxybutyrate from wheat bran hydrolysate using *Ralstonia eutropha* through microbial fermentation. J. Biotechnol. 237, 13–17.

Apolinar-Valiente, R., Romero-Cascales, I., Gómez-Plaza, E., López-Roca, J.M., Ros-García, J.M., 2015. Cell wall compounds of red grapes skins and their grape marcs from three different winemaking techniques. Food Chem. 187, 89–97.

Apprich, S., Tirpanalan, Ö., Hell, J., Reisinger, M., Böhmdorfer, S., Siebenhandl-Ehn, S., Novalin, S., Kneifel, W., 2014. Wheat bran-based biorefinery 2: valorization of products. LWT Food Sci. Technol 56, 222–231.

Arden, T.V., 1956. Improvements in the Recovery of Cream of Tartar from Wine Lees. Patent No. GB756854.

Argun, H., Onaran, G., 2015. Delignification of vineyard pruning residues by alkaline peroxide treatment. Ind. Crops Prod. 74, 697–702.

Asano, I., Fujii, S., Mutoh, K., Takao, I., Ozaki, K., Nakamuro, K., Matsushima, T., 2006. Mannooligosaccharide Composition for Body Fat Reduction. Patent WO 2006/036208.

Azabou, S., Najjar, W., Mohamed, B., Ghorbel, A., Sayadi, S., 2010. A compact process for the treatment of olive mill wastewater by combining wet hydrogen peroxide catalytic oxidation and biological techniques. J. Hazard Mater. 183 (1–3), 62–69. https://doi.org/10.1016/j.jhazmat.2010.06.104.

Azbar, N., Keskin, T., Ebru Cokay Catalkaya, 2008. Improvement in anaerobic degradation of Olive Mill Effluent (OME) by chemical pretreatment using batch systems. Biochem. Eng. J. 38 (3), 379–383. https://doi.org/10.1016/j.bej.2007.08.005.

Bagchi, D., Sen, C.K., Ray, S.D., Bagchi, M., Preuss, H.G., Vinson, J.A., 2003. Molecular mechanisms of cardioprotection by a novel grape seed proanthocyanidin extract. Mutat. Res. 523–524, 87–97.

Bakowska-Barczak, A.M., Schieber, A., Kolodziejczyk, P., 2009. Characterization of Canadian black currant (*Ribes nigrum* L.) seed oils and residues. J. Agric. Food Chem. 57, 11528–11536.

Balandran-Quintana, R.R., Mercado-Ruiz, J.N., Mendoza-Wilson, A.M., 2015. Wheat bran proteins: a review of their uses and potential. Food Rev. Int. 31, 279–293.

Ballesteros, L.F., Conde, T., Teixeira, J.A., Mussatto, S.I., 2011. Optimization of the extraction conditions for antioxidant phenolic compounds recovery from coffee silverskin. In: Book of Abstracts of the 11th International Chemical and Biological Engineering Conference. Lisabon, Portugal, pp. 223–224.

Barakat, N.A.M., Khil, M.S., Omran, A.M., Sheikh, F.A., Kim, H.Y., 2009. Extraction of pure natural hydroxyapatite from the bovine bones bio waste by three different methods. J. Mater. Process. Technol. 209 (7), 3408–3415.

Barba, F.J., Galanakis, C.M., Esteve, M.J., Frigola, A., Vorobiev, E., 2015. Potential use of pulsed electric technologies and ultrasounds to improve the recovery of high-added value compounds from blackberries. J. Food Eng. 167, 38–44.

Belscak-Cvitanovic, A., Komes, D., 2017. Chapter 4. Extraction and formulation of bioactive compounds. In: Galanakis, C.M. (Ed.), Handbook of Coffee Processing By-Products: Sustainable Applications. Elsevier-Academic Press.

Benamrouche, S., Crônier, D., Debeire, P., Chabbert, B., 2002. A chemical and histological study on the effect of (1 → 4)- β-endo-xylanase treatment on wheat bran. J. Cereal Sci. 36, 253–260.

Bettini, C.R., Bettini, D., 2011. Methods and Systems for Fuel Generation. Patent No. US20110185629.

Bicu, I., Mustata, F., 2011. Cellulose extraction from orange peel using sulfite digestion reagents. Bioresour. Technol. 102, 10013–10019.

Boles, J., Rathgeber, B., Shand, P., 2000. Recovery of proteins from beef bone and the functionality of these proteins in sausage batters. Meat Sci. 55 (2), 223–231.

Braga, F.G., Silva, F.A.L.E., Alves, A., 2002. Recovery of winery by-products in the douro demarcated region: production of calcium tartate and grape pigments. Am. J. Enol. Vitic. 53, 41–45.

Bravo, J., Juániz, I., Monente, C., Caemmerer, B., Kroh, L.W., De Peña, M., P, Cid, C., 2012. Evaluation of spent coffee obtained from the most common coffeemakers as a source of hydrophilic bioactive compounds. J. Agric. Food Chem. 60, 12565–12573.

Bravo, J., Monente, C., Juániz, I., De Peńa, M.P., Cid, C., 2013. Influence of extraction process on antioxidant capacity of spent coffee. Food Res. Int. 50, 610–616.

Brenes, A., Viveros, A., Chamorro, S., Arija, I., 2016. Use of polyphenol-rich grape by-products in monogastric nutrition. A review. Anim. Feed Sci. Technol. 211, 1–17.

Bressan, M., Liberatore, L., D'Alessandro, N., Tonucci, L., Belli, C., Ranalli, G., 2004. Improved combined chemical and biological treatments of olive oil mill wastewaters. J. Agric. Food Chem. 52 (5), 1228–1233. https://doi.org/10.1021/jf035128p.

Brijwani, K., Oberoi, S.H., Vadlani, P.V., 2010. Production of a cellulolytic enzyme system in mixed-culture solid-state fermentation of soybean hulls supplemented with wheat bran. Process Biochem. 45, 120–128.

Buranov, A.U., Mazza, G., 2009. Extraction and purification of ferulic acid from flax shives, wheat and corn bran by alkaline hydrolysis and pressurised solvents. Food Chem. 115 (4), 1542–1548.

Buratti, C., Barbanera, M., Lascaro, E., 2015. Ethanol production from vineyard pruning residues with steam explosion pretreatment. Environ. Prog. Sustain. Energy 34 (3), 802–809.

Burg, P., Vítěz, T., Turan, J., Burgová, J., 2014. Evaluation of grape pomace composting process. Acta U.Agr.Silvi. 62 (5), 875–888.

Caputo, A.C., Scacchia, F., Pelagagge, P.M., 2003. Disposal of by-products in olive oil industry: waste-to-energy solutions. Appl. Therm. Eng. 23 (2), 197–214. https://doi.org/10.1016/s1359-4311(02)00173-4.

Cassano, A., Galanakis, C.M., 2018. Chapter 6. Membrane technologies for the fractionation of compounds recovered from cereal processing by-products. In: Galanakis, C. (Ed.), Sustainable Recovery and Reutilization of Cereal Processing By-Products. M. Elsevier Inc., Waltham.

Cassano, A., Conidi, C., Giorno, L., Drioli, E., 2013. Fractionation of olive mill wastewaters by membrane separation techniques. J. Hazard Mater. 248–249, 185–193.

Castanas, E., Andricopoulos, N., Mposkou, G., Vercauteren, J., 2005. A Method for the Treatment of Olive Mill Waste Waters. World Intellectual Property Organization. WO/2005/003037.

Celiktas, M.S., Kirsch, C., Smirnova, I., 2014. Cascade processing of wheat bran through a biorefinery approach. Energy Convers. Manag. 84, 633–639.

Chang, S.W., Rho, Y.T., Shin, N.S., 2011. Powder Containing High Content of Polyphenolic Substances Using Grape Leaves, and a Producing Method Thereof. Patent No. KR1020110039756.

Chantaro, P., Devahastin, S., Chiewchan, N., 2008. Production of antioxidant high dietary fiber powder from carrot peels. LWT Food Sci. Technol. 41, 1987–1994.

Chaturvedi, V., Bhange, K., Bhatt, R., Verma, P., 2014. Production of kertinases using chicken feathers as substrate by a novel multifunctional strain of *Pseudomonas stutzeri* and its dehairing application. Biocatalysis Agric. Biotechnol. 3 (2), 167–174.

Chatzisymeon, E., Xekoukoulotakis, N.P., Mantzavinos, D., 2009. Determination of key operating conditions for the photocatalytic treatment of olive mill wastewaters. Catal. Today 144 (1–2), 143–148. https://doi.org/10.1016/j.cattod.2009.01.037.

Chedea, V.S., Kefalas, P., Socaciu, C., 2010. Patterns of carotenoid pigments extracted from two orange peel wastes (Valencia and Navel var.). J. Food Biochem. 34, 101–110.

Chen, L., Hu, J.Y., Wang, S.Q., 2012. The role of antioxidants in photoprotection: a critical review. J. Am. Acad. Dermatol. 67, 1013–1024.

Choi, S.U., Ryu, S.Y., Kim, Y.S., Choi, C.W., Choi, Y.H., Cha, M.R., Kim, Y.K., 2012. Anticancer Drug Composition Including Resveratrol Derivative Compound Separated from Grape Stem Shell. Patent No. KR1020120116115.

Correddu, F., Nudda, A., Manca, M.G., Pulina, G., Dalsgaard, T.K., 2015. Light-induced lipid oxidation in sheep milk: effects of dietary grape seed and linseed, alone or in combination, on milk oxidative stability. J. Agric. Food Chem. 63 (15), 3980–3986.

Costa, A.S.G., Alves, R.C., Vinha, A.F., Barreira, S.V.P., Nunes, M.A., Cunha, L.M., Oliveira, M.B.P.P., 2014. Optimization of antioxidants extraction from coffee silverskin, a roasting by-product, having in view a sustainable process. Ind. Crops Prod. 53, 350–357.

Cotana, F., Barbanera, M., Foschini, D., Lascaro, E., Buratti, C., 2015. Preliminary optimization of alkaline pretreatment for ethanol production from vineyard pruning. Energy Procedia 82, 389–394.

Cripwell, R., Favaro, L., Rose, S.H., Basaglia, M., Cagnin, L., Casella, S., Zyl, W.V., 2015. Utilisation of wheat bran as a substrate for bioethanol production using recombinant cellulases and amylolytic yeast. Appl. Energy 160, 610–617.

Da Ros, C., Cavinato, C., Bolzonella, D., Pavan, P., 2016. Renewable energy from thermophilic anaerobic digestion of winery residue: preliminary evidence from batch and continuous lab-scale trials. Biomass Bioenergy 91, 150–159.

Dalis, D., Anagnostidis, K., Lopez, A., Letsiou, I., Hartmann, L., 1996. Anaerobic digestion of total raw olive-oil wastewater in a two-stage pilot-plant (up-flow and fixed-bed bioreactors). Bioresour. Technol. 57 (3), 237–243. https://doi.org/10.1016/s0960-8524(96)00051-x.

Darine, S., Christophe, V., Gholamreza, D., 2010. Production and functional properties of beef lung protein concentrates. Meat Sci. 84 (3), 315–322.

De Magalhaes Nunes Da Ponte, M.L., Dos Santos, J.L.C., Matias, A.A.F., Nunes, A.V.M.M., Duarte, C.M.M., Crespo, J.P.S.G., 2008. Method of Obtaining a Natural Hydroxytyrosol-Rich Concentrate From Olive Tree Residues and Subproducts Using Clean Technologies. Patent US 2008/0179246.

de Mello Prado, R., Caione, G., Silva Campos, C.N., 2013. Filter cake and Vinasse as fertilizers contributing to conservation agriculture. Appl. Environ. Soil Sci. 8. Article ID 581984.

del Castillo, M.D., Ibáñez, E., Amigo, M., Herrero, M., Plaza, M., Ullate, M., 2013. Application of Products of Coffee Silverskin in Anti-aging Cosmetics and Functional Food. WO 2013/004873.

Delgado de la Torre, M.P., Priego-Capote, F., Luque de Castro, M.D., 2014. Comparative profiling analysis of woody flavouring from vine-shoots and oak chips. J. Sci. Food Agric. 94 (3), 504−514.

Delgado De La Torre, M.P., Priego-Capote, F., Luque De Castro, M.D., 2015. Tentative identification of polar and mid-polar compounds in extracts from wine lees by liquid chromatography-tandem mass spectrometry in high-resolution model. J. Mass Spectrom. 50 (6), 826−837.

De Martino, A., Michele, A., Iorio, M., Vinale, F., Lorito, M., Prenzler, P.D., Ryan, D., Obied, H.K., 2011. Detoxification of olive mill wastewaters by zinc-aluminium layered double hydroxides. Appl. Clay. Sci. 53 (4), 737−744. https://doi.org/10.1016/j.clay.2011.07.003.

Deng, Q., Zinoviadou, K.G., Galanakis, C.M., Orlien, V., Grimi, N., Vorobiev, E., Lebovka, N., Barba, F.J., 2015. The effects of conventional and non-conventional processing on glucosinolates and its derived forms, isothiocyanates: extraction, degradation and applications. Food Eng. Rev 7, 357−381.

Dewitt, C., Gomez, G., James, J., 2002. Protein extraction from beef heart using acid solubilization. J. Food Sci. 67 (9), 3335−3341.

Di Bernardini, R., Harnedy, P., Bolton, D., Kerry, J., O'Neill, E., Mullen, A.M., Hayes, M., 2011a. Antioxidant and antimicrobial peptidic hydrolysates from muscle protein sources and by-products. Food Chem. 124 (4), 1296−1307.

Di Bernardini, R., Rai, D.K., Bolton, D., Kerry, J., O'Neill, E., Mullen, A.M., Harnedy, P., Hayes, M., 2011b. Isolation, purification and characterization of antioxidant peptidic fractions from a bovine liver sarcoplasmic protein thermolysin hydrolyzate. Peptides 32 (2), 388−400.

Di Mauro, A., Fallico, B., Passerini, A., Rapisarda, P., Maccarone, E., 1999. Recovery of hesperidin from orange peel by concentration of extracts on styrene-divinylbenzene resin. J. Agric. Food Chem. 47, 4391−4397.

Dimou, C., Kopsahelis, N., Papadaki, A., Papanikolaou, S., Kookos, I.K., Mandala, I., Koutinas, A.A., 2015. Wine lees valorization: biorefinery development including production of a generic fermentation feedstock employed for poly(3-hydroxybutyrate) synthesis. Food Res. Int. 73, 81−87.

Directive 2008/98/EC of The European Parliament and of The Council of 19 November 2008 on waste and repealing certain directives, 2008. Available at: http://eur-lex.europa.eu/legal content/EN/TXT/PDF/?uri=CELEX:32008L0098&from=EN.

Domínguez-Perles, R., Martínez-Ballesta, M.C., Carvajal, M., García-Viguera, C., Moreno, D., 2010. Broccoli-derived by-productsea promising source of bioactive ingredients. J. Food Sci. 75, 383−392.

Duca, D., Toscano, G., Pizzi, A., Rossini, G., Fabrizi, S., Lucesoli, G., Servili, A., Mancini, V., Romanazzi, G., Mengarelli, C., 2016. Evaluation of the characteristics of vineyard pruning residues for energy applications: effect of different copper-based treatments. J. Agric. Eng. 47 (1), 22−27.

Ducournau, P., 1997. Method for Making Wine on Lees in Barrel or Vat and Implementing Device. Patent No. WO1998024878.

Eckert, P., Heinen, W., Knaudt, C., 2007. Grapeseed, Cold-Pressed Grapeoil, Crushed Grape and Grape Flour. Patent No. US7226627.

Ehsperester, A., Shehfer, E., Zakher, F., 2008. Filmed Tablet Containing Red Grape Leaves Extract. Patent No. RU02341279.

El Hajjouji, H., Fakharedine, N., Ait Baddi, G., Winterton, P., Bailly, J.R., Revel, J.C., Hafidi, M., 2007. Treatment of olive mill waste-water by aerobic biodegradation: an analytical study using gel permeation chromatography, ultraviolet-visible and Fourier transform infrared spectroscopy. Bioresour. Technol. 98 (18), 3513–3520. https://doi.org/10.1016/j.biortech.2006.11.033.

El-Abbassi, A., Hafidi, A., García-Payo, M.C., Khayet, M., 2009. Concentration of olive mill wastewater by membrane distillation for polyphenols recovery. Desalination 245, 670–674.

El-Abbassi, A., Kiai, H., Hafidi, A., Garcia-Payo, M.C., Khayet, M., 2012. Treatment of olive mill wastewater by membrane distillation using polytetrafluoroethylene membranes. Separ. Purif. Technol. 98, 55–61. https://doi.org/10.1016/j.seppur.2012.06.026.

El-Abbassi, A., Hafidi, A., Khayet, M., Garcia-Payo, M.C., 2013. Integrated direct contact membrane distillation for olive mill wastewater treatment. Desalination 323, 31–38. https://doi.org/10.1016/j.desal.2012.06.014.

El-Abbassi, A., Kiai, H., Raiti, J., Hafidi, A., 2014. Cloud point extraction of phenolic compounds from pretreated olive mill wastewater. J. Environ. Chem. Eng. (0) https://doi.org/10.1016/j.jece.2014.06.024.

El-Shishtawy, R.M., Mohamed, S.A., Asiri, A.M., Gomaa, A.M., Ibrahim, H.I., Al-Talhi, H.A., 2015. Saccharification and hydrolytic enzyme production of alkali pretreated wheat bran by *Trichoderma virens* under solid state fermentation. BMC Biotechnol. 15, 37.

European Commission. http://ec.europa.eu/environment/circular-economy/index_en.htm.

Fabbri, A., Bonifazi, G., Serranti, S., 2015. Micro-scale energy valorization of grape marcs in winery production plants. Waste Manag. 36, 156–165.

FAO, 2011. Global Food Losses and Food Waste: Extent, Causes and Prevention. Food and Agriculture Organization of the United Nations, Rome.

Farhat, A., Anne-Sylvie Fabiano-Tixier, A.-S., Maataoui, M.E., Maingonnat, J.-F., Romdhane, M., Chemat, F., 2011. Microwave steam diffusion for extraction of essential oil from orange peel: kinetic data, extract's global yield and mechanism. Food Chem. 125, 255–261.

Favaro, L., Basaglia, M., van Zyl, W.H., Casella, S., 2013. Using an efficient fermenting yeast enhances ethanol production from unfiltered wheat bran hydrolysates. Appl. Energy 102, 170–178.

Femenia, A., Lefebvre, A.-C., Thebaudin, J.-Y., Robertson, J.A., Bourgeois, C.-M., 1997. Physical and sensory properties of model foods supplemented with cauliflower fiber. Physical and sensory properties of model foods supplemented with cauliflower fiber. J. Food Sci. 62, 635–639.

Fia, G., Zanoni, B., Gori, C., 2016. A new technique for exploitation of wine lees. Agric. Agric. Sci. Procedia 8, 748–754.

Folts, J.D., Shanmuganayagam, D., Hutson, P.R., 2005. Polyphenol-containing Stem and Vine Extracts and Methods of Use. Patent No. WO2005048719.

Foresight Project, 2011. The Future of Food and Farming: Challenges and Choices for Global Sustainability. Government Office for Science, UK.

Frangi, E., Bertani, M., Mustich, G., Tuccini, G., 2007. Process for Preparing Grapeseed Extracts Enriched in Procyanidol Oligomers. Patent No. US5484594.

Galanakis, C.M., 2011. Olive fruit and dietary fibers: components, recovery and applications. Trends Food Sci. Technol. 22 (4), 175–184.

Galanakis, C.M., 2012. Recovery of high added-value components from food wastes: conventional, emerging technologies and commercialized applications. Trends Food Sci. Technol. 26, 68–87.

Galanakis, C.M., 2013. Emerging technologies for the production of nutraceuticals from agricultural by-products: a viewpoint of opportunities and challenges. Food Bioprod. Process. 91, 575–579.

Galanakis, C.M., 2015. Preface. In: Galanakis, C.M. (Ed.), Food Waste Recovery: Processing Technologies and Industrial Techniques.

Galanakis, C.M., 2015j. Chapter 3: development of a universal recovery strategy. In: Galanakis, C.M. (Ed.), Food Waste Recovery: Processing Technologies and Techniques. Elsevier Inc., Waltham.

Galanakis, C.M., 2016f. Innovation Strategies for the Food Industry: Tools for Implementation. Elsevier-Academic Press. ISBN: 9780128037515.

Galanakis, C.M., 2017f. Handbook of Coffee Processing By-Products: Sustainable Applications. Elsevier-Academic Press.

Galanakis, C.M., 2017g. Handbook of Grape Processing By-Products: Sustainable Solutions. Elsevier-Academic Press. ISBN: 9780128098707.

Galanakis, C.M., 2017h. Olive Mill Waste: Recent Advances for the Sustainable Management. Elsevier-Academic Press. ISBN: 9780128053140.

Galanakis, C.M., 2017i. Nutraceutical and Functional Food Components: Effects of Innovative Processing Techniques. Elsevier-Academic Press. ISBN: 9780128052570.

Galanakis, C.M., 2017q. Chapter 6. Membrane technologies for the separation of compounds recovered from grape processing by-products. In: Galanakis, C.M. (Ed.), Handbook of Grape Processing By-Products: Sustainable Solutions. Elsevier Inc., Waltham.

Galanakis, C.M., 2018f. Polyphenols: Properties, Recovery and Applications. Elsevier-Academic Press.

Galanakis, C.M., 2018g. Sustainable Recovery and Reutilization of Cereal Processing By-Products. Elsevier-Woodprint.

Galanakis, C.M., 2018h. Sustainable Food Systems from Agriculture to Industry: Improving Production and Processing. Elsevier-Academic Press.

Galanakis, C.M., 2018r. Phenols recovered from olive mill wastewater as additives in meat products. Trends Food Sci. Technol. 79, 98–105.

Galanakis, C.M., 2019a. Dietary Fiber: Properties, Recovery & Applications. Elsevier Inc.

Galanakis, C.M., 2019b. Proteins: Sustainable Source, Processing and Applications. Elsevier Inc.

Galanakis, C.M., 2019c. The Role of Alternative and Innovative Food Ingredients and Products in Consumer Wellness. C.M. Elsevier Inc.

Galanakis, C.M., 2019d. Saving Food: Production, Supply Chain, Food Waste and Food Consumption. Elsevier-Academic Press.

Galanakis, C.M., 2019e. Separation of Functional Molecules in Food by Membrane Technology (2019). Elsevier-Academic Press.

Galanakis, C.M., 2019f. In: Galanakis, C.M. (Ed.), Sustainable Meat Production and Processing (2018). Elsevier-Academic Press. ISBN: 9780128148747.

Galanakis, C.M., 2019g. Innovations in Traditional Foods. Elsevier-Woodhead Publishing. ISBN: 9780128148877.

Galanakis, C.M., Goulas, V., Tsakona, S., Manganaris, G.A., Gekas, V., 2013. A knowledge base for the recovery of natural phenols with different solvents. Int. J. Food Prop. 16, 382–396.

Galanakis, C.M., Kotsiou, K., 2017. Chapter 10. Recovery of bioactive compounds from olive mill waste. In: Galanakis, C.M. (Ed.), Olive Mill Waste: Recent Advances for the Sustainable Management. Elsevier Inc., Waltham.

Galanakis, C.M., Schieber, A., 2014. Editorial. Special Issue on Recovery and utilization of valuable compounds from food processing by-products. Food Res. Int. 65, 299–230.

Galanakis, C.M., Tornberg, E., Gekas, V., 2010a. A study of the recovery of the dietary fibres from olive mill wastewater and the gelling ability of the soluble fibre fraction. LWT Food Sci. Technol. 43 (7), 1009–1017.

Galanakis, C.M., Tornberg, E., Gekas, V., 2010b. Clarification of high-added value products from olive mill wastewater. J. Food Eng. 99 (2), 190–197.

Galanakis, C.M., Tornberg, E., Gekas, V., 2010c. Dietary fiber suspensions from olive mill wastewater as potential fat replacements in meatballs. LWT Food Sci. Technol. 43 (7), 1018–1025.

Galanakis, C.M., Tornberg, E., Gekas, V., 2010d. Recovery and preservation of phenols from olive waste in ethanolic extracts. J. Chem. Technol. Biotechnol. 85 (8), 1148–1155.

Galanakis, C.M., Tornberg, E., Gekas, V., 2010e. The effect of heat processing on the functional properties of pectin contained in olive mill wastewater. LWT Food Sci. Technol. 43 (7), 1001–1008.

Galanakis, C.M., Chasiotis, S., Botsaris, G., Gekas, V., 2014. Separation and recovery of proteins and sugars from halloumi cheese whey. Food Res. Int. 65, 477–483.

Galanakis, C.M., Martínez-Saez, N., del Castillo, M.D., Barba, F.J., Mitropoulou, V.S., 2015c. Chapter 15: patented and commercialized applications. In: Galanakis, C.M. (Ed.), Food Waste Recovery: Processing Technologies and Industrial Techniques. Elsevier-Academic Press.

Galanakis, C.M., Kotanidis, A., Dianellou, M., Gekas, V., 2015l. Phenolic content and antioxidant capacity of cypriot wines. Czech J. Food Sci. 33, 126–136.

Galanakis, C.M., Cvejic, J., Verardo, V., Segura-Carretero, A., 2016. Chapter 11. Food use for social innovation by optimizing food waste recovery strategies. In: Galanakis, C.M. (Ed.), Innovation Strategies for the Food Industry: Tools for Implementation. Elsevier Inc., Waltham.

Galanakis, C.M., Tsatalas, P., Charalambous, Z., Galanakis, I.M., 2018a. Control of microbial growth in bakery products fortified with polyphenols recovered from olive mill wastewater. Environ. Techno. Innov. 10, 1–15.

Galanakis, C.M., Tsatalas, P., Galanakis, I.M., 2018b. Implementation of phenols recovered from olive mill wastewater as UV booster in cosmetics. Ind. Crops Prod. 111, 30–37.

Galanakis, C.M., Tsatalas, P., Galanakis, I.M., 2018c. Phenols from olive mill wastewater and other natural antioxidants as UV filters in sunscreens. Environ. Technol. Innov. 9, 160–168.

Galanakis, C.M., Tsatalas, P., Charalambous, Z., Galanakis, I.M., 2018d. Polyphenols recovered from olive mill wastewater as natural preservatives in extra virgin olive oils and refined olive kernel oils. Environ. Technol. Innov. 10, 62–70.

Galbreath, J.S., 2001. Method for Preparing a Grape Derived Product. Patent No. US 6190716.

Garcia-Castello, E., Cassano, A., Criscuoli, A., Conidi, C., Drioli, E., 2010. Recovery and concentration of polyphenols from olive mill wastewaters by integrated membrane system. Water Res. 44, 3883–3892.

Georgacakis, D., Dalis, D., 1993. Controlled anaerobic digestion of settled olive-oil waste-water. Bioresour. Technol. 46 (3), 221–226. https://doi.org/10.1016/0960-8524(93)90124-t.

Giacobbo, A., Do Prado, J.M., Meneguzzi, A., Bernardes, A.M., De Pinho, M.N., 2015. Microfiltration for the recovery of polyphenols from winery effluents. Sep.Purif. Technol. 143, 12–18.

González-Vara, A., Pinelli, D., Rossi, M., Fajner, D., Magelli, F., Matteuzzi, D., 1996. Production of L(+) and D(−) lactic acid isomers by *Lactobacillus casei* subsp. Casei DSM 20011 and *Lactobacillus coryniformis* subsp. Torquens DSM 20004 in continuous fermentation. J. Ferment. Bioeng. 81 (6), 548–552.

Gortzi, O., Lalas, S., Chatzilazarou, A., Katsoyannos, E., Papaconstandinou, S., Dourtoglou, E., 2008. Recovery of natural antioxidants from olive mill wastewater using genapol-X080. J. Am. Oil Chem. Soc. 85, 8.

Gousterova, A., Nustorova, M., Paskaleva, D., Naydenov, M., Neshev, G., Vasileva-Tonkova, E., 2012. Assessment of feather hydrolysate from thermophilic actinomycetes for soil amendment and biological control application. Int. J. Environ. Res. 6 (2), 467–474.

Gül, H., Acun, S., Şen, H., Nayir, N., Türk, S., 2013. Antioxidant activity, total phenolics and some chemical properties of Öküzgözü and Narince grape pomace and grape seed flour. J. Food Agric. Environ. 11 (2), 28–34.

Gursoy-Haksevenler, Hande, B., Dogruel, S., Arslan-Alaton, I., 2014. Effect of ferric chloride coagulation, lime precipitation, electrocoagulation and the Fenton's reagent on the particle size distribution of olive mill wastewater. Int. J. Glob. Warming 6 (2–3), 194–211. https://doi.org/10.1504/ijgw.2014.061010.

Hachicha, S., Juan, C., Fatma, S., Ridha, H., Noureddine, D., Khaled, M., Ammar, E., 2009. Elimination of polyphenols toxicity from olive mill wastewater sludge by its co-composting with sesame bark. J. Hazard Mater. 161 (2–3), 1131–1139. https://doi.org/10.1016/j.jhazmat.2008.04.066.

Hadnađev, T.D., Hadnađev, M., Pojić, M., 2018. Chapter 2. The healthy components of cereal by-products and their functional properties. In: Galanakis, C.M. (Ed.), Sustainable Recovery and Reutilization of Cereal Processing By-Products. Elsevier-Woodprint.

Hajati, H., Hassanabadi, A., Golian, A., Nassiri-Moghaddam, H., Reza Nassiri, M., 2015. The effect of grape seed extract and vitamin C feed supplementation on some blood parameters and HSP70 gene expression of broiler chickens suffering from chronic heat stress. Ital. J. Anim. Sci. 14 (3), 1–9, 3273.

Hamza, M., Sayadi, S., 2015. Valorization of olive mill wastewater by enhancement of natural hydroxytyrosol recovery. Int. J. Food Sci. Technol. 50, 826–833.

Hanafi, F., Assobhei, O., Mountadar, M., 2010. Detoxification and discoloration of Moroccan olive mill wastewater by electrocoagulation. J. Hazard Mater. 174 (1–3), 807–812. https://doi.org/10.1016/j.jhazmat.2009.09.124.

Harris-Valle, C., Esqueda, M., Sánchez, A., Beltrán-García, M., Valenzuela-Soto, E.M., 2007. Polar vineyard pruning extracts increase the activity of the main ligninolytic enzymes in *Lentinula edodes* cultures. Can. J. Microbiol. 53 (10), 1150–1157.

Hartle, D.K., Greenspan, P., Hargrove, J.L., Hofacre, C.L., Bralley, E.E., 2006. Muscadine Grape Pomace in the Treatment of Intestinal Inflammation. Patent No. US20060121137.

He, X., Zhao, H., Sun, F., Liu, M., 2011. Production Process for Brewing Grape Fruit Vinegar by Using Grape Pomace. Patent No. CN102268358.

Heng, W.W., Xiong, L.W., Ramanan, R.N., Hong, T.L., Kong, K.W., Galanakis, C.M., Prasad, K.N., 2015. Two level factorial design for the optimization of phenolics and flavonoids recovery from palm kernel by-product. Ind. Crops Prod. 63, 238–248.

Heng, W.W., Yang, B., Prasad, K.N., Ramanan, R.N., Galanakis, C.M., Sun, J., Ismail, A., Ti, T.B., 2016. Extraction of phytochemicals using hydrotropic solvents. Separ. Sci. Technol. 51, 1151–1165.

Hsieh, Y.-L., 2013. Cellulose nanocrystals and self-assembled nanostructures from cotton, rice straw and grape skin: a source perspective. J. Mat. Sci. 48 (22), 7837–7846.

Hwang, S., Jonghwang, S.J., 2015. Manufacturing Method of Pickled Yacon Using Wine Lees. Patent No. KR1020150096062.

Ianiro, T., Fisher, L., 2016. Chardonnay Grape Seed Extract. Patent No. WO2015184291.

Ichimura, T., Yamanaka, A., Otsuka, T., Yamashita, E., Maruyama, S., 2009. Antihypertensive effect of enzymatic hydrolysate of collagen and Gly-Pro in spontaneously hypertensive rats. Biosci. Biotechnol. Biochem. 73 (10), 2317–2319.

Ignea, C., Dorobantu, C.M., Mintoff, C.P., Branza-Nichita, N., Ladomery, M.R., Kefalas, P., Chedea, V.S., 2013. Modulation of the antioxidant/pro-oxidant balance, cytotoxicity and antiviral actions of grape seed extracts. Food Chem. 14 (4), 3967–3976.

Incinur, H., Galanakis, C.M., 2018. Chapter 7. Recovery technologies and encapsulation techniques. In: Galanakis, C.M. (Ed.), Polyphenols: Properties, Recovery and Applications. Elsevier Inc., Waltham.

Ionescu, A., Aprodu, I., Zara, M., Vasile, A., Porneala, L., 2007. Evaluating of some functional properties of the myofibrillar protein concentrate from the beef heart. Sci. Study Res. VIII (2).

IOOC: Recycling of Vegetable Water & Olive Pomace on Agicultural Land (CFC/IOOC/04). Achievements of the Project CFC/IOOC/04 "Recycling of Vegetable Water & Olive Pomace on Agricultural Land" Good Practice in Vegetable Water and Compost Spreading on Agricultural Land: Case of Olive Growing, 2004. Project executing agency CFC/IOOC/04 Agro-pôle Olivier ENA Meknès.

Jaisan, C., Chase, S., Punbusayakul, N., 2015. Antioxidant and antimicrobial activities of various solvents extracts of arabica coffee pulp. J. Process. Energy Agric. 19, 224–227.

Jiménez-Quero, A., Pollet, E., Zhao, M., Marchioni, E., Avérous, L., Phalip, V., 2016. Itaconic and fumaric acid production from biomass hydrolysates by *Aspergillus* strains. J. Microbiol. Biotechnol. 26 (9), 1557–1565.

Jimenez-Zamora, A., Pastoriza, S., Rufian-Henares, J.A., 2015. Revalorization of coffee by-products. Prebiotic, antimicrobial and antioxidant properties. LWT Food Sci. Technol. 61, 12–18.

Juodeikiene, G., Basinskiene, L., Vidmantiene, D., Makaravicius, T., Bartkiene, E., 2011. Benefits of β-xylanase for wheat biomass conversion to bioethanol. J. Sci. Food Agric. 92, 84–89.

Kakaei, S., Shahbazi, Y., 2016. Effect of chitosan-gelatin film incorporated with ethanolic red grape seed extract and *Ziziphoraclinopodioides* essential oil on survival of *Listeria monocytogenes* and chemical, microbial and sensory properties of minced trout fillet. Food Sci. Technol LEB. 72, 432–438.

Kaparaju, P., Serrano, M., Thomsen, A.B., Kongjan, P., Angelidaki, I., 2009. Bioethanol, biohydrogen and biogas production from wheat straw in a biorefinery concept. Bioresour. Technol. 100, 2562–2568.

Kaplan, F., Hesenov, A., Gozmen, B., Erbatur, O., 2011. Degradations of model compounds representing some phenolics in olive mill wastewater via electro-Fenton and photoelectro-Fenton treatments. Environ. Technol. 32 (7), 685–692. https://doi.org/10.1080/09593330.2010.510536.

Kara, K., KocaoȷluGüçlü, B., Baytok, E., Şentürk, M., 2016. Effects of grape pomace supplementation to laying hen diet on performance, egg quality, egg lipid peroxidation and some biochemical parameters. J. Appl. Anim. Res. 44 (1), 303–310.

Karaoğlu, M.H., Zor, S., Uğurlu, M., 2010. Biosorption of Cr(III) from solutions using vineyard pruning waste. Chem. Eng. J. 159 (1–3), 98–106.

Ke, W.C., Yang, F.Y., Undersander, D.J., Guo, X.S., 2015. Fermentation characteristics, aerobic stability, proteolysis and lipid composition of alfalfa silage ensiled with apple or grape pomace. Anim. Feed Sci. Technol. 202, 12–19.

Kestioglu, K., Yonar, T., Azbar, N., 2005. Feasibility of physico-chemical treatment and Advanced Oxidation Processes (AOPs) as a means of pretreatment of Olive Mill Effluent (OME). Process Biochem. 40 (7), 2409–2416. https://doi.org/10.1016/j.procbio.2004.09.015.

Kew, S.J., Gwynne, J.H., Enea, D., Abu-Rub, M., Pandit, A., Zeugolis, D., Brooks, R.A., Rushton, N., Best, S.M., Cameron, R.E., 2011. Regeneration and repair of tendon and ligament tissue using collagen fibre biomaterials. Acta Biomater. 7 (9), 3237–3247.

Kilic, M.Y., Yonar, T., Kestioglu, K., 2013. Pilot-scale treatment of olive oil mill wastewater by physicochemical and advanced oxidation processes. Environ. Technol. 34 (12), 1521–1531. https://doi.org/10.1080/09593330.2012.758663.

Kim, W.C., Lee, D.Y., Lee, C.H., Kim, C.W., 2004. Optimization of narirutin extraction during washing step of the pectin production from citrus peels. J. Food Eng. 63, 191–197.

Kim, H., Kim, D.-H., Seo, K.-H., Chon, J.-W., Nah, S.-Y., Bartley, G.E., Arvik, T., Lipson, R., Yokoyama, W., 2015. Modulation of the intestinal microbiota is associated with lower plasma cholesterol and weight gain in hamsters fed chardonnay grape seed flour. J. Agric. Food Chem. 63 (5), 1460–1467.

Knoblich, M., Anderson, B., Latshaw, D., 2005. Analyses of tomato peel and seed byproducts and their use as a source of carotenoids. J. Sci. Food Agric. 85, 1166–1170.

Ko, J.M., Cha, H.C., An, N.R., 2009. Massive Extraction of Quercetin Glycoside Derived from Kyoho Grape Leaves Using Column Chromatography. Patent No. KR1020090125988.

Koutrotsios, G., Mountzouris, K.C., Chatzipavlidis, I., Zervakis, G.I., 2014. Bioconversion of lignocellulosic residues by *Agrocybecylindraceaand* *Pleurotusostreatus* mushroom fungi – assessment of theireffect on the final product and spent substrate properties. Food Chem. 161, 127–135.

Kovacevic, D.B., Barba, F.J., Granato, D., Galanakis, C.M., Herceg, Z., Dragovic-Uzelac, V., Prutnik, P., 2018. Pressurized hot water extraction (PHWE) for the green recovery of bioactive compounds from steviol glycosides from *Stevis rebaudiana* Bertoni leaves. Food Chem. 254, 150–157.

Kwon, S.I., Song, S.H., Cho, S.O., 2000. Distilled Liquor Using Wine Lees. Patent No. KR1020000049540.

Lavecchia, R., Zuorro, A., 2008. Process for the Extraction of Lycopene. World Intellectual Property Organization. WO/2008/055894.

Lavelli, V., Sri Harsha, P.S.C., Spigno, G., 2016. Modelling the stability of maltodextrin-encapsulated grape skin phenolics used as a new ingredient in apple puree. Food Chem. 209, 323–331.

Lee, S.-H., Song, K.B., 2009. Purification of an iron-binding nona-peptide from hydrolysates of porcine blood plasma protein. Process Biochem. 44 (3), 378–381.

Lee, S.-J., Kim, Y.-S., Hwang, J.-W., Kim, E.-K., Moon, S.-H., Jeon, B.-T., Jeon, Y.-J., Kim, J.M., Park, P.-J., 2012. Purification and characterization of a novel antioxidative peptide from duck skin by-products that protects liver against oxidative damage. Food Res. Int. 49 (1), 285–295.

Lee, W.A., Crockett, B.A., Pelle, E., Pernodet, N.A., Keupp, G.M., Georgena, M., 2016. Topical Preparations Comprising Grape Seed, Folic Acid, Biotin Bifidobacterium and Echinacea Purpurea. Patents No. WO2015050883.

Li, R., Li, X., 2010. Method for Extracting Procyanidin from Wine Lees. Patent No. CN101845036.

Li, L., Ma, S., Fan, L., Zhang, C., Pu, X., Zheng, X., Wang, X., 2016. The influence of ultrasonic modification on arabinoxylans properties obtained from wheat bran. Int. J. Food Sci. Technol. 51, 2338–2344.

Lin, D., 2016. K.D. BIO-BEAUTY. Patent No. AU2016100373.

Linde, M., Galbe, M., Zacchi, G., 2008. Bioethanol production from non-starch carbohydrate residues in process streams from a dry-mill ethanol plant. Bioresour. Technol. 99, 6505–6511.

Liu, Z., Ying, Y., Li, F., Ma, C., Xu, P., 2010. Butanol production by *Clostridium beijerinckii* ATCC 55025 from wheat bran. J. Ind. Microbiol. Biotechnol. 37 (5), 495–501.

Lota, M.L., de Rocca Serra, D., Tomi, F., Casanova, J., 2000. Chemical variability of peel and leaf essential oils of mandarins from *Citrus reticulata* Blanco. Biochem. Syst. Ecol. 28, 61–78.

Ma, L., Du, N., Li, T., 2002. Alcohol Extraction Method for Extracting Grape Skin Extract and its Grape Liver-Nourishing Capsule. Patent No. CN1380093.

Maes, C., Delcour, J.A., 2001. Alkaline hydrogen peroxide extraction of wheat bran non-starch polysaccharides. J. Cereal Sci. 34, 29–35.

Maes, C., Delcour, J.A., 2002. Structural characterisation of water-extractable and water-unextractable arabinoxylans in wheat bran. J. Cereal Sci. 35, 315–326.

Mandalaria, G., Faulds, C.B., Sancho, A.I., Saija, A., Bisignano, G., LoCurto, R., Waldron, K.W., 2005. Fractionation and characterisation of arabinoxylans from brewers' spent grain and wheat bran. J. Cereal Sci. 42, 205–212.

Masuda, K., Matsumoto, K., Okada, M., Takakhashi, K., 2009. Composition for Treating Chronic Venous Insufficiencies, Containing Aqueous Extract of Red Grape Leaves and Antiinflammatory Agent. Patent No. 02367464.

Mateo, J.J., Maicas, S., 2015. Valorization of winery and oil mill wastes by microbial technologies. Food Res. Int. 73, 13–25.

Marques, I.P., 2001. Anaerobic digestion treatment of olive mill wastewater for effluent re-use in irrigation. Desalination 137 (1–3), 233–239. https://doi.org/10.1016/s0011-9164(01)00224-7.

Martin-Cabrejas, M.A., Esteban, R.M., Lopez-Andreu, F.J., Waldron, K., Selvendran, R.R., 1995. Dietary fiber content of pear and kiwi pomaces. J. Agric. Food Chem. 43, 662–666.

Martinez, G.A., Rebecchi, S., Decorti, D., Domingos, J.M.B., Natolino, A., Del Rio, D., Bertin, L., Da Porto, C., Fava, F., 2015. Towards multi-purpose biorefinery platforms for the valorization of red grape pomace: production of polyphenols, volatile fatty acids, polyhydroxyalkanoates and biogas. Green Chem. 18 (1), 261–270.

Masi, F., Bresciani, R., Munz, G., Lubello, C., 2015. Evaporation-condensation of olive mill wastewater: evaluation of condensate treatability through SBR and constructed Wetlands. Ecol. Eng. 80, 156–161. https://doi.org/10.1016/j.ecoleng.2014.11.008.

Masmoudi, M., Besbes, S., Chaabouni, M., Robert, C., Paquot, M., Blecker, C., Attia, H., 2008. Optimization of pectin extraction from lemon by-product with acidified date juice using response surface methodology. Carbohydr. Polym. 74, 185–192.

Masowa, M.M., Kutu, F.R., Shange, P.L., Mulidzi, R., Vanassche, F.M.G., 2016. The effect of winery solid waste compost application on maize growth, biomass yield, and nutrient content under greenhouse conditions. Arch. Agron Soil Sci. 62 (8), 1082–1094.

Merali, Z., Marjamaa, K., Käsper, A., Kruus, K., Gunning, A.P., Morris, V.J., Waldron, K.W., 2016. Chemical Characterization of Hydrothermally Pretreated and Enzyme-Digested Wheat Straw: an.

Michael, P.T., 2001. Method for Making Grape Leaves with Meat. Patent No. US6268002.

EP Minerals®, 2013. Accessed on July 2016. http://blog.epminerals.com/blog/california-vineyards-recycle-de-spent-cake-from-filtration-into-compost.

Mironeasa, S., Codinə, G.G., Mironeasa, C., 2016. Optimization of wheat-grape seed composite flour to improve alpha-amylase activity and dough rheological behaviour. Int. J. Food Prop. 19 (4), 859–872.

Mizutani, K., Iemori, Y., Ikeda, K., 2000. Food or Pharmaceutical Composition Containing Wine Strained Lee Extract. Patent No. JP2000135071.

Mokochinski, J.B., Sovrani, V., Dalla Santa, H.S., Felsner, M.L., Sawaya, A.C.H.F., González-Borrero, P.P., Bataglion, G.A., Eberlin, M.N., Torres, Y.R., 2015. Biomass and sterol production from vegetal substrate fermentation using *Agaricusbrasiliensis*. J. Food Qual. 38 (3), 221–229.

Mullen, W., Nemzer, B., Stalmach, A., Ali, S., Combet, E., 2013. Polyphenolic and hydroxycinnamate contents of whole coffee fruits from China, India, and Mexico. J. Agric. Food Chem. 61, 5298–5309.

Mullen, A.M., Alvarez, C., Pojic, M., Hadnadev, T.D., Papageorgiou, M., 2015. Chapter 2. Classification and target compounds. In: Galanakis, C.M. (Ed.), Food Waste Recovery: Processing Technologies and Industrial Techniques.

Murphy, J.D., Power, N.M., 2008. How can we improve the energy balance of ethanol production from wheat? Fuel 87, 1799–1806.

Murthy, P.S., Naidu, M.M., 2012. Recovery of phenolic antioxidants and functional compounds from coffee industry by-products. Food Bioprocess Technol. 5, 897–903.

Mussatto, S.I., Ballesteros, L.F., Martins, S., Teixeira, J.A., 2011. Extraction of antioxidant phenolic compounds from spent coffee grounds. Separ. Purif. Technol. 83, 173–179.

Mussatto, S.I., Machado, E.M.S., Martins, S., Teixeira, J.A., 2011a. Production, composition, and application of coffee and its industrial residues. Food Bioprocess Technol. 4, 661–672.

Mussatto, S.I., Sampaio, A., Dragone, G.M., Teixeira, J.A., 2013. Bebida Destilada a Partir da Borra de Café e Respectivo Método de Produção (Distilled Beverage From Spent Coffee Grounds and Respective Production Method). Patent PT105346.

Mustafa, M.A., Misailidis, N., Mateos-Salvador, F., Du, C., Sadhukhan, J., 2007. Feasibility of Co- Producing Arabinoxylans and Ethanol in a Wheat Biorefinery. Project Report No. 425 2007.

Nair, R.B., Lundin, M., Brandberg, T., Lennartsson, P.R., Taherzadeh, M., 2014. Dilute phosphoric acid pretreatment of wheat bran for enzymatic hydrolysis and subsequent ethanol production by edible fungi *Neurospora* intermedia. Ind. Crops Prod. 69, 314–323.

Naveena, B.J., Altaf, M., Bhadriah, K., Reddy, G., 2005A. Selection of medium components by Plackett–Burman design for production of L (+) lactic acid by *Lactobacillus amylophilus* GV6 in SSF using wheat bran. Biosource Technol. 96, 485–490.

Naveena, B.J., Altaf, M., Bhadrayya, K., Madhavendra, S.S., Reddy, G., 2005B. Direct fermentation of starch to l (+) lactic acid in SSF by *Lactobacillus amylophilus* GV6 using wheat bran as support and substrate: medium optimization using RSM. Process Biochem. 40, 681–690.

Nedjar-Arroume, N., Dubois-Delval, V., Adje, E.Y., Traisnel, J., Krier, F., Mary, P., Kouach, M., Briand, G., Guillochon, D., 2008. Bovine hemoglobin: an attractive source of antibacterial peptides. Peptides 29 (6), 969–977.

Nieto, L.M., Hodaifa, G., Vives, S.R., Casares, J.A.G., 2010. Industrial plant for olive mill wastewater from two-phase treatment by chemical oxidation. J. Environ. Eng. Asce 136 (11), 1309–1313. https://doi.org/10.1061/(asce)ee.1943-7870.0000272.

Nofar, G., Cogan, U., 2016. Inhibition of Neurodegenerative Disease by Grape Seed Extract, Green Tea Extract and Probiotic Bacteria. Patents No. WO2014141265.

Odero-Marah, V., 2016. Muscadine Grape Skin Extract as Treatment for Bone Metastatic Cancer. Patent No. US2016067298.

Otles, S., Despoudi, S., Bucatariu, C., Kartal, C., 2015. Chapter 1, Food waste management, valorization, and sustainability in the food industry. In: Galanakis, C.M. (Ed.), Food Waste Recovery: Processing Technologies and Industrial Techniques.

Özvural, E.B., Vural, H., 2011. Grape seed flour is a viable ingredient to improve the nutritional profile and reduce lipid oxidation of frankfurters. Meat Sci. 88 (1), 179–183.

Pagan, J., Ibarz, A., Llorca, M., Coll, L., 1999. Quality of industrial pectin extracted from peach pomace at different pH and temperature. J. Sci. Food Agric. 79, 1038–1042.

Palmarola-Adrados, B., Chotěborská, P., Galbe, M., Zacchi, G., 2005. Ethanol production from non starch carbohydrates of wheat bran. Biosour. Technol. 96, 843–850.

Panusa, A., Zuorro, A., Lavecchia, R., Marrosu, G., Petrucci, R., 2013. Recovery of natural antioxidants from spent coffee grounds. J. Agric. Food Chem. 61, 4162–4168.

Paraskeva, C.A., Papadakis, V.G., Tsarouchi, E., Kanellopoulou, D.G., Koutsoukos, P.G., 2007. Membrane processing for olive mill wastewater fractionation. Desalination 213 (1–3), 218–229. https://doi.org/10.1016/j.desal.2006.04.087.

Pascoa, R.N.M.J., Magalhaes, L.M., Lopes, J.A., 2013. FT-NIR spectroscopy as a tool for valorization of spent coffee grounds: application to assessment of antioxidant properties. Food Res. Int. 51, 579–586.

Patsioura, A., Galanakis, C.M., Gekas, V., 2011. Ultrafiltration optimization for the recovery of β-glucan from oat mill waste. J. Membr. Sci. 373 (1–2), 53–63.

Pazmiño-Durán, A.E., Giusti, M.M., Wrolstad, R.E., Glória, M.B.A., 2001. Anthocyanins from banana bracts (*Musa X paradisiaca*) as potential food colorants. Food Chem. 73, 327–332.

Pedroza, M.A., Amendola, D., Maggi, L., Zalacain, A., De Faveri, D.M., Spigno, G., 2015. Microwave-assisted extraction of phenolic compounds from dried waste grape skins. Int. J. Food Eng. 11 (3), 359–370.

Pérez-Bibbins, B., Torrado-Agrasar, A., Salgado, J.M., Oliveira, R.P.D.S., Domínguez, J.M., 2015. Potential of lees from wine, beer and cider manufacturing as a source of economic nutrients: an overview. Waste Manag. 40, 72–81.

Pinelo, M., Arnous, A., Meyer, A.S., 2006. Upgrading of grape skins: significance of plant cell-wall structural components and extraction techniques for phenol release. Trends Food Sci. Technol. 17, 579–590.

Pizzichini, M., Russo, C., 2005. Process for Recovering the Components of Olive Mill Wastewater with Membrane Technologies. World Intellectual Property Organization. WO/2005/123603.

Potoglou, D., Kouzeli-Katsiri, A., Haralambopoulos, D., 2004. Solar distillation of olive mill wastewater. Renew. Energy 29 (4), 569–579. https://doi.org/10.1016/j.renene.2003.09.002.

Prinsen, P., Gutierrez, A., Faulds, C.B., del Río, J.C., 2014. Comprehensive study of valuable lipophilic phytochemicals in wheat bran. J. Agric. Food Chem. 62, 1664–1673.

Prisenžnáková, L., Nosálová, G., Hromádková, Z., Ebringerová, A., 2010. The pharmacological activity of wheat bran polysaccharides. Fitoterapia 81, 1037–1044.

Proietti, P., Federici, E., Fidati, L., Scargetta, S., Massaccesi, L., Nasini, L., Regni, L., Ricci, A., Cenci, G., Gigliotti, G., 2015. Effects of amendment with oil mill waste and its derived-compost on soil chemical and microbiological characteristics and olive (*Olea europaea* L.) productivity. Agric. Ecosyst. Environ. 207, 51–60.

Rahmanian, N., Jafari, S.M., Galanakis, C.M., 2014. Recovery and removal of phenolic compounds from olive mill wastewater. J. Am. Oil Chem. Soc. 91, 1–18.

Rajha, H.N., Boussetta, N., Louka, N., Maroun, R.G., Vorobiev, E., 2015. Electrical, mechanical, and chemical effects of high-voltage electrical discharges on the polyphenol extraction from vine shoots. Innov. Food Sci.Emerg. 31, 60–66.

Ramalakshmi, K., Rao, L.J.M., Takano-Ishikawa, Y., Goto, M., 2009. Bioactivities of low-grade green coffee and spent coffee in different in vitro model systems. Food Chem. 115, 79–85.

Razavi, S.-M., Gholamin, S., Eskandari, A., Mohsenian, N., Ghorbanihaghjo, A., Delazar, A., Rashtchizadeh, N., Keshtkar-Jahromi, M., Argani, H., 2013. Red grape seed extract improves lipid profiles and decreases oxidized low-density lipoprotein in patients with mild hyperlipidemia. J. Med. Food 16 (3), 255–258.

Regni, L., Giovanni, G., Luigi, N., Agrafioti, E., Galanakis, C.M., Primo, P., 2017. Chapter 5. Reuse of olive mill waste as soil amendment. In: Galanakis, C.M. (Ed.), Olive Mill Waste: Recent Advances for the Sustainable Management. Elsevier Inc., Waltham.

Reisinger, M., Tirpanalan, Ö., Huber, F., Kneifel, W., Novalin, S., 2014. Investigations on a wheat bran biorefinery involving organosolv fractionation and enzymatic treatment. Bioresour. Technol. 170, 53–61.

Roberts, P.J., Simmonds, D.H., Wootton, M., Wrigley, C.W., 1985. Extraction of protein and solids from wheat bran. J. Sci. Food Agric. 36, 5–10.

Rogalinski, T., Herrmann, S., Brunner, G., 2005. Production of amino acids from bovine serum albumin by continuous sub-critical water hydrolysis. J. Supercrit. Fluids 36 (1), 49–58.

Romain, C., Gaillet, S., Carillon, J., Vidé, J., Ramos, J., Izard, J.-C., Cristol, J.-P., Rouanet, J.-M., 2012. Vineatrol and cardiovascular disease: beneficial effects of a vine-shoot phenolic extract in a hamster atherosclerosis model. J. Agric. Food Chem. 60 (44), 11029–11036.

Romo Sánchez, S., Gil Sánchez, I., Arévalo-Villena, M., Briones Pérez, A., 2015. Production and immobilization of enzymes by solid-state fermentation of agroindustrialwaste. Bioproc. Biosyst. Eng. 38 (3), 587–593.

Roselló-Soto, E., Barba, F.J., Parniakov, O., Galanakis, C.M., Grimi, N., Lebovka, N., Vorobiev, E., 2015a. High voltage electrical discharges, pulsed electric field and ultrasounds assisted extraction of protein and phenolic compounds from olive kernel. Food Bioprocess Technol. 8, 885–894.

Roselló-Soto, E., Galanakis, C.M., Brncic, M., Orlien, V., Trujillo, F.J., Mawson, R., Knoerzer, K., Tiwari, B.K., Francisco, J., Barba, F.J., 2015b. Clean recovery of antioxidant compounds from plant foods, byproducts and algae assisted by ultrasounds processing. Modeling approaches to optimize processing conditions. Trends Food Sci. Technol. 42, 134–149.

Rozzi, A., Malpei, F., 1996. Treatment and disposal of olive mill effluents. Int. Biodeterior. Biodegrad. 38 (3–4), 135–144. https://doi.org/10.1016/s0964-8305(96)00042-x.

Rubio, R., Pérez-Murcia, M.D., Agulló, E., Bustamante, M.A., Sánchez, C., Paredes, C., Moral, R., 2013. Recycling of agro-food wastes into vineyards by composting: agronomic validation in field conditions. Commun. Soil Sci. Plan. 44 (1–4), 502–516.

Russo, C., 2007. A new membrane process for the selective fractionation and total recovery of polyphenols , water and organic substances from vegetation waters (VW). J. Membr. Sci. 288 (1–2), 239–246. https://doi.org/10.1016/j.memsci.2006.11.020.

San José, M.J., Alvarez, S., García, I., Peñas, F.J., 2013. A novel conical combustor for thermal exploitation of vineyard pruning wastes. Fuel 110, 178–184.

Sánchez-Gómez, R., Garde-Cerdán, T., Zalacain, A., Garcia, R., Cabrita, M.J., Salinas, M.R., 2016b. Vine-shoot waste aqueous extract applied as foliar fertilizer to grapevines: effect on amino acids and fermentative volatile content. Food Chem. 197, 132–140.

Sánchez-Gómez, R., Zalacain, A., Alonso, G.L., Salinas, M.R., 2016z. Effect of vine-shoots toasting on the generation of high added value volatiles. Flavour Fragrance J. https://doi.org/10.1002/ffj.3313.

Sarika, R., Kalogerakis, N., Mantzavinos, D., 2005. Treatment of olive mill effluents Part II. Complete removal of solids by direct flocculation with poly-electrolytes. Environ. Int. 31 (2), 297–304. https://doi.org/10.1016/j.envint.2004.10.006.

Sato, T., Morikawa, H., 2011. Silver Skin-Containing Paper and Method for Producing the Same. Patent US 7927460.
Schieber, A., Stintzing, F.C., Carle, A., 2001. By-products of plant food processing as a source of functional compounds – recent developments. Trends Food Sci. Technol. 12, 401–413.
Schönnenbeck, C., Trouvé, G., Valente, M., Garra, P., Brilhac, J.F., 2016. Combustion tests of grape marc in a multi-fuel domestic boiler. Fuel 180, 324–331.
Servili, M., Esposto, S., Veneziani, G., Urbani, S., Taticchi, A., Di Maio, I., Selvaggini, R., Sordini, B., Montedoro, G., 2011. Improvement of bioactive phenol content in virgin olive oil with an olive-vegetation water concentrate produced by membrane treatment. Food Chem. 124, 1308–1315.
Seçmeler, Ö., Galanakis, C.M., 2019. Chapter 8. Olive fruit and olive oil. In: Galanakis, C.M. (Ed.), Innovations in Traditional Foods. Elsevier Inc., Waltham.
Shanbrom, E., 2002. Antimicrobial Lees. Patent No. US20020102287.
Sharma, P.C., Tilakratne, B.M., Anil, G., 2010. Utilization of wild apricot kernel press cake for extraction of protein isolate. J. Food Sci. Technol. Mysore 47, 682–685.
Sharma, A.K., Kumar, R., Azad, Z.R.A.A., Adsule, P.G., 2015. Use of fine wine lees for value addition in ice cream. J. Food Sci. Technol. 52 (1), 592–596.
Shi, G., 2016. Traditional Chinese Medicine Toothpaste Adopting Grape Seed Matrix Abrasive and Preparation Method of Traditional Chinese Medicine Toothpaste. Patent No. CN105342912.
Shin, H.J., Kim, U.J., An, Y.H., 2009. Method for Extracting Resveratrol Using Grape Stems, Capable of Using the By-Product of Grape as a Material. Patent No. KR1020090092405.
Shrikhande, A.J., Race, E.J., Wightman, J.D., Sambueso, R.D., 2003. Process for Extraction, Purification and Enrichment of Polyphenolic Substances from Whole Grapes, Grape Seeds and Grape Pomace. Patent No. US6544581.
Siorou, S., Vgenis, T.T., Dareioti, M.A., Vidali, M.S., Efthimiou, I., Kornaros, M., Vlastos, D., Dailianis, S., 2015. Investigation of Olive Mill Wastewater (OMW) ozonation efficiency with the use of a battery of selected ecotoxicity and human toxicity assays. Aquat. Toxicol. 164, 135–144. https://doi.org/10.1016/j.aquatox.2015.04.017.
Skendi, A., Harasym, J., Galanakis, C.M., 2018. Chapter 7. Recovery of high added value compounds from brewing and distillate processing by-products. In: Galanakis, C.M. (Ed.), Sustainable Recovery and Reutilization of Cereal Processing By-Products. Elsevier Inc., Waltham.
Smaali, I., Rémond, C., Skhiri, Y., O'Donohue, M., 2009. Biocatalytic conversion of wheat bran hydrolysate using an immobilized GH43 b-xylosidase. Bioresour. Technol. 100, 338–344.
Socaci, S.A., Fărcaş, A.C., Galanakis, C.M., 2019. Chapter 2. Introduction in functional components for membrane separations. In: Galanakis, C.M. (Ed.), Separation of Functional Molecules in Food by Membrane Technology. Elsevier Inc., Waltham.
Song, H., Lee, S.Y., 2006. Production of succinic acid by bacterial fermentation. Enzym. Microb. Technol. 39 (3), 352–361.
Souilem, S., El-Abbassi, A., Kiai, H., Hafidi, A., Sayadi, S., Galanakis, C.M., 2017. Chapter 1. Olive oil production sector: environmental effects and sustainability challenges. In: Galanakis, C.M. (Ed.), Olive Mill Waste: Recent Advances for the Sustainable Management. Elsevier Inc., Waltham.
Sousa, C., Gabriel, C., Cerqueira, F., Manso, M.C., Vinha, A.F., 2015. Coffee industrial waste as a natural source of bioactive compounds with antibacterial and antifungal activities. In: Méndez-Vilas, A. (Ed.), The Battle Against Microbial Pathogens: Basic Science, Technological Advances and Educational Programs. Formatex Research Center, Badajoz, Spain, pp. 131–136.

Spigno, G., Maggi, L., Amendola, D., Dragoni, M., De Faveri, D.M., 2013a. Influence of cultivar on the lignocellulosic fractionation of grape stalks. Ind. Crops Prod. 35, 178–184.

Spigno, G., Donsì, F., Amendola, D., Sessa, M., Ferrari, G., De Faveri, D.M., 2013b. Nanoencapsulation systems to improve solubility and antioxidant efficiency of a grape marc extract into hazelnut paste. J. Food Eng. 114 (2), 207–214.

Spigno, G., Amendola, D., Dahmoune, F., Jauregi, P., 2015. Colloidal gas aphrons based separation process for the purification and fractionation of natural phenolic extracts. Food Bioprod. Process. 94, 434–442.

Spigno, G., Marinoni, L., Garrido, G.D., 2017. Chapter 1. State of the art in grape processing byproducts. In: Galanakis, C.M. (Ed.), Handbook of Grape Processing By-Products: Sustainable Solutions. Elsevier Inc., Waltham.

Spinelli, R., Nati, C., Pari, L., Mescalchin, E., Magagnotti, N., 2012. Production and quality of biomass fuels from mechanized collection and processing of vineyard pruning residues. Appl.Energy 89 (1), 374–379.

Stoller, M., 2008. Technical optimization of a dual ultrafiltration and nanofiltration pilot plant in batch operation by means of the critical flux theory: a case study. Chem. Eng. Process: Process Intensification 47, 1165–1170.

Stoller, M., Angelo, C., 2006. Technical optimization of a batch olive wash wastewater treatment membrane plant. Desalination 200 (1–3), 734–736. https://doi.org/10.1016/j.desal.2006.03.501.

Sueyoshi, Y., Hashimoto, T., Yoshikawa, M., Watanabe, K., 2011. Transformation of intact chicken feathers into chiral separation membranes. Waste Biomass Valorization 2 (3), 303–307.

Sun, Y., Cheng, J., 2002. Hydrolysis of lignocellulosic materials for ethanol production: a review. Bioresour. Technol. 83, 1–11.

Sun, Q., Shen, H., Luo, Y., 2011. Antioxidant activity of hydrolysates and peptide fractions derived from porcine hemoglobin. J. Food Sci. Technol. 48 (1), 53–60.

Sung, K.J., Jeong, Y.S., Jung, H.K., Cho, K.H., Hong, J.H., Shim, S.P., Jung, J.E., 2013. Anti-obesity Composition Based on Grape Skin Extracts as an Effective Component and a Manufacturing Method of the Same. Patent No. KR1020130064595.

Takao, I., Asano, I., Fujii, S., Kaneko, M., Nielson, J.R., Steffen, D.G., Hatzold, T., 2009. Composition Having Blood Pressure Reducing And/or Elevation Suppressing Effect and Food and Drink Containing the Same. US2009/0005342. PCT/JP2006/301025.

Takase, H., Ohashi, H., Okoshi, T., Kobayashi, H., Takata, R., Kon-No, T., 2011. Soft Drink, Grape Skin Extract, and Methods of Preparation of Same. Patent No. US20110311680.

Talebnia, F., Karakashev, D., Angelidaki, I., 2010. Production of bioethanol from wheat straw: an overview on pretreatment, hydrolysis and fermentation. Bioresour. Technol. 101, 4744–4753.

Tan, B., 2015. Grape Skin Pigment Extraction Technology. Patent No. CN104448902.

Tao, Y., Wu, D., Zhang, Q.-A., Sun, D.-W., 2014. Ultrasound-assisted extraction of phenolics from wine lees: modeling, optimization and stability of extracts during storage. Ultrason. Sonochem. 21 (2), 706–715.

Tirpanalan, Ö., Reisinger, M., Smerilli, M., Huber, F., Neureiter, M., Kneifel, W., Novalin, S., 2015. Wheat bran biorefinery – an insight into the process chain for the production of lactic acid. Bioresour. Technol. 180, 242–249.

Toldrà, M., Parés, D., Saguer, E., Carretero, C., 2011. Hemoglobin hydrolysates from porcine blood obtained through enzymatic hydrolysis assisted by high hydrostatic pressure processing. Innov. Food Sci. Emerg. Technol. 12 (4), 435–442.

Tsagaraki, E., Lazarides, H.N., Petrotos, K.B., 2007. Olive mill wastewater treatment. In: Vasso, O., Winfried, R. (Eds.), Utilization of By-Products and Treatment of Waste in the Food Industry. Springer, pp. 133–157.

Turano, E., Curcio, S., De Paola, M.G., Calabro, V., Iorio, G., 2002. An integrated centrifugation-ultrafiltration system in the treatment of olive mill wastewater. J. Membr. Sci. 209 (2), 519–531. https://doi.org/10.1016/s0376-7388(02)00369-1.

Ucak, I., Gokoglu, N., Toepfl, S., Galanakis, C.M., 2018. Inhibitory Effects of High pressure processing on *Photobacterium phosphoreum* and *Morganella psychrotolerans* in vacuum packed herring (*Clupea harengus*). J. Food Safety 38 (6). December 2018e12519.

Ucak, I., Gokoglu, N., Kiessling, M., Toepfl, S., Galanakis, C.M., 2019. Inhibitory effects of high pressure treatment on microbial growth and biogenic amine formation in marinated herring (*Clupea harengus*) inoculated with *Morganella psychrotolerans*. LWT Food Sci. Technol. 99, 50–56.

Ucar, R.C., Sngul, A., Celictas, M.S., 2018. Chapter 4. Biorefineries for wheat bran: sustainable utilization and market potential. In: Galanakis, C. (Ed.), Sustainable Recovery and Reutilization of Cereal Processing By-Products. M. Elsevier Inc., Waltham.

Umesh Hebbar, H., Sumana, B., Raghavarao, K.S.M.S., 2008. Use of reverse micellar systems for the extraction and purification of bromelain from pineapple wastes. Bioresour. Technol. 99, 4896–4902.

Velioglu, S.G., Curi, K., Camlilar, S.R., 1987. Laboratory experiments on the physical treatment of olive oil wastewater. International journal for development technology.

Velioglu, S.G., Curi, K., Camlilar, S.R., 1992. Activated-sludge treatability of olive oil-bearing waste-water. Water Res. 26 (10), 1415–1420.

Villanova, L., Villanova, L., Fasiello, G., Merendino, A., 2006. Process for the Recovery of Tyrosol and Hydroxytyrosol from Oil Mill Wastewaters and Catalytic Oxidation Method in Order to Convert Tyrosol in Hydroxytyrosol. Patent US 2006/0070953.

Vincenzi, S., Dinnella, C., Recchia, A., Monteleone, E., Gazzola, D., Pasini, G., Curioni, A., 2013. Grape seed proteins: a new fining agent for astringencyreduction in red wine. Aust. J. Grape Wine Res. 19, 153–160.

Vulic, J., Saponjac, V.T., Cetkovic, G., Canadanovic-Brunet, J., Djilas, S., Stajcic, S., 2017. Chapter 8. Patent survey. In: Galanakis, C.M. (Ed.), Handbook of Grape Processing By-Products: Sustainable Solutions. Elsevier Inc., Waltham.

Wang, Y.Q., Schuchardt, F., 2010. Effect of C/N ratio on the composting of vineyard pruning residues. Landbauforsch Volk 60 (3), 131–138.

Wang, S., Chen, F., Wu, J., Wang, Z., Liao, X., Hu, X., 2007. Optimization of pectin extraction assisted by microwave from apple pomace using response surface methodology. J. Food Eng. 78, 693–700.

Wang, J., Sun, B., Liu, Y., Zhang, H., 2013. Optimisation of ultrasound-assisted enzymatic extraction of arabinoxylan from wheat bran. Food Chem. 150, 482–488.

Wang, H., Liang, Y., Li, H., Yang, C., 2015. Method for Making Grape Pomace Dietary Fiber Nutritious Biscuits. Patent No. CN104798854.

Wild, H.P., Sass, M., 2006. Concentrate Comprising Green Tea, Grape Skin Extract and Grape Extract, the Production Thereof and Use of the Same. Patent No. US7087259.

Wood, I.P., Cook, N.M., Ryden, P.R., Robertson, J.A., Waldron, K.W., 2016. Ethanol from a biorefinery waste stream: saccharification of amylase, protease and xylanase treated wheat bran. Food Chem. 198, 125–131.

Xing, Y., Che, Z., Jiang, L., Lin, H., Liu, H., Cao, D., Chen, Y., Li, X., Wang, R., Deng, A., 2015. Preparation Method of Grape Pomace Essence and Fruit and Vegetable Powder Combined Blending Agent. Patent No. CN104287048 (A).

Xu, Z., 2016. Application of Soft Capsule Containing Sheep Placenta, Radix Salviae Miltiorrhizae and Grape Seeds in Preparing Cough Medicine. Patent No. CN105343353.
Yang, Q., 2012. Grape Skin Pigment Extracting Method. Patent No. CN102675904.
Yaragalla, S., Rajendran, R., Jose, J., Almaadeed, M.A., Kalarikkal, N., Thomas, S., 2016. Preparation and characterization of green graphene using grape seed extract for bioapplications. Mater. Sci. Eng. C Mater. Biol. Appl. 65, 345–353.
Yuan, X., Wang, J., Yao, H., 2005. Antioxidant activity of feruloylated oligosaccharides from wheat bran. Food Chem. 90, 759–764.
Zagklis, D.P., Paraskeva, C.A., 2014. Membrane filtration of agro-industrial wastewaters and isolation of organic compounds with high added values. Water Sci. Technol. 69, 202–207.
Zagklis, D.P., I Vavouraki, A., Kornaros, M.E., Paraskeva, C.A., 2015. Purification of olive mill wastewater phenols through membrane filtration and resin adsorption/desorption. J. Hazard Mater. 285, 69–76.
Zeugolis, D.I., Paul, R.G., Attenburrow, G., 2008. Factors influencing the properties of reconstituted collagen fibers prior to self-assembly: animal species and collagen extraction method. J. Biomed. Mater. Res. A 86A (4), 892–904.
Zhang, D., Hamauzu, Y., 2004. Phenolic compounds and their antioxidant properties in different tissues of carrots. Food Agric. Environ. 2, 95–100.
Zhang, L., Sun, X., 2016. Improving green waste composting by addition of sugarcane bagasse and exhausted grape marc. Bioresour. Technol. 218, 335–343.
Zhang, G., Zhao, S., 1990. Method of Making Vinegar Using Grape Skin. Patent No. CN1039841.
Zhang, B., Wang, X., Zheng, Q., Huang, F., Yang, X., 2010. Method for Extracting Resveratrol from Grape Leaves. Patent No. CN101913995.
Zhang, B., Wang, M., Chu, Y., 2016. Method for Preparing Wine from Grape Juice Precipitation Solution. Patent No. CN105349297.
Zhu, F., Du, B., Zheng, L., Li, J., 2015. Advance on the bioactivity and potential applications of dietary fibre from grape pomace. Food Chem. 186, 207–212.
Zhu, Q., Wu, J., Wang, L., Yang, G., Zhang, X., 2016. Adsorption characteristics of Pb^{2+} onto wine lees-derived biochar. Bull. Environ. Contam. Toxicol. https://doi.org/10.1007/s00128-016-1760-4.
Zinoviadou, K.G., Galanakis, C.M., 2017. Glucosinolates and respective derivatives. In: Puri, M. (Ed.), Food Bioactives: Extraction and Biotechnology Applications. Springer, New York.
Zinoviadou, K.G., Barba, F.J., Galanakis, C.M., Brnčić, M., Trujillo, F., Mawson, R., Knoerzer, K., 2015. Fruit juice sonication: implications on food safety and physicochemical and nutritional properties. Food Res. Int. 77 (4), 743–752.
Zuorro, A., Lavecchia, R., 2011. Polyphenols and energy recovery from spent coffee grounds. Chem. Eng. Trans. 25, 285–290.

Further reading

Cherubini, F., Jungmeier, G., Wellisch, M., Willke, T., Skiadas, I., Van Ree, R., Jong, E., 2009. Toward a common classification approach for biorefinery systems. Biofuels Bioprod. Biorefining 3, 534–546.
Chotěborská, P., Palmarola-Adrados, B., Galbe, M., Zacchi, G., Melzoch, K., Rychtera, M., 2004. Processing of wheat bran to sugar solution. J. Food Eng. 61, 561–565.
del Castillo, M.D., Martinez-Saez, N., Ullate, M., 2014. Healthy Bakery Products with High Level of Dietary Antioxidant Fibre. PCT/ES2014/070062.

D'Alessandro, F., Marucchini, C., Minuti, L., Zadra, C., Tatocchi, A., 2005. GC/MS-SIM analysis of phenolic compounds in olive oil waste waters. Ital. J. Food Sci. 17, 83–88.

Galanakis, C.M., 2016j. High Value-Added Compounds From Food Waste. Reference Module in Food Sciences. Elsevier, pp. 1–8. https://doi.org/10.1016/B978-0-08-100596-5.03510-1.

HLPE, 2014. Food Losses and Waste in the Context of Sustainable Food Systems. A Report by the High Level Panel of Experts on Food Security and Nutrition of the Committee on World Food Security. Rome 2014. http://www.fao.org/fileadmin/user_upload/hlpe/hlpe_documents/HLPE_Reports/HLPE-Report-8_EN.pdf.

Stoeglehner, G., Narodoslawsky, M., 2009. How sustainable are biofuels? Answer and further questions arising from an ecological footprint perpective. Bioressour. Technol. 16, 3825–38030.

Index

Note: 'Page numbers followed by "f" indicate figures and "t" indicates tables'.

A

Abiotic Depletion Potential, 67
Absorption, 296–297
Acidification Potential (AP), 67
Activated carbon assisted biofilters, 301–302
Adsorbate, 295
Adsorption, 295–296
Aerobic processes, 250–252
Agricultural commodity, 230
Agricultural water productivity, 234
Agriculture 4.0. Smart Farming, 41
Agrifood industry, 2
Agri-footprint, 89–90, 91t–98t
Alliance for Water Stewardship (AWS) Standard, 43
Alliance Shippers, 51–52
Alternative energy sources, 17–18
3-Alternative forced-choice (3-AFC), 294
Anaerobic processes, 252–253
Animal-derived waste, 157–158
Apple pomace, 177
Artisanal cheeses, 121
Avoidable fraction, 18

B

Baking industry, 285–286
Beer, 284
Bennett's Law, 38
"Biobased" bioplastics, 203–204
Biobased plastic materials, 202–208
Biobased products, 330–331
Biodegradability, 204–206
Biofilter technology, 301–304
  activated carbon assisted biofilters, 301–302
  engineered biofilter media, 303–304
  multistaged high performance modules, 302–303
  nonthermal plasma assisted biofilters, 302
Biological trickling filters (BTFs)
  and olive mill wastewater, 261–264
  and second cheese whey, 265
  and table olive wastewater, 264–265
  treatment using, 261, 262t–263t
Biopolymer cellulose, 208
Biorefineries based on food processing waste
  biowaste-based biorefineries, 176–178
  brewer's spent grain, 165–169
  olive waste, 169–174
  potato peels, 174–176
  wine grape waste, 160–165
Biorefinery, 22
Biotechnologies, 299–301
Biowaste-based biorefineries
  apple pomace, 177
  citrus waste, 177–178
  dairy by-products, 178
  tomato pomace, 176
Blood proteins, 367–370
Blue (virtual) water, 227–231
Brainstorming sessions, 7–8
Brewer's spent grain, biorefinery based on, 165–169
Brewery industry, 284–285
Brewing, 284

C

Caciocavallo Palermitano cheese, 130
Catalytic oxidation, 297–298
Caves, cheese ripening in, 130–131
Cellulose nanoparticles (CNs), 212
Cereal by-products, 348t–350t
Cereal processing by-products, 347–351, 348t–350t
Cereals, 347
Cheese ripening
  in caves, 130–131

Cheese ripening (*Continued*)
in goat's skin bag, 130
in wooden vats, 122–130
Cheeses microbial ecology studies, 123t
Cheeses with house microbiota
cheese ripening in caves, 130–131
cheese ripening in goat's skin bag, 130
cheeses microbial ecology studies, 123t
cheeses ripening in wooden vats, 122–130
pit cheese, 122
Cheese terroir, 121
Chili grinding, 282–283
Chitosan, 211–212
Ciauscolo, 139
Circular Economy Package, 19
Circular supply chains, 54–55
Citrus waste, 177–178
Cleaner production (CP), 3–9, 5f
assessment, 7–8
environmental improvement in, 9
energy efficiency, 14–18
raw material efficiency, 18–22
water efficiency, 10–13
evaluation and feasibility study, 8–9
methodology, 3
opportunities, implementation of viable, 9
planning and organizing, 5–6
preassessment, 6–7
Cleaner Production (CP) methodology, 3
Climate change, 232
global vital flow impacted by, 231–232
Coagulase-negative cocci (CNC), 131
Coffee processing by-products, 358–366, 359t–362t, 364t–365t
Constructed wetlands, 258–259, 265–270, 266t–267t, 269f
and olive mill wastewater, 268
and second cheese whey, 268–270
and table olive wastewater, 268
Consumer behavior, 196
Control of odors
chili grinding, 282–283
example sources of
baking industry, 285–286
brewery industry, 284–285
palm oil industry, 286–288
poultry industry, 288–289
food processing industry, 282–283
maintenance requirement, 305–306
modifications of biofilter technology, 301–304
activated carbon assisted biofilters, 301–302
engineered biofilter media, 303–304
multistaged high performance modules, 302–303
nonthermal plasma assisted biofilters, 302
selection based
on investment cost, 304–305
on operational requirements, 305
on preceding environmental impacts, 306
technologies
absorption, 296–297
adsorption, 295–296
biotechnologies, 299–301
catalytic oxidation, 297–298
nonthermal plasma injection, 298–299
thermal oxidation, 297
through odor management plan
documentation, 289–290
identification of odor sources from processing activities, 290–294
validation of odor exposure, 294–295
Conventional extraction methods, 383
"Conventional" food products, 326
Cosmetics, 363
Countryside Stewardship (CS) scheme, 40–41
CPC code, 108–109, 108f
Crop cultivation, 75–78
Crude palm oil, 287f
Cultural innovations, 318–320

D

Dairy by-products, 178
Dairy industry, house microbiota of, 121–122
Design for the environment (DfE), 43–44
3D food printing, 324–325
Dietary fiber, 342–343
Digital gastronomy, 324–325
Dimethyl sulfide (DMS), 304
Disruptive technologies, 324–325
Divle Cave (DC) cheese, 131
Dried Grains with Solubles (DDGS), 351
Dual purpose product concept, 44

E

Eco-design of food products, 43–44
Eco-innovation, 316
Ecoinvent, 82–89
Eco-labeling of food products, 44–47
Economic viability, 8
Electrochemical methods, 257–258
Empty fruit bunches (EFB), 286–287
End-of-life of a product, 42
Energy, 16–17
  consumption at source, reducing, 16
  efficiency, 14–18
  recovery techniques, 17
Engineered biofilter media, 303–304
Environmental degradation, 36
Environmental/ecoefficiency analysis, 7
Environmental footprint reduction, 51
Environmental impact indicator, 23–26, 24t–25t
Environmental impacts, 306
Environmentally Preferable Purchasing (EPP) policies, 48
Environmental performance evaluation, 8
Environmental product declaration (EPD), 100–101, 102t–107t
Environmental Protection Agency (EPA), 283
Environmental sustainability of logistics, 51
EU organic logo, 98–99
European Bioplastics Organization (EUBP), 202–206
European Recycling Platform (ERP), 54–55
Eutrophication Potential (EP), 67
EverGreen refrigerator, 51–52

F

Fermented foods, microbial ecology of, 139–141
Fermented sausages microbial ecology studies, 135t–137t
Flow chart of production activity, 2
"Food and Beverages" category, 100–101, 102t–107t
Food by-products, 20
Food chains, sustainability of, 196–199
Food contamination, 4–5
Food database, 81
Food-derived by-products, 159–160
Food incremental innovation, 323–324
Food industry
  performance and environmental indicators, 22–26, 24t–25t
  principles and methodology for, 3–9
Food industry 4.0, 26–29, 30f
Food industry wastewaters, 246–247, 272–273
  and their characteristics, 247, 248t–249t
Food industry wastewater treatment, 260–261
  biological trickling filters (BTFs)
    and olive mill wastewater, 261–264
    and second cheese whey, 265
    and table olive wastewater, 264–265
    treatment using, 261, 262t–263t
  constructed wetlands, 266t–267t
    constructed wetlands, 258–259, 265–270, 266t–267t, 269f
    and second cheese whey, 268–270
    and table olive wastewater, 268
  electrocoagulation treatment, 270–271, 271f
Food Layered Manufacture, 325
Food market breakthrough, 324
Foodomics, 324–327
Food processing industry, 282–283
Food processing plant, 3
Food processing waste, 156–157
  biorefineries based on
    biorefinery based on brewer's spent grain, 165–169
    biorefinery based on olive waste, 169–174
    biorefinery based on potato peels, 174–176
    biorefinery based on wine grape waste, 160–165
    biowaste-based biorefineries, 176–178
  legislation on waste management, 178–179
  wastes and valuable by-products
    animal-derived waste, 157–158
    market potentials from food-derived by-products, 159–160
    plant-derived waste, 158
Food products, 43
  stewardship, 42–43, 42f
Food radical innovation, 326–328
Food sanitation, 331

Food Supply Chain (FSC), 37, 57
Food system, 282
Food technological breakthrough, 324–325
Food use, packaging materials for, 192–194
Food wasted by-products, 343
Food waste management, 53–54
Food waste recovery, 327–331, 343–346
Food waste reduction, 20
Food wastes on overall sustainability, 194–196
Food waste valorization opportunities, 341–342
  cereal processing by-products, 347–351, 348t–350t
  coffee processing by-products, 358–366, 359t–362t, 364t–365t
  grape processing by-products, 377–402, 378t–382t, 384t–386t
  innovation challenges and commercialization aspects of, 344–346
  meat processing co-products and by-products, 366–370, 368t–369t
  olive mill processing by-products, 370–377, 371t–374t
  universal recovery strategy, 343–344, 344f
Frequency–Intensity–Duration–Offensiveness–Location (FIDOL), 282
Fruit processing by-products, 387–402, 400t–401t
Functional food (FF), 318–320, 326–327
Functional unit, 68–69

G

Genetically modified organisms (GMOs), 40
Global policy agenda, 225–227
Global vital flow impacted by climate change, 231–232
Global Warming Potential (GWP), 67
"Goal and scope definition" phase, 66
Goat's skin bag, cheese ripening in, 130
Gram-positive catalase-positive cocci (GCC), 132
Grape by-products, 387, 388t–399t
Grape marc/pomace, 163–164
Grape processing by-products, 377–402, 378t–382t, 384t–386t
Grape seeds, 164
Gray (virtual) water, 227–231
Greece, 138
Greencore, 44
Green evolution, 36
Green farming, 40–41
Green food
  definition of, 39–40
  logistics, 51–52
  product design, 44, 45f
    design for the environment (DfE), 43–44
    eco-design of food products, 43–44
    eco-labeling of food products' and product certification, 44–47
    food product stewardship, 42–43, 42f
    life cycle assessment and life cycle costing, 47
  sourcing and procurement, 47–49
  supply chains
    and circular food supply chains, 52–55
    definition of, 39–40
Green purchasing, 47–48
Green supply chain, 35–36
  definition of green food and green food supply chain, 39–40
  green farming, 40–41
  green food logistics, 51–52
  green food product design
    eco-design of food products or design for the environment (DfE), 43–44
    eco-labeling of food products' and product certification, 44–47
    food product stewardship, 42–43, 42f
    life cycle assessment and life cycle costing, 47
  green food sourcing and procurement, 47–49
  green food supply chains and circular food supply chains
    internal environmental management, 52–53
    reverse logistics and circular supply chains, 54–55
    waste management and food waste management, 53–54
  green warehousing, 49–51, 50f
Green Supply Chain Management (GSCM) approach, 36–37
Green warehousing, 49–51, 50f
Green (virtual) water, 227–231

H

Heat recovery, 16–17
High degree of innovativeness, 321
High-throughput sequencing (HTS) techniques, 139–140
House microbiota
  cheeses with characteristic in
    cheese ripening in caves, 130–131
    cheese ripening in goat's skin bag, 130
    cheeses microbial ecology studies, 123t
    cheeses ripening in wooden vats, 122–130
    pit cheese, 122
  of dairy industry, 121–122
  of meat industry, 131–133
Hybrid systems, 259–260
Hydrogen sulfide/mercaptans, 283–284

I

ILCD method, 3
In-field odor assessment, 294
Innovation efforts in food industry, 316
Innovation management and sustainability, 315–316
  innovation in food industry, 317–321
    food incremental innovation, 323–324
    food market breakthrough, 324
    food radical innovation, 326–327
  research and development (R&D) activities, 316
  sustainability driven foods innovation, 327–331
Integrated Water Resources Management (IWRM) framework, 232
Intended Nationally Determined Contributions (INDCs), 232
Internal environmental management, 52–53
International standards of odor assessment, 293–294
Investment cost, selection based on, 304–305
Italy, 138–139

J

"Joint Scorecard and Business Plan", 48

K

Key performance indicators (KPI), 22–23
Koyoto protocol, 36

L

Labeling schemes, 90–98
Lactic acid bacteria (LAB), 122–130, 132–133
Land Use, 67
Leadership in Energy and Environmental Design (LEED), 49–50
Legislation on waste management, 178–179
"Less than full container load" (LCL), 51
Levacka, 134
Life Cycle Analysis framework, 232
Life cycle assessment (LCA), 23, 47, 63–65, 66f, 193
  agrifootprint, 89–90, 91t–98t
  application of
    allocation, 71–74
    crop cultivation, 75–78
    functional unit, 68–69
    manure management, 78
    packaging, 79–80
    system boundary, 69–71
    wastewater management, 78–79
  databases
    agri-footprint, 89–90
    ecoinvent, 82–89
    food database, 81
    "market activity" processes in food sector, 83t–89t
    World food life cycle assessment database (WFLDB), 81–82
  environmental product declaration (EPD), 100–101, 102t–107t
  EU organic logo and standards on organic farming, 98–99
  "Food and Beverages" category, 100–101, 102t–107t
  "goal and scope definition" phase, 66
  labeling schemes, 90–98
  PEFCRs, 99–100, 101t
  product category rules (PCR) related to food, 107–109, 108f, 109t–111t
  product environmental footprint (PEF), 99–100
Life cycle costing, 47
"Life cycle interpretation" phase, 68
Life Cycle Inventory analysis (LCI), 66–67
Lightweighting design of packaging, 202
"Limits to growth model", 38
Listeria monocytogenes, 132–133

Low degree of innovativeness, 321
Low-fat margarine, 323–324

M

Manure management, 78
"Market activity" processes in food sector, 83t–89t
Material Sourcing, 47–48
"Meat by-product", 367
Meat industry, 366–367
  house microbiota of, 131–133
  interaction of fermented sausages with, 131
Meat processing co-products and by-products, 366–370, 368t–369t
Medium degree of innovativeness, 321
"Me-too products", 320–321, 323–324
Microalgal-based processes, 253–255
Microbes and environment, 119–120
  cheeses with characteristic in house microbiota
    cheese ripening in caves, 130–131
    cheese ripening in goat's skin bag, 130
    cheeses microbial ecology studies, 123t
    cheeses ripening in wooden vats, 122–130
    pit cheese, 122
  house microbiota of dairy industry, 121–122
  house microbiota of meat industry, 131–133
  interaction of cheese with dairy environment, 120–121, 120f
  interaction of fermented sausages with meat industry, 131
  traditional sausages with characteristic house microbiota
    fermented sausages microbial ecology studies, 135t–137t
    Greece, 138
    Italy, 138–139
    microbial ecology of fermented foods, 139–141
    Portugal, 134
    Serbia, 134
    Spain, 134–138
Microbial communities, 121–122
Microbial ecology of fermented foods, 139–141
Multistaged high performance modules, 302–303

N

Nanoencapsulation, 363
Nanotechnology-based packaging, 213
Nasal Ranger Field Olfactometer, 295
New Product Development (NPD), 43
Nonthermal plasma assisted biofilters, 302
Nonthermal plasma injection, 298–299

O

Odor assessment, international standards of, 293–294
Odor emission, quantification of, 292–293
Odor management plan
  documentation, 289–290
  odor sources from processing activities, 290–294
  validation of odor exposure, 294–295
Odor removal efficiency, 296
Olfactometer, 292–293
Olive leaves, 172–173
Olive mill processing by-products, 370–377, 371t–374t
Olive mill wastewater (OMW), 173, 342, 370–375
  biological trickling filters (BTFs) and, 261–264
  constructed wetlands and, 268
Olive oil industries, 375
Olive pomace, 169–172, 170f
Olive stones, 172
Olive tree pruning, 173–174
Olive waste, biorefinery based on
  olive leaves, 172–173
  olive mill wastewater, 173
  olive pomace, 169–172, 170f
  olive stones, 172
  olive tree pruning, 173–174
Operational requirements, 305
Organic dairy products, 46
Organic farming, 40–41
  standards on, 98–99
Organizational stakeholders, 36–37

P

Packaging, 79–80
  improvement, nanotechnologies for, 211–214
  lightweighting, 200–201

materials, environmental responsibility of, 193
Packaging and food sustainability, 191
food wastes on overall sustainability, 194–196
packaging materials for food use, 192–194
sustainability of food chains, 196–199
sustainability-oriented strategies
correct shelf life assignment, 214–215
maximize packaging protective role, 210–211
nanotechnologies for packaging improvement, 211–214
packaging lightweighting, 200–201
packaging materials with focus on recyclability, 201–202
redesign product-packaging configurations, 210
shift to biobased plastic materials, 202–208
Packaging Relative Environmental Impact (PREI), 197–198, 199f
Palm kernel oil, 287f
Palm oil industry, 286–288
Palm oil mill effluent (POME), 288
Paris Agreement, 234
"Pasta-filata" product, 130
PEFCRs, 99–100, 101t
Penicillium roqueforti, 130–131
Periodic monitoring, 9
Photochemical Ozone Creation Potential (POCP), 67
Physicochemical methods, 255–257
Pit cheese, 122
Plant-derived waste, 158
Plastic packaging, 192
"Port Centric Logistics", 51
Portugal, 134
Postharvest Food Losses (PHFL), 37
Potato peels, biorefinery based on, 174–176
Poultry industry, 288–289
Processing aid, 11
Process innovations, 318–320
Process integration techniques, 16–17
Product category rules (PCR) related to food, 107–109, 108f, 109t–111t
Product certification, 44–47
Product environmental footprint (PEF), 99–100
Product innovations, 318–320
Product stewardship, 42, 56
Protection Designation of Origin (PDO), 121

Q

"Quasi-new products", 320–321, 323

R

Ragusano cheese, 122–130
Raw material efficiency, 18–22
Raw milk, 121–122
Recovery and recycling, 20–22
Redesign product-packaging configurations, 210
Research and development (R&D) activities, 316
Resource efficiency, 4, 36
"Responsive packaging", 211
"Return on Investment", 246
Reverse logistics, 54–55

S

Scrubber, 297
Secondary shelf life, 215
Second cheese whey
biological trickling filters (BTFs) and, 265
constructed wetlands and, 268–270
Serbia, 134
Slaughter by-products, 367
"Slow food movement", 325
Smart farming, 41
Soppressata molisana, 138
Soxhlet extraction, 358–363
Soy-based food, 324
Spain, 134–138
Strawberry surpluses, 387–402
Streptococcus thermophilus, 122–130
Sulfur-containing compounds, 284–285
Sustainability-driven foods innovation, 327–331
Sustainability-oriented strategies
correct shelf life assignment, 214–215
maximize packaging protective role, 210–211
nanotechnologies for packaging improvement, 211–214
packaging lightweighting, 200–201

Sustainability-oriented strategies (*Continued*)
packaging materials with focus on recyclability, 201–202
redesign product-packaging configurations, 210
shift to biobased plastic materials, 202–208
System boundary, 69–71

T

Table olive wastewater
biological trickling filters (BTFs) and, 264–265
constructed wetlands and, 268
Thermal oxidation, 297
Thermoplastic starches (TPS), 206
Tina, 122–130
Tomato pomace, 176
Traditional CP strategies, 29, 29f
Traditional pinch analysis, 17
Traditional sausages with house microbiota
fermented sausages microbial ecology studies, 135t–137t
Greece, 138
Italy, 138–139
microbial ecology of fermented foods, 139–141
Portugal, 134
Serbia, 134
Spain, 134–138
Traditional supply chains, 35–36
Treatment methods, 247–260
biological methods
aerobic processes, 250–252
anaerobic processes, 252–253
microalgal-based processes, 253–255
case studies of food industry wastewater treatment, 260–261
biological trickling filters (BTFs), 261–265
constructed wetlands, 265–270, 266t–267t
electrocoagulation treatment, 270–271, 271f
constructed wetlands, 258–259
electrochemical methods, 257–258
hybrid systems, 259–260
physicochemical methods, 255–257
Tulum cheese, 130

U

Unavoidable fraction, 18
UN Decade of Action on Nutrition, 235–236
Universal recovery strategy, 343–344, 344f
US organic products, 46–47

V

Valorization of food processing waste, 156–157
biorefineries based on food processing waste
biowaste-based biorefineries, 176–178
brewer's spent grain, 165–169
olive waste, 169–174
potato peels, 174–176
wine grape waste, 160–165
legislation on waste management, 178–179
wastes and valuable by-products
animal-derived waste, 157–158
market potentials from food-derived by-products, 159–160
plant-derived waste, 158
Valuable by-products, wastes and
animal-derived waste, 157–158
market potentials from food-derived by-products, 159–160
plant-derived waste, 158
Vine shoots, 160–163
by-products, 161–162
Virtual water
blue, green, and gray (virtual) water, 227–231
to global food and agricultural system water requirement, 233–236
global vital flow impacted by climate change, 231–232
water (blue, green, gray), 224–225
water on global policy agenda, 225–227
water security, 223
Volatile organic compounds (VOCs), 284–285

W

Waste Framework Directive, 19
Waste management, 53–54
legislation on, 178–179

Wastes and valuable by-products
  animal-derived waste, 157–158
  market potentials from food-derived by-products, 159–160
  plant-derived waste, 158
Waste-to-energy process, 375–376
Wastewater generation, 12–13
Wastewater management, 78–79
Wastewater pollution control, 13, 14t
Wastewater treatment plant (WWTP), 12, 299–300, 300f
Wastewater treatment technologies, 247–250
Water consumption, 9–13, 14t
Water efficiency, 10–13
Water security, 223
Water sustainability, 228
Wheat (bran-straw) based biorefinery concept, 351, 352t–357t
Wine grape waste, biorefinery based on
  grape marc or pomace, 163–164
  grape seeds, 164
  vine shoots, 160–163
  wine lees, 164–165
Wine lees, 164–165
Winemaking-derived ingredients, 387
Wine production, simplified process for, 160, 161f
Wooden vats, cheeses ripening in, 122–130
World food life cycle assessment database (WFLDB), 81–82

Z

"Zero waste", 53–54
  economy, 351

MIX
Papier | Fördert gute Waldnutzung
FSC® C083411

Kolibri 360 GmbH
Ferdinand-Jühlke-Straße 7
99095 Erfurt, Deutschland
produktsicherheit@kolibri360.de